T0310128

Mechanics of Microsystems

The Wiley Microsystem and Nanotechnology Series

Series Editors – Ronald Pethig and Horacio Espinosa

Mechanics of Microsystems
Corigliano, Ardito, Comi, Frangi, Ghisi, and Mariani, December 2017

Advanced Computational Nanomechanics
Silvestre, February 2016

Micro-Cutting: Fundamentals and Applications
Cheng and Huo, August 2013

Nanoimprint Technology: Nanotransfer for Thermoplastic and Photocurable Polymers
Taniguchi, Ito, Mizuno, and Saito, August 2013

Nano and Cell Mechanics: Fundamentals and Frontiers
Espinosa and Bao, January 2013

Digital Holography for MEMS and Microsystem Metrology
Asundi, July 2011

Multiscale Analysis of Deformation and Failure of Materials
Fan, December 2010

Fluid Properties at Nano/Meso Scale: A Numerical Treatment
Dyson, Ransing, Williams, and Williams, September 2008

Introduction to Microsystem Technology: A Guide for Students
Gerlach and Dötzel, March 2008

AC Electrokinetics: Colloids and Nanoparticles
Morgan and Green, January 2003

Microfluidic Technology and Applications
Koch, Evans, and Brunnschweiler, November 2000

Mechanics of Microsystems

Alberto Corigliano, Raffaele Ardito, Claudia Comi, Attilio Frangi, Aldo Ghisi, and Stefano Mariani

Politecnico di Milano
Italy

This edition first published 2018

Registered Offices
John Wiley & Sons, Inc., 111 River Street, Hoboken, NJ 07030, USA
John Wiley & Sons Ltd, The Atrium, Southern Gate, Chichester, West Sussex, PO19 8SQ, UK

Editorial Office
The Atrium, Southern Gate, Chichester, West Sussex, PO19 8SQ, UK

For details of our global editorial offices, customer services and more information about Wiley products visit us at www.wiley.com.

Wiley also publishes its books in a variety of electronic formats and by print-on-demand. Some content that appears in standard print versions of this book may not be available in other formats.

Library of Congress Cataloging-in-Publication Data

Name: Corigliano, Alberto, author.
Title: Mechanics of microsystems / by Alberto Corigliano, Dr. Raffaele Ardito, Dr. Claudia Comi, Dr. Attilio Frangi, Dr. Aldo Ghisi, Dr. Stefano Mariani.
Description: Hoboken : Wiley, [2018] | Series: The Wiley microsystem and nanotechnology series ; 7646 | Includes bibliographical references and index. | Identifiers: LCCN 2017033389 (print) | LCCN 2017052694 (ebook) | ISBN 9781119053804 (ePDF) | ISBN 9781119053811 (ePUB) | ISBN 9781119053835 (cloth)
Subjects: LCSH: Microelectromechanical systems.
Classification: LCC TK7875 (ebook) | LCC TK7875 .C665 2018 (print) | DDC 621.381–dc23
LC record available at https://lccn.loc.gov/2017033389

Cover Design: Wiley
Cover Images: (First 3 images) Courtesy of Alberto Corigliano; (Abstract) © LuxEterna/Gettyimages; (Large image) Courtesy of Alberto Corigliano; (Lower left corner) © Fuse/Gettyimages

Set in 10/12pt WarnockPro by SPi Global, Chennai, India

Printed and bound by CPI Group (UK) Ltd, Croydon, CR0 4YY

10 9 8 7 6 5 4 3 2 1

To our families

Contents

Series Preface

The Microsystem and Nanotechnology book series provides a thorough contextual summary of the current methods used in micro- and nano-technology research and how these advances are influencing many scientific fields of study and practical application. Readers of these books are guided to learn the fundamental principles necessary for the topic while finding many examples that are representative of the application of these fundamental principles. This approach ensures that the books are appropriate for readers with varied backgrounds and useful for self-study or as classroom materials.

Micro- and nano-scale systems, fabrication techniques, and metrology methods are the basis for many modern technologies. Several books in this series, including *Introduction to Microsystem Technology* by Gerlach and Dotzel, *Microfluidic Technology and Applications* edited by Koch, Evans, and Brunnschweiler, and *Fluid Properties at Nano/Meso Scale* by Dyson, Ransing, P. Williams and R. Williams, provide a resource for building a scientific understanding of the field. Multiscale modeling, an important aspect of microsystem design, is extensively reviewed in *Multiscale Analysis of Deformation and Failure of Materials* by Jinghong Fan. Emergent nanofabrication techniques are presented in *Nanoimprint Technology: Nanotransfer for Thermoplastic and Photocurable Polymer* by Jan Taniguchi. Modern topics in mechanics are covered in *Nano and Cell Mechanics: Fundamentals and Frontiers* edited by Espinosa and Bao. Specific implementations and applications are presented in *AC Electrokinetics: Colloids and Nanoparticles* by Morgan and Green, and *Digital Holography for MEMS and Microsystem Metrology* edited by Asundi.

This book, by A. Corigliano, R. Ardito, C. Comi, A. Frangi, A. Ghisi, and S. Mariani, presents in-depth mechanics analyses and powerful numerical modeling strategies particularly useful to the field of microsystems. Topics such as mechanical characterization at the microscale, fracture and fatigue of microsystems, intrinsic and extrinsic dissipative mechanisms, and surface interactions are treated in a unique perspective and depth not found in other MEMS books. Given the focus and depth of the topics covered, the reader will find this book complementary to classic books in MEMS such as *Fundamentals of Microfabrication*, by M. Madou, and *Microsystem Design*, by S. Senturia.

The book starts by covering the fundamentals needed to design and anlayze microsystems. An extensive chapter is dedicated to the analytical and numerical treatment of linear and nonlinear mechanics, which is essential to understand design principles exploiting geometric and electrostatic stiffness variations in various sensing modalities.

The authors have made numerous and unique contributions to the field by working closely with STS engineers on microsystems, which designs were translated into commercial products. Hence the book contains many examples of novel microsystems, e.g., three-axis resonant accelerometers with optimized performance and integration. In addition to accelerometers, the book also discusses Coriolis-based gyroscopes and vibrating Lorentz force magnetometers. As such, the reader will find in one book all the knowledge needed to gain insight into the use of microsystems in inertial measurement unites typically found in modern cell phones, unmanned or remotely operated vehicles, space satellites, and civil and military aviation. Other applications include the use of piezoelectric materials in resonators and energy harvesters.

The pedagogic treatment given by the authors makes this book suitable for inclusion in undergraduate and graduate courses in engineering. Students who are familiar with undergraduate dynamics, strength of materials, and electricity and magnetisms, should be able to grasp the majority of the book content. Some chapters may require background in continuum mechanics and finite element modeling to take full advantage of their contents. The book as a whole should be considered as recommended reading for researchers across a wide range of disciplines including materials science, mechanical engineering, electrical engineering, and applied physics.

Horacio Espinosa
Series co-editor

Preface

This book originated from the experience that our research group, working on the mechanics of solids and structures, had the opportunity of accumulating in the fascinating field of microsystems.

Microsystems are multidisciplinary devices that are creating new opportunities in many markets, such as the consumer sector and the automotive industry, and are now undergoing a great expansion, owing to the emerging Internet of Things.

When one looks inside these very small machines, one soon realizes that, in addition to many fundamental electronic components and read-out circuits, in many cases their core is based on mechanics and related complex interactions in other physical domains.

Another thing that newcomers to the microsystem world very quickly learn is that microsystems design is related to the whole process of fabrication, with related high technology processes, up to the final packaging and assembly, and to many reliability issues that can arise along the long journey from the initial idea to the final product. *Design for reliability* is a necessity if one is to create successful commercial products.

Mechanics can greatly help in understanding the basic working principles of microsystems, in proposing new designs and in understanding many reliability issues. To put it briefly: good mechanics is of paramount importance for good microsystems.

The purpose of this book is precisely to give researchers, students and designers a mechanical perspective of the world of microsystems and to focus on those mechanically related aspects that, in our view, are most important for the study and design of microsystems. Therefore, this is not a book about the whole world of microsystems; many important features of microsystem research and design are not dealt with here, such as microfabrication technologies, electronics and control theory.

The book is divided into three main parts.

The first part, 'Fundamentals', gives the reader fundamental knowledge for understanding the microsystems world; it contains notions that are traditionally given in various disciplines, such as solid and structural mechanics, fluid mechanics, electrostatics, thermal analysis and a first introduction to typical coupled and nonlinear problems encountered in microsystems.

The second part, 'Devices', contains examples of typical microsystems interpreted from the point of view of mechanics. Important inertia devices such as accelerometers and gyroscopes are discussed, together with other less common but important devices like micropumps and energy harvesters.

The third part, 'Reliability and dissipative phenomena' has the purpose of illustrating, in some detail, fundamental reliability issues in microsystems: fracture and fatigue, accidental drops, residual stresses, damping phenomena and stiction.

In our opinion, this book could be used in various ways. It can be a useful reference for microsystems researchers and designers looking for precise information on many mechanical issues in microsystems and for help in understanding and modelling some reliability issues. It can be used in graduate courses, either as an introduction to the mechanics of microsystems (Parts One and Two), or in more advanced courses (Parts Two and Three).

Milano, March 2017

Alberto Corigliano
Raffaele Ardito
Claudia Comi
Attilio Frangi
Aldo Ghisi
Stefano Mariani

Acknowledgements

Our group started working in the world of microsystems thanks to a collaboration with STMicroelectronics in 2002; the experience accumulated over the last 15 years would not have been possible without the intense academia–industry collaborations that consolidated during these years. A special acknowledgement goes to the whole MEMS group of STMicroelectronics, which includes mechanical and electronic designers and experts in fabrication processes.

Another key ingredient behind our expertise in microsystems is the active collaboration with other research groups on the Politecnico di Milano, working at the Departments of Electronics, Materials, Mechanics and Mathematics. We are perfectly aware that without close multidisciplinary collaborations, many features of microsystems are simply intractable and therefore we would like to acknowledge the importance of their contribution behind many of the results discussed in this book.

Many results and experiences described in this book are also based on the work of many young collaborators, post-docs, Ph.D. and M.S. students; to them our thanks for their fundamental contribution and for giving us the opportunity of introducing them to the strange and fascinating world of microsystems.

Acknowledgements

Notation

Latin symbols

a	crack length
a_i	unknown variables in the method of weighted residuals
$\mathbf{a}$	acceleration
A	cross-sectional area of beam
$\mathbf{b}$	body force
B	beam width; magnetic field intensity
c	damping coefficient
c_h	specific heat
$\mathbf{c}$	distance vector between two origins; material compliance matrix
C	concentrated couple in a beam
C	capacitance
$\mathbf{d}$	material stiffness matrix
$\mathbf{D}$	electric displacement
E	Young's modulus
$\mathbf{E}$	electric field
f_0	frequency
$\mathbf{f}$	surface force
F_elec	electrostatic force
F_v	viscous force
$\mathbf{F}$	concentrated force
$\mathbf{F}_\mathrm{e}$	elastic force
g	acceleration due to gravity
g_0	initial gap in a capacitor
G	energy release rate
G_c	critical energy release rate
h	convection constant; function in Weibull approach
H	beam height
$\mathbf{i}, \mathbf{j}, \mathbf{k}$	unit vectors of Cartesian axes
J	moment of inertia of beam
J_T	geometrical torsional stiffness of beam
k	stiffness of linear elastic spring; thermal conductivity
k_elec	electrostatic stiffness
K_I	mode I stress intensity factor

K_{II}	mode II stress intensity factor
K_{III}	mode III stress intensity factor
K_{IC}	material fracture toughness in mode I
$\mathbf{K}$	stiffness matrix
l	length of beam; length of other device
l_{pz}	length of fracture process zone
L	length of beam
$\mathcal{L}$	Lagrangian functional
$\mathbf{L}_O$	angular momentum with respect to point O
m	point mass; Weibull material parameter
M	bending moment in a beam
M_{T}	torsional moment in a beam
$\mathbf{M}$	mass matrix
n	axially distributed load in a beam
N	axial force in a beam
O	origin of a reference system
p	hydrostatic part of a stress tensor or distributed load in a beam orthogonal to beam axis; gas pressure
P_{e}	external power
P_{f}	probability of failure in Weibull approach
P_{i}	internal power
P_0	assigned axial load in a beam
q	electric charge; flow rate
q_{int}	electric charge contained in a volume
$\mathbf{q}$	heat flux
Q_T	internal source of thermal power per unit volume
$\mathbf{Q}$	linear momentum
r	radial coordinate in linear elastic fracture mechanics
R	error; residual in the method of weighted residuals
R	stress ratio in fatigue
s	out-of-plane beam thickness
$\mathbf{s}$	deviatoric stress
S	surface in a capacitor
$\mathcal{S}$	action functional
t	time; thickness
T	temperature
$\mathcal{T}$	kinetic energy
u	displacement function
$\tilde{u}$	approximating function in the method of weighted residuals
$\mathbf{u}$	displacement vector
U	displacement amplitude
v	displacement function; stroke volume
$\mathbf{v}$	velocity vector
w_i	weight functions in the method of weighted residuals
W	shear force in a beam
W_{nc}	work of nonconservative forces
x, y, z	Cartesian axes

$\mathbf{x}$	position vector
Y	time-dependent amplitude of free vibration of beam

Greek symbols

α	coefficient of thermal expansion
β	parameter in Weibull approach
Γ	boundary of a deformable body
Γ_{f}	film fracture energy
Γ_f	free boundary of a deformable body
Γ_u	constrained boundary of a deformable body
δ	displacement jump across a crack
ϵ_0	permittivity of free space
ϵ_{r}	relative permittivity
$\boldsymbol{\varepsilon}$	column matrix of strains
η	axial deformation of beam
θ	angle of torsional rotation per unit length in a beam; angle of rotation in a rotational microactuator
ϑ	Euler angle; circumferential coordinate in linear elastic fracture mechanics
λ	first Lamé constant
μ	second Lamé constant
μ_{f}	dynamic viscosity
ν	Poisson's ratio
ν_{f}	kinematic viscosity
ξ	nondimensional constant for damping
Π	total potential energy
ρ	mass density per unit volume
ρ_{ql}	electric charge density per unit length
$\rho_{q\Gamma}$	electric charge density per unit surface area
$\rho_{q\Omega}$	electric charge density per unit volume
$\sigma_{\mathrm{I}}, \sigma_{\mathrm{R}}, \sigma_{\mathrm{T}}$	incident, reflected, transmitted stress in impacts
σ_{nom}	nominal stress in Weibull approach
$\sigma_{\mathrm{p}i}$	principal stresses
σ_{s}	nominal stress at rupture
σ_u	Weibull material parameter
σ_0	Weibull material parameter
$\tilde{\sigma}$	equivalent stress in Weibull approach
$\boldsymbol{\sigma}$	column matrix of stresses
$\tau(\delta)$	cohesive crack law
φ	Euler angle; phase angle; electrostatic potential
φ_i	displacement function at crack tip in linear elastic fracture mechanics
φ_{p}	constant polarization voltage
χ	beam curvature
ψ	Euler angle
ψ_i	basis functions in the method of weighted residuals

ψ_{ij}	stress functions at crack tip in linear elastic fracture mechanics
ω	undamped circular natural frequency
ω_{d}	damped circular natural frequency
$\boldsymbol{\omega}$	angular velocity
Ω	volume of deformable body
Ω_{r}	equivalent volume in Weibull approach

Acronyms

dof	degree of freedom; degrees of freedom
MEMS	microelectromechanical system; microelectromechanical systems
NEMS	nanoelectromechanical system; nanoelectromechanical systems
RH	relative humidity
1D	one-dimensional
2D	two-dimensional
3D	three-dimensional

Mathematical operators and symbols

C^1	continuous function with continuous first derivative
$\dot{u}$	first derivative of u with respect to time
$\ddot{u}$	second derivative of u with respect to time
$u', u'', u''', u^{\mathrm{IV}}$	first, second, third, fourth derivatives of u with respect to x
d	infinitesimal increment
Δ	finite increment
δ	infinitesimal variation
grad	gradient operator: grad $a = \frac{\partial a}{\partial x}\mathbf{i} + \frac{\partial a}{\partial y}\mathbf{j} + \frac{\partial a}{\partial z}\mathbf{k}$
div	divergence operator: div $\mathbf{a} = \frac{\partial a_x}{\partial x} + \frac{\partial a_y}{\partial y} + \frac{\partial a_z}{\partial z}$
curl	curl operator: curl $\mathbf{a} = \left(\frac{\partial a_z}{\partial y} - \frac{\partial a_y}{\partial z}\right)\mathbf{i} + \left(\frac{\partial a_x}{\partial z} - \frac{\partial a_z}{\partial x}\right)\mathbf{j} + \left(\frac{\partial a_y}{\partial x} - \frac{\partial a_x}{\partial y}\right)\mathbf{k}$
∇^2	Laplace operator: $\nabla^2 a = \frac{\partial^2 a}{\partial x^2} + \frac{\partial^2 a}{\partial y^2} + \frac{\partial^2 a}{\partial z^2}$
∇^4	double Laplace operator: $\nabla^4 a = \frac{\partial^4 a}{\partial x^4} + \frac{\partial^4 a}{\partial y^4} + \frac{\partial^4 a}{\partial z^4} + 2\frac{\partial^4 a}{\partial x^2 \partial y^2} + 2\frac{\partial^4 a}{\partial x^2 \partial z^2} + 2\frac{\partial^4 a}{\partial y^2 \partial z^2}$
$\wedge$	vector product: $\mathbf{a} \wedge \mathbf{b}$
$\cdot$	scalar product: $\mathbf{a} \cdot \mathbf{b}$
$\otimes$	tensor product: $(\mathbf{a} \otimes \mathbf{b}) \cdot \mathbf{c} = \mathbf{a}(\mathbf{b} \cdot \mathbf{c})$

About the Companion Website

This book is accompanied by a companion website, which hosts supplementary material:

www.wiley.com/go/corigliano/mechanics 

The website includes:

- Biographical notes
- Slides for books contents presentation
- Images and videos

Scan this QR code to visit the companion website.

1

Introduction

The purpose of this first chapter is to give, in a brief and simple way, preliminary information on what are currently termed microsystems, how they are fabricated and the role of mechanics in their design and overall behaviour.

This preliminary information helps in understanding the purpose, organization and limitations of this book. At the end of this introductory chapter, a short description of the contents of each chapter is given.

1.1 Microsystems

It is usually recognized that the history of microsystems started with the talk given by Feynman on December 1959 (Feynman, 1960) and with the creation of the resonant gate transistor (Nathanson *et al.*, 1967).

After a period in which microsystems were the subject of intense research activity in academic laboratories, the MEMS industry started with a large number of spin-offs and small and medium enterprises. More recently MEMS production expanded, mainly thanks to the wide diffusion of inertial MEMS (microaccelerometers and microgyroscopes) in consumer market products.

The 'more than Moore' trend driven by MEMS was able to compensate a partial saturation of ICT product growth and opened new possibilities for distributed sensing and monitoring. Nowadays, the new popular phrases 'Internet of Things' and 'Internet of Everything' are paradigms of an increasingly connected world in which many kinds of information can be automatically transmitted by everyday life smart products such as phones, tablets, watches, glasses and cars.

We are now in a 'more than MEMS' era in which, after a first phase mainly driven by microsystem miniaturization, augmented capabilities are added to microsystems; for example, multiaxis sensor units, which combine accelerometers, gyroscopes, pressure and magnetic field sensors. In the near future, we will witness a new phase, in which microsystems inserted in wireless sensor networks will have extra abilities and performances, such as energy autonomy and high reliability, even in extremely aggressive environments.

A first glance at microsystems can be obtained examining the popular acronym (MEMS) which stands for *micro electro mechanical system*.

Mechanics of Microsystems, First Edition. Alberto Corigliano, Raffaele Ardito, Claudia Comi, Attilio Frangi, Aldo Ghisi, and Stefano Mariani.

Companion website: www.wiley.com/go/corigliano/mechanics

Micro means that we are speaking of small devices, in which the single smallest dimension can also be at the submicrometre scale. A complete MEMS containing mechanical and electronic parts can be of the order of millimetres. These dimensions tell us that we are not yet in the field of the nanoscopic world, but that we are very near to it. Some phenomena, e.g. interaction forces between surfaces almost in contact, must in fact be studied and modelled taking into account the nanoscale.

To have an immediate feeling of the microsystems' dimensions, let us observe Figure 1.1, which shows schematically a commercial product. The fully packaged device has dimensions 3.5 mm × 3 mm × 1 mm; it has, therefore, a volume of 10.5 mm^3. Inside the device there are three sensors: a three-axis accelerometer, a three-axis gyroscope and a three-axis magnetometer. This is why the device is called a *nine-axis module.*

If we now make the exercise of superposing the device of Figure 1.1 on a 1 euro cent coin, we realize that more than 16 nine-axis MEMS can be arranged on its surface, as shown in Figure 1.2. This gives the impressive result that on a surface equivalent to 1 euro cent, one can, in principle have, $9 \times 16 = 144$ sensing signals!

This remark gives a clear idea of the potentialities behind the microsystems technology; these small machines can really be placed everywhere!

Electro means that inside these small devices there are electric and electronic components; these are needed to create connections between the MEMS and the external world, to transform physical information (inertia forces, pressure, ...) in electric signals and also to activate movements inside the device. In addition to the electronic components needed to make the MEMS core work, it is always necessary to add read-out electronic circuits. In Figure 1.1, electric and electronic components are represented by the small wires that connect various components and by the thin lower plate that represents the application-specific integrated circuit for read-out control. Definitely, electronics form a very important part of every microsystem.

Mechanical means that microsystems, in addition to the fact that they are very small and contain electronic components, have some portion of their architecture that works thanks to mechanical principles. For instance, inside the MEMS there can be very small beams or plates that are loaded by inertia forces caused by the overall acceleration of the device and that can be considered as structural components exactly in the same way as beams and plates are structural components of large structures, such as the buildings in

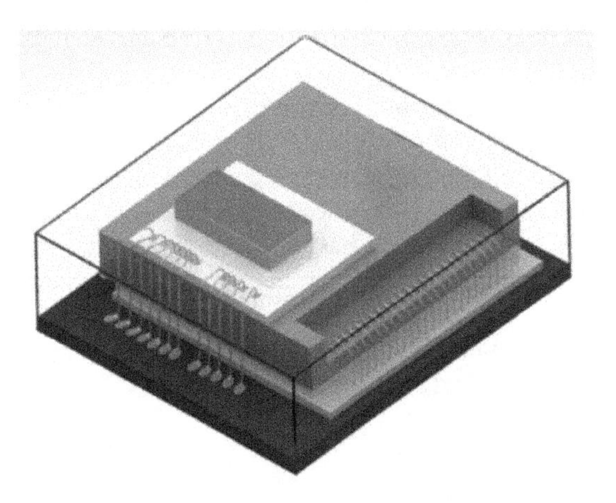

Figure 1.1 Typical commercial microsystem: a nine-axis accelerometer, gyroscope, magnetometer.

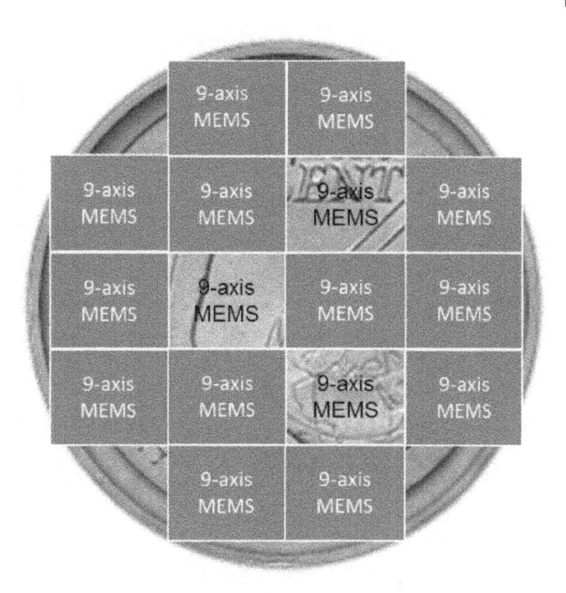

Figure 1.2 16 nine-axis MEMS superposed on a 1 euro cent coin.

which we all live. In addition to this immediate evidence of mechanics in microsystems, there are many other implications, as briefly discussed in Section 1.3.

System means that the microdevices are not simple; they must be interpreted as complex systems combining various components: the electric and electronic parts, the mechanical parts, and possibly other parts, such as optical components. Moreover, they may be produced by means of complex fabrication processes and complex integration of different portions.

1.2 Microsystems Fabrication

The fabrication of microsystems is a complex process that was mainly adapted from those used in the fabrication of integrated circuits. Here, only a few concepts are given with reference to one of the possible fabrication processes; for a full understanding of microsystems fabrication processes, the reader must consult specialized textbooks, e.g. Madou (2002).

The starting point for current microsystems is a *wafer* of monocrystalline silicon, which is used as a support on top of which various materials can be subsequently deposited by means of various techniques. The wafer is a flat disc very similar to those used in CDs for music or data storage.

Microsystems industries with high volume production currently use 8 inch wafers, i.e. discs with a diameter of 20.32 cm.

Various materials are added on top of the wafer, such as metals or other kinds of silicon, e.g. polycrystalline silicon. Each added layer has a pre-specified thickness and role in the design and fabrication of the device.

Beside the stratification, or *deposition*, of various materials, microsystems fabrication has another recurrent and fundamental step, which is the selective elimination of portions of one or more deposited layers. This is the so-called *lithography* process, which means *writing on stone* (*lithos* = stone, in Greek).

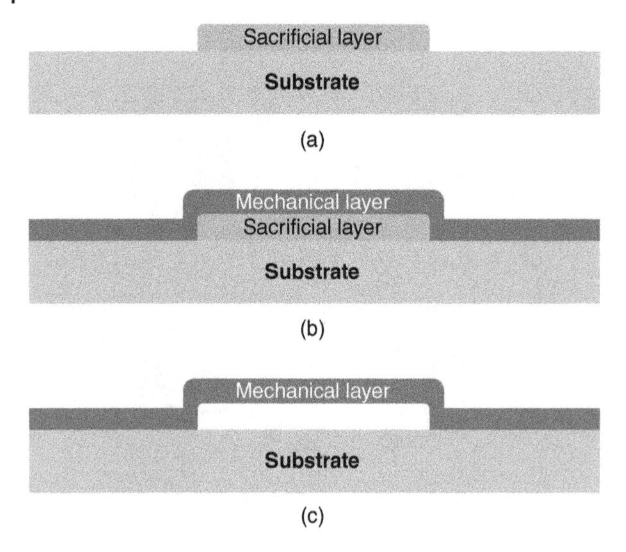

Figure 1.3 Lithographic process for MEMS.

Lithography is used in the production of electronic circuits and is conceptually similar to artistic lithography, in which drawings are chemically engraved on stone surfaces.

During a lithographic phase, a thin film, called resist, is first exposed to light only on some portions of the surface, depending on the pattern or *drawings* to be reproduced on the wafer. The exposure of thin films to light modifies them in such a way that when chemical substances are subsequently deposited on the resist, portions of it are eliminated and therefore can be used as a *mask* that covers the wafer surface, following the desired drawing. At this point, the mask can be used on top of the wafer and additional chemical substances can selectively eliminate portions of the pre-deposited layers on the wafer surface, again depending on the desired pattern.

In addition to classical lithography, in the microsystem production, additional phases are added to enable the selective elimination of portions of materials below some pre-deposited layers. In this way, there is the possibility not only to dig but also to create suspended structures, beams and plates, as shown in Figure 1.3.

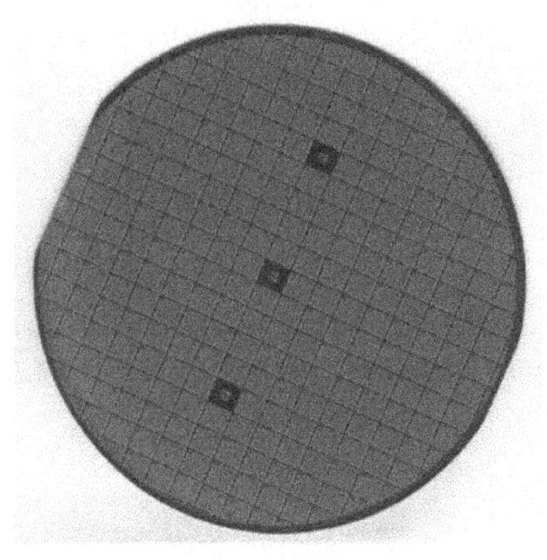

Figure 1.4 Patterned wafer.

With a sequence of deposition and selective elimination, which can involve various lithographic phases, the whole wafer is patterned as desired, as shown in Figure 1.4.

Because microsystems are very small, it is possible to pattern many small devices on the same wafer (e.g. with a 20.32 cm diameter); this is the real advantage of silicon-related technology: to produce hundreds of identical devices on the same wafer! This means low unit cost and also high yield in mass production, since the differences among various devices are really reduced.

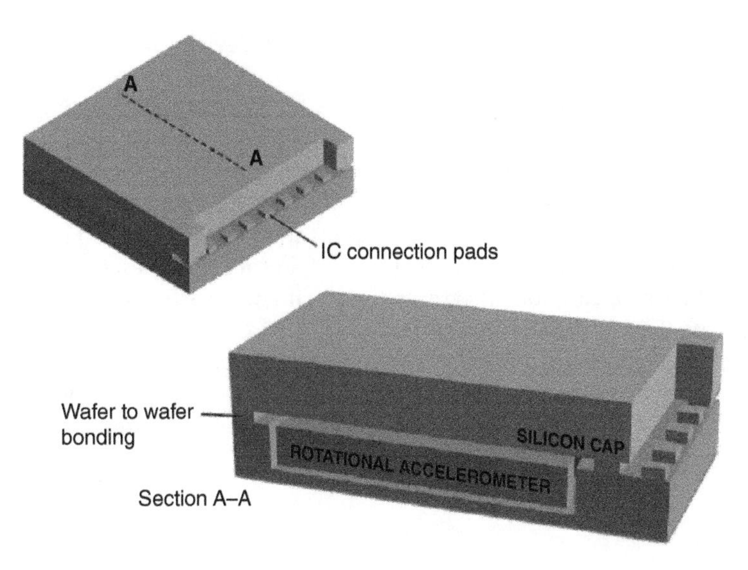

Figure 1.5 Single die and cross-sectional view. IC, integrated circuit.

The fabrication process does not end with the patterned wafer, which is only part of the story. A second wafer must be prepared, usually with much simpler patterning than the first one, to be used as a *cap* for the whole set of devices patterned on the first wafer.

A procedure called *wafer-to-wafer bonding* is thus put in place; basically, some kind of *glue* is deposited on strips of the wafers that separate each device, then the two wafers are pressed towards each other by a machine that also controls the processing temperature. At the end of the thermocompressive bonding process, the two wafers are glued along the corresponding separation strips.

To obtain single devices, the two glued wafers must be cut, e.g. with hydro-jets, corresponding to the separation strips. A so-called *die* is thus obtained, as shown in Figure 1.5.

At this point in the fabrication process, one has obtained the portion of the final MEMS that is represented in Figure 1.1 as the central grey box.

To obtain the final product, the die must be connected to the electronic circuit and possibly to other dies, as shown in Figure 1.1, and everything must be protected by the final *package.*

The package is of crucial importance in protecting the device; in many cases, it is made of black plastic, exactly like the one used for a processing unit in a PC.

1.3 Mechanics in Microsystems

As anticipated by these remarks, mechanics is a fundamental ingredient in microsystems. It can be present in the realization of microsystems from the first idea, e.g. for mechanically based inertia MEMS, to the fabrication process.

Solid and structural mechanics is necessary to interpret the behaviour of small structures on board microsystems. The mechanical or structural design can give the right direction to increasing the performance of the device.

Fluid mechanics is the tool to study and design micro-fluidic devices, like micropumps.

Mechanical behaviour coupled with other physical fields, e.g. electric, magnetic and thermal, can be another way in which mechanics enters microsystems. Coupled problems are intrinsic in MEMS, which could be more realistically and precisely named *micro multiphysics systems*, but the resulting acronym would not be very smart!

In addition to the initial idea and to the design, mechanics enters the fabrication process. First because design needs materials with specific mechanical properties, but also because the wafer and the single components are stressed in many ways by thermomechanical processes during the fabrication phase.

A third area in which mechanics has an important role is *reliability*. Many microsystems reliability issues are related to mechanics, for example, accidental drop or spontaneous adhesion or stiction. Without good mechanical interpretations, it is impossible to obtain complete control of the device reliability.

1.4 Book Contents

After these introductory thoughts on microsystems, it is now possible to describe the motivations behind this book and the specific choices made in selecting its contents.

This is not a book on the microsystems world in general. This world today is too complex and many good books have already appeared dealing with all aspects of MEMS; here we mention Senturia (2001), for its quality and for its important historical role. Many other MEMS books can nowadays be found on the market that cover all aspects of MEMS; examples are those of Adams and Layton (2010), Hartzell, da Silva and Shea (2011), Korvink and Paul (2006) and Pelesko and Bernstein (2003).

This book does not contain chapters on microfabrication technology, microsystem electronics, control theory for circuits or applications of microsystems. The main focus of this book, as declared in the title, is the role of mechanics in the study, design and use of microsystems. We tried to transfer our experience in this book that we believe can be useful for students, researchers and designers. The various chapters composing this book therefore contain rigorous descriptions of mechanical or mechanical coupled problems in MEMS.

Reference lists are given at the end of each chapter. They cannot be considered as separate state of the art for each subject; they are useful references to act as a good starting point to obtain more insight in each specific subject.

The book is divided into three main parts: 'Fundamentals', 'Devices' and 'Reliability'.

The purpose of the first part, containing Chapters 2 and 3, is to give the reader fundamental knowledge for the understanding of the microsystems world from a mechanical perspective. In Chapter 2, a simple yet rigorous overview of the mechanics of rigid bodies and deformable solid bodies and of fluid mechanics is proposed, thus setting the stage for the understanding of basic mechanical design. Owing to the intrinsic multiphysics character of microsystems, Chapter 2 also contains introductions to electrostatics and the coupled electromechanical problem and to heat conduction and the relevant coupled thermomechanical problem. Smart piezoelectric materials are now becoming more important in new microsystems products; this is why a brief discussion on piezoelectric materials is inserted in Chapter 2. Chapter 3 contains a first guide on how to build computing models for MEMS and also a discussion on some particular nonlinear mechanical problems that can be met in microsystems.

The second part is focused on devices and contains Chapters 4 to 10. The purpose of the second part is to show how various devices, like accelerometers, gyroscopes and resonators, work, also thanks to mechanics, and to explain in some detail which kind of mechanics must be used in different cases.

The third part deals with reliability and dissipative phenomena and goes from Chapter 11 to the final Chapter 16. The purpose of the third part is to enter with some detail into mechanical-related reliability problems in MEMS. It starts with Chapter 11 in which techniques for the *on-chip* mechanical characterization of microsystems are discussed and in which the Weibull approach for brittle materials is explained. The subsequent two chapters, Chapters 12 and 13, deal with the important issues of fracture and fatigue and of the related consequences in accidental drop events. Chapter 14 contains an overview of how residual stresses can originate in microsystems and how they can be computed. Damping in microsystems is discussed in depth in Chapter 15, while the final chapter, Chapter 16, is dedicated to the important reliability problem caused by spontaneous adhesion or *stiction*.

References

Adams, T. and Layton, R. (2010) *Introductory MEMS: Fabrication and Applications*, Springer.

Feynman, R.P. (1960) There's plenty of room at the bottom, in *Miniaturization* (ed. H.D. Gilbert), Reinhold, p. 282.

Hartzell, A., da Silva, M. and Shea, H. (2011) *MEMS Reliability*, Springer.

Korvink, J. and Paul, O. (eds) (2006) *MEMS: A Practical Guide to Design, Analysis and Applications*, Springer.

Madou, M. (2002) *Fundamentals of Microfabrication*, CRC Press.

Nathanson, H., Newell, W.E., Wickstrom, R.A. and Ransford Davis, Jr., J. (1967) The resonant gate transistor. *IEEE Transactions on Electron Devices*, **14** (3), 117–133.

Pelesko, J. and Bernstein, D. (2003) *Modeling MEMS and NEMS*, CRC Press.

Senturia, S. (2001) *Microsystem Design*, Springer.

Part I

Fundamentals

2

Fundamentals of Mechanics and Coupled Problems

2.1 Introduction

The purpose of this preliminary chapter is to recall fundamental concepts, formulations and equations that are at the basis of the majority of problems encountered in everyday microsystems mechanical design.

Many formulations will be given without proof; it is, in fact, out of the scope of this book to make a complete treatise on each subject.

We start recalling in Section 2.2 the equations that govern the movement of a rigid body subject to a set of external actions, which are fundamental for the understanding of the mechanical principles at the basis of typical inertial MEMS, such as accelerometers and gyroscopes. The chapter then focuses on the schematization of solids and fluids as continua; basic notions concerning the response of elastic deformable bodies subject to external actions in static and dynamic regimes are given in Section 2.3, while introductory concepts of fluid mechanics are discussed in Section 2.4. These are needed, for example, to understand the mechanical responses of deformable parts in MEMS, such as suspension beams and their interaction with the surrounding air.

The subsequent sections of the chapter are devoted to the most important and most frequently found multiphysics problems in microsystems; these are the electromechanical and the thermomechanical problems, respectively described in Sections 2.5 and 2.7, after the introduction of the electrostatic and thermal problems separately. Also these notions are at the basis of a microsystem design because they govern the behaviour of electrostatic sensors and actuators and of electrothermomechanical actuators. Moreover, thermomechanical coupling is frequently at the origin of fabrication-induced residual eigenstresses.

Section 2.6 presents basic concepts on piezoelectric materials and on the way in which their intrinsic electromechanical coupled response can be described in terms of constitutive equations. Piezoelectric sensing and actuation is becoming more and more important in microsystems and the concepts given here are fundamental for the understanding of real devices.

For the reader who would like to know more on each of the subjects dealt with in this chapter, some major references are suggested at the beginning of each section and listed at the end of the chapter.

Mechanics of Microsystems, First Edition. Alberto Corigliano, Raffaele Ardito, Claudia Comi, Attilio Frangi, Aldo Ghisi, and Stefano Mariani.

Companion website: www.wiley.com/go/corigliano/mechanics

The expert researcher or microsystems engineer can skip this chapter and go directly to the thematic chapters of the book.

2.2 Kinematics and Dynamics of Material Points and Rigid Bodies

This section starts with the basic concepts of position and motion description referred to a so-called *material point* and concludes with the description of *rigid-body* dynamics. The concepts and formulae here recalled are fundamental for the understanding of inertial microsystems and their design (see Part Two). To have a complete discussion on what is presented in this section, please refer to textbooks on classical mechanics and classical dynamics of systems, e.g. Goldstein, Poole and Safko (2002) or Thornton and Marion (2003).

2.2.1 Basic Notions of Kinematics and Motion Composition

Let us consider a point P referred to an orthogonal reference frame with origin O, coordinates x, y, z (or x_1, x_2, x_3) and unit vectors $\mathbf{i}$, $\mathbf{j}$, $\mathbf{k}$, as in Figure 2.1; the position of point P is given by the vector $\mathbf{x}$:

$$\mathbf{x} = x\mathbf{i} + y\mathbf{j} + z\mathbf{k}. \tag{2.1}$$

The position of point P can be represented in a different reference frame, with origin O', coordinates x', y', z' (or x'_1, x'_2, x'_3) and unit vectors $\mathbf{i}'$, $\mathbf{j}'$, $\mathbf{k}'$ (Figure 2.2):

$$\mathbf{x}' = x'\mathbf{i}' + y'\mathbf{j}' + z'\mathbf{k}'. \tag{2.2}$$

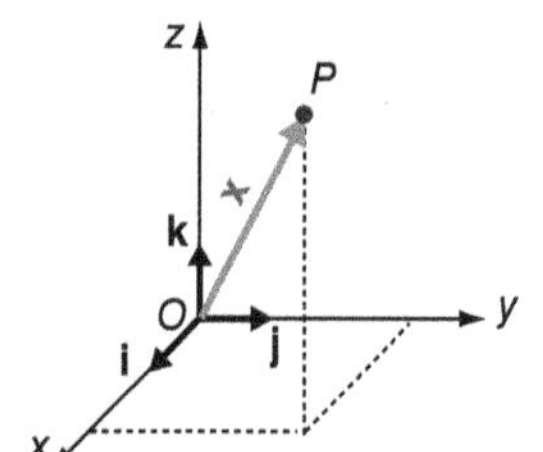

Figure 2.1 Reference system.

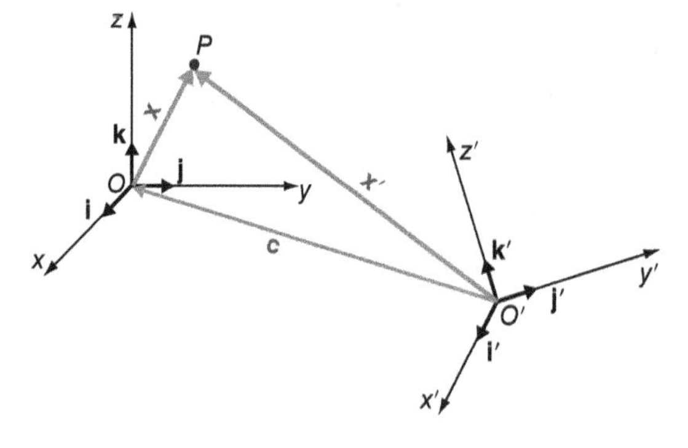

Figure 2.2 Two reference systems.

Denoting by $\mathbf{c}$ the position vector of origin O in the new reference frame, as shown in Figure 2.2, the relations between the position vectors in the two reference frames is:

$$\mathbf{x}' = \mathbf{c} + \mathbf{x}. \tag{2.3}$$

Equation 2.3 can be written in terms of components x_i', $i = 1, 2, 3$, in the new reference frame. Denoting the coordinates of vector $\mathbf{c}$ by c_i, $i = 1, 2, 3$, and the components of unit vectors $\mathbf{i}, \mathbf{j}, \mathbf{k}$ in the new reference frame by n_{ij}, $i, j = 1, 2, 3$, one has:

$$x_i' = c_i + \sum_{j=1}^{3} n_{ij} x_j, \quad i = 1, 2, 3. \tag{2.4}$$

It is worth noting that, in view of the previous notation, n_{ij} is the ith component in the new reference frame of the jth unit vector of the initial reference frame, where unit vectors are ordered from 1 to 3 as $\mathbf{i}, \mathbf{j}, \mathbf{k}$.

The movement of point P in space can be represented by the following three functions of time t:

$$x = x(t), \;\; y = y(t), \;\; z = z(t), \tag{2.5}$$

or, in vectorial notation,

$$\mathbf{x} = \mathbf{x}(t). \tag{2.6}$$

If represented in the four-dimensional $\mathbf{x}$–t space, Equations 2.5 are the so-called *world line* of point P, while if represented in three-dimensional $\mathbf{x}$ space, Equations 2.5 represent the *trajectory* of point P. Knowing the trajectory of a point P means being able to follow its movement in space. The derivatives with respect to time of Equation 2.5 are the components v_x, v_y, v_z of the velocity vector $\mathbf{v}$ of point P:

$$v_x = \frac{dx}{dt}, \;\; v_y = \frac{dy}{dt}, \;\; v_z = \frac{dz}{dt}, \;\; \mathbf{v} = \frac{d\mathbf{x}}{dt}, \;\; \mathbf{v} = v_x\mathbf{i} + v_y\mathbf{j} + v_z\mathbf{k}. \tag{2.7}$$

With an additional derivation with respect to time, we obtain the components a_x, a_y, a_z of the acceleration vector $\mathbf{a}$ of point P:

$$a_x = \frac{dv_x}{dt}, \;\; a_y = \frac{dv_y}{dt}, \;\; a_z = \frac{dv_z}{dt}, \;\; \mathbf{a} = \frac{d\mathbf{v}}{dt}, \;\; \mathbf{a} = a_x\mathbf{i} + a_y\mathbf{j} + a_z\mathbf{k}. \tag{2.8}$$

The position vector $\mathbf{x}$, together with the velocity vector $\mathbf{v}$ and the acceleration vector $\mathbf{a}$ are functions of time; their physical dimensions are, respectively, length, length/time and length/time^2.

In many situations it is important to be able to interpret the movement of a point P as seen from different *observers*, i.e. as described in different reference frames, like the two already considered in Figure 2.2.

Let us first consider the following situation: the position of point P in the reference frame with origin O is fixed, while this reference frame is travelling in space with respect to the reference frame with origin O'. Starting from Equations 2.3 and 2.4, and computing the time derivatives, one obtains the velocity of point P as seen by the observer O':

$$\mathbf{v}' = \frac{\mathrm{d}\mathbf{x}'}{\mathrm{d}t} = \frac{\mathrm{d}\mathbf{c}}{\mathrm{d}t} + \frac{\mathrm{d}\mathbf{x}}{\mathrm{d}t}. \tag{2.9}$$

$$v_i' = \frac{\mathrm{d}x_i'}{\mathrm{d}t} = \frac{\mathrm{d}c_i}{\mathrm{d}t} + \sum_{j=1}^{3} \frac{\mathrm{d}n_{ij}}{\mathrm{d}t} x_j, \quad i = 1, 2, 3. \tag{2.10}$$

The first term in the right-hand side of Equations 2.9 and 2.10 is the velocity $\mathbf{v}_0$ of point O, while the second term depends directly on how the directions of axes x, y, z vary in time with respect to axes x', y', z'. This last term therefore contains the information related to the rotation of axes x, y, z and it can be given a compact expression introducing the rotation vector or angular velocity $\boldsymbol{\omega}$; Equation 2.9 can thus be substituted by

$$\mathbf{v}' = \mathbf{v}_0 + \boldsymbol{\omega} \wedge (P - O), \tag{2.11}$$

where the symbol $\wedge$ denotes a vectorial product. The vector $\boldsymbol{\omega}$ is the angular velocity of the frame with origin O with respect to the frame with origin O'; its components are:

$$\omega_1 = \sum_{j=1}^{3} \frac{\mathrm{d}n_{3j}}{\mathrm{d}t} n_{2j}, \quad \omega_2 = \sum_{j=1}^{3} \frac{\mathrm{d}n_{1j}}{\mathrm{d}t} n_{3j}, \quad \omega_3 = \sum_{j=1}^{3} \frac{\mathrm{d}n_{2j}}{\mathrm{d}t} n_{1j}. \tag{2.12}$$

It is now possible to obtain an expression analogous to Equation 2.11 for the acceleration, as seen by observer O'; it is given by the sum of three contributions, as

$$\mathbf{a}' = \mathbf{a}_0 + \frac{\mathrm{d}\boldsymbol{\omega}}{\mathrm{d}t} \wedge (P - O) + \boldsymbol{\omega} \wedge (\boldsymbol{\omega} \wedge (P - O)). \tag{2.13}$$

Equations 2.11 and 2.13 are very important for the description of motion of material points and of rigid bodies, as will be described in Section 2.2.4.

We now continue the discussion concerning the way in which a movement is seen by different *observers* and remove the hypothesis that the position of point P is fixed with respect to the observer O. We consider two observers O and O' and the point P travelling in space. In this case the observer O' will measure the following velocities and acceleration of point P:

$$\begin{aligned} \mathbf{v}' &= \mathbf{v}_\mathrm{r} + \mathbf{v}_\mathrm{d} \\ &= \mathbf{v}_\mathrm{r} + \mathbf{v}_0 + \boldsymbol{\omega} \wedge (P - O), \end{aligned} \tag{2.14}$$

$$\begin{aligned} \mathbf{a}' &= \mathbf{a}_\mathrm{r} + \mathbf{a}_\mathrm{d} + \mathbf{a}_\mathrm{c} \\ &= \mathbf{a}_\mathrm{r} + \left[\mathbf{a}_0 + \frac{\mathrm{d}\boldsymbol{\omega}}{\mathrm{d}t} \wedge (P - O) + \boldsymbol{\omega} \wedge (\boldsymbol{\omega} \wedge (P - O)) \right] + 2\boldsymbol{\omega} \wedge \mathbf{v}_\mathrm{r}. \end{aligned} \tag{2.15}$$

In Equation 2.14, the additional term $\mathbf{v}_\mathrm{r}$ has appeared with respect to Equation 2.11; this is the relative velocity vector, i.e. the velocity of point P as seen by the observer O. The other two terms have been collected in vector $\mathbf{v}_\mathrm{d}$, which is the *drag* velocity, i.e. the velocity of point P as *dragged* by the reference frame with origin in O.

The acceleration in Equation 2.15 has three terms: the relative $\mathbf{a}_\mathrm{r}$ and the *drag* $\mathbf{a}_\mathrm{d}$ acceleration vectors, with meanings analogous to the relative and drag velocities, and the term $\mathbf{a}_\mathrm{c}$, which is the Coriolis acceleration, given by twice the vectorial product of the angular velocity and of the relative velocity vectors. The Coriolis acceleration is zero if vectors $\boldsymbol{\omega}$ and $\mathbf{v}_\mathrm{r}$ are parallel or if vectors $\boldsymbol{\omega}$ or $\mathbf{v}_\mathrm{r}$ are zero.

Equations 2.14 and 2.15 are the starting point for the mechanical design of inertial microsystems such as accelerometers and gyroscopes (see Chapters 4 and 5). These relations are used by imagining that the observer O' is the one with respect to which we would like to measure, e.g., a linear acceleration in a uniaxial accelerometer or a component of the angular velocity vector in a gyroscope.

2.2.1.1 Summary

The main points discussed in this section are the following.

- The description of movement of a point in space has been given in terms of trajectory, velocity and acceleration.
- The description of the movement of a point given by different observers has been discussed.
- The concepts of relative and drag velocities and of relative, drag and Coriolis accelerations have been discussed.

2.2.2 Basic Notions of Dynamics and Relative Dynamics

After the discussion concerning the pure description of the movement in space of a single point P, it is necessary to recall the physical principles that are at the origin of the movement. This involves answering the following questions. Why does a point move? Which are the actions and rules that provoke and govern the movement? This is the field of dynamics.

Classical dynamics is ruled by Newton's famous and fundamental laws; before introducing them, we need to define two important concepts.

- *Inertial reference frame.* All inertial frames are in a state of constant rectilinear motion with respect to one another; they are not accelerating.
- *Material point.* A material point is a point with a given mass m. An isolated material point is in uniform rectilinear movement with respect to an inertial reference frame.

Given these definitions, Newton's three laws of dynamics can be stated as follows.

- *Newton's first law.* In an inertial reference frame, a material point remains in a state of constant velocity unless acted on by an external unbalanced force.
- *Newton's second law.* A material point of mass m subject to a net force $\mathbf{F}$ undergoes an acceleration $\mathbf{a}$ that has the same direction as the force and a magnitude that is directly proportional to the force and inversely proportional to the mass:

$$m\mathbf{a} = \mathbf{F}. \tag{2.16}$$

- *Newton's third law.* The forces of action and reaction between the bodies are equal, opposite and collinear (action–reaction law).

Equation 2.16 tells us how a certain force or sum of forces will cause a material point to move. As in Section 2.2.1, it is also possible to re-interpret the movement on changing the observer. In this case, we must rely on the equations describing the composition of movements; in particular, we can substitute the acceleration in Equation 2.16 into the expression given by Equation 2.15, thus obtaining

$$\begin{aligned} m\mathbf{a} &= m(\mathbf{a}_\mathrm{r} + \mathbf{a}_\mathrm{d} + \mathbf{a}_\mathrm{c}) \\ &= m\left\{\mathbf{a}_\mathrm{r} + \left[\mathbf{a}_0 + \frac{\mathrm{d}\boldsymbol{\omega}}{\mathrm{d}t} \wedge (P - O) + \boldsymbol{\omega} \wedge (\boldsymbol{\omega} \wedge (P - O))\right] + 2\boldsymbol{\omega} \wedge \mathbf{v}_\mathrm{r}\right\} = \mathbf{F}. \end{aligned} \tag{2.17}$$

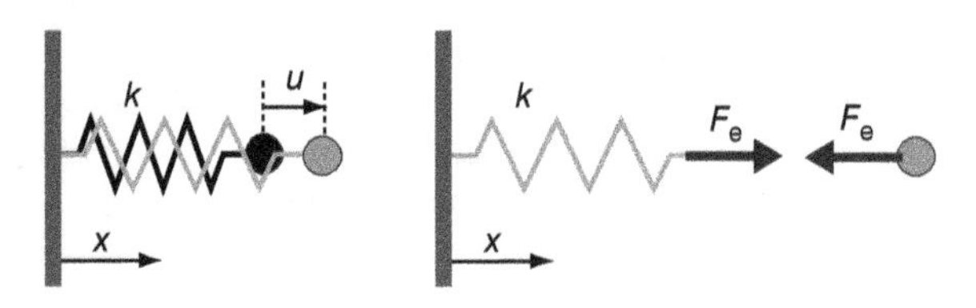

Figure 2.3 Linear elastic spring.

Equation 2.17 governs the motion of a material point as seen by a *noninertial observer*, i.e. an observer that does not move with respect to the inertial observer with rectilinear motion with constant velocity. This is the case of a MEMS containing material points in motion inside it: the microsystem will move in a very general way with accelerations and rotations; hence, an observer in motion with it is noninertial.

Starting from Equation 2.17, it is possible to write:

$$m\mathbf{a}_{\mathrm{r}} = \mathbf{F} - m\mathbf{a}_{\mathrm{d}} - m\mathbf{a}_{\mathrm{c}} \equiv \mathbf{F} + \mathbf{F}_{\mathrm{d}} + \mathbf{F}_{\mathrm{c}}, \tag{2.18}$$

where the two terms $\mathbf{F}_{\mathrm{d}}$ and $\mathbf{F}_{\mathrm{c}}$ have been defined; these are called the *apparent drag force* and *apparent Coriolis force*, respectively.

We can therefore conclude that it is also possible to study the movement of a material point with respect to a noninertial reference frame, provided that in Newton's second law the external forces and the apparent forces are included.

As meaningful examples of forces that can act on a material point inside a MEMS, we mention here the elastic restoring force, the weight and the viscous damping force.

The first is a manifestation of the elasticity of deformable bodies, which is discussed in more detail in Section 2.3 and in the simplest case can be represented by a linear elastic spring, like that shown in Figure 2.3: when the spring is elongated by a displacement u, the elastic spring reacts with a force equal to the displacement multiplied by a proportionality constant k named the *elastic stiffness* of the spring, which is dimensionally equal to a force divided by a length, i.e.:

$$F_{\mathrm{e}} = -ku. \tag{2.19}$$

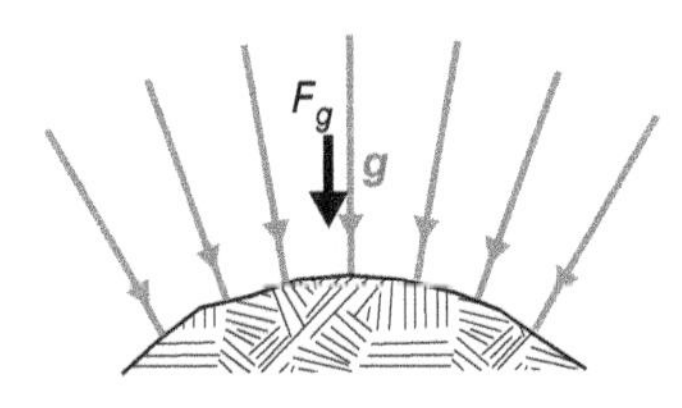

Figure 2.4 Weight force.

The weight force is due to gravity: every mass inside the gravity field is accelerated by gravity towards the centre of the Earth (see Figure 2.4), this depends on latitude and altitude, with a conventional standard value equal to $g = 9.81\,\mathrm{m/s^2}$. When we limit our considerations to dimensions that are very small with respect to the Earth's radius, then we can consider the gravitational field as locally orthogonal to the Earth's surface, which is locally considered flat. The weight force is therefore:

$$F_g = mg. \tag{2.20}$$

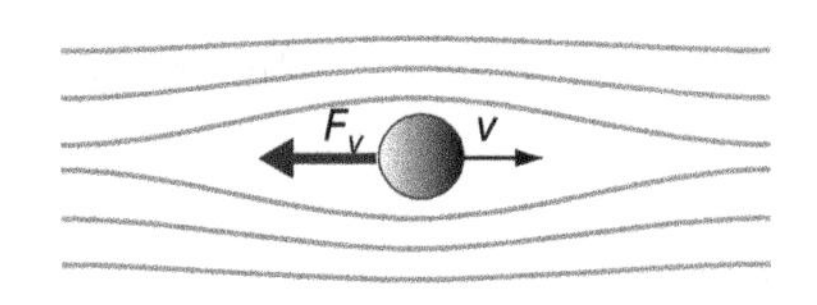

Figure 2.5 Viscous force.

The third example concerns the viscous damping force; this is a manifestation of the fluid–structure interaction arising whenever a solid body moves inside a fluid (see Figure 2.5), which is typically gaseous in microsystems. This is the same kind of force exerted on a car travelling on a road by the surrounding air that opposes the movement and must be studied with the tools of fluid mechanics, as briefly recalled in Section 2.4.

Under simplifying hypotheses, the viscous force can be assumed to be directly proportional to the relative velocity v of the solid moving inside the fluid; the proportionality constant is named c and has the dimensions of a force divided by a velocity:

$$F_v = -cv. \tag{2.21}$$

2.2.2.1 Summary

In this section, the fundamental concepts that govern the dynamics and relative dynamics have been given:

- The concepts of inertial reference frame and of material point;
- Newton's three fundamental laws;
- The equation of motion written for an inertial observer and for a noninertial observer, together with the notion of apparent forces;
- The elastic, weight and viscous forces, which are fundamental for the understanding of material point dynamics in microsystems.

2.2.3 One-Degree-of-Freedom Oscillator

In this section, the case of a one-degree-of-freedom (dof) oscillator is discussed in some detail. This example contains many fundamental features related to the dynamic response of microsystems and is also relevant for its practical importance, since many real-life microsystems can be described in a simple way by means of equivalent one-dof mass–spring oscillators.

Let us consider the device shown in Figure 2.6: a material point of mass m, constrained by an ideal massless extensional spring of stiffness $k \geq 0$ and subject to a viscous damping force $c\dot{u}$ and to an additional external force F.

The equation of motion reads:

$$m\ddot{u} + c\dot{u} + ku = F, \tag{2.22}$$

and must be supplemented by initial conditions on displacement and velocity, i.e.

$$u(0) = u_0, \quad \dot{u}(0) = \dot{u}_0. \tag{2.23}$$

In Equation 2.22, first and second derivatives with respect to the time variable t have been denoted with a superposed dot and two superposed dots, respectively. This is a very common notation, more compact than the use of partial differentiation symbols with respect to the time variable, and will be used throughout the whole book.

Equation 2.22 is linear; with constant coefficients, its general solution can be obtained as the sum of the general integral of the homogeneous equation $u_\mathrm{h}(t)$, for $F = 0$, and of

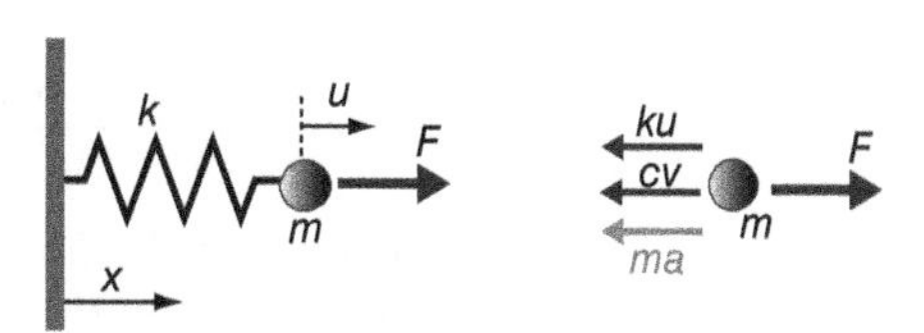

Figure 2.6 One-degree-of-freedom oscillator.

a particular solution of the whole equation $u_\mathrm{p}(t)$:

$$u(t) = u_\mathrm{h}(t) + u_\mathrm{p}(t). \tag{2.24}$$

where $u_\mathrm{h}(t)$ represents the *free vibration* of the ideal system for assigned initial conditions. Considering the undamped case corresponding to $c = 0$, Equation 2.22 for $F = 0$ can be rewritten as

$$\ddot{u} + \omega^2 u = 0, \tag{2.25}$$

with

$$\omega \equiv 2\pi f_0 \equiv \sqrt{\frac{k}{m}}. \tag{2.26}$$

The positive parameter ω is called the undamped circular natural *frequency*, with units of radians per second [rad/s], and f_0 is the frequency, expressed in cycles per second or Hertz. The solution of the equation of motion (2.25) is given in terms of trigonometric functions as

$$u(t) = A_1 \cos \omega t + A_2 \sin \omega t, \tag{2.27}$$

where A_1 and A_2 are constants that can be directly determined from the initial conditions:

$$u(0) = u_0 = A_1, \quad \dot{u}(0) = \dot{u}_0 = A_2 \omega. \tag{2.28}$$

Another common version of Equation 2.27 is the phased response form:

$$u(t) = U \cos(\omega t - \alpha). \tag{2.29}$$

The *amplitude* U and the *phase angle* α are linked to the initial conditions by

$$U = \sqrt{u_0^2 + \left(\frac{\dot{u}_0}{\omega}\right)^2}, \quad \tan \alpha = \frac{\dot{u}_0}{u_0 \omega}. \tag{2.30}$$

When considering damped free oscillations, the equation of motion is usually written in the following way:

$$\ddot{u} + 2\xi\, \omega \dot{u} + \omega^2 u = 0, \tag{2.31}$$

where $\xi = c/(2\omega m)$ is the nondimensional damping factor. The magnitude of the damping factor ξ compared with unity can be used to distinguish three cases: for $\xi < 1$, the underdamped case, the motion is oscillatory with decreasing amplitude; for $\xi \geq 1$, critically damped and overdamped cases, the motion is nonoscillatory and its amplitude decays monotonically, except possibly for one zero crossing.

In the underdamped case, the damped circular natural frequency is given by

$$\omega_\mathrm{d} = \omega \sqrt{1 - \xi^2} \tag{2.32}$$

and the solution reads

$$u(t) = \mathrm{e}^{-\xi \omega t} \left(u_0 \cos \omega_\mathrm{d} t + \frac{\dot{u}_0 + \xi \omega u_0}{\omega_\mathrm{d}} \sin \omega_\mathrm{d} t \right). \tag{2.33}$$

Figure 2.7 shows the free oscillation of an underdamped system for $\dot{u}_0 = 0$. An important parameter characterizing the response is the so-called logarithmic damping

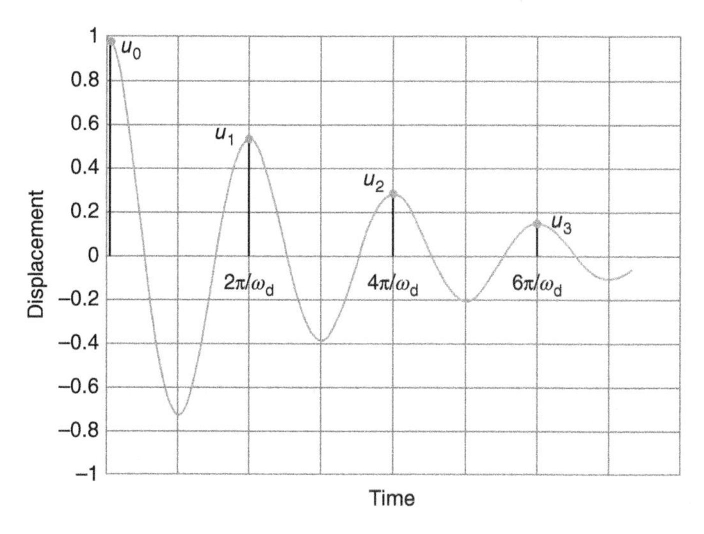

Figure 2.7 Damped oscillation of a one-dof system.

decrement δ, defined as the natural logarithm of the ratio between two successive positive peaks, such as u_n and u_{n+1},

$$\delta \equiv \ln \frac{u_n}{u_{n+1}} = \frac{2\pi\xi}{\sqrt{1-\xi^2}}. \tag{2.34}$$

For small damping values, the logarithmic decrement can be approximated by

$$\delta \cong 2\pi\xi. \tag{2.35}$$

The reciprocal of δ is proportional to the *quality factor*, denoted by Q, as discussed in Chapter 15.

For $\xi = 1$, the critically damped system response is given by

$$u(t) = \mathrm{e}^{-\omega t}(u_0 + (\dot{u}_0 + \omega u_0)t). \tag{2.36}$$

Finally, the solution in the overdamped case is

$$u(t) = \mathrm{e}^{-\xi\omega t}\left(u_0 \cosh \omega_* t + \frac{\dot{u}_0 + \xi\omega u_0}{\omega_*} \sinh \omega_* t\right), \tag{2.37}$$

with $\omega_* = \omega\sqrt{\xi^2 - 1}$.

As already remarked, the response of the one-dof oscillator governed by Equation 2.22 to a given external force $F(t)$ can be found by superposing a particular integral to the solution of the homogeneous equation.

We only consider here the case of harmonic forces, owing to their importance in MEMS applications; hence, the given external force is:

$$F(t) = F \sin(\omega_F t), \tag{2.38}$$

where F is the amplitude of the assigned force and ω_F its angular frequency.

It can be verified that the particular integral in this case is an oscillating function of amplitude u_F, which has the same angular frequency of the assigned force but has a phase shift φ:

$$u(t) = u_F \sin(\omega_F t + \varphi), \tag{2.39}$$

$$u_F = \frac{F}{k\sqrt{\left(1-\left(\frac{\omega_F}{\omega}\right)^2\right)^2 + 4\xi^2\left(\frac{\omega_F}{\omega}\right)^2}}, \tag{2.40}$$

$$\tan\varphi = \frac{2\xi\left(\frac{\omega_F}{\omega}\right)}{\left(\left(\frac{\omega_F}{\omega}\right)^2 - 1\right)}. \tag{2.41}$$

Finally, the complete solution of Equation 2.22, for a given external force and assigned initial conditions, is obtained by superposing the general solution of the homogeneous equation to the particular integral given in Equation 2.39; in the underdamped case,

$$u(t) = e^{-\xi\omega t}(B_1 \cos\omega_d t + B_2 \sin\omega_d t) + u_F \sin(\omega_F t + \varphi), \tag{2.42}$$

where B_1 and B_2 are constants to be determined by enforcing the initial conditions (2.23). In this way one obtains

$$B_1 = u_0 - u_F \sin\varphi, \quad B_2 = \frac{1}{\omega_d}(\dot{u}_0 + \xi\omega u_0) - \frac{u_F}{\omega_d}(\omega_F \cos\varphi + \xi\omega \sin\varphi). \tag{2.43}$$

It is now interesting to examine the solution obtained. Plots of the amplitude, Equation 2.40, and of the phase, Equation 2.41, of the forced response, are given in Figures 2.8 and 2.9, respectively.

From the plot of the amplitude, it can be noticed that for every damping level below the critical one there is a maximum; this is obtained when the frequency of the external force is near to the frequency of the oscillator. When the damping is zero, the maximum becomes infinite. The phenomenon that brings a high increase in the amplitude for a given frequency of the external input is called *resonance* and the corresponding

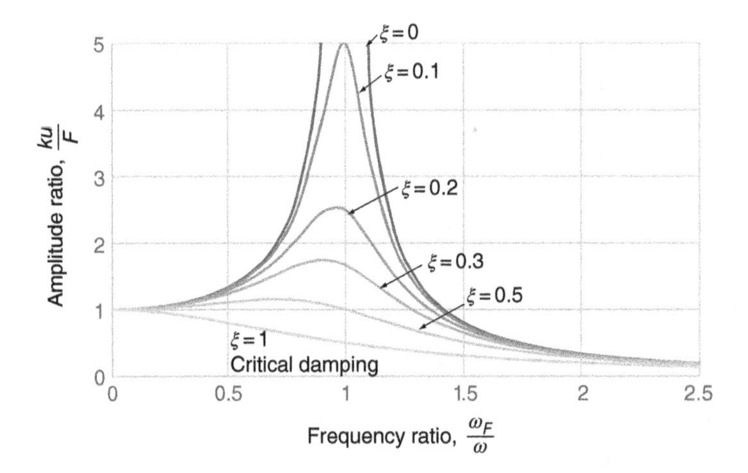

Figure 2.8 One-dof oscillator: amplitude as a function of the ratio between the frequency of the external force and the frequency of the oscillator for varying damping factors.

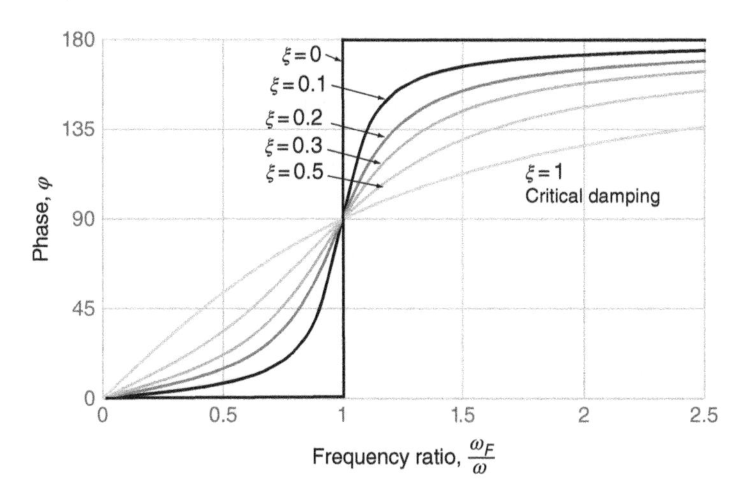

Figure 2.9 One-dof oscillator: phase as a function of the ratio between the frequency of the external force and the frequency of the oscillator for varying damping factors.

frequency is called the *resonance frequency*. This is very important in microsystems in which a device must be kept oscillating with a minimum amount of power consumption and is discussed in Chapter 6, which is devoted to resonators.

Another interesting feature of the one-dof forced response is the fact that when the frequency of the external input goes to zero or it is near to zero, the amplitude of the forced response is equal to F/k, which corresponds to the static response of the elastic spring subject to a force with intensity F, i.e. no dynamic effects are felt by the oscillator. This is also why the term multiplying the value F/k in Equation 2.40 is simply called the amplification factor. This particular situation is exploited in many practical examples concerning microsystems, typically in accelerometers (see Chapter 4).

Considering an external force with a frequency much higher than the frequency of the oscillator, a de-amplification with respect to the purely static response F/k is obtained until the amplitude can be considered negligible; this circumstance is also exploited in the design of microsystems.

From the phase diagram of Figure 2.9, it can be noticed that in the case with zero damping the phase angle is either zero or π, while in the other situations there is a continuous variation of the phase angle.

Considering now the complete response given by Equation 2.42, it can be observed that the first part goes to zero whenever the damping factor is nonzero; this means that, after a sufficient amount of time, the prevailing term will always be the one given by the forced oscillation.

2.2.3.1 Summary

It is now possible to summarize the most important features of the one-dof oscillator dynamics discussed in this section.

- Free vibration is obtained when the oscillator is activated by the initial conditions only, with zero external force.
- Free vibrations are characterized by the frequency of the oscillator.

- Depending on the damping level, the dynamic response can be oscillating or not.
- When a damped oscillator is loaded by an external force varying with a given frequency, the dynamic response will vary with the same frequency of the external force after a sufficient amount of time.
- When the frequency of the external signal equals the frequency of the oscillator, the resonance phenomenon occurs, characterized by a high amplification of the oscillation amplitude.

2.2.4 Rigid-Body Kinematics and Dynamics

Let us consider a continuous distribution of material points with fixed relative distances; in this way, we reach the concept of a continuous *rigid body*. In other words, a continuous rigid body is considered as a continuous distribution in the space of infinitesimal rigid volumes with given mass.

The fact that we can speak of infinitesimal volumes and of continuity in space enables us to define functions of the position in space; in particular, the mass density per unit volume ρ can be defined as a continuous function of the position $\mathbf{x}$ and the total mass of a continuous rigid body of volume Ω can be computed as

$$m_\Omega = \int_\Omega \rho(\mathbf{x})\, \mathrm{d}\Omega. \tag{2.44}$$

As for material points, in the case of rigid bodies it is also important to recall how the possible movements can be described and to understand how these movements can be originated by external causes; in other words, it is important to recall the kinematics and the dynamics of a rigid body.

The scheme of a rigid body is often used for objects that cannot be considered as simple material points because their shape has an influence on their motion. As we will see in the next section, another possible scheme is that of a deformable solid body, in which not only the position in space of the object changes in time but also its shape can be modified.

Depending on the scale and on the kind of problem to be considered, the same object can sometimes be considered as a point mass, as a rigid body or as a deformable body. Consider, as an example, an aeroplane flying. If we want overall information on the aeroplane's flight and its expected trajectory, we can, to a first approximation, consider it as a material point. If we want to know something about how the aeroplane is interacting with the surrounding air and its flight attitude, we must consider also the fact that the aeroplane rotates around its axis and that the inclination of wings counts; hence, we need to consider the overall aeroplane as a rigid body. If we want to know something about how the wings are internally loaded during the flight, we will have to consider the wings as deformable bodies.

2.2.4.1 Rigid-Body Kinematics

As far as the description of rigid-body movements is concerned, we can exploit the results recalled in Section 2.2.1 directly because we can describe the movement of any point P of the rigid body as seen by an external inertial observer similarly to that done for a single point with a fixed position with respect to a noninertial reference frame. To this purpose, it is sufficient to define a reference frame attached to the rigid

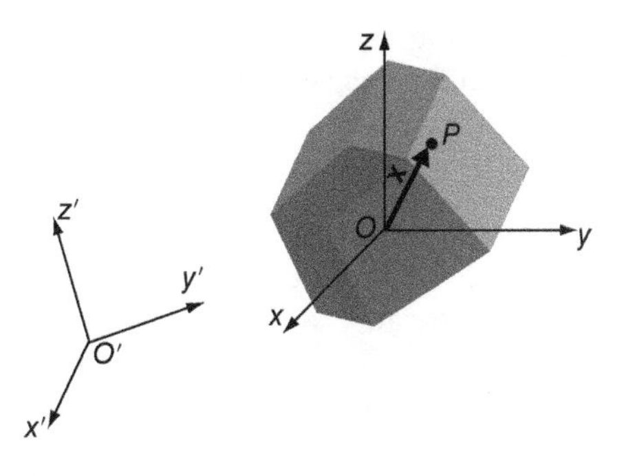

Figure 2.10 Rigid body.

body with origin O, as shown schematically in Figure 2.10; from Equations 2.11 and 2.13 we have the velocity and acceleration vectors of any point P seen by an external inertial observer. At varying position of point P, we obtain the velocity and acceleration of any point of the rigid body:

$$\mathbf{v}_P = \mathbf{v}_0 + \boldsymbol{\omega} \wedge (P - O), \tag{2.45}$$

$$\mathbf{a}_P = \mathbf{a}_0 + \frac{\mathrm{d}\boldsymbol{\omega}}{\mathrm{d}t} \wedge (P - O) + \boldsymbol{\omega} \wedge (\boldsymbol{\omega} \wedge (P - O)). \tag{2.46}$$

Equations 2.45 and 2.46 show that any instantaneous movement of a rigid body can be described starting from the velocity $\mathbf{v}_0$ and acceleration $\mathbf{a}_0$ of a reference point O attached to the rigid body and from the rotation velocity vector $\boldsymbol{\omega}$. Hence, only the velocity and rotation velocity and their derivatives are necessary; this means that we need only six scalar quantities and that a rigid body has six degrees of freedom.

It is interesting to remark that if we consider rigid bodies that are contained in a plane, with movements all contained in the same plane, only three degrees of freedom are necessary: the velocity components of a reference point and one rotation component around an axis orthogonal to the rigid-body plane.

2.2.4.2 Rigid-Body Dynamics

Starting from the last remark of the previous section, we understand that six scalar equations are needed to govern the dynamics of a rigid body. These are obtained by generalizing the equations governing the dynamics of a rigid system of material points, i.e. points kept at fixed distances, to a continuous distribution of mass. The expressions of these equations simplify if we make reference to a particular point of the rigid body, which is the *centre of mass G*, for which coordinates x_G, y_G, z_G with respect to an inertial reference frame are given by

$$x_G = \frac{\int_\Omega x\rho \,\mathrm{d}\Omega}{\int_\Omega \rho \,\mathrm{d}\Omega}, \quad y_G = \frac{\int_\Omega y\rho \,\mathrm{d}\Omega}{\int_\Omega \rho \,\mathrm{d}\Omega}, \quad z_G = \frac{\int_\Omega z\rho \,\mathrm{d}\Omega}{\int_\Omega \rho \,\mathrm{d}\Omega}. \tag{2.47}$$

When the rigid body is homogeneous, i.e. when its mass density ρ does not depend on position, the coordinates of G become a purely geometric property of the rigid body.

The *linear momentum* $\mathbf{Q}$ of the whole rigid body can now be expressed as the product of the total mass of the rigid body and the velocity of the centre of mass, as

$$\mathbf{Q} = m_G \mathbf{v}_G. \tag{2.48}$$

Another quantity that must be defined to study the rigid body dynamics is the *angular momentum* with respect to a point O, computed starting from the velocity $\mathbf{v}_P$ of any point P of the rigid body:

$$\mathbf{L}_O = \int_\Omega (P - O) \wedge \mathbf{v}_P \rho \,\mathrm{d}\Omega, \tag{2.49}$$

Given these definitions of centre of mass and linear and angular momentum, the rigid body dynamics is governed by the following two vector differential equations:

$$\frac{d\mathbf{Q}}{dt} = \mathbf{F}, \quad \frac{d\mathbf{L}_O}{dt} = \mathbf{M}_O, \tag{2.50}$$

where $\mathbf{F}$ is the resultant of all external forces acting on the rigid body, while $\mathbf{M}_O$ is the resultant of moments exerted by the external forces with respect to point O, which can be a fixed point or a point in motion with a velocity parallel to that of the centre of mass.

Combining the first of Equations 2.50 with Equation 2.48, we can equivalently write:

$$m_G \mathbf{a}_G = \mathbf{F}, \tag{2.51}$$

where $\mathbf{a}_G$ is the acceleration vector of the centre of mass. This vector equation states that the movement of the centre of mass can be studied as that of a material point that has a mass equivalent to the whole mass of the rigid body. Solving this equation, we obtain the trajectory of the centre of mass of the rigid body but we do not know how the rigid body is changing its orientation in space. This second piece of information comes from the second of Equations 2.50.

The simplest way to interpret the second of Equations 2.50 is to make the hypothesis that point O coincides with the centre of mass of the rigid body. In this case, the angular momentum $\mathbf{L}_O$ can be given a simple expression as a function of the components of the rotation vector $\boldsymbol{\omega}$ with respect to a special noninertial reference frame with coordinates x, y, z and unit vectors $\mathbf{i}, \mathbf{j}, \mathbf{k}$ centred on G and fixed to the rigid body, called the *principal reference frame*:

$$\mathbf{L}_O = I_x \omega_x \mathbf{i} + I_y \omega_y \mathbf{j} + I_z \omega_z \mathbf{k}. \tag{2.52}$$

In Equation 2.52, the quantities I_x, I_y, I_z are the *moments of inertia* of the continuous rigid body with respect to axes x, y, z, respectively:

$$I_x = \int_\Omega \rho \left(y^2 + z^2\right) \, d\Omega, \tag{2.53}$$

$$I_y = \int_\Omega \rho \left(x^2 + z^2\right) \, d\Omega, \tag{2.54}$$

$$I_z = \int_\Omega \rho \left(x^2 + y^2\right) \, d\Omega. \tag{2.55}$$

Notice that if the reference frame fixed to the rigid body is not the principal reference frame, six additional terms appear in general in Equation 2.52, containing products of components of the rotation vector $\boldsymbol{\omega}$ and other inertia terms, which are called *products of inertia*, of the kind:

$$I_{xy} = \int_\Omega \rho xy \, d\Omega, \tag{2.56}$$

The property of the principal reference frame is that all products of inertia are zero with respect to it.

The time derivative of the angular momentum can now be computed from Equation 2.52, as

$$\frac{d\mathbf{L}_O}{dt} = I_x \dot{\omega}_x \mathbf{i} + I_y \dot{\omega}_y \mathbf{j} + I_z \dot{\omega}_z \mathbf{k} + \boldsymbol{\omega} \wedge \mathbf{L}_O. \tag{2.57}$$

In this equation, terms like $\dot{\omega}_x$ denote the time derivative of the components of the angular velocity vector, while the last term $\boldsymbol{\omega} \wedge \mathbf{L}_O$ comes from the fact that the reference frame is moving with the rigid body and therefore the unit vector directions also change with time.

The expression (2.57) of the time derivative of the angular momentum inserted in the second of the equations of motion (2.50) gives the additional vector equation, which is necessary to study the rigid body dynamics. Equations 2.50 can be projected onto the three axes, x, y, z, giving the six scalar equations governing the dynamics of the rigid body.

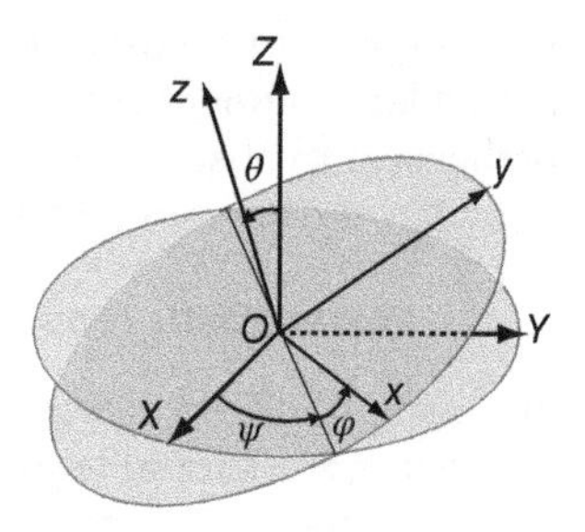

Figure 2.11 Euler angles.

The resulting equations are nonlinear in the components of the rotation vector $\boldsymbol{\omega}$, as can be verified by recalling the definitions of $\mathbf{L}_O$ given in Equation 2.49 and the expression of velocity $\mathbf{v}_P$ given by Equation 2.45.

Usually, the components of the angular velocity vector are expressed as functions of the so-called Euler angles, ϑ, ψ, φ, as follows:

$$\begin{aligned}\omega_x &= \dot{\psi} \sin\varphi \sin\vartheta + \dot{\vartheta} \cos\varphi \\ \omega_y &= \dot{\psi} \cos\varphi \sin\vartheta - \dot{\vartheta} \sin\varphi \\ \omega_z &= \dot{\psi} \cos\varphi + \dot{\varphi}.\end{aligned} \tag{2.58}$$

The Euler angles, ϑ, ψ, φ, are used to give the orientation of one reference frame with respect to another with the same origin. They are represented in Figure 2.11 and are called the *nutation* $\vartheta, (0 \leq \vartheta \leq \pi)$, *precession* $\psi, (0 \leq \psi < 2\pi)$ and *intrinsic rotation* $\varphi, (0 \leq \varphi < 2\pi)$.

2.2.4.3 Summary

In this section, we have discussed the equations governing rigid-body kinematics and dynamics; it is now useful to summarize the main results.

- The whole motion of any rigid body can be described by means of six degrees of freedom.
- In the case of a plane rigid body, only three degrees of freedom are necessary.
- Rigid-body dynamics is governed by two vector equations (2.50), which express linear momentum and angular momentum conservation.
- The linear momentum equation can be interpreted as the equation of motion of the centre of mass, having a mass equal to the mass of the whole rigid body, subject to the resultant of forces acting on it.
- The angular momentum equation can be projected onto a noninertial reference frame with its origin in the centre of mass and expressed in terms of rotation velocity components and Euler angles.

2.3 Solid Mechanics

Deformable bodies are the objects studied in the discipline called *solid mechanics*. For the understanding of what can happen to slender deformable parts inserted in microsystems and subject to various loadings that can cause their movement and deformation, it is important to have a clear idea of the fundamentals of solid mechanics and of the particular treatment reserved to deformable beams, given in this section. Many comprehensive treatises on solid mechanics can be found in the scientific literature, e.g. Marsden and Hughes (1993) and Timoshenko and Goodier (1970).

2.3.1 Linear Elastic Problem for Deformable Solids

The assumption of a *continuum* made for rigid bodies is also applied to deformable bodies. Hence, a deformable solid is seen as a continuum distribution in space of infinitesimal volumes. This fundamental assumption is also applied to fluids (see Section 2.4) and gives rise to the so-called *mechanics of deformable continua*, i.e. the study of solid and fluid continua subject to many kinds of external loadings that can cause their movement and deformation. The material point discussed in Section 2.2.2 can now be made coincident with an infinitesimal volume with mass that can move and deform in space. The most important difference with respect to what has been previously discussed is that now the solid can deform and the rigid-body assumption is abandoned.

Let us consider a deformable body, as shown in Figure 2.12, described in the given Cartesian orthogonal reference frame with coordinates x, y, z.

The solid has volume Ω and external surface Γ; it is totally or partially constrained by kinematic constraints that limit its movement possibilities to the portion of the external surface named Γ_u.

On the complementary portion of the external surface, named Γ_f, the body is loaded by external surface forces $\mathbf{f}$, which are dimensionally forces per unit surface (like an external pressure), while volume or body forces $\mathbf{b}$ are distributed inside the volume; these last are forces per unit volume (like unit weight).

On the constrained external surface Γ_u, displacements $\bar{\mathbf{u}}$ can also be assigned.

The vector functions $\mathbf{f}$ and $\mathbf{b}$ are considered as functions of the position $\mathbf{x}$; together with the assigned displacements $\bar{\mathbf{u}}$ they can also vary with time t, depending on the considered situation.

The movements of a continuum body like the one shown in Figure 2.12 are described in terms of displacements of all points P of coordinates $\mathbf{x}$ inside the body. The displacement field $\mathbf{u}$ (dimensionally, a length) is defined as the vectorial function collecting the three components of displacements u_x, u_y, u_z in directions x, y, z, respectively. The displacement is assumed to be a continuous function of the position $\mathbf{x}$, possibly varying with time t, depending on the considered situation.

In view of the goals of this book, we limit the discussion here by assuming the hypothesis of *small strain and displacements*; this means that the displacements and the shape and volume variations of the solid body are very small with respect to the body dimensions. This assumption implies that the final configuration of the body is considered geometrically coincident with the initial one, i.e. material points after displacements and deformation will have the same coordinates! Moreover, all the relations that govern the deformation process in practice coincide with those governing an infinitesimal movement of the body. This hypothesis for deformable solids will be removed

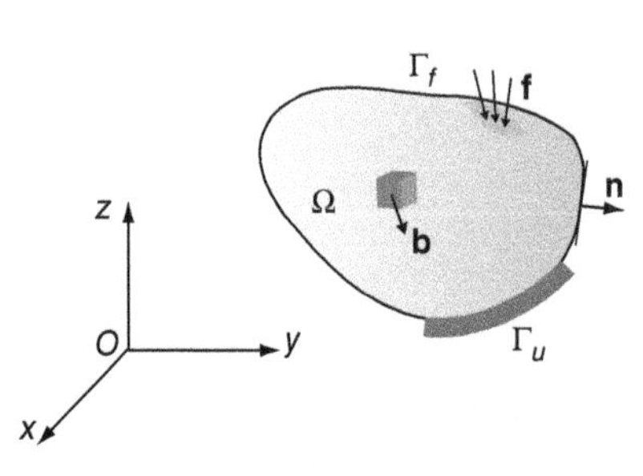

Figure 2.12 Deformable body.

only in some parts of the book and specific remarks and formulations will be given when necessary.

Under the hypothesis of *small strains and displacements*, the whole description of the movements of the body and of its deformation process is highly simplified, and the equations relating the displacement field to the functions describing the deformation process become linear.

The deformation process is described by means of a second-order symmetric *strain tensor* ε.

The strain tensor, as all second-order tensors, is defined by nine functions of the position $\mathbf{x}$ and possibly of the time t, which reduce to six functions if the tensor is symmetric. A common representation in the Cartesian reference frame of Figure 2.12 is as follows:

$$\varepsilon = \begin{bmatrix} \varepsilon_{xx} & \varepsilon_{yx} & \varepsilon_{zx} \\ \varepsilon_{xy} & \varepsilon_{yy} & \varepsilon_{zy} \\ \varepsilon_{xz} & \varepsilon_{yz} & \varepsilon_{zz} \end{bmatrix}. \tag{2.59}$$

Owing to the symmetry of the strain tensor, it is possible and convenient to define a six-component vector that contains the six independent components of the strain tensor, i.e.

$$\varepsilon^{\mathrm{T}} = \begin{bmatrix} \varepsilon_{xx} & \varepsilon_{yy} & \varepsilon_{zz} & 2\varepsilon_{xy} & 2\varepsilon_{yz} & 2\varepsilon_{xz} \end{bmatrix}. \tag{2.60}$$

The reason why, in Equation 2.60, the fourth, fifth and sixth strain components appear multiplied by two is related to the necessity of correctly defining the scalar product of the strains and stresses, as will be explained in the following, after the introduction of the stress tensor.

An alternative notation for the six independent components of the strain tensor is

$$\varepsilon^{\mathrm{T}} = \begin{bmatrix} \varepsilon_{x} & \varepsilon_{y} & \varepsilon_{z} & \gamma_{xy} & \gamma_{yz} & \gamma_{xy} \end{bmatrix}. \tag{2.61}$$

The strain components ε_x, ε_y, ε_z, called longitudinal strains, represent the elongation or reduction in length of unit material fibres directed as x, y, z, respectively, in the original undeformed configuration of the body. A strain component like γ_{xy} represents the variation of the angle between directions x and y in the original undeformed configuration of the body and similarly for components γ_{yz} and γ_{xz}. The corresponding strain components, ε_{xy}, ε_{yz}, ε_{xz}, are called angular strains.

It is important to remark that the strain tensor and all its components in the various possible notations are nondimensional numbers. As an example, declaring that $\varepsilon_x = 0.02$, uniformly, for a specimen with original length 100 cm under tensile loading means that the length increased up to 102 cm.

Having defined the displacement and strain fields, it is now possible to recall the linear relations that hold between the derivatives of the displacement vector components and the strain components under the assumed hypothesis of small strain and displacement:

$$\varepsilon_{xx} = \frac{\partial u_x}{\partial x}, \quad \varepsilon_{yy} = \frac{\partial u_y}{\partial y}, \quad \varepsilon_{zz} = \frac{\partial u_z}{\partial z},$$

$$2\varepsilon_{xy} = \left(\frac{\partial u_x}{\partial y} + \frac{\partial u_y}{\partial x} \right),$$

$$2\varepsilon_{yz} = \left(\frac{\partial u_y}{\partial z} + \frac{\partial u_z}{\partial y} \right),$$

$$2\varepsilon_{xz} = \left(\frac{\partial u_x}{\partial z} + \frac{\partial u_z}{\partial x} \right). \tag{2.62}$$

These equations are called the *kinematic compatibility* or, briefly, *compatibility* relations; the whole set of kinematic relations is supplemented by boundary conditions on the constrained boundary Γ_u of the body, with $\bar{\mathbf{u}}$ the assigned displacement:

$$\mathbf{u} = \bar{\mathbf{u}} \quad \text{on } \Gamma_u. \tag{2.63}$$

Volume forces $\mathbf{F}$ and surface forces $\mathbf{f}$ load the deformable body; their effect is felt in terms of reaction forces in correspondence of the constrained boundary and of an internal field of forces per unit surface, which are called *stresses*. The stress field is defined by means of another second-order and symmetric tensor, usually denoted $\boldsymbol{\sigma}$.

A possible representation in the Cartesian reference frame of Figure 2.12 is

$$\boldsymbol{\sigma} = \begin{bmatrix} \sigma_{xx} & \sigma_{yx} & \sigma_{zx} \\ \sigma_{xy} & \sigma_{yy} & \sigma_{zy} \\ \sigma_{xz} & \sigma_{yz} & \sigma_{zz} \end{bmatrix}. \tag{2.64}$$

Owing to the symmetry of the stress tensor, similar to that done for the strain tensor, it is possible and convenient to define a six-component vector that contains the six independent components of the stress tensor, i.e.

$$\boldsymbol{\sigma}^{\mathrm{T}} = \begin{bmatrix} \sigma_{xx} & \sigma_{yy} & \sigma_{zz} & \sigma_{xy} & \sigma_{yz} & \sigma_{xz} \end{bmatrix}. \tag{2.65}$$

An alternative notation for the six independent components of the stress tensor is

$$\boldsymbol{\sigma}^{\mathrm{T}} = \begin{bmatrix} \sigma_{x} & \sigma_{y} & \sigma_{z} & \tau_{xy} & \tau_{yz} & \tau_{xz} \end{bmatrix}. \tag{2.66}$$

With these definitions of the vectors containing the six independent components of the strain and stress tensors, the scalar product of the two vectors correctly gives the tensorial product of the two tensors; this is important because the scalar product is the internal work per unit volume during a deformation process.

A physical interpretation of the stress tensor components is as follows: consider an infinitesimal cubic volume inside the body at position $\mathbf{x}$, as shown in Figure 2.13. A stress tensor component, like σ_x, is the internal force per unit surface exerted on the surface of the infinitesimal cube orthogonal to the axis x, in the direction of axis x. A stress component, like σ_{xy}, is the force per unit surface area exerted on the same surface orthogonal to axis x but in the direction of axis y. The components σ_x, σ_y, σ_z are also called normal stresses, while the components σ_{xy}, σ_{yz}, σ_{xz} are also called shear stresses.

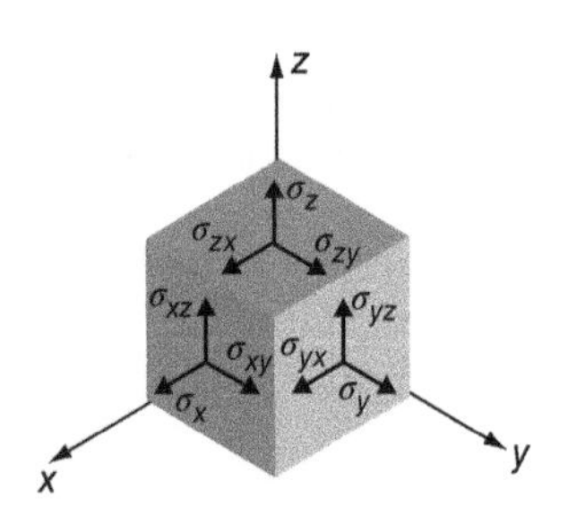

Figure 2.13 Infinitesimal volume with stress tensor components.

Given a stress tensor, the mean of the normal components $p \equiv (\sigma_x + \sigma_y + \sigma_z)/3$ is called the *hydrostatic* portion of the stress, while the stress tensor without the hydrostatic component is called the *deviatoric* part of the stress.

The following decomposition can therefore be always introduced, which in turn defines the deviatoric part of the stress **s**:

$$\mathbf{s} \equiv \boldsymbol{\sigma} - p\mathbf{I} = \begin{bmatrix} \sigma_{xx} - p & \sigma_{yx} & \sigma_{zx} \\ \sigma_{xy} & \sigma_{yy} - p & \sigma_{zy} \\ \sigma_{xz} & \sigma_{yz} & \sigma_{zz} - p \end{bmatrix}, \tag{2.67}$$

where **I** represents the second-order identity tensor. The hydrostatic and deviatoric portions of the stress are important because materials react differently to them.

The values of the stress tensor components vary as the orientation of the infinitesimal cube inside the solid body is varied with respect to a fixed reference frame. The whole set of values assumed by the stress tensor components at varying orientation represents the *stress state* at point **x**. This property of the variation of tensor components is also shown by the strain tensor and we can speak analogously of the *strain state*.

The rules that govern the variation of tensor components at varying reference frames define the tensorial character of the strain and stress and can be found in any book of continuum mechanics, e.g. Marsden and Hughes (1993).

For any given stress state, it is possible to find a reference frame with respect to which the infinitesimal cube is subject to normal stress components only, as shown in Figure 2.14. When the cube is oriented in this way, we say that the cube axes are the *principal directions* and the normal stress components acting on the cube are the *principal stresses*. Principal directions and stresses are important because they give a clearer vision of what is happening inside the body in terms of stresses and because the three principal stress values contain the maximum and minimum values of all stress components that can be found at varying orientations of the reference frame.

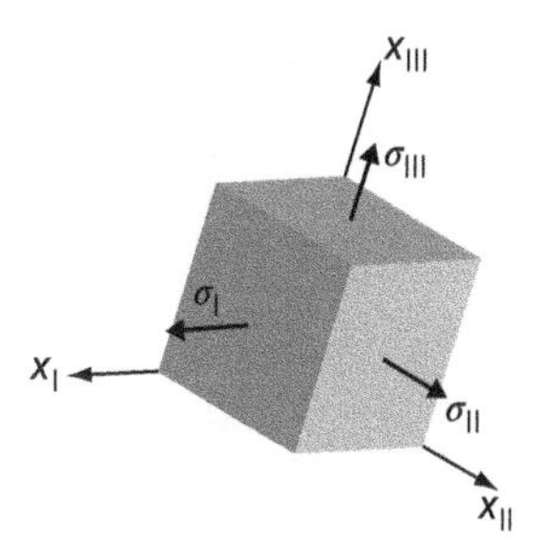

Figure 2.14 Infinitesimal volume with principal stress tensor components.

Similar properties also hold for the strain state; hence, it is possible to find principal strain directions characterized by the fact that only fibre elongations in the direction of the reference axis are found and no angle variations are observed in the principal strain reference frame.

Depending on the kind of loading applied to the deformable body, the problem in hand can be considered as *quasi-static* or *dynamic*. In the first case, inertia forces do not influence the mechanical response; stresses, volume forces and surface forces must then satisfy the static equilibrium conditions inside the volume and on the loaded boundary Γ_f:

$$\left.\begin{aligned} \frac{\partial \sigma_{xx}}{\partial x} + \frac{\partial \sigma_{yx}}{\partial y} + \frac{\partial \sigma_{zx}}{\partial z} + b_x = 0, \\ \frac{\partial \sigma_{xy}}{\partial x} + \frac{\partial \sigma_{yy}}{\partial y} + \frac{\partial \sigma_{zy}}{\partial z} + b_y = 0, \\ \frac{\partial \sigma_{xz}}{\partial x} + \frac{\partial \sigma_{yz}}{\partial y} + \frac{\partial \sigma_{zz}}{\partial z} + b_z = 0, \end{aligned}\right\} \text{in} \quad \Omega \tag{2.68}$$

$$\left.\begin{aligned}\sigma_{xx}n_x + \sigma_{yx}n_y + \sigma_{zx}n_z &= f_x,\\ \sigma_{xy}n_x + \sigma_{yy}n_y + \sigma_{zy}n_z &= f_y,\\ \sigma_{xz}n_x + \sigma_{yz}n_y + \sigma_{zz}n_z &= f_z.\end{aligned}\right\}\text{on } \Gamma_f \tag{2.69}$$

In Equation 2.69, n_x, n_y, n_z are the components of the unit vector $\mathbf{n}$ orthogonal to the external surface of the solid body, directed outside the surface, as shown in Figure 2.12.

When the external loading is *rapidly varying* in time, inertia forces must be taken into consideration in the equilibrium equations in the volume and these transform into equations of motion. These are the equivalent for a deformable solid of the equations governing the dynamics of a material point or of a rigid body, shown in Section 2.2. It is important to remark that the notion of *rapidly varying* loading can be correctly interpreted only with respect to the dynamic properties of the deformable body or structure, as already discussed with reference to the single-dof oscillator presented in Section 2.2.

Introducing the inertia force per unit volume in Equation 2.68, given by the product of the mass density ρ per unit volume (dimensionally, mass divided by length cubed) and acceleration $\ddot{u}_x$, $\ddot{u}_y$, $\ddot{u}_z$, one obtains the continuum version of the Newton's second law or of the conservation of linear momentum:

$$\left.\begin{aligned}\frac{\partial\sigma_{xx}}{\partial x} + \frac{\partial\sigma_{yx}}{\partial y} + \frac{\partial\sigma_{zx}}{\partial z} + b_x &= \rho\ddot{u}_x,\\ \frac{\partial\sigma_{xy}}{\partial x} + \frac{\partial\sigma_{yy}}{\partial y} + \frac{\partial\sigma_{zy}}{\partial z} + b_y &= \rho\ddot{u}_y,\\ \frac{\partial\sigma_{xz}}{\partial x} + \frac{\partial\sigma_{yz}}{\partial y} + \frac{\partial\sigma_{zz}}{\partial z} + b_z &= \rho\ddot{u}_z.\end{aligned}\right\}\quad \text{in} \quad \Omega. \tag{2.70}$$

In the dynamic case, initial conditions must also be provided, i.e. displacements and velocities must be assigned at every point of the body at the initial time instant $t = 0$, in addition to initial values of stresses and strains:

$$\mathbf{u} = \mathbf{u}_0, \quad \dot{\mathbf{u}} = \dot{\mathbf{u}}_0, \quad \text{in} \quad \Omega. \tag{2.71}$$

Up to now we have a set of kinematic compatibility relations with relevant boundary conditions (2.62, 2.63) and a set of equilibrium or motion equations with relevant boundary and possibly initial conditions (2.68 to 2.71). The available equations are nine in total, while the problem in hand contains 15 unknowns: three displacement components, six independent strain tensor components and six independent stress tensor components. This means that we still need six additional independent equations relating the problem unknowns. The missing equations are found by introducing the specific material that composes the deformable body; this is why we usually speak of *material constitutive laws.*

A constitutive law for a deformable material introduces a relation between the strain and the stress components. Given a strain state, by means of a constitutive law it is possible to find the stress components, and vice versa if the law is invertible. We expect, for instance, that, for the same stress level, a material like rubber shows much higher strain levels than a material like steel. The differences among various materials that can be used to fabricate objects is therefore captured by the constitutive law. In the case of microsystems, the material used most often is silicon, either monocrystalline or polycrystalline; other materials typically used in microsystems fabrication are metals like gold and aluminium.

A typical way to capture the material response and therefore to formulate a possible constitutive law, is by experiment. The simplest one is the uniaxial tension test, as shown in Figure 2.15, during which a material specimen of cylindrical shape, of length ℓ and cross-sectional area A, is subject to an ideally uniform and uniaxial stress state by applying two loads F at the two ends of the cylinder in the direction of the cylinder axis. In this ideal test, inside the material we have only a stress component of the normal kind $\sigma = F/A$ in the direction of the axis. Let us observe that in this case σ is a principal stress and that the specimen axis is a principal direction.

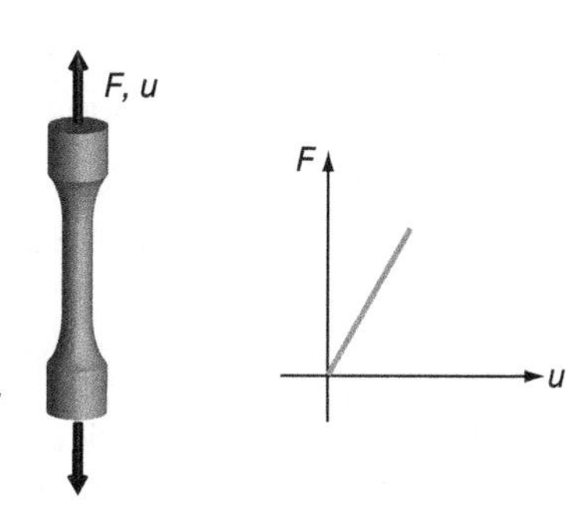

Figure 2.15 Tensile test on material specimen.

In a typical uniaxial test, the majority of materials, when loaded at sufficiently low levels of stress, show a linear relation between stress and strain. This linearity is lost when a sufficient high level of stress is reached; let us call it the critical stress level or σ_c. Depending on the kind of material, what happens after reaching σ_c is different: in some cases, the material breaks; in other cases, it enters a nonlinear regime without immediate rupture.

The two most important examples are *ductile* behaviour, e.g. as shown by metals and depicted in Figure 2.16a and *brittle* behaviour, e.g. as shown by silicon and depicted in Figure 2.16b.

Ductile behaviour means that when the critical stress is reached, the material enters a so-called elastoplastic regime characterized by nonlinearity and irreversibility, owing to the presence of irreversible or *plastic strains* that start to develop just after the beginning of the nonlinear regime. In this case, the critical stress is called the *yield stress* and denoted σ_y.

Brittle behaviour means that when the critical stress is reached, the material undergoes rupture. In this case, the critical stress can be called the nominal rupture stress or *nominal strength* of the material and denoted σ_s. It is important to observe that rupture can be almost instantaneous in a tensile test but that in reality the final separation of the two parts of the specimen is a consequence of a complex phenomenon that occurs at the microscale, in which one (or more than one) crack starts propagating at a certain high but finite velocity. More insight into fracture phenomena in microsystems will be given in Chapter 12; a useful reference on fracture mechanics is Broberg (1999).

Inspired by the uniaxial tensile test of Figure 2.15, we can now focus on the initial linear regime that is defined as *linear elastic*; during this regime, the linearity of the response is also accompanied by full reversibility, i.e. if we invert the loading sign and start unloading, we go back following the same path of the initial loading phase. The

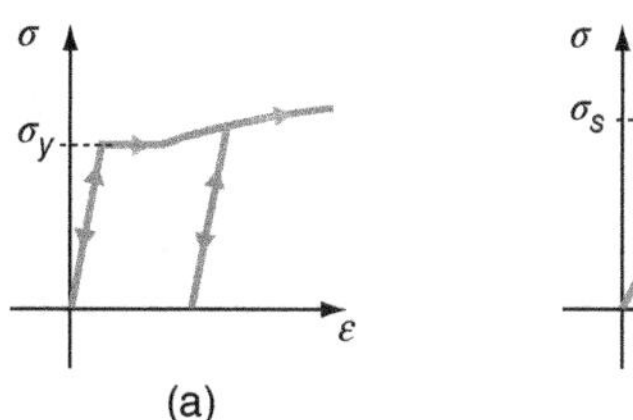

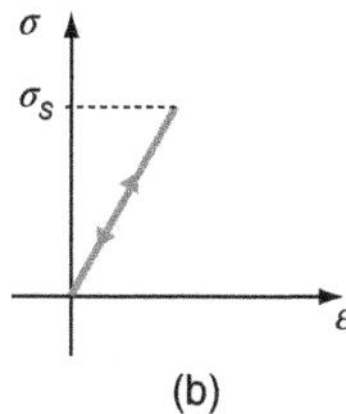

Figure 2.16 Typical experimental responses of materials under a uniaxial tensile test: (a) ductile; (b) brittle.

linear elastic behaviour can be mathematically represented by the simplest possible constitutive law, as a linear relation between stress and strain:

$$\sigma = E\varepsilon. \tag{2.72}$$

Equation 2.72 is known as *Hooke's law* for a linear elastic material; the parameter E is positive and dimensionally a force per unit surface (like the stress) and is called the *Young's modulus* of the material, or the material *stiffness*. To determine the Young's modulus experimentally, a tensile test is sufficient.

Typical values of E for bulk metals are 210 GPa for steel, 117 GPa for copper, 80 GPa for gold and 70 GPa for aluminium; in the case of polycrystalline silicon, an average value is 150 GPa.

We know that in a solid body stress and strain states are usually nonuniaxial and represented at any point $\mathbf{x}$ of the volume Ω by two second-order tensors. The uniaxial linear elastic constitutive law given by Equation 2.72 can be generalized to any stress and strain state by the introduction of a linear relation between the two vectors containing the six independent components of the stress and strain tensors; we thus obtain the so-called *generalized Hooke's law*:

$$\sigma = \mathbf{d}\varepsilon, \tag{2.73}$$

where a six by six matrix $\mathbf{d}$ has been introduced, which contains elastic material constants. Matrix $\mathbf{d}$ is called the *material stiffness matrix*; it can be shown that for linear elastic materials for which an elastic potential can be defined, $\mathbf{d}$ is symmetric and positive definite. This in turn means that we can always invert Equation 2.73 to obtain a fully equivalent description of the linear elastic constitutive law:

$$\varepsilon = \mathbf{c}\sigma. \tag{2.74}$$

Matrix $\mathbf{c}$, which is the inverse of matrix $\mathbf{d}$, is called the *material compliance matrix* and is another six by six, symmetric and positive definite matrix.

Owing to the symmetry of matrices $\mathbf{d}$ and $\mathbf{c}$, we need at most 21 parameters to completely define a linear elastic constitutive law of the kind of Equations 2.73 or 2.74. This is not a simple task, because we should invent clever experimental tests to determine the 21 unknown parameters and define matrices $\mathbf{d}$ and $\mathbf{c}$.

Luckily enough, the number of independent parameters in matrices $\mathbf{d}$ and $\mathbf{c}$ can be strongly reduced if we take into account material symmetries; in other words, 21 is the number of independent parameters that define matrices $\mathbf{d}$ and $\mathbf{c}$ only if the material response does not show any symmetry and the response in terms of strains for an assigned stress state depends on the directions along which the various stress tensor components are applied; this is the case of a fully *anisotropic* material.

The most simple situation is the *isotropic case*. This means that the strain for assigned stress does not depend at all on the direction of loading; the material response is always the same, independent of the loading direction. In the isotropic case, it can be shown that matrices $\mathbf{d}$ and $\mathbf{c}$ are defined starting from two material parameters only. Typically, the two independent parameters are the Young's modulus, already defined in the case of a uniaxial tensile test, and *Poisson's ratio*, ν.

This second parameter is a nondimensional number, which can again be obtained from a uniaxial test, in which the material not only deforms in the direction of loading but also transversely: the Poisson's ratio is minus the ratio between a strain component in the direction orthogonal to the specimen axis and the strain component in the direction of the specimen axis. Typical values of Poisson's ratio are 0.3 for steel, 0.33

for copper, 0.4 for gold and 0.32 for aluminium; in the case of polycrystalline silicon, an average value is 0.22.

Given E and ν, the material compliance matrix $\mathbf{c}$ can be written as

$$\mathbf{c} = \frac{1}{E}\begin{bmatrix} 1 & -\nu & -\nu & 0 & 0 & 0 \\ -\nu & 1 & -\nu & 0 & 0 & 0 \\ -\nu & -\nu & 1 & 0 & 0 & 0 \\ 0 & 0 & 0 & 2(1+\nu) & 0 & 0 \\ 0 & 0 & 0 & 0 & 2(1+\nu) & 0 \\ 0 & 0 & 0 & 0 & 0 & 2(1+\nu) \end{bmatrix}. \tag{2.75}$$

By inverting matrix $\mathbf{c}$ in Equation 2.75, matrix $\mathbf{d}$ can also be obtained, expressed in terms of E and ν. Nevertheless, another representation for $\mathbf{d}$ is usually preferred, introducing a second couple of material parameters λ and μ (μ is also named the shear stiffness modulus and denoted G), named Lamé constants, which can be defined starting from E and ν:

$$\lambda = \frac{E\nu}{(1+\nu)(1-2\nu)}, \quad \mu = \frac{E}{2(1+\nu)}. \tag{2.76}$$

Matrix $\mathbf{d}$ can then be expressed as follows, as a function of Lamé constants:

$$\mathbf{d} = \begin{bmatrix} \lambda+2\mu & \lambda & \lambda & 0 & 0 & 0 \\ \lambda & \lambda+2\mu & \lambda & 0 & 0 & 0 \\ \lambda & \lambda & \lambda+2\mu & 0 & 0 & 0 \\ 0 & 0 & 0 & \mu & 0 & 0 \\ 0 & 0 & 0 & 0 & \mu & 0 \\ 0 & 0 & 0 & 0 & 0 & \mu \end{bmatrix}. \tag{2.77}$$

Matrices $\mathbf{d}$ and $\mathbf{c}$ are both positive definite; this property implies that parameters E and ν must satisfy some constraints, namely:

$$E > 0, -1 < \nu < 0.5. \tag{2.78}$$

Usual values of ν are positive but materials with negative Poisson's ratios can be obtained, starting from particular microstructures; these are called *auxetic materials*. In the case of positive ν, parameters λ and μ are both strictly positive.

Between the simplest case of a linear elastic isotropic material (two material parameters) and the most complex case of a fully anisotropic linear elastic material (21 material parameters), other situations are found, depending on the kind of symmetries shown by the material in hand.

Here are recalled only the meaningful cases of *cubic*, *transversely isotropic* and *orthotropic* linear elastic materials, in view of their relevance in microsystems design.

2.3.1.1 Linear Elastic Cubic Material

The elastic stiffness of silicon monocrystals can be given in this format, defined starting from three independent parameters, and can be written as in Equation 2.79:

$$\mathbf{d} = \begin{bmatrix} d_{11} & d_{12} & d_{12} & 0 & 0 & 0 \\ d_{12} & d_{11} & d_{12} & 0 & 0 & 0 \\ d_{12} & d_{12} & d_{11} & 0 & 0 & 0 \\ 0 & 0 & 0 & d_{44} & 0 & 0 \\ 0 & 0 & 0 & 0 & d_{44} & 0 \\ 0 & 0 & 0 & 0 & 0 & d_{44} \end{bmatrix}. \tag{2.79}$$

2.3.1.2 Linear Elastic Transversely Isotropic Material

This case can be assumed for polycrystalline silicon with a columnar structure. The behaviour in the plane orthogonal to the columns axis can be considered isotropic, but distinguished from that in the direction of the column axis. The material stiffness matrix is defined starting from five independent parameters, and can be written as in Equation 2.80:

$$\mathbf{d} = \begin{bmatrix} d_{11} & d_{11} - 2d_{66} & d_{13} & 0 & 0 & 0 \\ d_{11} - 2d_{66} & d_{11} & d_{13} & 0 & 0 & 0 \\ d_{13} & d_{13} & d_{33} & 0 & 0 & 0 \\ 0 & 0 & 0 & d_{44} & 0 & 0 \\ 0 & 0 & 0 & 0 & d_{44} & 0 \\ 0 & 0 & 0 & 0 & 0 & d_{66} \end{bmatrix}. \tag{2.80}$$

2.3.1.3 Linear Elastic Orthotropic Material

Orthotropic behaviour can be shown by layered materials in which the single layers are anisotropic, e.g. by microbeams obtained by depositing piezoelectric layers on top of polycrystalline silicon beams. The stiffness matrix in this case is defined starting from nine independent constants, as in Equation 2.81:

$$\mathbf{d} = \begin{bmatrix} d_{11} & d_{12} & d_{13} & 0 & 0 & 0 \\ d_{12} & d_{22} & d_{23} & 0 & 0 & 0 \\ d_{13} & d_{23} & d_{33} & 0 & 0 & 0 \\ 0 & 0 & 0 & d_{44} & 0 & 0 \\ 0 & 0 & 0 & 0 & d_{55} & 0 \\ 0 & 0 & 0 & 0 & 0 & d_{66} \end{bmatrix}. \tag{2.81}$$

The linear elastic isotropic constitutive law is used in most practical cases, owing to its simplicity, even when the material under study is not rigorously isotropic. This is the case of metals, which are usually considered isotropic; this assumption is a good idealization of reality at the macroscale and is also applied in the world of microsystems.

Nevertheless, it should always be taken into account that for a polycrystalline material the overall isotropicity of the response highly depends on the internal crystal structure and the overall dimensions of the device under study with respect to the crystal grain dimensions.

Consider again polycrystalline silicon: it is made of face-centred-cubic single crystals, nevertheless on average it can be considered transversely isotropic, owing to the random distribution of crystal axes, or even isotropic. This last is usually the drastic assumption made in many design phases for microsystems; its implications and possible over-simplifications must be understood and controlled for, depending on the specific case considered.

The linear elastic problem for deformable bodies is the simplest one among the various possible schematizations and mathematical treatments for the study of solids under various external actions. Notwithstanding its simplicity, closed-form solutions can be obtained in only a few cases; to solve the problem for general body shapes and loading conditions, one should use numerical approaches. Among the available numerical treatments, the most popular is the *finite element method*, which allows for the obtainment of solutions not only for the linear elastic problem but also for nonlinear solid mechanics problems and, more generally, for problems governed by partial differential equations in space, quasi-static or evolving in time, referred to single physics or to multiphysics situations. Good references on the finite element method are Bathe (1996), Bonnet, Frangi and Rey (2014) and Zienkiewicz and Taylor (2000). In this book, we will refer to finite element method solutions in various sections and relevant remarks will be made when necessary.

2.3.1.4 Summary

It is now useful to sum up the main points discussed in this section.

- The mechanical response of deformable bodies subject to external actions is governed by compatibility, equilibrium (motion) and constitutive equations.
- Depending on the kind of external loading, we can define a *quasi-static problem*, mathematically a boundary value problem, or a *dynamic problem*, mathematically an initial boundary value problem.
- The nature and properties of the material enter the constitutive law.
- The simplest problem for deformable solids is the linear elastic one, in which the hypotheses of small strain and displacements and linear elastic constitutive law are introduced.
- When the material is isotropic, the linear elastic constitutive law can be defined in terms of two parameters: the Young's modulus and Poisson's ratio.
- The linear elastic problem for deformable bodies can be solved analytically only in a few cases. General solutions can be obtained by means of numerical procedures, such as the finite element method.

2.3.2 Linear Elastic Problem for Beams

In many practical situations, a deformable body has particular shapes that allow for a simplified treatment of the linear elastic problem.

The most meaningful example is that of an *elastic beam*, like that shown in Figure 2.17. A beam is a 3D body that can be created by starting from a plane shape, the beam *cross-section*, and making its centre of mass G move in the space following an open trajectory, the length of which is much larger than the maximum dimension along the cross-section. The trajectory becomes the *beam axis*, and the resulting body has a typically slender shape.

Beams are the simplest example of structural elements and are widely used to schematize real structures, e.g. in buildings, cars, aeroplanes and satellites. In microsystems, the beam schematization is typically used for elastic suspension parts, simply called *springs* in the MEMS literature, and for many vibrating elements.

The fact that one dimension of the beam, the axis length, is much larger than all the other dimensions, makes it possible to introduce drastic simplifications and to reduce the linear elastic problem for the beam to a 1D problem in which all input data and unknowns depend only on a coordinate along the beam axis.

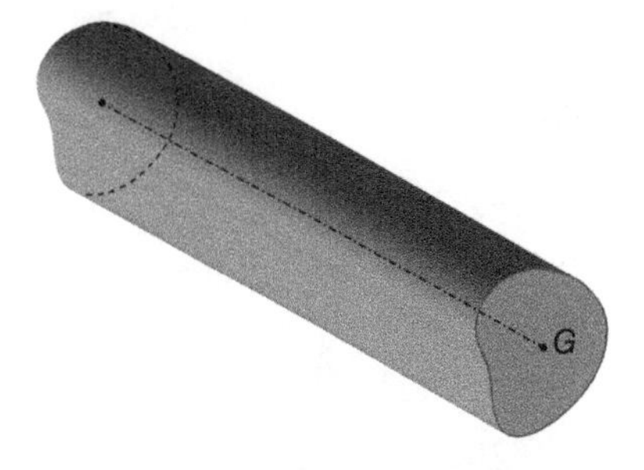

Figure 2.17 Geometry of a beam.

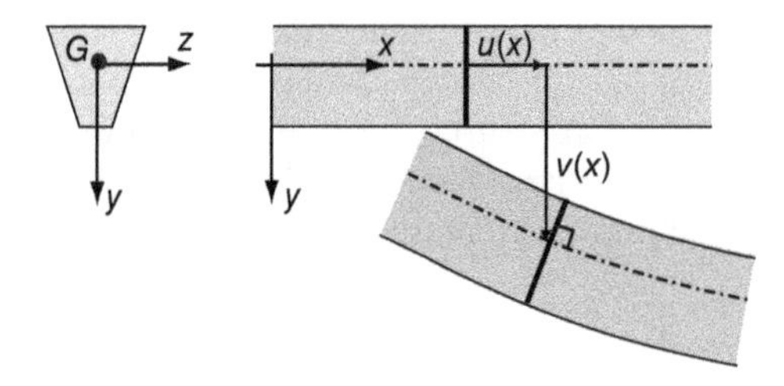

Figure 2.18 Deformed beam.

The formulation here presented is known as the *Euler–Bernoulli beam theory* and is discussed here for the simplest case of a rectilinear beam whose axis x is therefore contained in a plane. In addition to this, we make the hypothesis that the beam cross-section has at least one symmetry axis which, together with the beam axis, defines the plane x, y where we study the beam movement and deformation, as shown in Figure 2.18. Moreover, we assume that the material is homogeneous, linear elastic and isotropic.

In the Euler–Bernoulli beam theory, the beam is considered as a set of rigid cross-sections, which move in space, keeping their position orthogonal to the beam axis. This strong assumption implies that no shear strain is considered in the beam; the only strain component considered is the one in the direction of the beam axis. It is important to remark that zero shear strain does not mean zero shear stress inside the beam! It only means that the beam is considered to be infinitely stiff with respect to shear strain.

On the basis of these simplifying hypotheses, a reformulation of the elastic problem for the deformable beam can be obtained; this means that kinematic compatibility, equilibrium and the linear elastic constitutive law are rewritten in new variables, which are only functions of the coordinate x along the beam axis.

Let us consider the schematic representation of a beam in the initial and deformed configurations given in Figure 2.18; the displacement of the beam is governed by the functions $u(x)$ and $v(x)$, which coincide with the displacement components of a point at position x along the beam axis in the directions parallel and orthogonal to the beam axis, respectively.

Starting from the displacements $u(x)$ and $v(x)$, and taking into account the reference frame of Figure 2.18, the only strain component ε_x in the Euler–Bernoulli beam can be obtained as

$$\varepsilon_x = \frac{\mathrm{d}u}{\mathrm{d}x} - y\frac{\mathrm{d}^2 v}{\mathrm{d}x^2} \equiv \eta + y\chi. \tag{2.82}$$

In Equation 2.82, two new functions $\eta(x)$ and $\chi(x)$ have been defined: η is the axial deformation of the beam, i.e. the strain ε_x computed at $y = 0$; χ is called the beam *curvature*, dimensionally the reciprocal of length, and is the inverse of the *radius of curvature* of the beam axis at a certain position.

From Equation 2.82, applying the uniaxial linear elastic constitutive law, the stress σ_x follows:

$$\sigma_x = E(\eta + y\chi). \tag{2.83}$$

It can be seen from Equations 2.82 and 2.83 that both ε_x and σ_x are linear functions of the coordinate y.

Integrating σ_x along the cross-sectional area A, we obtain the axial force N:

$$N = \int_A \sigma_x \,\mathrm{d}A. \tag{2.84}$$

Integrating the product $\sigma_x y$ along the cross-section, we obtain the bending moment M:

$$M = \int_A \sigma_x y \, \mathrm{d}A. \tag{2.85}$$

In the Euler–Bernoulli beam description, N and M are the static variables corresponding to the kinematic variables η and χ, respectively.

We can now obtain a constitutive law in the beam variables, by introducing Equation 2.83 into Equations 2.84 and 2.85:

$$\begin{aligned} N &= \int_A \sigma_x \, \mathrm{d}A = \int_A E(\eta + y\chi) \, \mathrm{d}A = EA\eta, \\ M &= \int_A \sigma_x y \, \mathrm{d}A = \int_A E(\eta + y\chi)y \, \mathrm{d}A = EJ\chi. \end{aligned} \tag{2.86}$$

In Equation 2.86, it has been taken into account that the material is homogeneous and therefore E is a constant and that $\int_A y \, \mathrm{d}A = 0$ because it coincides with the first moment of inertia of the beam cross-section computed with respect to an axis containing the centre of gravity. Finally the second moment of inertia $J = \int_A y^2 \, \mathrm{d}A$ has been introduced.

The last ingredient that is needed for the reformulation of the linear elastic problem in the beam variables is equilibrium, in quasi-static or dynamic conditions.

Equilibrium conditions can be directly obtained by enforcing equilibrium of an infinitesimal portion of the beam, like that depicted in Figure 2.19, where we see the axial force and its increment, the bending moment and its increment and an additional internal force W and its increment. W is called the shear force and is the integral along the cross-section of the shear stress component distributed along the beam cross-section in the direction of axis y:

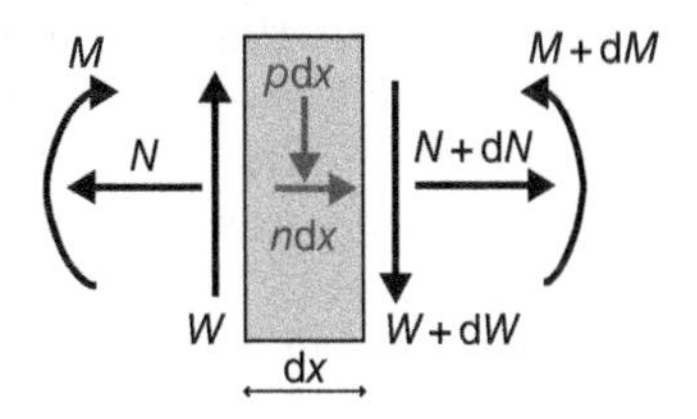

Figure 2.19 Infinitesimal element of an Euler–Bernoulli beam.

$$W = \int_A \tau_{xy} \, \mathrm{d}A. \tag{2.87}$$

The infinitesimal portion of the beam is also loaded with loads per unit length n and p, directed, respectively, along the beam axis x and in the direction of axis y.

Enforcing equilibrium in quasi-static conditions of the infinitesimal portion of the beam in Figure 2.19 in the direction of the beam axis, in the direction orthogonal to the beam axis and at the rotation around an axis orthogonal to the beam plane, it is possible to obtain

$$\begin{aligned} \frac{\mathrm{d}N}{\mathrm{d}x} + n &= 0, \\ \frac{\mathrm{d}W}{\mathrm{d}x} + p &= 0, \\ \frac{\mathrm{d}M}{\mathrm{d}x} &= W. \end{aligned} \tag{2.88}$$

It is now possible to combine the compatibility (2.82), constitutive (2.86) and equilibrium (2.88) equations and obtain two differential equations in the unknowns u and v, governing, separately, the axial and the bending behaviour, respectively:

$$\frac{\mathrm{d}}{\mathrm{d}x}\left(EA\frac{\mathrm{d}u}{\mathrm{d}x}\right)+n=0,$$
$$-\frac{\mathrm{d}^2}{\mathrm{d}x^2}\left(EJ\frac{\mathrm{d}^2 v}{\mathrm{d}x^2}\right)+p=0. \tag{2.89}$$

To solve Equations 2.89, it is also necessary to define suitable boundary conditions.

As far as the second-order equation in the unknown u is concerned, boundary conditions must be given for the value of the displacement u, possibly assigned due to the presence of kinematic constraints in some point of the beam, and for its first derivative. This second condition involves assigning the value of the axial force or of the axial deformation.

Concerning the fourth-order equation in the unknown v, we need to find four conditions expressed on the displacement v and on its derivatives up to the third order. The mechanical meaning is as follows: an assigned value of the displacement v means that a kinematic constraint is assigned in some point of the beam in the direction orthogonal to the beam axis; assigning the first derivative of v means that the rotation is known; the second derivative is directly related to the curvature or bending moment value, while the third derivative is proportional to the shear force.

The transformation of the quasi-static beam problem into a dynamic problem can be obtained by inserting inertia forces in the equilibrium conditions. This involves considering the unknown functions u and v as also functions of the time variable t and considering accelerations in the direction of the beam axis $\ddot{u}$ and in the direction orthogonal to the beam axis $\ddot{v}$. Inertia forces per unit length that act on the infinitesimal beam element of Figure 2.19 are obtained by multiplying accelerations by the mass density per unit volume ρ and by the area A:

$$\frac{\mathrm{d}N}{\mathrm{d}x}+n=\rho A\ddot{u},$$
$$\frac{\mathrm{d}W}{\mathrm{d}x}+p=\rho A\ddot{v},$$
$$\frac{\mathrm{d}M}{\mathrm{d}x}=W. \tag{2.90}$$

Differential equations in the unknowns u and v can be obtained from Equations 2.90 by making use of compatibility and constitutive laws for the beam:

$$\frac{\mathrm{d}}{\mathrm{d}x}\left(EA\frac{\mathrm{d}u}{\mathrm{d}x}\right)+n=\rho A\ddot{u},$$
$$-\frac{\mathrm{d}^2}{\mathrm{d}x^2}\left(EJ\frac{\mathrm{d}^2 v}{\mathrm{d}x^2}\right)+p=\rho A\ddot{v}. \tag{2.91}$$

Equations 2.91 can be solved by adding boundary conditions as for Equation 2.89 and initial conditions on displacements u and v and velocities $\dot{u}$ and $\dot{v}$.

It is important to remark that, in view of the small strain and displacement and homogeneous beam, the axial and the bending-shear responses of the beam are fully decoupled.

This description of beam mechanics is restricted to a 2D case. More generally, a beam can be seen as a 1D body in a 3D space; in this case, the previous discussion must be generalized; other possible generalizations concern heterogeneous beams, such as those

Figure 2.20 Beam under torsion.

used in layered composites, and different hypotheses on the deformation possibility given to the beam. Complete treatments of beam mechanics can be found in specialized books, see, for example, Hjelmstad (2005).

Before concluding this section devoted to beam mechanics, it is worth mentioning another recurrent situation in MEMS design; the case of a beam subject to a torque M_T, i.e. a moment around the beam axis, as shown in Figure 2.20.

This case must be studied by considering the beam in a 3D space; here, we mention only that an equation analogous to Equation 2.86, holding for axial and bending responses, can be proved. In the case of torsion, the kinematic variable is the angle θ of relative rotation in torsion per unit length and the relation can be written as

$$M_T = GJ_T\theta, \tag{2.92}$$

where G is the material elastic shear modulus and J_T is the geometrical torsional momentum (dimensionally a length to the 4th power), which depends on the shape of the beam cross-section. A meaningful case, which is often met in MEMS design, is the one in which the cross-section is rectangular, as in Figure 2.20, with one side H much longer than the other one t. In this case, the geometrical torsional stiffness J_T can be simply computed as:

$$J_T = \frac{1}{3}Ht^3. \tag{2.93}$$

2.3.2.1 Examples

It is now possible to discuss some simple examples, which show how the beam formulation can be used.

Clamped Beam Subject to a Distributed Axial Force Let us consider the beam shown in Figure 2.21: a homogeneous beam of length l, constant cross-sectional area A, is completely constrained at its left end, free at its right end and loaded by a uniform load per unit length n in the axial direction.

In quasi-static conditions, the governing equation of the problem is the first of Equations 2.89, while the boundary conditions are

$$u(0) = 0, \quad \left.\frac{\mathrm{d}u}{\mathrm{d}x}\right|_l = 0. \tag{2.94}$$

The solution of the problem is:

$$\begin{aligned} u(x) &= \frac{nl^2}{EA}\left[-\frac{1}{2}\left(\frac{x}{l}\right)^2 + \frac{x}{l}\right], \\ \eta &= \frac{\mathrm{d}u}{\mathrm{d}x} = \frac{nl}{EA}\left[-\frac{x}{l} + 1\right], \end{aligned} \tag{2.95}$$

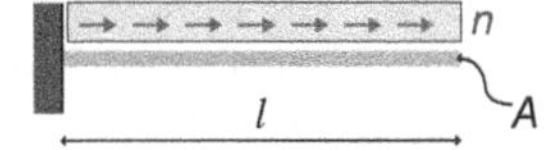

Figure 2.21 Clamped beam subject to a distributed axial force.

$$N = EA\eta = \frac{\mathrm{d}u}{\mathrm{d}x} = nl\left[-\frac{x}{l} + 1\right]. \tag{2.96}$$

The displacement is therefore quadratic, while the axial strain and the axial force are linear functions of the position. From Equation 2.96, it can be obtained that the value of the axial force at the left, clamped, end is $N = nl$, which coincides with the resultant of the distributed load n along the beam.

It is now possible to estimate the maximum stress inside the beam; this will be found in the clamped end and will be uniformly distributed along the cross-section:

$$\sigma_x(0) = \frac{N(0)}{A} = \frac{nl}{A}. \tag{2.97}$$

By means of the value of the maximum stress inside the beam it is possible to check whether the beam is able to sustain the load remaining in the elastic regime or whether it reaches the critical value σ_c.

Clamped Beam Subject to a Concentrated Shear Force at the Free End Let us now consider the beam shown in Figure 2.22: a homogeneous beam of length l, constant cross-sectional area and hence constant J, is completely constrained at its left end ($x = 0$), and is loaded by a concentrated load F in the direction orthogonal to the beam axis at the free end ($x = l$).

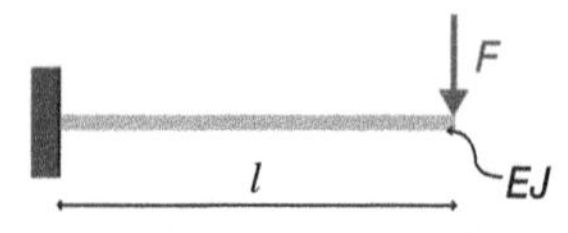

Figure 2.22 Clamped beam subject to a concentrated shear force at the free end.

In quasi-static conditions, the governing equation of the problem is the second of Equations 2.89, while the boundary conditions are now

$$v(0) = 0, \quad \left.\frac{\mathrm{d}v}{\mathrm{d}x}\right|_0 = 0, \quad \left.\frac{\mathrm{d}^2v}{\mathrm{d}x^2}\right|_l = 0, -EJ\left.\frac{\mathrm{d}^3v}{\mathrm{d}x^3}\right|_l = F. \tag{2.98}$$

The solution of the problem is as follows:

$$v(x) = \frac{Fl^3}{2EJ}\left[-\frac{1}{3}\left(\frac{x}{l}\right)^3 + \left(\frac{x}{l}\right)^2\right], \tag{2.99}$$

$$\frac{\mathrm{d}v}{\mathrm{d}x} = \frac{Fl^2}{EJ}\left[-\frac{1}{2}\left(\frac{x}{l}\right)^2 + \frac{x}{l}\right],$$

$$M = -EJ\chi = -EJ\frac{\mathrm{d}^2v}{\mathrm{d}x^2} = Fl\left[\frac{x}{l} - 1\right],$$

$$W = -EJ\frac{\mathrm{d}^3v}{\mathrm{d}x^3} = F. \tag{2.100}$$

In this second example, the displacement v is cubic, the rotation quadratic, the curvature and the bending moment linear and the shear force constant and equal to the applied external force F. The displacement at the free end, where the concentrated load is applied, reads:

$$v(l) = \frac{Fl^3}{3EJ}. \tag{2.101}$$

Finally, it is possible to obtain the maximum value of the normal stress σ_x inside the beam, which is found in the clamped end; it depends directly on the value of the bending moment in that section. The value of σ_x can be found by combining Equations 2.83

and 2.86 to obtain the stress at the clamped end:

$$\sigma_x(0,y) = \frac{M(0)}{J}y = -\frac{Fl}{J}y. \tag{2.102}$$

The maximum stress value is found at maximum y, i.e. on the upper or lower part of the cross-section; if the cross-section is rectangular with height H and width B, the maximum normal tensile stress in the beam of Figure 2.22 is on the upper side at the clamped cross-section and is given by

$$\sigma_x(0,-H/2) = \frac{FlH}{2J} = 6\frac{Fl}{BH^2}. \tag{2.103}$$

Doubly Clamped Beam Subject to an Imposed Displacement at One End The case of Figure 2.23 is now considered: it is again a homogeneous beam with constant cross-sectional area, clamped at both ends, and subject to a displacement U at one end, as shown in the figure. To solve the problem, we can again refer to the second of Equations 2.89 with the following boundary conditions:

$$v(0) = 0, \quad \left.\frac{\mathrm{d}v}{\mathrm{d}x}\right|_0 = 0,$$
$$v(l) = U, \quad \left.\frac{\mathrm{d}v}{\mathrm{d}x}\right|_l = 0. \tag{2.104}$$

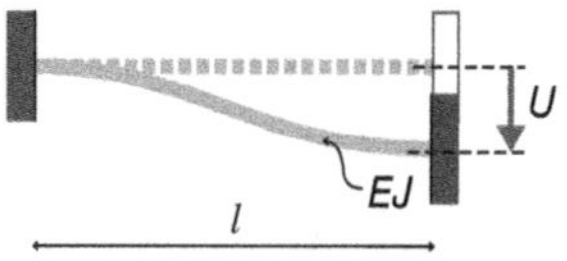

Figure 2.23 Doubly clamped beam subject to imposed displacement at one end.

The solution in terms of the displacement v is:

$$v(x) = U\left[-2\left(\frac{x}{l}\right)^3 + 3\left(\frac{x}{l}\right)^2\right]. \tag{2.105}$$

From Equation 2.105, it is possible to compute the shear force, which turns out to be constant:

$$W = -EJ\frac{\mathrm{d}^3v}{\mathrm{d}x^3} = \frac{12EJ}{l^3}U. \tag{2.106}$$

The ratio $W/U = 12EJ/l^3$ can be named the beam stiffness in the sense that it represents the force necessary to impose a unit displacement at one of the two ends of the beam; this value is used on many occasions when designing microsystems anchored to a substrate by means of flexible slender beams.

Doubly Clamped Beam Subject to an Imposed Rotation at One End Consider the case shown in Figure 2.24: a homogeneous beam with constant cross-sectional area, clamped at the left and right end, where a rotation Θ is assigned.

To solve the problem, we refer as usual to the second of Equations 2.89 with the following boundary conditions:

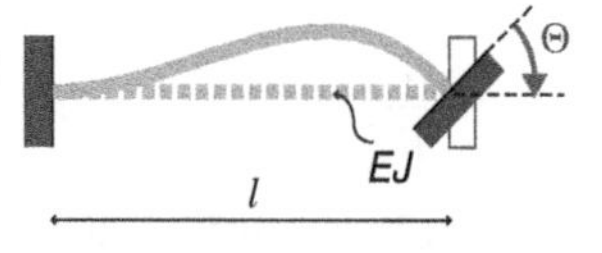

Figure 2.24 Doubly clamped beam subject to an imposed rotation at one end.

$$v(0) = 0, \quad \left.\frac{\mathrm{d}v}{\mathrm{d}x}\right|_0 = 0, \quad v(l) = 0, \quad \left.\frac{\mathrm{d}v}{\mathrm{d}x}\right|_l = \Theta. \tag{2.107}$$

The solution in terms of the displacement v is:

$$v(x) = \Theta l\left[\left(\frac{x}{l}\right)^3 - \left(\frac{x}{l}\right)^2\right]. \tag{2.108}$$

From Equation 2.108, it is possible to compute the bending moment at the right beam end, where the rotation is imposed:

$$M(l) = -EJ\left.\frac{d^2 v}{dx^2}\right|_l = -4\frac{EJ}{l}\Theta. \tag{2.109}$$

The ratio $M/\Theta = 4EJ/l$ can be named the beam stiffness in the sense that it represents the moment necessary to impose a unit rotation at one of the two ends of the beam; this value is also used on many occasions when designing microsystems anchored to the substrate by means of flexible slender beams.

Folded Beams The last example of this section is a case that is often found in microsystems design, called *folded beams*. The case is shown in Figure 2.25: a homogeneous beam with constant cross-sectional area clamped on one end is fabricated in a folded shape by clamping together a certain number of straight beams, as shown in the figure.

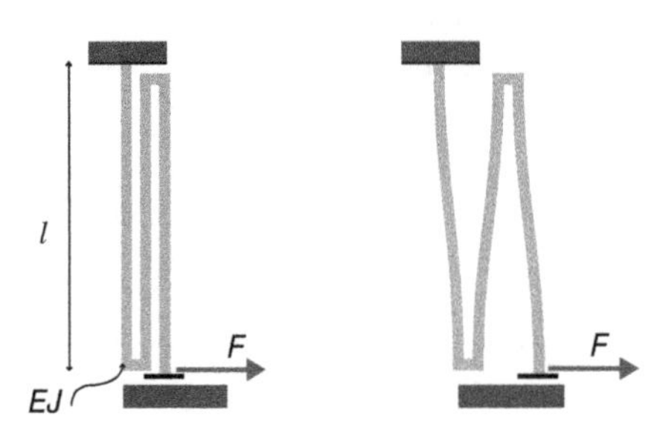

Figure 2.25 Three-folded beam.

This case can be simply resolved by making use of the solution presented in the third example of this section. By assuming that the short beam portions that connect each fold of the beam are rigid, it can be observed that the deformed shape of the folded beam is as shown in Figure 2.25. As a consequence, each fold of the beam deforms, as in the case of a doubly clamped beam subject to an imposed displacement at an end. Moreover, owing to local equilibrium, the force applied to the free end is transmitted unaltered to the final clamped end; hence, with the same applied force, we obtain a lateral displacement that is the one of the single fold multiplied by the number of folds n_f and the global stiffness relating the force to the global displacement is given by

$$\frac{12EJ}{n_f l^3}. \tag{2.110}$$

A similar reasoning holds for folded beams subject to torsion, in this case the global stiffness relating the applied torque to the total rotation angle is given by

$$\frac{GJ_T}{n_f l}. \tag{2.111}$$

From these results, it can be observed that folded beams in bending and in torsion can be used to increase the compliance, while keeping the in-plane space occupation limited.

2.3.2.2 Summary

We can now summarize the main points discussed in this section devoted to beam mechanics.

- Straight beams have been defined as special 3D prismatic solids with one dimension prevailing with respect to the others.
- Beams can be studied by introducing only one spatial dimension.

- The linear elastic problem for deformable bodies has been reformulated in terms of beam variables.
- The governing equations for beams have been given in quasi-static and dynamic conditions.
- Examples have been proposed and solved for meaningful cases.

2.4 Fluid Mechanics

Fluid mechanics studies the motion of gaseous or liquid fluids under a wide spectrum of conditions; fundamentally, the same equations can be used to describe waves, water flowing in pipes, gases in the atmosphere or fluids in microsystems. As for rigid bodies and deformable solids, the hypothesis of the continuum is usually assumed.

The contents of this section are limited to an introduction to the standard formulation of the equations governing fluid dynamics, called Navier–Stokes equations, and to some specializations that are particularly meaningful for microsystem design. The interested reader can find a complete description of fluid dynamics and its computational aspects in, for example, Versteeg and Malalasekera (2007).

2.4.1 Navier–Stokes Equations

The hypothesis of continuum allows all given and unknown quantities to be defined as fields, i.e. as functions of position and time. The description of fluids in motion is based on the velocity field $\mathbf{v}$ and not on the displacement one, as is usually done in solids. Another important point is that the mass density of the fluid ρ is, in many cases, variable with time and not only with position. In general, density is another unknown of the problem.

The equations governing the fluid motion are based on conservation laws, constitutive relations that characterize, as in solid mechanics, the kind of fluid we are dealing with and on state equations, which are typically different for gases and liquids.

The first conservation law concerns the mass, i.e. the mass of the fluid can move with fluid particles but, in the standard hypothesis considered, cannot disappear or be created. This is usually called the *mass conservation law* or *continuity condition* and is written as

$$\frac{\partial \rho}{\partial t} + \mathrm{div}(\rho \mathbf{v}) = 0. \tag{2.112}$$

In the continuity condition, the symbol div means the *divergence* operator applied to any vector field $\mathbf{a}$:

$$\mathrm{div}\, \mathbf{a} = \frac{\partial a_x}{\partial x} + \frac{\partial a_y}{\partial y} + \frac{\partial a_z}{\partial z}. \tag{2.113}$$

The second conservation law is that of *linear momentum conservation*, a vector relation, which is basically the application of Newton's second law of dynamics to an infinitesimal portion of fluid in motion and is conceptually analogous to the dynamic equilibrium equations (2.70) written for a deformable solid.

Before writing the linear momentum conservation equation for a fluid, it is important to remark that, in the typical fluid description, the velocity field gives the value of the fluid velocity in a certain position in space and at a certain time instant. If we compute the derivative of the velocity with respect to the time variable, we simply compute how the velocity in a certain position in space is varying. If we want to compute the acceleration of a fluid particle that is moving in space, we must also consider, in the computation of the velocity time derivative, the variation in space of the coordinates. In this way, we compute what is called the *material* or *advective* time *derivative*, i.e.:

$$\frac{\mathrm{D}\mathbf{v}}{\mathrm{D}t} = \frac{\partial \mathbf{v}}{\partial t} + \frac{\partial \mathbf{v}}{\partial x}\frac{\partial x}{\partial t} + \frac{\partial \mathbf{v}}{\partial y}\frac{\partial y}{\partial t} + \frac{\partial \mathbf{v}}{\partial z}\frac{\partial z}{\partial t} = \frac{\partial \mathbf{v}}{\partial t} + v_x\frac{\partial \mathbf{v}}{\partial x} + v_y\frac{\partial \mathbf{v}}{\partial y} + v_z\frac{\partial \mathbf{v}}{\partial z}. \tag{2.114}$$

Taking into account this concept of the material time derivative, assuming also that for fluids a stress tensor field $\boldsymbol{\sigma}$ can be defined and that possible body forces $\mathbf{b}$, such as gravity, act on the fluid, we can rewrite the dynamic equilibrium equations (2.70) and obtain the linear momentum conservation equations for a fluid:

$$\begin{aligned}
&\frac{\partial \sigma_{xx}}{\partial x} + \frac{\partial \sigma_{yx}}{\partial y} + \frac{\partial \sigma_{zx}}{\partial z} + b_x = \rho\frac{\mathrm{D}v_x}{\mathrm{D}t},\\
&\frac{\partial \sigma_{xy}}{\partial x} + \frac{\partial \sigma_{yy}}{\partial y} + \frac{\partial \sigma_{zy}}{\partial z} + b_y = \rho\frac{\mathrm{D}v_y}{\mathrm{D}t},\\
&\frac{\partial \sigma_{xz}}{\partial x} + \frac{\partial \sigma_{yz}}{\partial y} + \frac{\partial \sigma_{zz}}{\partial z} + b_z = \rho\frac{\mathrm{D}v_z}{\mathrm{D}t}.
\end{aligned} \tag{2.115}$$

This is not the standard format in which the linear momentum equation is written in fluid mechanics. To obtain this, we can make use of the additive decomposition of a stress tensor in its hydrostatic p and deviatoric $\mathbf{s}$ parts, given in Equation 2.67, and rewrite Equation 2.115 as

$$\begin{aligned}
&\rho\frac{\mathrm{D}v_x}{\mathrm{D}t} = \frac{\partial p}{\partial x} + \frac{\partial s_{xx}}{\partial x} + \frac{\partial s_{yx}}{\partial y} + \frac{\partial s_{zx}}{\partial z} + b_x,\\
&\rho\frac{\mathrm{D}v_y}{\mathrm{D}t} = \frac{\partial p}{\partial y} + \frac{\partial s_{xy}}{\partial x} + \frac{\partial s_{yy}}{\partial y} + \frac{\partial s_{zy}}{\partial z} + b_y,\\
&\rho\frac{\mathrm{D}v_z}{\mathrm{D}t} = \frac{\partial p}{\partial z} + \frac{\partial s_{xz}}{\partial x} + \frac{\partial s_{yz}}{\partial y} + \frac{\partial s_{zz}}{\partial z} + b_z.
\end{aligned} \tag{2.116}$$

Notice that the hydrostatic part p is here assumed as positive when tensile, while in standard fluid mechanics it is the contrary and therefore a minus sign would appear in front of the relevant terms; the unknown is simply called *pressure*.

Here, p is usually considered as one of the problem unknowns, while the components of the deviatoric stress tensor $\mathbf{s}$ are eliminated by means of a constitutive law for the fluid. The most meaningful cases are described in the following.

2.4.1.1 Newtonian Fluids

In the case of *Newtonian fluids*, the basic hypotheses concerning fluid behaviour are that the fluid is considered isotropic and that a linear relation exists between the components of the stress tensor and those of the strain rate tensor. In the 1D case, this relation reads:

$$\sigma = \mu_\mathrm{f}\frac{\mathrm{d}\epsilon}{\mathbf{d}t} = \mu_\mathrm{f}\frac{\mathrm{d}v}{\mathrm{d}x}, \tag{2.117}$$

where a new material parameter μ_f called *dynamic viscosity* has been introduced. Parameter μ_f has the dimension of a force per unit surface multiplied by time

(N·s/m^2 = Pa·s). Typical values of viscosity at ambient temperature (25°C) are: 35×10^{-4} for blood, 8.9×10^{-4} for water and 0.19×10^{-4} for air. Referring now to a 3D case and considering again the decomposition of stress into deviatoric and hydrostatic parts, it is possible to obtain the following relations:

$$
\begin{aligned}
s_{xx} &= 2\mu_{\mathrm{f}}\frac{\partial v_x}{\partial x} - \frac{2}{3}\mu_{\mathrm{f}}\mathrm{div}\,\mathbf{v},\\
s_{yy} &= 2\mu_{\mathrm{f}}\frac{\partial v_y}{\partial y} - \frac{2}{3}\mu_{\mathrm{f}}\mathrm{div}\,\mathbf{v},\\
s_{zz} &= 2\mu_{\mathrm{f}}\frac{\partial v_z}{\partial z} - \frac{2}{3}\mu_{\mathrm{f}}\mathrm{div}\,\mathbf{v},\\
s_{xy} &= \mu_{\mathrm{f}}\left(\frac{\partial v_x}{\partial y} + \frac{\partial v_y}{\partial x}\right),\\
s_{yz} &= \mu_{\mathrm{f}}\left(\frac{\partial v_y}{\partial z} + \frac{\partial v_z}{\partial y}\right),\\
s_{xz} &= \mu_{\mathrm{f}}\left(\frac{\partial v_x}{\partial z} + \frac{\partial v_z}{\partial x}\right).
\end{aligned}
\tag{2.118}
$$

Substituting Equations 2.118 into the momentum equation (2.116), and considering the continuity equation (2.112), we obtain a set of four partial differential relations in the unknowns density ρ, velocity $\mathbf{v}$ and pressure $-p$.

When the viscosity does not depend on spatial coordinates, the form of the dynamic equilibrium equations is

$$
\begin{aligned}
\rho\frac{\mathrm{D}v_x}{\mathrm{D}t} &= \frac{\partial p}{\partial x} + \mu_{\mathrm{f}}\nabla^2 v_x + \frac{1}{3}\mu_{\mathrm{f}}\frac{\partial}{\partial x}\mathrm{div}\,\mathbf{v} + b_x,\\
\rho\frac{\mathrm{D}v_y}{\mathrm{D}t} &= \frac{\partial p}{\partial y} + \mu_{\mathrm{f}}\nabla^2 v_y + \frac{1}{3}\mu_{\mathrm{f}}\frac{\partial}{\partial y}\mathrm{div}\,\mathbf{v} + b_y,\\
\rho\frac{\mathrm{D}v_z}{\mathrm{D}t} &= \frac{\partial p}{\partial z} + \mu_{\mathrm{f}}\nabla^2 v_z + \frac{1}{3}\mu_{\mathrm{f}}\frac{\partial}{\partial z}\mathrm{div}\,\mathbf{v} + b_z,
\end{aligned}
\tag{2.119}
$$

where ∇^2 is the Laplace operator.

Since we now have four differential equations (the continuity condition and the momentum conservation) in five unknowns, we have to find an additional equation that is a state equation. A typical example is the gas law in the case of gaseous fluids. If non-isothermal processes are considered, temperature appears as an additional unknown and a final additional relation coming from a conservation of energy statement is needed.

To complete the formulation, initial and boundary conditions must be specified on the unknown variables, as briefly discussed at the end of this section.

The obtained partial differential equations are called the Navier–Stokes equations. These equations are nonlinear and very complex to solve. Solutions in general conditions can be obtained only by means of computational methods and are the main goal of computational fluid dynamics.

A key parameter, allowing the identification of different regimes with different types of solution, is the nondimensional number called the *Reynolds number Re* defined as

$$
Re = \frac{\rho V L}{\mu_{\mathrm{f}}},
\tag{2.120}
$$

where V is a characteristic velocity of the fluid and L a characteristic length. The Reynolds number can be interpreted as the ratio between inertia and viscous forces: a low Reynolds number characterizes *laminar flow*, while high Reynolds numbers characterize *turbulent flow*; for intermediate Reynolds numbers, the flow is in the so-called *transition regime*.

2.4.1.2 Incompressible Fluids

In many situations, it is possible to consider the fluid as incompressible, with constant density ρ. In this case, the continuity condition gives

$$\text{div}\,\mathbf{v} = 0. \tag{2.121}$$

This condition allows the Navier–Stokes equations to be simplified; in the case of constant viscosity, these can now be written as follows:

$$\begin{aligned}
\rho\frac{\mathrm{D}v_x}{\mathrm{D}t} &= \frac{\partial p}{\partial x} + \mu_\mathrm{f}\nabla^2 v_x + b_x,\\
\rho\frac{\mathrm{D}v_y}{\mathrm{D}t} &= \frac{\partial p}{\partial y} + \mu_\mathrm{f}\nabla^2 v_y + b_y,\\
\rho\frac{\mathrm{D}v_z}{\mathrm{D}t} &= \frac{\partial p}{\partial z} + \mu_\mathrm{f}\nabla^2 v_z + b_z.
\end{aligned} \tag{2.122}$$

These equations are usually rewritten dividing by the mass density and defining a new parameter called the *kinematic viscosity* $\nu_\mathrm{f} \equiv \mu_\mathrm{f}/\rho$:

$$\begin{aligned}
\frac{\mathrm{D}v_x}{\mathrm{D}t} &= \frac{1}{\rho}\frac{\partial p}{\partial x} + \nu_\mathrm{f}\nabla^2 v_x + \frac{1}{\rho}b_x,\\
\frac{\mathrm{D}v_y}{\mathrm{D}t} &= \frac{1}{\rho}\frac{\partial p}{\partial y} + \nu_\mathrm{f}\nabla^2 v_y + \frac{1}{\rho}b_y,\\
\frac{\mathrm{D}v_z}{\mathrm{D}t} &= \frac{1}{\rho}\frac{\partial p}{\partial z} + \nu_\mathrm{f}\nabla^2 v_z + \frac{1}{\rho}b_z.
\end{aligned} \tag{2.123}$$

Since the condition of constant density is an equation of state that also eliminates the density as an unknown, Equations 2.121 and 2.123 are now a system of four equations in four unknowns, and only the initial and boundary conditions must be specified to complete the problem.

2.4.1.3 Perfect Fluids

A further strong approximation considers a fluid that is not able to sustain shear stress components; hence, only the pressure or hydrostatic term survives in the momentum equations and these read:

$$\begin{aligned}
\rho\frac{\mathrm{D}v_x}{\mathrm{D}t} &= \frac{\partial p}{\partial x} + b_x,\\
\rho\frac{\mathrm{D}v_y}{\mathrm{D}t} &= \frac{\partial p}{\partial y} + b_y,\\
\rho\frac{\mathrm{D}v_z}{\mathrm{D}t} &= \frac{\partial p}{\partial z} + b_z,
\end{aligned} \tag{2.124}$$

Note that the same set of equations can be obtained by making the hypothesis that the fluid is inviscid, i.e. with zero viscosity. If we combine the hypothesis of inviscid fluid with that of incompressibility, Equations 2.121 and 2.124 constitute a set of four equations in four unknowns, called the *Euler equations*.

2.4.1.4 Stokes Flow

A very particular and simple way to describe fluid motion is the one governed by the so-called *Stokes equations*. Basically, a fluid is in the Stokes regime when we can neglect the inertial terms in the momentum conservation equations, thus obtaining equations that coincide with those of static equilibrium in deformable solids:

$$\begin{aligned}
&\frac{\partial \sigma_{xx}}{\partial x} + \frac{\partial \sigma_{yx}}{\partial y} + \frac{\partial \sigma_{zx}}{\partial z} + b_x = 0, \\
&\frac{\partial \sigma_{xy}}{\partial x} + \frac{\partial \sigma_{yy}}{\partial y} + \frac{\partial \sigma_{zy}}{\partial z} + b_y = 0, \\
&\frac{\partial \sigma_{xz}}{\partial x} + \frac{\partial \sigma_{yz}}{\partial y} + \frac{\partial \sigma_{zz}}{\partial z} + b_z = 0.
\end{aligned} \tag{2.125}$$

Combining these equations with the conservation of mass, we obtain the Stokes equations.

Stokes flow is also called creeping flow; it is realized when the fluid velocity is small and the viscosity high or the length scale very small. Typical examples are the flow of polymers and particular situations that are realized in MEMS (see Chapters 10 and 15).

Stokes equations are usually combined with the properties of a Newtonian fluid and the condition of incompressibility, hence they read:

$$\begin{aligned}
&\frac{\partial p}{\partial x} + \mu_f \nabla^2 v_x + b_x = 0, \\
&\frac{\partial p}{\partial y} + \mu_f \nabla^2 v_y + b_y = 0, \\
&\frac{\partial p}{\partial z} + \mu_f \nabla^2 v_z + b_z = 0,
\end{aligned} \tag{2.126}$$

$$\text{div } \mathbf{v} = 0. \tag{2.127}$$

This Stokes regime is also obtained as the limit of incompressible Navier–Stokes equations when the Reynolds number *Re* goes towards zero.

Stokes equations are linear in the unknowns velocity and pressure and therefore much more simple than the complete Navier–Stokes equations.

2.4.1.5 Initial and Boundary Conditions

For formulations in which velocity and pressure are the unknowns, the values of the pressure and of the velocity vector in the whole domain at the initial time instant should be assigned (initial conditions). Boundary conditions must be dealt with with some care. In general terms we can have fluid–solid boundaries, as for a fluid flowing in a pipe, and fluid–fluid boundaries, as when two fluids are mixed together. Only the first case will be discussed briefly here. The conditions to assign at a fluid–solid boundary depend on the nature of the solid surface and on the kind of fluid. Generally, in the direction orthogonal to the solid wall, a *no-penetration condition* must be enforced, stating that the component of the fluid velocity orthogonal to the wall must be zero. In addition to this, a *no-slip boundary condition* can be assigned in many cases; this is based on experimental evidence and involves assigning zero tangent velocity near the solid wall. In particular situations that can be found in microsystems, the no-slip boundary condition does not correctly represent reality and must be substituted by other conditions (see Chapter 15).

2.4.1.6 Summary

The main results discussed in this section devoted to fluid dynamics are now summarized.

- The governing equations of fluid dynamics are: conservation of mass, conservation of linear momentum, constitutive law, the state equation and energy conservation.
- The concept of the material time derivative has been introduced.
- The Navier–Stokes equations are written with density, pressure and velocity as unknowns.
- In a Newtonian fluid, pressure and deviatoric stress components appear, together with a material parameter called viscosity.
- The Reynolds number can be used to define different flow regimes: laminar, transition and turbulent.
- In an incompressible fluid, the mass density is constant and the Navier–Stokes equations simplify.
- In an inviscid or perfect fluid, no deviatoric stresses appear.
- Stokes flow is obtained as the limit of the incompressible Navier–Stokes equation when the Reynolds number goes to zero.

2.4.2 Fluid–Structure Interaction

In general terms, a fluid–structure interaction problem is met whenever a fluid in motion interacts with a deformable solid in such a way that the mechanical response of the solid depends on the behaviour of the fluid and the fluid motion in the region surrounding the solid is influenced by the mechanical response of the solid.

Solving a fluid–structure interaction problem thus means solving simultaneously the equations governing the solid and those governing the fluid. In many cases, this is a very difficult task and many numerical procedures have been formulated *ad hoc*. In a more general view, the fluid–structure interaction problem can be considered as a multiphysics or coupled problem; other examples of coupled problems are briefly discussed in Section 2.5.3, for electromechanics, and Section 2.7.2, for thermomechanics.

In microsystems, a typical fluid–structure interaction problem is the one governing the behaviour of a micropump (see Chapter 10), where a deformable membrane interacts with a fluid with the purpose of moving the fluid. A precise evaluation of the whole system behaviour needs the solution of the fully coupled fluid–structure interaction problem.

A situation somewhat similar to that of micropumps concerns ultrasound microtransducers or piezo ultrasound microtransducers, which send or receive acoustic waves in the surrounding fluid (air or liquid) domain.

Fluid–structure interaction can also play a role in the evaluation of damping in vibrating microstructures, as discussed in Section 15.2, devoted to the so-called *fluid damping* phenomena, in which vibration kinetic energy is lost through interaction with the surrounding fluid.

2.5 Electrostatics and Electromechanics

Electrostatics and its many implications play a major role in the study and design of MEMS, also in a mechanically oriented vision as put forward in this book.

This section contains the basic concepts of electrostatics and electromechanics and their application for the creation of electrostatic (or capacitive) sensors and actuators, which are among the most used devices in real-life microsystems.

First, basic notions of electrostatics are recalled in Section 2.5.1, then a simple electromechanical problem is discussed in Section 2.5.2. Finally, a complete formulation for the coupled electromechanical problem is discussed in Section 2.5.3. For readers interested in more information on electrostatics, a good reference is Jonassen (2002).

2.5.1 Basic Notions of Electrostatics

Electrostatics is the part of physics that studies all phenomena related to stationary (or slowly moving) electric charges.

First of all, we must recall the fundamental Coulomb's law, which gives the interaction force $\mathbf{F}_{12}$ of two electrically charged particles or *point charges* q_1 and q_2, with positive or negative sign, in vacuum at a given relative distance r (see Figure 2.26):

$$\mathbf{F} = \frac{1}{4\pi\epsilon_0}\frac{q_1 q_2}{r^2}\mathbf{r}_{12}. \tag{2.128}$$

In Equation 2.128, $\mathbf{r}_{12}$ is the nondimensional unit vector along the direction connecting the two charges, while ϵ_0 is the *permittivity* or *dielectric* constant of free space. Since electric charges are measured in coulombs [C], forces in newtons [N] and distances in metres [m], the permittivity constant has units of coulombs squared per newton metre squared [C^2/Nm^2]; its value is $\epsilon_0 = 8.854\ 10^{-12}\ C^2/Nm^2$. When we consider the effect of charges in a specific material and not vacuum, the permittivity constant is given by the product of ϵ_0 and a relative, nondimensional, permittivity constant ϵ_r.

Typical values of ϵ_r at 20°C are slightly above 1 for air, 3.9 for silicon dioxide, 11.68 for silicon, 80 for water, and between 500 and 6000 for lead zirconate titanate (PZT), which is an important piezoelectric material (see Section 2.6).

Coulomb's law also states that if the electric charges have the same sign, the electrostatic force is repulsive, while it is attractive if the two charges have opposite signs. Conventionally, from Equation 2.128 when the force is positive, charges are with the same sign and the force is repulsive.

A concept strictly related to the Coulomb's law is that of the *electric field* $\mathbf{E}$, defined as

$$\mathbf{E} = \frac{1}{4\pi\epsilon_0}\frac{q}{r^2}\mathbf{r}, \tag{2.129}$$

where $\mathbf{r}$ is the nondimensional unit vector pointing from the point charge q to the point at distance r. The electric field is therefore a measure of the influence of a point charge q on the surrounding space; its unit is newtons per coulomb [N/C].

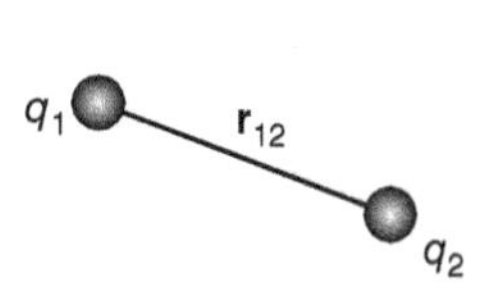

Figure 2.26 Two electric charges in empty space.

Given the electric field, the force that acts on a small test charge q, that is unable to influence the existing electric field, placed in a certain point of the space where no other charges exist, is given by:

$$\mathbf{F}_{\mathrm{elec}} = q\mathbf{E}. \tag{2.130}$$

Real applications concern the presence of many electric charges and the resultant electric field comes from the superposition of the contributions of the electric field generated by single point charges.

As in solid and fluid mechanics, in electrostatics it is also convenient to consider continuous distributions of quantities, in this case of electric charges, along lines, on surfaces or inside volumes. Charge densities per unit length ρ_{ql}, surface area $\rho_{q\Gamma}$ and volume $\rho_{q\Omega}$ are therefore defined such that an infinitesimal charge dq can be obtained in the three cases as:

$$\mathrm{d}q = \rho_{ql}\mathrm{d}l, \quad \mathrm{d}q = \rho_{q\Gamma}\,\mathrm{d}\Gamma, \quad \mathrm{d}q = \rho_{q\Omega}\,\mathrm{d}\Omega, \tag{2.131}$$

where dl, dΓ and dΩ are infinitesimal portions of length, surface area or volume, respectively.

From distributions of charges on lines, surfaces or volumes and relevant electrostatic forces, it is possible to create electrostatic actuators in microsystems. At small scales, electrostatic forces are high enough to create movement and deform beams, plates and other microstructures. This is not true at the macroscale: electrostatic forces are not used to create electric engines or deform large structures!

Given a distribution of charges in a given body, the problem of finding the electric field, the electrostatic forces and other relevant quantities, is the so-called *electrostatic problem*. This is governed by a specialization of the whole set of *Maxwell's equations* that govern all electromagnetic phenomena.

The first of Maxwell's equation relevant for electrostatics, also named Gauss's equation, states that the flux of the electric field through a closed surface Γ is equal to the sum of electric charges q_{int} contained inside Γ, divided by the permittivity constant $\epsilon_r\epsilon_0$:

$$\oint_{\Gamma} \mathbf{E}\cdot\mathbf{n}\,\mathrm{d}\Gamma = \frac{\sum q_{\mathrm{int}}}{\epsilon_r\epsilon_0}, \tag{2.132}$$

where $\mathbf{n}$ is the unit vector orthogonal to the surface, pointing outside.

By considering a distribution of charges in the volume Ω with volumetric density $\rho_{q\Omega}$, from Equation 2.132 one obtains

$$\oint_{\Gamma} \mathbf{E}\cdot\mathbf{n}\,\mathrm{d}\Gamma = \frac{1}{\epsilon_r\epsilon_0}\int_{\Omega} \rho_{q\Omega}\,\mathrm{d}\Omega. \tag{2.133}$$

Applying the divergence theorem, from Equation 2.133, the local expression of Maxwell's first equation is obtained:

$$\mathrm{div}\,\mathbf{E} = \frac{\rho_{q\Omega}}{\epsilon_r\epsilon_0}. \tag{2.134}$$

The second Maxwell's equation relevant for electrostatics, also named Faraday's law, specialized to the case of stationary conditions, states that the electric field is irrotational, with null curl:

$$\mathrm{curl}\,\mathbf{E} = \mathbf{0}. \tag{2.135}$$

As a direct consequence of Equation 2.135, the electric field is conservative and a potential function φ can be defined, such that

$$\mathbf{E} = -\text{grad } \varphi. \tag{2.136}$$

The electric potential φ is measured in volts [V].

It can be observed that the electric potential φ can also be interpreted as the amount of work per unit charge required to move a charge from point A to point B in the space:

$$-\int_A^B \mathbf{E} \cdot \mathrm{d}\mathbf{u} = \varphi(A) - \varphi(B). \tag{2.137}$$

The combination of Maxwell's first law (2.134) and Equation 2.136 gives rise to the fundamental equation governing the electrostatic problem, in which the Laplace operator ∇^2 appears:

$$\nabla^2 \varphi = -\frac{\rho_{q\Omega}}{\epsilon_{\mathrm{r}}\epsilon_0}. \tag{2.138}$$

The electrostatic problem is governed by Equation 2.138 together with suitable boundary conditions, which can be given on the potential φ, or on the flux of the electric field.

In the first case, the potential is assigned on a portion Γ_φ of the external boundary of the domain Ω:

$$\varphi = \overline{\varphi} \quad \text{on} \quad \Gamma_\varphi. \tag{2.139}$$

In the second case, a surface charge density is assigned on a portion Γ_E of the external boundary of the domain Ω:

$$\text{grad } \varphi \cdot \mathbf{n} = -\frac{\rho_{q\Gamma}}{\epsilon_{\mathrm{r}}\epsilon_0} \quad \text{on} \quad \Gamma_E. \tag{2.140}$$

The solution of the electrostatic problem gives the electric potential as a function of the position; from its gradient we can compute the electric field.

It is now useful to introduce two additional quantities: the *electrostatic energy* and the *capacitance.*

The electrostatic energy can be defined as the work necessary to bring electric charges in the considered configuration, starting from infinite distance.

In terms of the electric field, the electrostatic energy can be expressed as

$$U = \frac{1}{2}\int_{\text{Space}} \epsilon_{\mathrm{r}}\epsilon_0\, |\mathbf{E}|^2 \,\mathrm{d}\Omega = \frac{1}{2}\int_{\text{Space}} \epsilon_{\mathrm{r}}\epsilon_0 \text{grad } \varphi \cdot \text{grad } \varphi \,\mathrm{d}\Omega. \tag{2.141}$$

The electrostatic energy is also useful in computing forces originating from the charge distributions (electrostatic forces) that act on solid parts. These can be obtained by computing the infinitesimal work δW done by the system as a consequence of an infinitesimal displacement $\delta\mathbf{u}$ of the solid part, which can be written as

$$\delta W = \mathbf{F}_{\text{elec}} \cdot \delta\mathbf{u} = \delta U = \frac{\partial U}{\partial \mathbf{u}} \cdot \delta\mathbf{u}. \tag{2.142}$$

From this expression, once the electrostatic energy is known, we can compute the force acting on the solid part as its infinitesimal variation due to an infinitesimal variation of the position of a solid part (the infinitesimal displacement).

The capacitance can be considered as the property of a conductive body of being able to hold an electric charge. When a total charge Q is given to a conductive body, this will be distributed on its surface with a surface charge density $\rho_{q\Gamma}$ in such a way that the electric potential φ will be constant on the volume.

The capacitance C is the ratio

$$C = \frac{Q}{\varphi}. \tag{2.143}$$

and is constant at varying charge Q, i.e. if the charge Q is doubled, the potential φ also doubles. The capacitance is a geometrical property of the conductive body and its unit is the farad [F = C/V].

A *capacitor* can be defined as a device in which electrostatic energy can be stored. It can be built with two conductive parts, kept at a certain voltage difference $\overline{\varphi}$, separated by a dielectric, nonconductive material.

The electrostatic energy stored in a capacitor is equal to the work done to charge it. The capacitance of a capacitor is given by the ratio of the charge distributed on each conductive part, divided by the potential difference between the two parts.

By considering the charging process as a sequence of infinitesimal increments of charges dq at potential difference $\overline{\varphi}$, the infinitesimal work is equal to $\overline{\varphi}$ dq and therefore the total electrostatic energy of the capacitor can be computed as

$$U = \int_0^Q \overline{\varphi}\, \mathrm{d}q = \int_0^Q \frac{q}{C}\, \mathrm{d}q = \frac{1}{2}\frac{Q^2}{C} = \frac{1}{2}C\overline{\varphi}^2. \tag{2.144}$$

Capacitors in real microsystems are often used in groups; this is done to increase the total force that can be exerted on a solid portion or the total capacitance of the system. Typical electric connections of capacitors are *in parallel* and *in series*. Capacitors connected in parallel are all kept at the same electric potential difference; in this case the total capacitance is the sum of the single capacitances. Capacitors connected in series are such that neighbouring capacitor plates have equal and opposite charges. In this case, the total capacitance is equal to the reciprocal of the sum of the reciprocals of the single capacitances.

2.5.1.1 Examples

Two meaningful examples are now discussed, which are found in many real-life microsystems.

Parallel-Plate Capacitor Let us consider the capacitor shown in Figure 2.27. Two conductive plates with surface S are placed parallel to each other at a distance d (the so-called *gap* in microsystems), separated by a dielectric material.

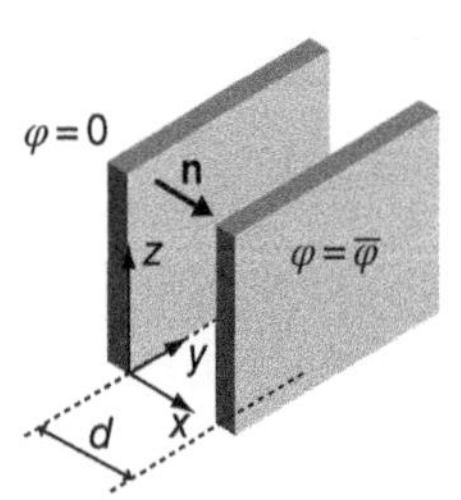

Figure 2.27 Parallel-plate capacitor.

The potential is constant on each conductive plate and the total potential difference is $\overline{\varphi}$. It can be shown that the potential φ varies linearly in the dielectric; it can therefore be given by the expression

$$\varphi = \overline{\varphi}\frac{x}{d}. \tag{2.145}$$

Notice that this expression satisfies the governing equation (2.138) with zero volume charge density in the dielectric and boundary conditions of the kind of Equation 2.139 on the two parallel plates.

Given the electric potential, it is possible to compute the electric field, applying Equation 2.136:

$$E_x = -\frac{\partial \varphi}{\partial x} = -\frac{\overline{\varphi}}{d},$$
$$E_y = -\frac{\partial \varphi}{\partial y} = 0,$$
$$E_z = -\frac{\partial \varphi}{\partial z} = 0. \quad (2.146)$$

From Equation 2.141 it is then possible to compute the electrostatic energy:

$$U = \frac{1}{2}\int_{\text{Space}} \epsilon_r \epsilon_0 \, |\mathbf{E}|^2 \, \mathrm{d}\Omega = \frac{\epsilon_r \epsilon_0 S}{2} \frac{\overline{\varphi}^2}{d}. \quad (2.147)$$

Comparing Equations 2.144 and 2.147, it can be deduced that the capacitance for an ideal parallel-plate capacitor is

$$C = \frac{\epsilon_r \epsilon_0 S}{d}. \quad (2.148)$$

Finally, exploiting Equation 2.142, the resultant force acting on one plate, which is attracted to the other plate, owing to the uniform distribution of charges, can be computed in the direction orthogonal to the plates as

$$F_{\text{elec}} = \frac{\partial U}{\partial d} = -\frac{\epsilon_r \epsilon_0 S}{2} \left(\frac{\overline{\varphi}}{d}\right)^2. \quad (2.149)$$

Owing to the important implications in the design of electrostatic MEMS, it must be noticed that the capacitance and the electrostatic force are both nonlinear functions of the relative distance between the capacitor plates: the capacitance is inversely proportional to the relative distance of the two plates, while the electrostatic force is inversely proportional to the second power of the relative distance. This issue will be discussed further in Section 2.5.2.

The results given in Equations 2.148 and 2.149 are valid with reference to an ideal parallel-plate capacitor, disregarding the variation of the electric field at the boundary of the plates.

Comb-Finger Capacitor Let us now consider the capacitor shown in Figure 2.28.

The configuration shown in the figure concerns one element of the so-called *interdigitated comb-finger capacitor*, where the name comes from the fact that many unit cells like the one shown in the figure are used in practice; these, all together, create a device which has the shape of a comb.

A plate is placed symmetrically at distance d with respect to the other two parallel plates. It is assumed that the distance l of Figure 2.28 is large enough and that the electric field of interest is only at the sides of the central plate, created by a voltage difference $\overline{\varphi}$.

In this case, the capacitance of the system can be approximated by the capacitance of two parallel-plate capacitors connected in parallel, with the same potential difference,

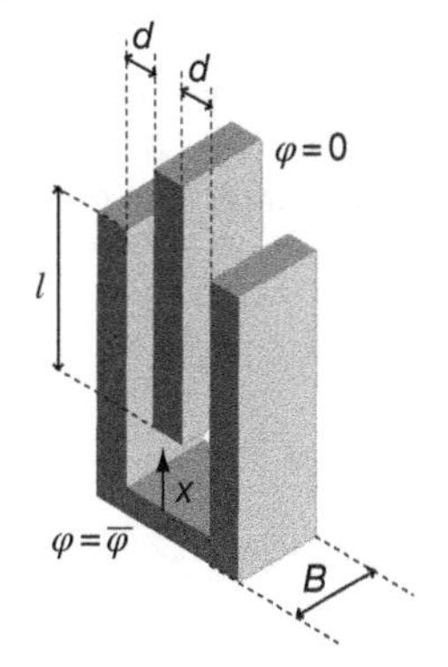

Figure 2.28 Comb-finger capacitor.

hence:

$$C = 2\frac{\epsilon_r \epsilon_0 Bl}{d}. \tag{2.150}$$

The electrostatic energy can then be computed as

$$U = \frac{1}{2}\frac{2\epsilon_r \epsilon_0 Bl}{d}\overline{\varphi}^2. \tag{2.151}$$

If we imagine that the symmetric configuration of the capacitor is maintained, the only possible movement that we can consider is the one in the direction x of Figure 2.28; hence, it is useful to compute the electrostatic force in the same direction, as

$$F_{\text{elec}} = \frac{\partial U}{\partial x} = \frac{\partial}{\partial x}\left(-\frac{1}{2}\frac{2\epsilon_r \epsilon_0 B(l-x)}{d}\overline{\varphi}^2\right) = \frac{\epsilon_r \epsilon_0 B}{d}\overline{\varphi}^2. \tag{2.152}$$

It is now interesting to remark that, unlike the parallel-plate configuration, the ideal comb-finger capacitor shows an electrostatic force that does not depend on the coordinate in the direction of possible relative movements between the two conductive portions of the device.

The simple results obtained are valid only for ideal conditions.

2.5.1.2 Summary

It is now useful to sum up the main contents of this section.

- Coulomb's law, which gives the interaction force of two point charges, has been recalled.
- The concepts of electric field, continuous distributions of charges and electric potential have been introduced.
- The equations governing electrostatics have been shown as a particular case of Maxwell's relations.
- The concepts of electrostatic energy, capacitance and the capacitor have been introduced.
- The important examples of parallel-plate and comb-finger capacitors have been discussed.

2.5.2 Simple Electromechanical Problem

After the discussion on solid mechanics given in Section 2.3 and the introduction to electrostatics given in the previous section, it is now possible to examine how the fields of mechanical and electrostatic variables can influence each other in coupled electromechanical problems.

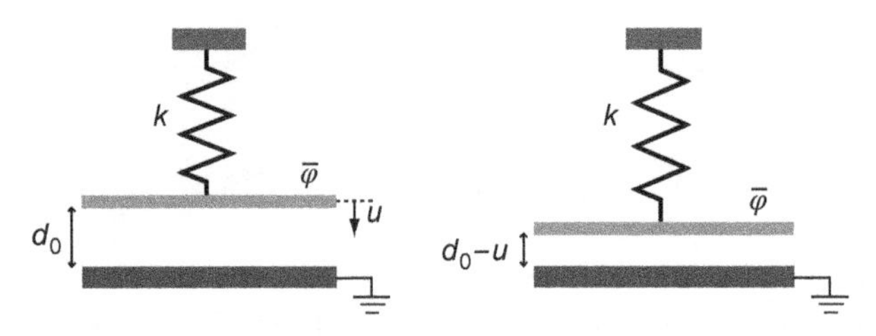

Figure 2.29 Parallel-plate capacitor connected with an elastic spring.

Before entering into a general formulation (see Section 2.5.3), a simple example is here discussed, combining the results for a one-dof mechanical oscillator, as presented in Section 2.2.3, and simple capacitors, as presented at the end of Section 2.5.1.

This is the first example of a coupled problem met in this book. The situation shown in Figure 2.29 is considered: a movable plate is connected through a linear elastic spring with stiffness k to a fixed point and is kept parallel to a second fixed plate.

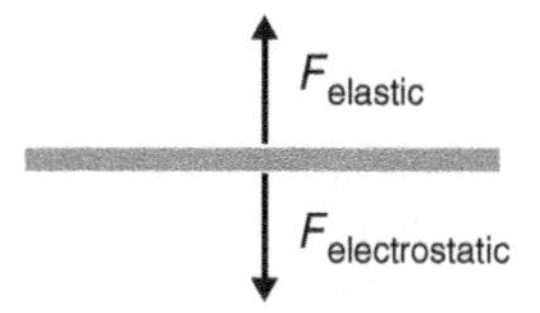

Figure 2.30 Parallel-plate capacitor connected with an elastic spring: equilibrium of forces.

2.5.2.1 Static Equilibrium

The static equilibrium of the plate connected with the elastic spring is guaranteed by the presence of elastic and electrostatic forces, as shown in Figure 2.30.

The static equilibrium equation therefore reads:

$$ku - \frac{\epsilon_r \epsilon_0 S}{2} \left(\frac{\overline{\varphi}}{d_0 - u} \right)^2 = 0. \tag{2.153}$$

The obtained equilibrium equation is nonlinear in the displacements: coupling two linear problems, the elastic and the electrostatic ones, the so-called *coupled problem* becomes nonlinear!

Equation 2.153 relates the applied voltage difference $\overline{\varphi}$ to the displacement u. It is possible to solve this relation with respect to $\overline{\varphi} = \overline{\varphi}(u)$, for $u \in [0 - d_0)$, obtaining

$$\overline{\varphi} = \sqrt{\frac{2k}{\epsilon_r \epsilon_0 S} u (d_0 - u)^2}. \tag{2.154}$$

The plot corresponding to Equation 2.154 is qualitatively shown in Figure 2.31, where the variables have been nondimensionalized. The plot has a maximum corresponding to the following values of displacement and potential:

$$\begin{aligned} u \equiv u_{\text{pi}} &= \frac{d_0}{3}, \\ \overline{\varphi} \equiv \overline{\varphi}_{\text{pi}} &= \sqrt{\frac{8}{27} \frac{k}{\epsilon_r \epsilon_0 S} (d_0)^3}. \end{aligned} \tag{2.155}$$

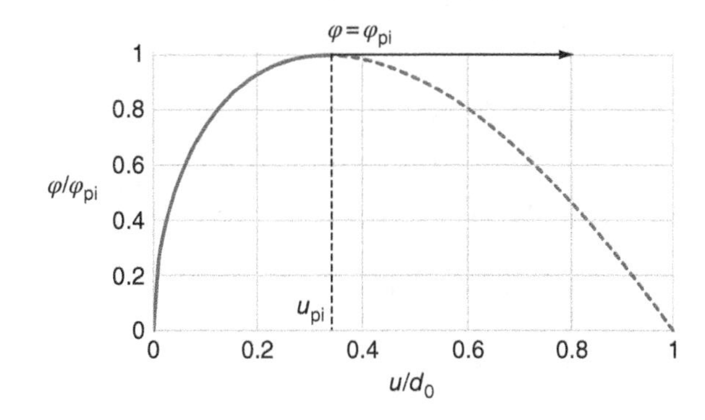

Figure 2.31 Parallel-plate capacitor connected with an elastic spring: voltage–displacement plot.

The index 'pi' in Equation 2.155 means *pull-in*, and the corresponding displacement and voltage are *pull-in values*, which characterize an important situation in parallel-plate capacitors. The pull-in phenomenon can be understood by imagining that the device is controlled at increasing voltage difference between parallel plates. It can be seen from Figure 2.31 that when the voltage and displacement reach the pull-in values, the static equilibrium can no longer be guaranteed and the movable parallel plate moves suddenly towards the opposite plate, closing the gap and creating a short circuit; in other words the elastic force of the spring is no longer able to equilibrate the nonlinear electrostatic force and the situation becomes unstable.

Notice that the pull-in value for the displacement u_{pi} means that a parallel-plate capacitor connected with a linear elastic spring can, in practice, be used until the gap reduces to one-third of the initial value.

A deeper insight, also in view of the subsequent dynamic study, can be obtained by developing the electrostatic load in a Taylor's series as a function of the displacement u, around a reference position u_0; the static equilibrium equation of the movable parallel plate thus becomes

$$\begin{aligned} ku - \frac{\epsilon_r\epsilon_0 S}{2}\left(\frac{\overline{\varphi}}{d_0 - u}\right)^2 & \\ \approx ku - \frac{\epsilon_r\epsilon_0 S\overline{\varphi}^2}{2}&\left[\frac{1}{(d_0-u_0)^2} + \frac{2(u-u_0)}{(d_0-u_0)^3} + \frac{3(u-u_0)^2}{(d_0-u_0)^4} + \frac{4(u-u_0)^3}{(d_0-u_0)^5} + \dots\right] \\ = 0. & \end{aligned} \tag{2.156}$$

By choosing the terms in the development up to the linear one in the displacement u, Equation 2.156 can be rewritten as

$$\begin{aligned} \left(k - \frac{\epsilon_r\epsilon_0 S\overline{\varphi}^2}{(d_0-u_0)^3}\right)u - \frac{\epsilon_r\epsilon_0 S\overline{\varphi}^2}{2}\left[\frac{1}{(d_0-u_0)^2} - \frac{2u_0}{(d_0-u_0)^3}\right] & \\ \equiv (k - k_{elec})u - \frac{\epsilon_r\epsilon_0 S\overline{\varphi}^2}{2}\left[\frac{(d_0-3u_0)}{(d_0-u_0)^3}\right] & \\ = 0. & \end{aligned} \tag{2.157}$$

The coefficient of u is given by the difference of the elastic stiffness k and of a second term called the *electrostatic stiffness* k_{clcc}, which is always positive in the considered range of displacement values. This means that the electromechanical interaction has the effect of progressively reducing the equivalent global stiffness $(k - k_{elec})$ at increasing voltage difference between the two plates.

2.5.2.2 Dynamic Response

We can now study the previous problem in dynamic conditions, considering the movable plate with a mass m; we come up with a one-dof equation in the unknown displacement u, similar to the one studied in Section 2.2.3:

$$m\ddot{u} + ku - \frac{\epsilon_r\epsilon_0 S}{2}\left(\frac{\overline{\varphi}}{d_0 - u}\right)^2 = 0. \tag{2.158}$$

The obtained equation is second-order, ordinary differential and must be supplemented with initial conditions on displacement and velocity. As for the static case, the obtained equation is nonlinear in displacement; hence, we must now deal with a nonlinear oscillator. It can be studied with the tools of nonlinear dynamics.

In practical applications, Equation 2.158 is studied with the voltage difference considered as a given function of time, e.g. periodically variable in time, and a large variety of possible responses can be obtained. Here, we limit the discussion to basic features only. Further insight into nonlinear oscillators will be given in other chapters of the book, in particular Chapters 3, 6, 7, 8 and 9.

As for the static case, it is useful to develop the electrostatic force as a function of the displacement; in this case, the reference value for the displacement is chosen to be equal to the solution of the static equilibrium (2.153):

$$u_0 = u_{eq} : \quad ku_{eq} - \frac{\epsilon_r \epsilon_0 S}{2}\left(\frac{\overline{\varphi}}{d_0 - u_{eq}}\right)^2 = 0. \tag{2.159}$$

The displacement u can then be considered as the sum of u_{eq} and an increment Δu:

$$\Delta u \equiv (u - u_0) = (u - u_{eq}), \quad u = u_{eq} + \Delta u. \tag{2.160}$$

The equation of motion can be rewritten as follows:

$$\begin{gathered} m\ddot{u} + ku - \frac{\epsilon_r \epsilon_0 S}{2}\left(\frac{\overline{\varphi}}{d_0 - u}\right)^2 = 0 \approx \\ m\Delta\ddot{u} + \left(k - \frac{\epsilon_r \epsilon_0 S\overline{\varphi}^2}{(d_0 - u_{eq})^3}\right)\Delta u - \frac{\epsilon_r \epsilon_0 S\overline{\varphi}^2}{2}\left[\frac{3\Delta u^2}{(d_0 - u_{eq})^4} + \frac{4\Delta u^3}{(d_0 - u_{eq})^5} + \dots\right] = 0. \end{gathered} \tag{2.161}$$

The obtained equation is now expressed in the unknown Du, which is the variation of the displacement with respect to the static equilibrium position.

Depending on the chosen order for the development of the electrostatic force, we obtain a linear or a nonlinear oscillator of second, third or higher order.

As a general remark, from Equation 2.161, it can be seen that the nonlinear effects are amplified by the voltage, considered as a given quantity. Depending on the electromechanical properties of the device, it will be possible to detect a maximum value of given voltage below which the nonlinear effects become negligible.

As will be discussed in the following chapters, the optimal response of a microsystem in terms of the whole device controlled by an electronic circuit is usually obtained in the linear regime; hence, the study of nonlinearities such as those described in Equation 2.161 is of paramount importance in understanding the working regime of a device and in forecasting possible malfunctions. In some particular cases, a nonlinear response can also be advantageous to obtain better performances, as in the case of *energy-harvesting* devices, as discussed in Chapter 9.

If we consider only the terms up to the linear one in Equation 2.161, we obtain the linear oscillator equation:

$$m\Delta\ddot{u} + (k - k_{elec})\Delta u = m\Delta\ddot{u} + \left(k - \frac{\epsilon_r \epsilon_0 S\overline{\varphi}^2}{(d_0 - u_{eq})^3}\right)\Delta u = 0. \tag{2.162}$$

The linear oscillator described by Equation 2.162 has a frequency that depends on the assigned voltage difference between the capacitor plates, i.e.:

$$\omega = \sqrt{\frac{(k - k_{\text{elec}})}{m}} = \sqrt{\frac{\left(k - \frac{\epsilon_r \epsilon_0 S \overline{\varphi}^2}{(d_0 - u_{\text{eq}})^3}\right)}{m}}. \tag{2.163}$$

As observed for the static response, from this expression we understand that the electromechanical interaction has the effect of reducing the effective stiffness. This, in turn, reduces the natural frequency of the device and has important implications in the design of resonant microsystems or inertial microsystems in which resonant devices are inserted.

If the static equilibrium position u_{eq} is the one at pull-in (see Equation 2.155), the natural frequency of vibration around the static equilibrium position goes to zero; this in turn implies that the vibration motion transforms into a divergent one and a dynamic pull-in phenomenon is exhibited. More generally, dynamic pull-in should be considered for the electromechanical coupled problem, applying the notion of stable or unstable motion and looking for situations that transform a stable vibrating motion into an unstable one.

2.5.2.3 Summary

Here are the major points discussed in this section:

- A simple example of electromechanical coupled behaviour has been analysed.
- The concept of static pull-in has been introduced.
- The nonlinear dynamic response typical of electromechanical coupling has been demonstrated.

2.5.3 General Electromechanical Coupled Problem

In the previous section, we discussed a first example of an electromechanical coupled problem.

In more general terms, coupled problems arise when different physical phenomena interact with reciprocal influence; these situations are characteristic of microsystems in which the designers exploit the coupling among physics to obtain smart devices.

In coupled problems, two or more physical problems or *physical domains* cannot be resolved independently of each other. Two main classes of coupled problems can be defined.

- Coupling through boundary conditions imposed at the interface between nonoverlapping physical domains; this is the case of fluid–structure interaction, briefly mentioned in Section 2.4.2, and of the most common electromechanical couplings.
- The domains in which the solution is looked for are partially or totally overlapping and the coupling comes from the differential equations governing the various physical phenomena; this is the case of thermomechanical coupling, as discussed in Section 2.7.2.

Here, we briefly discuss the *electromechanical coupled problem*, in which the mechanical response is linearly elastic, as presented in Section 2.3.1, and the electrical response is electrostatic, as presented in Section 2.5.

Other examples of electromechanical coupling are those briefly discussed in Section 2.7.2, concerning the interaction among the Joule effect due to current flowing in a wire and relevant heat exchange, which in turn is a cause of deformation, and in Section 2.6, where an introduction to piezoelectric behaviour is presented.

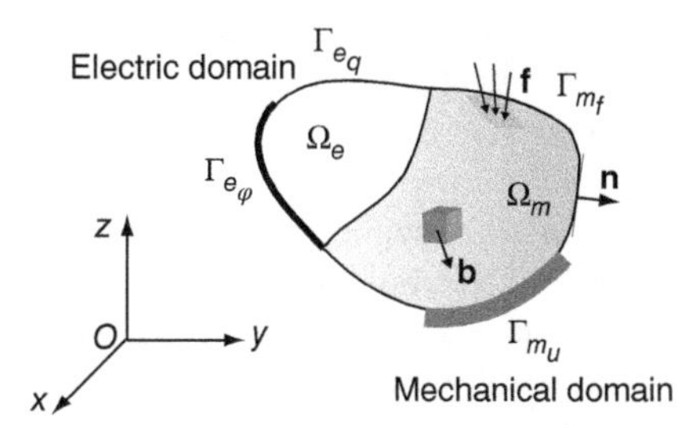

Figure 2.32 General scheme for a coupled electromechanical problem.

We consider here the general situation shown in Figure 2.32, where a deformable solid is subject to loads and is partially constrained on its boundary, while a nonoverlapping, neighbouring electric domain is subject to electric boundary conditions; typically, an electric potential difference is applied, and electric charges are possibly assigned.

The two physical domains influence each other in the following way: at the interface between the solid and the electric domain, the mechanical domain is subject to electrostatic forces, which are in addition to the forces assigned on the mechanical domain; at the same time, the shape of the electric domain changes due to the deformation of the mechanical domain, which moves the interface between the two physical domains. As a consequence, it is not possible to have the complete solution of the elastic solid if we do not know the electrostatic forces acting at the interface between the two domains, while it is not possible to solve the electrostatic problem without knowing how much the solid body deformed and moved the interface between the two physical domains.

This is a typical situation of a fully coupled problem, the two problems can only be solved together. Moreover, the considered electromechanical coupled problem is nonlinear, even if the individual physics problems are linear. The occurrence of nonlinearity in the coupled problem was already observed with reference to the simple problem studied in Section 2.5.2 and comes from the the expression of electrostatic forces and from the variation of the electrostatic response at varying solution domains.

An important remark concerns the computation of the electrostatic forces, which must be considered in the mechanical response of the elastic solid.

In Section 2.5.1, Equation 2.142 was given, which allows for the computation of an electrostatic force starting from a derivative of the electrostatic energy. More generally, electrostatic interaction forces must be computed starting from the Maxwell stress tensor, which, in the absence of magnetic field, can be written as function of the electric field $\mathbf{E}$ only, as follows:

$$\boldsymbol{\sigma} = \epsilon_0 \epsilon_\mathrm{r} \begin{bmatrix} \frac{1}{2}(E_x^2 - E_y^2 - E_z^2) & E_x E_y & E_x E_z \\ E_y E_x & \frac{1}{2}(E_y^2 - E_x^2 - E_z^2) & E_y E_z \\ E_z E_x & E_z E_y & \frac{1}{2}(E_z^2 - E_x^2 - E_y^2) \end{bmatrix}. \tag{2.164}$$

Notice that if Equation 2.164 is applied to the example of the parallel-plate capacitor, with the electric field given by Equation 2.146, the Maxwell stress component in the direction orthogonal to the plate is

$$\sigma_x = \frac{\epsilon_0 \epsilon_\mathrm{r}}{2} \left(\frac{\overline{\varphi}}{d} \right)^2, \tag{2.165}$$

which gives the electrostatic force of Equation 2.149 if multiplied by the surface area S.

Electromechanical coupling is pervasive in microsystem study and design. Whenever electrostatics is used, either for sensing or actuation, the coupled behaviour must be carefully studied and understood. Meaningful examples are discussed in Chapters 6, 7, 8, 9 and 10, devoted to the description of devices. Electromechanical coupling is also important for the design of microsystems used for mechanical characterization at the microscale, like those discussed in Chapter 11.

In general, to obtain a solution of an electromechanical coupled problem is not a trivial task and *ad-hoc* numerical approaches, mainly based on the finite element method or the boundary element method, have been developed and are still being developed.

General principles, like those discussed in Chapter 3, can be adopted to formulate general numerical procedures and to obtain simplified solutions. These, based on reduced-order models, are governed by a reduced number of dof and are very useful for design purposes.

2.5.3.1 Summary

The most relevant concepts presented in this section are here summarized.

- The general fully coupled electromechanical problem has been presented.
- Coupling terms in the electromechanical problem are a source of nonlinearity.
- The definition of Maxwell's stresses has been recalled.

2.6 Piezoelectric Materials in Microsystems

Piezoelectric materials are used nowadays in commercial microsystems to obtain highly efficient sensing and actuating properties. In this section, the basic notions on piezoelectric materials are presented, together with a description of mathematical models focusing on piezoelectric constitutive equations applied to thin films.

2.6.1 Piezoelectric Materials

Two major phenomena characterize the piezoelectric materials: by means of the *direct piezoelectric effect*, the material is able to transform mechanical energy into electrical energy; through the *converse piezoelectric effect*, the material is able to transform electrical energy into mechanical energy.

The direct piezoelectric effect makes the material, coupled with some appropriate electrodes, work as a transducer, which produces charge (voltage difference) at the electrodes on the application of strain (or stress). This effect is usually employed in inertial sensors (e.g. accelerometers, gyroscopes, pressure sensors) to measure the motion of a moving mass. Other applications include strain sensors, which measure the change in resistivity or even the voltage output of piezoelectric gauges, and microphones, in which input acoustic waves excite piezoelectric membranes.

The converse piezoelectric effect manifests as strain (or stress) on the application of an electrical potential difference between the electrodes. Piezoelectric actuators use this phenomenon to electrically control and induce movements and strains in mechanical elements, such as robots' artificial muscles or to activate the vibrations of membranes in acoustic wave transducers (e.g. for sonographic instruments).

Piezoelectric actuators are also the basis of a number of everyday applications, such as the ignition source for cigarette lighters and push-starts for propane barbecues.

Piezoelectricity was first observed in the late 1800s by Pierre and Jacques Curie, who demonstrated the direct piezoelectric effect in quartz and in some other crystalline materials in the natural state.

Piezoelectricity is due to the particular crystalline structure of such materials, which include asymmetric ions or charged molecular groups. Some piezoelectric materials belong to the class of ferroelectric materials that are observed to spontaneously polarize by decreasing the temperature to below the so-called Curie temperature. During the cooling, each electric dipole moment tends to be in electric equilibrium with all the other dipoles, so the global polarization is null. However, by imposing an external electric field, it is possible to align all the dipole moments in the same direction, obtaining a global residual polarization.

The name ferroelectricity is used because this polarization process is very similar to the magnetization process used to magnetize iron or to obtain ferromagnetic materials. In the ferroelectric case, it is common to use a similar name even if iron is not involved. The internal electric dipoles of a ferroelectric material are coupled to the material lattice, so anything that changes the lattice also changes the strength of the dipoles (in other words, a change in the spontaneous polarization). The change in the spontaneous polarization results in a change in the surface charge that can eventually be collected on electrodes. Two stimuli that modify the lattice dimensions of a material are force and temperature. Thus, ferroelectric materials are also piezoelectric and pyroelectric (materials that change polarization in response to a temperature variation).

The most common piezoelectric materials used nowadays are ceramics. In MEMS, piezoelectric materials are employed in thin films deposited on the micromachined silicon substrate or even directly micromachined. Owing to their piezoelectric and manufacturability qualities, the two most used in applications are aluminium nitride and lead zirconate titanate (PZT).

Aluminium nitride is a piezoelectric material with a wurtzite crystal structure which is an example of a hexagonal crystal system. Many binary compounds (compounds that contain exactly two different chemical elements, e.g. zinc oxide) are assembled in wurtzite structures, which lends them piezoelectric properties because this structure is noncentrosymmetric. Owing to its structure, the polar axis of this material cannot be oriented by the application of an electric field. Therefore, the development of any deposition process for these films must result in films with a well-oriented polarization axis. Although its piezoelectric coefficients are nine or ten times smaller than those of PZT, aluminium nitride has been used in many applications, such as bulk acoustic wave resonators. Aluminium nitride is mainly used in sensors because it has electromechanical coupling coefficients (see Section 2.6.2) too low for actuation. The great success of this material in microscale devices results from its extremely good manufacturability properties, which assure high-quality materials.

PZT or $Pb(Zr_x,Ti_{1-x})O_3$ is the most widely used ferroelectric material, owing to its high piezoelectric coefficients. It is an oxide of lead, zirconium and titanium disposed in a perovskite crystalline structure.

Zirconium and titanium can be present in different proportions; this influences the final properties of the material.

It is worth noting that significant effort is currently being made to develop new piezoelectric materials with improved characteristics. The research is now focused on piezoelectric materials that do not contain lead (which is toxic) and are suitable for bio-applications, such as polyvinylidene fluoride or barium titanate (Saito *et al.*, 2004; Shrout and Zhang, 2007; Takenaka, Nagata and Hiruma, 2008), and phase boundary transition piezoelectrics, such as relaxors, which show extremely high piezoelectric properties.

2.6.2 Piezoelectric Modelling

Piezoelectric materials can be modelled taking into account fully coupled electromechanical behaviour on the same domain Ω: we therefore have a coupled problem characterized by completely overlapped domains.

Dynamic equilibrium and kinematic compatibility together with static and kinematic boundary conditions govern the mechanical response (see Section 2.3.1), while the electric behaviour is governed by the electrostatic relations described in Section 2.5.

The coupling terms between the two physics domains are contained in the constitutive law, which must be written accounting for the electromechanical coupling due to the piezoelectric effect.

The elastic constitutive law given by Equation 2.73 is a relation between stresses and strains; also, in electrostatics (more generally in electromagnetism) a constitutive law is introduced as a relation between the electric field **E** and another vector field called the *electric displacement* **D**:

$$\mathbf{D} = \epsilon_0 \epsilon_\mathrm{r} \mathbf{E}. \tag{2.166}$$

In the piezoelectric constitutive law, stresses depend not only on strains but also on the electric field, while the electric displacement depends not only on the electric field but also on strains.

In writing the piezoelectric constitutive law, a peculiar notation is used in the literature (ANSI/IEEE 176-1987, 1988), different from that used for standard continuum mechanics: the second-order stress and strain tensor components are denoted T_{ij} and S_{ij}, respectively; their six-component vector forms are therefore **T** and **S**. The elastic stiffness tensor is denoted **c**, while the compliance tensor is denoted **a**. This notation will be used for piezoelectric materials only in the remainder of the book.

Generalizing Equation 2.166, a matrix of the dielectric constant is defined as $\boldsymbol{\epsilon}$. Finally, a new matrix of electromechanical coupling coefficients with units of newtons per volt metre [N/Vm] is defined and denoted as **e**.

The coupled piezoelectric constitutive law, taking into account the new notation, can be written in matrix-vector form as

$$\begin{aligned} \mathbf{T} &= \mathbf{c}^E \, \mathbf{S} - \mathbf{e}^T \mathbf{E}, \\ \mathbf{D} &= \mathbf{e} \, \mathbf{S} + \boldsymbol{\epsilon}^S \, \mathbf{E}. \end{aligned} \tag{2.167}$$

These relations (2.167) represent the so-called *e-form* of the piezoelectric constitutive equations, in which the strain and the electric field are used as coupling variables. The superscript E to matrix **c** denotes that the elastic stiffness coefficients must be evaluated at constant electric field, while the superscript S to matrix $\boldsymbol{\epsilon}$ denotes that the dielectric coefficients must be evaluated at constant strain.

Another format of the piezoelectric constitutive equations that is used in the literature, is the so-called *d-form*, in which stresses instead of strains are used as coupling variables; in this case, the matrix of coupling coefficients is replaced with the matrix **d**, containing coefficients with units of metres per volt [m/V]:

$$\mathbf{S} = \mathbf{a}^E\,\mathbf{T} + \mathbf{d}^T\mathbf{E}$$
$$\mathbf{D} = \mathbf{d}\,\mathbf{T} + \boldsymbol{\epsilon}^T\,\mathbf{E}. \tag{2.168}$$

The superscript E above matrix **a** denotes that the elastic compliance coefficients must be evaluated at constant electric field, while the superscript T above matrix $\boldsymbol{\epsilon}$ denotes that the dielectric coefficients must in this case be evaluated at constant stress.

Matrices **d** and **e** must satisfy the following relation:

$$\mathbf{d}^T = \mathbf{a}\,\mathbf{e}^T. \tag{2.169}$$

Piezoelectric constitutive equations are a combination of the classical Hooke's law employed in continuum mechanics and standard linear constitutive relations between electric field and displacement. The peculiarity of piezoelectricity lies in the coupling terms contained in matrices **d** and **e**. These coefficients refer to three main coupling mechanisms, which are collected in Figure 2.33, using the standard assumption that direction 3 is always the polarization direction:

- 3–3 mode: when applying an electric field along the polarization axis, the piezoelectric element stretches in this same direction (and vice versa).
- 3–1 mode: when applying an electric field along the polarization axis, the piezoelectric element shrinks in the orthogonal plane (and vice versa).
- Shear mode: when applying an electric field orthogonal to the polarization axis, a shear occur in the element plane (and vice versa).

According to the physics of piezoelectric coupling, only a few coupling constants are nonzero. Matrix **e** therefore contains many zeros and can usually be represented as

$$\mathbf{e} = \begin{bmatrix} 0 & 0 & 0 & 0 & e_{15} & 0 \\ 0 & 0 & 0 & e_{24} & 0 & 0 \\ e_{31} & e_{32} & e_{33} & 0 & 0 & 0 \end{bmatrix}. \tag{2.170}$$

Table 2.1 shows meaningful values of piezoelectric constitutive parameters.

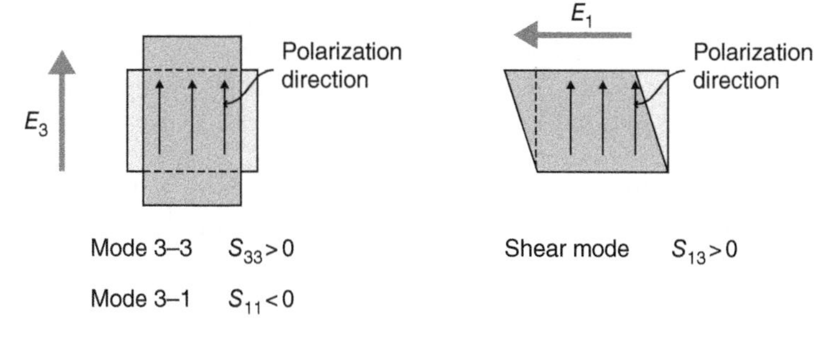

Figure 2.33 Piezoelectric modes.

Table 2.1 Piezoelectric material parameters. $\epsilon_0 = 8.85 \times 10^{-12}$ F/m.

	E[GPa]	ν[−]	e_{31}[N/mV]	e_{33}[N/mV]	$\epsilon_{33}^S[\epsilon_0]$
Lead zirconate titanate	60/100	0.3	$-8/-18$	18/25	300/2000
Aluminium nitride	260/350	0.28	$-0.8/-2$	1.8/2.4	8/10.5
Zinc oxide	30/35	0.34	$-0.6/-1$	1.2/1.5	10/11
Polyvinylidene fluoride	2/8	0.3	$1.1/1.25 \times 10^{-2}$	-2.9×10^{-2}	10/12

2.6.2.1 Summary

In this section devoted to piezoelectric materials, the following main points have been made.

- The main properties of piezoelectric materials have been discussed.
- The piezoelectric constitutive relation has been given in the *e-form* and in the *d-form*.
- Piezoelectric coupling parameters have been defined.

2.7 Heat Conduction and Thermomechanics

The purpose of this section is to introduce heat conduction and the coupled thermomechanical problem.

Many sources of heat exchange are present in MEMS, primarily the variations of temperature related to the external environment in which the microsystems must operate; in addition, there is the Joule effect, which transforms electric energy into thermal energy.

Heat conduction is also present in MEMS in the form of coupled effects; in particular, the study of thermomechanical coupling is very important in understanding solid damping (see Chapter 15), in which vibration kinetic energy is partially transformed into thermal energy.

As for the electromechanical problem, the first part of this section is devoted to a description of the thermal problem, then the coupled problem is described with reference to a purely elastomechanical behaviour.

As with the other parts of this chapter, the introductory treatment presented here can be usefully supplemented by specialized reading, e.g. Ozisik (1980).

2.7.1 Heat Problem

The heat conduction problem is presented here in the simple case of a continuous and homogeneous solid, like the one shown in Figure 2.34. The fields that play a central role in the heat problem are the temperature T and the heat flux $\mathbf{q}$: T is a scalar function of position, and possibly time, and is measured in kelvins [K] in SI units (or in other units, such as degrees Celsius, [° C]); $\mathbf{q}$ is a vectorial function of position, and possibly of time, and is measured in units of power divided by units of surface area (watts divided by square metres in SI units).

The heat problem is governed by two main equations: Fourier's law of conduction and the power balance. Fourier's law introduces a relation between the spatial gradient of the temperature grad T and the heat flux:

$$\mathbf{q} = -k \text{ grad } T. \tag{2.171}$$

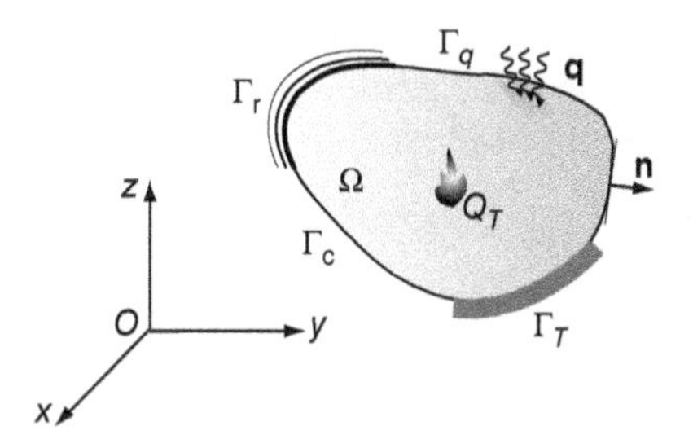

Figure 2.34 Solid subject to thermal boundary conditions.

In Equation 2.171, k is the thermal conductivity and is considered a positive material parameter; it is measured in units of power divided by units of length and units of temperature (watts per metres kelvin in SI units); Fourier's law states that heat is transferred (conducted) inside a solid from zones at higher temperatures to zones at lower temperatures. In reality, k depends on the temperature and therefore to consider it as a parameter in the heat conduction problem is a strong simplification. Typical values of k at ambient temperature (i.e. 293 K or 20 ° C) in watts per kelvin metre [W/K·m] are: 35 for lead, 149 for monocrystalline silicon, 220 for aluminium, 315 for gold and 390 for copper.

The second important relation comes from a power balance on an infinitesimal volume of the continuum body, in which the heat flux and its spatial variations, possible internal sources of thermal energies and the heat exchanged due to variation of temperature with time must be overall balanced, i.e. the power entering the infinitesimal body must be balanced by the power going out from it. The resulting equation is

$$-\text{div } \mathbf{q} + Q_T = c_\text{h}\rho\frac{\partial T}{\partial t}. \tag{2.172}$$

In Equation 2.172, Q_T is the internal source of thermal power per unit volume (in SI units, watts divided by cubic metres), c_h is a new parameter called *specific heat capacity* or, briefly, *specific heat*, measured in units of energy divided by mass and temperature (in SI units, joules per kilogram kelvin), while ρ is the previously introduced mass density per unit volume.

Typical values of c_h at 298 K (or 25 ° C) in joules per kilogram kelvin are 129 for gold and lead, 385 for copper, 712 for monocrystalline silicon and 890 for aluminium.

Equations 2.171 and 2.172 can be combined to obtain a differential relation in the unique unknown temperature, which is known as the *heat equation*:

$$\text{div } k \text{ grad } T + Q_T = c_\text{h}\rho\frac{\partial T}{\partial t}. \tag{2.173}$$

With the hypothesis of a homogeneous body, k does not depend on position and the heat equation (2.173) can be rewritten as

$$k\nabla^2 T + Q_T = c_\text{h}\rho\frac{\partial T}{\partial t}. \tag{2.174}$$

When the problem is stationary, there is a perfect analogy between Equation 2.174 and Equation 2.138 governing the electrostatic problem in the unknown voltage φ.

Equation 2.174 must be solved with the addition of initial and boundary conditions. The initial condition simply states that the temperature must be known at the initial time instant inside the whole body:

$$T(\mathbf{x}, 0) = T_0(\mathbf{x}), \quad \text{in } \Omega. \tag{2.175}$$

The simplest boundary condition states that the temperature is assigned on a portion of the external surface of the body:

$$T(\mathbf{x}, t) = \overline{T}(\mathbf{x}, t), \quad \text{on } \Gamma_T. \tag{2.176}$$

Another possibility is to assign the component of the heat flux projected in the direction of the normal to the external surface:

$$\mathbf{q}(\mathbf{x}, t) \cdot \mathbf{n} = \overline{q}(\mathbf{x}, t), \quad \text{on } \Gamma_q. \tag{2.177}$$

Other kinds of boundary conditions are obtained by studying the way in which the heat is exchanged in the proximity of the external surface Γ between the solid body and the external world. The precise determination of what is happening in these cases is a complex problem, which can involve fluid dynamics, if the heat exchange uses gases, or the study of radiation processes, if gases are not present. Here, we condense the results considering *convection* and *radiation* boundary conditions given in the standard format.

Convection boundary conditions relate the component of the heat flux projected in the direction of the normal to the external surface to the temperature difference between the surface and a reference value T_∞ far from the boundary:

$$\mathbf{q}(\mathbf{x}, t) \cdot \mathbf{n} = h(T - T_\infty), \quad \text{on } \Gamma_c. \tag{2.178}$$

In Equation 2.178, h is a parameter named the *convection constant* that depends on the nature of the surface and of the fluids circulating nearby it and on the relevant fluid dynamics. Values of h can be found in specialized textbooks on heat exchange.

Radiation boundary conditions relate the component of the heat flux projected in the direction of the normal to the external surface to the difference between the fourth power of the surface temperature and that of the reference value T_∞:

$$\mathbf{q}(\mathbf{x}, t) \cdot \mathbf{n} = \epsilon\sigma(T^4 - T_\infty^4), \quad \text{on } \Gamma_r. \tag{2.179}$$

Equation 2.179 comes from the Stefan–Boltzmann law, which rules the radiation emission from a surface with a certain temperature; ϵ is the emissivity of the body ($\epsilon = 1$ for a perfect *black body*, otherwise it is less than 1), σ is the Stefan–Boltzmann constant, and is equal to $5.67 \times 10^{-8}\ \mathrm{Wm^{-2}K^{-4}}$.

Combining Equation 2.173 with suitable initial and boundary conditions of the kind described, the heat problem is fully formulated. The problem remains linear if radiation conditions are neglected and if all parameters are assumed to be independent of temperature.

2.7.1.1 Example: Unidimensional Thermal Problem

As a simple example, let us consider a unidimensional heat problem of a bar with length L, in the absence of an internal heat source ($Q_T = 0$).

The heat equation reduces to

$$k\frac{\partial^2 T}{\partial x^2} = c_\mathrm{h}\rho\frac{\partial T}{\partial t}. \tag{2.180}$$

The problem is considered here with the following initial and boundary conditions:

$$T(x,0) = T_0(x),$$
$$T(0,t) = T_A, \quad T(L,t) = T_B. \tag{2.181}$$

The solution of Equation 2.180 with these initial and boundary conditions can be found by considering the temperature in a separate variable format, i.e.

$$T(x,t) = f(x)g(t). \tag{2.182}$$

Substituting this into Equation 2.180, two separate ordinary differential equations are obtained for functions $f(x)$ and $g(t)$, which can be separately solved to give the final solution in the following format:

$$\begin{aligned} T(x,t) = {} & \left(1 - \frac{x}{L}\right) T_A \\ & + \frac{2}{l} \sum_{1}^{\infty} \left\{ \int_0^L \left[T_0(x) - \left(1 - \frac{x}{L}\right) T_A \right] \sin\left(\frac{n\pi x}{L}\right) \mathrm{d}x \right\} \\ & \times \sin\left(\frac{n\pi x}{L}\right) \exp\left(-\frac{k}{c_{\mathrm{h}}\rho} \frac{n^2\pi^2}{L^2} t\right). \end{aligned} \tag{2.183}$$

The first term of this solution represents the stationary response of the thermal problem, which is eventually reached after a transient phase, as the second term is time-dependent and decays to zero.

2.7.1.2 Summary

It is now possible to summarize the main points discussed in this section concerning the introduction to heat problems.

- The governing equations for the heat problem have been presented, together with the main material parameters involved.
- The heat equation has been obtained for the unknown function temperature.
- Boundary conditions have been discussed; these are on temperature and heat flux and are of the convective and radiation kinds.
- A simple example concerning the solution of the 1D heat problem has been presented.

2.7.2 Thermomechanical Coupled Problem

After the presentation of the reference mechanical problem for a deformable body given in Section 2.3 and the introduction to the thermal problem discussed in Section 2.7, it is now possible to combine the two physics domains to obtain the coupled thermomechanical problem.

The thermomechanical coupled problem is an example of a situation in which the domains where the solution is looked for are partially or totally overlapping.

This kind of coupling is often present in quasi-static and dynamic situations; it is very important in microsystem design and can have various causes.

The first comes from the fact that a solid subject to a temperature variation deforms and therefore thermal strains must be considered in the mechanical problem. This means that, given a temperature distribution, the mechanical problem is modified.

Recalling the vectorial representation of strains given in Equation 2.60, Section 2.3, and assuming that the material has an isotropic thermal behaviour, thermal strains can be given as

$$\varepsilon_T^{\mathrm{T}} = \left[\alpha(T - T_0) \quad \alpha(T - T_0) \quad \alpha(T - T_0) \quad 0 \quad 0 \quad 0\right], \tag{2.184}$$

where $(T - T_0)$ denotes a variation of the temperature with respect to a reference value T_0 and α is a new material parameter called the linear coefficient of thermal expansion, with physical dimensions equal to the inverse of temperature (inverse of kelvin or degree Celsius).

Typical values of α at ambient temperature (20 ° C), measured in 10^{-6} K^{-1} are 29 for lead, 23.1 for aluminium, 18 for silver, 17 for copper, 14 for gold, 2.56 for silicon and 1.2 for Invar. This last material is a nickel-iron alloy; its name comes from the fact that it has a very low coefficient of thermal expansion and therefore is almost geometrically invariant at varying temperatures.

Given the value of the coefficient of thermal expansion, it is possible to compute, numerically, the thermal strain caused by a given increment of temperature. As an example, let us consider a cylindrical specimen with axis x; the thermal strain in the direction of x for $T - T_0 = 100$ K will be 23.1×10^{-4} for aluminium and 2.56×10^{-4} for silicon.

It is to be noticed that differences in coefficients of thermal expansion for various materials cause differences in thermal strains and therefore in the geometrical modifications driven by temperature variations.

The linear elastic problem for deformable solids of Section 2.3 must be modified taking into account thermal strains, simply adding them to strains of an elastic origin, named **e**, i.e. linked to stresses by the elastic constitutive law:

$$\varepsilon = \mathbf{e} + \varepsilon_T, \tag{2.185}$$

$$\sigma = \mathbf{d}\,\mathbf{e} = \mathbf{d}(\varepsilon - \varepsilon_T), \tag{2.186}$$

keeping the other equations (kinematic compatibility and equilibrium) unmodified.

The problem thus obtained is a one-way thermomechanical coupled problem; as already observed, given the temperature distribution, it is possible to compute what happens to displacements, strains and stresses in an elastic solid.

It can now be observed that temperature distributions in an elastic solid that create thermal strain distributions that are not compatible with internal and external kinematic constraints are at the origin of stress modifications. In other words, elastic strains **e** arise to make the total strain ε kinematically compatible and, as a consequence, stresses arise inside the body because these are related to **e** by the constitutive law.

A typical situation in which thermal stresses arise in a deformable elastic body is when the body is inhomogeneous and made with different parts having different coefficients of thermal expansion; in this case uniform temperature distributions also cause incompatible thermal strains owing to the coefficient of thermal expansion mismatch in neighbouring materials.

Always related to thermal strains and thermomechanical coupling are thermal stresses that remain inside the body after a full loading cycle has been completed, i.e. after temperature variations, possibly combined with mechanical load variations, have occurred and the body returns to the initial conditions in terms of overall temperature distribution and the mechanical loads return to zero. Stresses of this kind are named *residual* or *self-equilibrated* because they are in equilibrium with zero mechanical loads.

Residual thermal stresses originate from a combination of temperature gradients, coefficient of thermal expansion mismatch, kinematic constraints and the way in which different parts of the body are warmed or cooled. Residual stresses are of paramount importance in thin-film production and must be estimated and taken into consideration during the design of microsystems; Chapter 14 is devoted to this important issue.

A second source of thermomechanical coupling comes from the fact that portions of a deformable body that undergo fast volumetric deformation processes exchange heat with the surrounding parts, i.e. they behave similarly to gases in which expansion or compression causes heat exchange and variation in temperature. Compressed parts increase the temperature while extended parts cool down.

In terms of equations, this phenomenon can be taken into account by introducing an internal source of thermal power proportional to the velocity of volumetric deformation $\dot{\epsilon}_{\mathrm{vol}}$ into the heat Equation 2.173:

$$\mathrm{div}\,(k\,\mathrm{grad}\,T) + Q_T - \alpha T_0 \frac{E}{(1-2\nu)}\,\dot{\epsilon}_{\mathrm{vol}} = c_{\mathrm{h}}\rho \frac{\partial T}{\partial t}. \tag{2.187}$$

After taking into consideration the coupling terms inserted in Equations 2.185 and 2.187, one obtains the fully coupled thermoelastic problem simply by combining the whole set of equations governing the linear elastic problem and the heat problem, together with suitable boundary conditions.

The thermoelastic problem is of paramount importance for a careful evaluation of dissipative phenomena in vibrating microsystems, e.g. resonators. In this case, the energy exchange from mechanical to thermal represents a loss of kinetic energy, interpreted as damping, in the vibration. A discussion of this and other damping phenomena is given in Chapter 15.

A third source of thermomechanical coupling comes from the fact that parameters governing the purely mechanical response may depend on the temperature. A typical example is the dependence of the Young's modulus on temperature, which can provoke important drifts in mechanical response; this must be carefully evaluated in the MEMS design. If the temperature dependence of mechanical parameters is known, this can be taken into consideration in the governing equations that can be solved for any temperature level.

Of course parameters governing the heat problem can also vary with temperature; in this case the standard heat equation becomes nonlinear in the unknown temperature.

More generally, if one has to solve the fully coupled thermomechanical problem, the possible variation of parameters with temperature must be evaluated and introduced whenever necessary, at the price of an important increase in computational difficulties.

The previous considerations are valid in the hypothesis of elastic material behaviour; such inelastic phenomena as plasticity and viscosity are accompanied by dissipative phenomena, which induce local temperature increments, thermal strains and mechanical property variations, which should be described by suitable constitutive laws, as in the case of thermoviscoplasticity. At the limit, at very high temperatures, an elastoplastic solid can behave like a fluid; in this case, there is a very strong thermomechanical coupling, as, for example, in metal forming processes.

Before concluding this section, it is important to mention the Joule effect, which is another important coupling phenomenon, and which results from the transformation of electrical energy into thermal energy whenever a current flows in a conductor. The power dissipated in heat is given by the product of the electric current and the voltage variation across the conductive element; taking into account Ohm's law, it can also be expressed as potential difference to the second power divided by electric resistance.

Electrothermal coupling coming from Joule effect can be taken into consideration in the thermal problem by inserting an internal heat source given by Joule effect into the thermal power balance. In the case in which electric parameters, such as resistivity, can be considered as independent of temperature, the electric and the thermal problem can be solved in sequence. More generally, the electric and the thermal problem must be solved together as fully coupled.

The Joule effect can be exploited in MEMS to create electrothermomechanical actuators that are capable of exerting a high force with respect to electrostatic actuators; an example is discussed in Chapter 11, where electrothermomechanical actuators are used to perform mechanical rupture tests on silicon microspecimens.

2.7.2.1 Summary

The major points discussed in this section dedicated to thermomechanical coupling are here recalled:

- The general thermomechanical coupled problem has been introduced.
- The thermoelastic problem has been formulated.
- The various sources of thermomechanical coupling have been discussed.
- The concept of residual stresses has been introduced.
- The Joule effect has been mentioned.

References

ANSI/IEEE 176-1987 (1988) *IEEE standard on piezoelectricity*, American National Standards Institute.

Bathe, K. (1996) *Finite Element Procedures*, Prentice Hall.

Bonnet, M., Frangi, A. and Rey, C. (2014) *The Finite Element Method in Solid Mechanics*, McGraw-Hill.

Broberg, K. (1999) *Cracks and Fracture*, Academic Press.

Goldstein, H., Poole, C.P. and Safko, J.L. (2002) *Classical Mechanics*, 3rd edn, Addison Wesley.

Hjelmstad, K.D. (2005) *Fundamentals of Structural Mechanics*, Springer.

Jonassen, N. (2002) *Electrostatics*, Springer.

Marsden, J.E. and Hughes, T.J. (1993) *Mathematical Foundations of Elasticity*, Dover.

Ozisik, M.N. (1980) *Heat Conduction*, John Wiley & Sons, Inc.

Saito, Y., Takao, H., Tani, T. *et al.* (2004) Lead-free piezoceramics. *Nature*, **432**, 84–87.

Shrout, T.R. and Zhang, S.J. (2007) Lead-free piezoelectric ceramics: alternatives for PZT? *Journal of Electroceramics*, **19**, 113–126.

Takenaka, T., Nagata, H. and Hiruma, Y. (2008) Current developments and prospective of lead-free piezoelectric ceramics. *Japanese Journal of Applied Physics*, **47**, 3787–3801.

Thornton, S.T. and Marion, J.B. (2003) *Classical Dynamics of Particles and Systems*, 5th edn, Brooks Cole.
Timoshenko, S. and Goodier, J.N. (1970) *Theory of Elasticity*, 3rd edn, McGraw-Hill.
Versteeg, H.K. and Malalasekera, W. (2007) *An Introduction to Computational Fluid Dynamics*, Pearson Education.
Zienkiewicz, O. and Taylor, R. (2000) *The Finite Element Method*, 5th edn, Butterworth-Heinemann.

3
Modelling of Linear and Nonlinear Mechanical Response

3.1 Introduction

The purpose of this chapter is to show how the formulations of the various fundamental problems presented in the previous chapter (solid and fluid mechanics, electrostatics, thermal,...) can be suitably transformed into simpler sets of relations that can be used for design purposes or to obtain numerical procedures that give approximate but reliable solutions.

Also, in view of the character of the whole book, the choice has been made to focus in this chapter on solid mechanics problems in statics and dynamics and to show, for some reference problems, how to set the stage for the development of practical solutions, which are, in many cases, everyday tools for microsystems designers.

The starting point of the chapter is the presentation of the fundamental principles of virtual power and total potential energy and Hamilton's principle. Next, in Section 3.2, these are also specialized to the case of beam mechanics discussed in the previous chapter.

Section 3.3 provides an introduction to the method of weighted residuals, which allows approximate solutions to be obtained through models having a reduced number of degrees of freedom.

In Section 3.4, approximate and exact solutions referred to the dynamic response of beams are given explicitly, since these are very useful in the study and design of many microsystems, as shown in subsequent chapters of the book.

Finally, Section 3.5 contains a practical example of applications to a bistable element, which shows a characteristic highly nonlinear mechanical behaviour.

This chapter does not in any way aim at presenting an exhaustive list of nonlinear problems that can arise in MEMS. An additional example of dynamic instability is addressed in Chapter 7, where the parametric resonance of electrostatically actuated micromirrors will be analysed. Useful and more complete references on the topics of this chapter are Landau and Lifshitz (1969) and Marsden and Hughes (1993). Specialized books on numerical approaches for the solution of multiphysics problems in MEMS include Korvink and Paul (2006) and Bechtold, Schrag and Feng (2013).

Mechanics of Microsystems, First Edition. Alberto Corigliano, Raffaele Ardito, Claudia Comi, Attilio Frangi, Aldo Ghisi, and Stefano Mariani.

Companion website: www.wiley.com/go/corigliano/mechanics

3.2 Fundamental Principles

3.2.1 Principle of Virtual Power

The principle of virtual power plays a key role in continuum mechanics as it provides a framework for the formulation of structural theories; it also constitutes the foundation of the finite element method for solid mechanics. The finite element method is the most popular method for the numerical solution of problems governed by sets of partial differential equations, such as those described in Chapter 2.

The notion of *virtual displacements* or of *virtual motion* must first be introduced: virtual motions are imaginary movements of material points described at any instant by a velocity field $\delta\dot{\mathbf{u}}(\mathbf{x})$, corresponding, through kinematic compatibility, to strain rates $\delta\dot{\varepsilon}$ and satisfying homogeneous kinematic boundary conditions.

The principle of virtual power is a necessary and sufficient condition for the equilibrium of a material body of volume Ω and surface area Γ that is subjected to body forces $\mathbf{b}$ and surface forces $\mathbf{f}$ on the free part of the boundary, see Figure 2.12 and Section 2.3; it involves the so-called *internal* δP_{i} and *external* δP_{e} *virtual powers*, exerted respectively by stresses inside the body multiplied by virtual strain rates and external applied forces and inertia forces multiplied by virtual velocities:

$$\delta P_{\mathrm{i}} \equiv \int_{\Omega} \boldsymbol{\sigma}^{\mathrm{T}} \delta\dot{\varepsilon}\, \mathrm{d}\Omega; \qquad \delta P_{\mathrm{e}} \equiv \int_{\Omega} \mathbf{b}^{\mathrm{T}} \delta\dot{\mathbf{u}}\, \mathrm{d}\Omega + \int_{\Gamma_f} \mathbf{f}^{\mathrm{T}} \delta\dot{\mathbf{u}}\, \mathrm{d}\Gamma - \int_{\Omega} \rho\ddot{\mathbf{u}}^{\mathrm{T}} \delta\dot{\mathbf{u}}\, \mathrm{d}\Omega. \tag{3.1}$$

Consider now the following equation:

$$\delta P_{\mathrm{i}} - \delta P_{\mathrm{e}} = \int_{\Omega} \boldsymbol{\sigma}^{\mathrm{T}} \delta\dot{\varepsilon}\, \mathrm{d}\Omega - \int_{\Omega} \mathbf{b}^{\mathrm{T}} \delta\dot{\mathbf{u}}\, \mathrm{d}\Omega - \int_{\Gamma_f} \mathbf{f}^{\mathrm{T}} \delta\dot{\mathbf{u}}\, \mathrm{d}\Gamma + \int_{\Omega} \rho\ddot{\mathbf{u}}^{\mathrm{T}} \delta\dot{\mathbf{u}}\, \mathrm{d}\Omega = 0. \tag{3.2}$$

The *principle of virtual powers* can then be expressed as follows:

> A necessary and sufficient condition for the dynamic equilibrium of a deformable body is that Equation 3.2 holds for any virtual motion.

It is worth remarking that dynamic equilibrium means that the conservation of linear momentum (2.70) and equilibrium boundary conditions (2.69), recalled in Chapter 2, are both satisfied at the considered time instant.

In the case of quasi-static evolution problems, Equation 3.2 specializes to

$$\int_{\Omega} \boldsymbol{\sigma}^{\mathrm{T}} \delta\dot{\varepsilon}\, \mathrm{d}\Omega - \int_{\Omega} \mathbf{b}^{\mathrm{T}} \delta\dot{\mathbf{u}}\, \mathrm{d}\Omega - \int_{\Gamma_f} \mathbf{f}^{\mathrm{T}} \delta\dot{\mathbf{u}}\, \mathrm{d}\Gamma = 0. \tag{3.3}$$

3.2.2 Total Potential Energy Principle

Consider the quasi-static evolution of a linear elastic body subject to conservative forces. In this case, the solution can be characterized by a variational principle. To this purpose, let us first define the functional (i.e. a mathematical operator that takes as its input a function and returns a single number, a scalar) *total potential energy* Π, which is the sum of the elastic energy and the potential of external loads:

$$\Pi(\mathbf{u}) \equiv \frac{1}{2} \int_{\Omega} \varepsilon^{\mathrm{T}} \boldsymbol{d} \varepsilon\, \mathrm{d}\Omega - \int_{\Omega} \mathbf{b}^{\mathrm{T}} \mathbf{u}\, \mathrm{d}\Omega - \int_{\Gamma_f} \mathbf{f}^{\mathrm{T}} \mathbf{u}\, \mathrm{d}\Gamma. \tag{3.4}$$

Note that the potential energy Π is defined within the class of kinematically admissible fields $\mathbf{u}$ and ε, i.e. with $\mathbf{u} = \overline{\mathbf{u}}$ on Γ_u and $\varepsilon = \frac{1}{2}(\text{grad } \mathbf{u} + \text{grad}^{\mathrm{T}}\mathbf{u})$ in Ω (see Equations 2.62 and 2.63).

The following theorem, known as the *theorem of minimum potential energy*, holds:

> Among all kinematically admissible solutions, the real, equilibrated one, minimizes the total potential energy.

3.2.3 Hamilton's Principle

In classical mechanics, the *Lagrangian* function is defined as the kinetic energy $\mathcal{T}$ of the system minus its potential energy, Π, in symbols, $\mathcal{L} = \mathcal{T} - \Pi$.

Hamilton's principle states that the true evolution of a system between two specified states $\mathbf{u}_1 = \mathbf{u}(t_1)$ and $\mathbf{u}_2 = \mathbf{u}(t_2)$ at two specified time instants t_1 and t_2 is a stationary point (a point where the first variation is zero), of the *action functional*:

$$\mathcal{S}[\mathbf{u}] \quad \equiv \int_{t_1}^{t_2} \mathcal{L}(\mathbf{u}(t), \dot{\mathbf{u}}(t), t)\, \mathrm{d}t. \tag{3.5}$$

In other words, any first-order perturbation of the true evolution results in (at most) second-order changes in $\mathcal{S}$. Hamilton's principle states that the true evolution of a physical system is a solution of the functional equation

$$\int_{t_1}^{t_2} [\delta\mathcal{T}(\dot{\mathbf{u}}) - \delta\Pi(\mathbf{u})]\, \mathrm{d}t = 0. \tag{3.6}$$

When nonconservative forces are also considered, this principle states that the sum of the time variation of the difference between kinetic and potential energies and the work done by the nonconservative forces over any time interval t_1 and t_2 equals zero, for any varied path $\delta\mathbf{u}$ from time t_1 and t_2 with $\delta\mathbf{u}(t_1) = \delta\mathbf{u}(t_2) = 0$:

$$\int_{t_1}^{t_2} \delta[\mathcal{T}(\dot{\mathbf{u}}) - \Pi(\mathbf{u})]\, \mathrm{d}t + \int_{t_1}^{t_2} \delta W_{\mathrm{nc}}(t)\, \mathrm{d}t = 0, \tag{3.7}$$

where δW_{nc} is the virtual work of nonconservative forces. If the Lagrangian of a system is known, the equations of motion of the system may be obtained by a direct substitution of the expression for the Lagrangian into the Euler–Lagrange equation:

$$\frac{\partial\mathcal{L}}{\partial\mathbf{u}} - \frac{\mathrm{d}}{\mathrm{d}t}\frac{\partial\mathcal{L}}{\partial\dot{\mathbf{u}}} = 0. \tag{3.8}$$

The application of Hamilton's principle to a particle will clarify the method. Suppose we have a 3D space in which a particle of mass m moves under the influence of a conservative force $\mathbf{F}$. Since the force is conservative, it corresponds to a potential energy function $\Pi(\mathbf{u})$, such that $F_i = -\partial\Pi/\partial u_i$. The Lagrangian of the particle can be written as

$$\mathcal{L}(\mathbf{u}, \dot{\mathbf{u}}) = \frac{1}{2}m\dot{\mathbf{u}}^2 - \Pi(\mathbf{u}). \tag{3.9}$$

The equation of motion for the particle (see Equation 2.16) is found by applying the Euler–Lagrange equation

$$\frac{\partial\mathcal{L}}{\partial u_i} - \frac{\mathrm{d}}{\mathrm{d}t}\left(\frac{\partial\mathcal{L}}{\partial\dot{u}_i}\right) = 0, \text{where} \quad i = 1, 2, 3. \tag{3.10}$$

with

$$\frac{\partial \mathcal{L}}{\partial u_i} = F_i, \quad \frac{\mathrm{d}}{\mathrm{d}t}\left(\frac{\partial \mathcal{L}}{\partial \dot{u}_i}\right) = m\ddot{u}_i, \tag{3.11}$$

and reads in scalar components:

$$F_i = m\ddot{u}_i, \quad i = 1, 2, 3. \tag{3.12}$$

3.2.4 Specialization of the Principle of Virtual Powers to Beams

The principle of virtual power, presented in Section 3.2.1, is here specialized to beams. For the sake of simplicity, a single-span straight beam of length L is considered. The beam is loaded in the x–y plane by distributed forces $n(x)$ and $p(x)$ in the axial and transverse directions, respectively, and by concentrated forces F_i and couples W_j. Shear strain is neglected. The principle of virtual power is expressed as

$$\int_0^L N\delta\dot{\eta}\,\mathrm{d}x + \int_0^L M\delta\dot{\chi}\,\mathrm{d}x - \int_0^L n\delta\dot{u}\,\mathrm{d}x - \int_0^L p\delta\dot{v}\,\mathrm{d}x - \sum_i (F_x\delta\dot{u} + F_y\delta\dot{v})_i - \sum_j W_j\delta\dot{\varphi}_j + \int_0^L \rho A(\ddot{u}\delta\dot{u} + \ddot{v}\delta\dot{v})\,\mathrm{d}x = 0. \tag{3.13}$$

If Equation 3.13 holds for any virtual motion $\delta\dot{u}(x)$ and $\delta\dot{v}(x)$, the system is in dynamic equilibrium, i.e. the equations of motion (2.90) are locally satisfied.

In the case of quasi-static evolution problems, Equation 3.13 specializes to:

$$\int_0^L N\delta\dot{\eta}\,\mathrm{d}x + \int_0^L M\delta\dot{\chi}\,\mathrm{d}x - \int_0^L n\delta\dot{u}\,\mathrm{d}x - \int_0^L p\delta\dot{v}\,\mathrm{d}x - \sum_i (F_x\delta\dot{u} + F_y\delta\dot{v})_i - \sum_j W_j\delta\dot{\varphi}_j = 0. \tag{3.14}$$

If Equation 3.14 holds for any virtual motion $\delta\dot{u}(x)$ and $\delta\dot{v}(x)$, the system is in static equilibrium, i.e. the equations of equilibrium (2.88) are locally satisfied.

3.3 Approximation Techniques and Weighted Residuals Approach

The method of weighted residuals is an approximate technique for solving differential equations, such as those governing the equilibrium of an elastic body or of a beam, and can be used to derive the equations of the finite element method.

The unknown functions (for instance the displacements $\mathbf{u}(\mathbf{x})$) are approximated with functions $\tilde{\mathbf{u}}$, which are a linear combination of n basis functions $\boldsymbol{\psi}_i(\mathbf{x})$ chosen from a linearly independent set. That is:

$$\mathbf{u}(\mathbf{x}) \cong \tilde{\mathbf{u}}(\mathbf{x}) = \sum_i a_i \boldsymbol{\psi}_i(\mathbf{x}), \quad i = 1, \dots, n. \tag{3.15}$$

Now, when substituted into the differential equations, an error or residual $\mathbf{R}$ will exist. The notion in the method of weighted residuals is to force the residual to zero in some average sense over the domain. That is:

$$\int_{\Omega} \mathbf{R}(x) w_i \mathrm{d}\mathbf{x} = 0, \quad i = 1, 2, \dots, n, \tag{3.16}$$

where the number of equations of the kind of Equation 3.16 and the number of weight functions w_i is exactly equal to the number of unknowns a_i in $\tilde{\mathbf{u}}$; notice that a_i can be parameters or functions of time in time-dependent problems. The result is a set of n algebraic equations for the unknown constants a_i. There are several different submethods for the method of weighted residuals, according to the choices of weight or test functions w_i. The *Galerkin* method uses the basis functions themselves, $\psi_i(\mathbf{x})$, as test functions.

Let us now exemplify the method for a linear elastic beam in bending subject to a transverse distributed load p. The differential equation that governs the dynamic equilibrium of the beam, see Equation 2.91, reads:

$$\rho A \ddot{v} + \frac{\mathrm{d}^2}{\mathrm{d}x^2}\left(EJ \frac{\mathrm{d}^2 v}{\mathrm{d}x^2}\right) = p. \tag{3.17}$$

The transverse displacement is approximated by

$$v(x,t) \cong \tilde{v}(x,t) = \sum_i a_i(t)\psi_i(x), \quad i = 1, 2, \dots, n, \tag{3.18}$$

where the functions $\psi_i(x)$ are C^1 and fulfil the boundary conditions. Substituting Equation 3.18 into Equation 3.17, multiplying by the same functions $\psi_j(x)$ and integrating, one obtains

$$\sum_i \ddot{a}_i \int_0^L \rho A \psi_i \psi_j \,\mathrm{d}x + \sum_i a_i \int_0^L \frac{\mathrm{d}^2}{\mathrm{d}x^2}\left(EJ \frac{\mathrm{d}^2 \psi_i}{\mathrm{d}x^2}\right) \psi_j \,\mathrm{d}x$$
$$= \int_0^L p \psi_j \,\mathrm{d}x, \quad j = 1, \dots, n. \tag{3.19}$$

Integrating the second term by parts, taking into account the boundary conditions and rearranging, one obtains the system of equations (algebraic in space) from which the time-dependent coefficients a_i can be obtained:

$$\mathbf{M}\ddot{\mathbf{a}} + \mathbf{K}\mathbf{a} = \mathbf{F}, \tag{3.20}$$

with

$$[\mathbf{M}]_{ij} = \int_0^L \rho A \psi_i \psi_j \,\mathrm{d}x; \quad [\mathbf{K}]_{ij} = \int_0^L EJ \frac{\mathrm{d}^2 \psi_i}{\mathrm{d}x^2} \frac{\mathrm{d}^2 \psi_j}{\mathrm{d}x^2} \,\mathrm{d}x;$$
$$[\mathbf{F}]_j = \int_0^L p \psi_j \,\mathrm{d}x; \quad [\mathbf{a}]_i = a_i. \tag{3.21}$$

Matrix $\mathbf{M}$ is the so-called *mass matrix* and matrix $\mathbf{K}$ is the *stiffness matrix*, while vector $\mathbf{F}$ is the *load vector* of the beam mechanical model, which is discretized in space.

Equations 3.20 are a set of ordinary differential equations in time, in the unknown functions $a_i(t)$. This system of equations is conceptually analogous to the relation governing the one-dof undamped oscillator (see Section 2.2.3) and is a simplified and

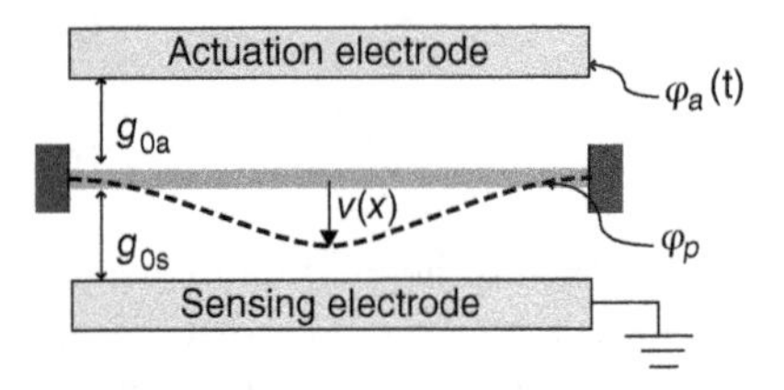

Figure 3.1 Electrostatically actuated resonator.

approximate description of beam dynamics, transformed into a n-dof system by means of the weighted residuals method.

Remark 1 *When quasi-static evolutions are considered, the contribution depending on the mass matrix* **M** *disappears.*

Remark 2 *An equation system similar to Equation 3.20 can be obtained by applying the finite element method; the interested reader can refer to specialized textbooks on the finite element method, such as Bonnet, Frangi and Rey (2014).*

Remark 3 *The method can also be used when the applied forces depend on the displacement (as in the case of electrostatic forces) or on the velocity (as in the case of viscous damping forces). A simple example is the electrostatically actuated resonator sketched in Figure 3.1, in its usual working condition: an actuation voltage is applied to the driving electrode to sustain the oscillation; the central beam is biased by a DC voltage φ_p, and the sensing electrode is kept at virtual ground for current read-out. The electrostatic force per unit length acting on the beam in this configuration is (see Section 2.5):*

$$p(x,t) = \frac{\epsilon_0 s}{2}\left[\frac{\mathrm{d}}{\mathrm{d}v}\frac{(\varphi_\mathrm{p}-\varphi_\mathrm{a})^2}{(g_{0\mathrm{a}}+v(x,t))} + \frac{\mathrm{d}}{\mathrm{d}v}\frac{(\varphi_\mathrm{p})^2}{(g_{0\mathrm{s}}-v(x,t))}\right]. \tag{3.22}$$

In Equation 3.22, ϵ_0 is the dielectric permittivity, s is the out-of-plane beam thickness, g_{0a} and g_{0s} are the gaps at rest between the central beam and the driving or sensing electrodes (see Figure 3.1).

For small displacement and small actuation voltage, the electrostatic force can be expressed as

$$p(x,t) \cong \overline{p}(t) + \epsilon_0 s\varphi_\mathrm{p}^2\left(\frac{1}{g_{0\mathrm{a}}^3}+\frac{1}{g_{0\mathrm{s}}^3}\right)v(x,t). \tag{3.23}$$

Substituting into the weighted residuals equation 3.19, one obtains the following algebraic equation:

$$\mathbf{M}\ddot{\mathbf{a}} + (\mathbf{K}+\mathbf{K}_\mathrm{e})\mathbf{a} = \mathbf{F} \tag{3.24}$$

with

$$[\mathbf{K}_\mathrm{e}]_{ij} = \int_0^L \epsilon_0 s\varphi_\mathrm{p}^2\left(\frac{1}{g_{0\mathrm{a}}^3}+\frac{1}{g_{0\mathrm{s}}^3}\right)\psi_i\psi_j\,\mathrm{d}x. \tag{3.25}$$

Remark 4 *The weighted residuals method is very general; it applies directly to differential equations (e.g. to the equilibrium equation in strong form) and allows to obtain an approximate solution of their weak form. The substitution of approximate functions in* *the principle of virtual power gives the same results as the Galerkin–weighted residuals method. In cases when the problem to be solved is endowed with a variational principle, the static equilibrium equations obtained through the Galerkin method can also be obtained from the stationarity of the total potential energy. This method is more general inasmuch as it does not require the existence of an energy principle.*

3.4 Exact and Approximate Solutions for Dynamic Problems

3.4.1 Free Flexural Linear Vibrations of a Single-span Beam

Consider a prismatic elastic beam with constant cross-sectional area A, momentum of inertia J and length l; its free vibrations are governed by Equation 3.17 with $p = 0$. In this case, the closed-form solution can be obtained by separation of variables using

$$v(x, t) = Y(t)\psi(x), \tag{3.26}$$

where $\psi(x)$ is the shape of the free-vibration motion while $Y(t)$ is its time-dependent amplitude. Notice that Equation 3.26 is analogous to Equation 3.15 with $n = 1$; hence, it is an approximation that reduces the beam dynamics to an equivalent one-dof oscillator, where the dof is the function $Y(t)$.

Substituting Equation 3.26 into Equation 3.17, one obtains:

$$\rho A\ddot{Y}(t)\psi(x) + EJY(t)\psi^{\mathrm{IV}}(x) = 0. \tag{3.27}$$

In Equation 3.27, the symbol $\psi^{\mathrm{IV}}(x)$ denotes the fourth-order derivative of $\psi(x)$ with respect to the spatial variable x; similar notation will be used for brevity in the remainder of the book for first- to fourth-order spatial derivatives.

Let us now consider the free vibration of an undamped one-dof oscillator, governed by

$$\ddot{Y}(t) + \omega^2 Y(t) = 0, \tag{3.28}$$

where ω is the angular frequency of the oscillation. From Equation 3.28, $\ddot{Y}(t) = -\omega^2 Y(t)$, hence, substituting in Equation 3.27:

$$-\rho A\omega^2\psi(x) + EJ\psi^{\mathrm{IV}}(x) = 0, \tag{3.29}$$

Equation 3.28 governs the free-vibration of a single-dof system, it has the form of Equation (2.25) with the solution (2.27) depending on initial conditions, i.e. with the present notation,

$$Y(t) = Y(0)\cos\omega t + \frac{\dot{Y}(0)}{\omega}\sin\omega t. \tag{3.30}$$

Equation 3.27 can be solved in closed form for different boundary conditions. The solution in terms of eigenvalues $\omega = \omega_k$ and eigenfunctions $\psi = \psi_k$ is

$$\begin{aligned}
\omega_k &= \lambda_k^2\sqrt{\frac{EJ}{\rho A}}, \quad k = 1, 2, \dots, \\
\psi_k &= A_k\,(\cos\lambda_k x + \mathrm{Ch}\lambda_k x) + B_k\,(\cos\lambda_k x - \mathrm{Ch}\lambda_k x) \\
&\quad + C_k\,(\sin\lambda_k x + \mathrm{Sh}\lambda_k x) + D_k\,(\sin\lambda_k x - \mathrm{Sh}\lambda_k x).
\end{aligned} \tag{3.31}$$

Table 3.1 collects the solutions of the first three modes for different boundary conditions.

Table 3.1 Eigenfrequencies and eigenfunctions of a single-span beam.

Constraint	k	A_k	B_k	C_k	D_k	$\lambda_k L$
Clamped–clamped	1	0	1	0	−0.9826	4.730
	2	0	1	0	−1.0008	7.853
	3	0	1	0	−1.0000	10.996
Clamped–free	1	0	1	0	−0.7341	1.875
	2	0	1	0	−1.0185	4.694
	3	0	1	0	−0.9992	7.885
Hinged–hinged	1	1	0	0.5	0.5	3.142
	2	1	0	0.5	0.5	6.283
	3	1	0	0.5	0.5	9.425

The constants A_k, B_k, C_k, D_k (see Equation 3.31) are defined up to a common amplification factor.

These analytical solutions are also useful for finding approximate solutions for many other problems for which exact solutions are not readily available. In fact they can be used as trial functions for numerical methods, such as the Galerkin–weighted residuals method. A practical application of these concepts is detailed in Section 8.2.2 of Chapter 8 on Magnetometers.

3.4.2 Nonlinear Vibration of an Axially Loaded Beam

Consider a prismatic elastic beam with cross-sectional area A, momentum of inertia J and length L subject to constant axial load P_0 and to a transversely distributed dynamic load $p(t, x)$. Let the longitudinal and transverse displacements of the beam axis be described by $u(x, t)$ and $v(x, t)$, respectively. The beam is axially constrained at both ends, possibly with elastic constraints represented by an axial spring of stiffness k_a at one end (see Figure 3.2).

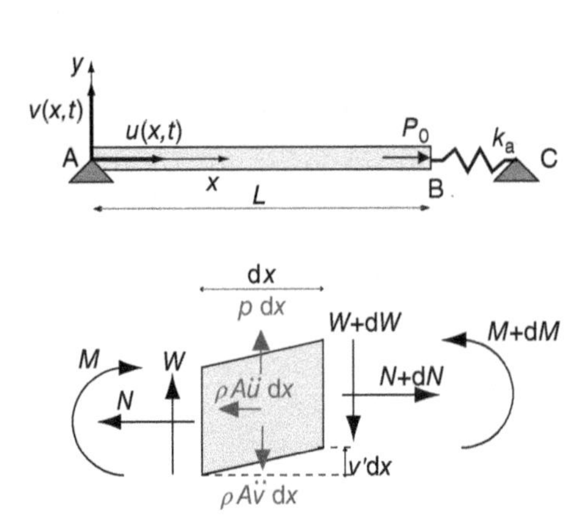

Figure 3.2 Vibrating beam and free body diagram.

The model here adopted for the description of the dynamic response of the beam is restricted by several hypotheses: (i) the beam is modelled by the Euler–Bernoulli theory (see Chapter 2, Section 2.3.2); (ii) the variation in cross-section during vibration is neglected; (iii) the stretching of the beam is small but finite. With these hypotheses, the axial strain along the beam axis can be expressed as

$$\eta_0(x, t) = u' + \frac{1}{2}(v')^2. \tag{3.32}$$

The axial force N and the bending moment M can be expressed as

$$N(x,t) = P_0 + EA\left(u' + \frac{1}{2}(v')^2\right),$$
$$M(x,t) = EJv''. \tag{3.33}$$

The equations of motion are derived by considering the equilibrium of forces acting on the differential segment of mass $\rho A\,\mathrm{d}x$ of the deflected beam (see Figure 3.2). Notice that this method of enforcing equilibrium is different from that used in Chapter 2, Section 2.3.2, where equilibrium is enforced in the undeformed configuration of the beam:

$$N' - \rho A\ddot{u} = 0,$$
$$M'' - (Nv')' + \rho A\ddot{v} = p. \tag{3.34}$$

Introducing Equations 3.33 into Equations 3.34, one obtains a coupled differential system for the axial and transverse displacements u and v.

If inertial effects in the axial direction can be neglected, the problem decouples, leading to a differential equation in v only, as explained in the following. Setting to zero the longitudinal inertial force, from Equation 3.34a, the axial force turns out to be a function of time only. From Equation 3.33a it follows that:

$$u' = \frac{N(t) - P_0}{EA} - \frac{1}{2}(v')^2. \tag{3.35}$$

Integrating Equation 3.35 in space, with the boundary conditions

$$u(0,t) = 0, \quad u(l,t) = -\frac{N(t) - P_0}{k_a}, \tag{3.36}$$

one obtains the axial force in the form

$$N(t) = P_0 + \frac{k_a EA}{k_a L + EA}\int_0^L \frac{1}{2}(v')^2\,\mathrm{d}x. \tag{3.37}$$

Note that the axial force has two contributions: the first is due to a pre-stress P_0/A, acting on the beam, constant in time and independent of the beam transverse displacement v, while the second is generated by the elongation of the beam induced by its finite deflection. This second contribution is present only in axially constrained beams. Substituting the axial force into the transverse dynamic equilibrium (3.34b), the following nonlinear equation for the transverse beam oscillation is obtained:

$$(EJv'')'' - P_0 v'' - \frac{k_a EA}{k_a L + EA}v''\int_0^L \frac{1}{2}(v')^2\,\mathrm{d}x + \rho A\ddot{v} = p. \tag{3.38}$$

If the axial elongation is not constrained (i.e. $k_a = 0$), the governing differential equation for the transverse displacement simplifies to

$$(EJv'')'' - P_0 v'' + \rho A\ddot{v} = p. \tag{3.39}$$

Equation 3.39, which generalizes Equation 3.17 to the present case, has been widely studied and several solutions are available in the literature. The presence of a constant axial load P_0 changes the natural angular frequencies of the beam oscillation. For a single-span beam, frequencies increase in the case of a tensile load and decrease in the case of a compressive load. The first resonant angular frequency ω can be expressed as

$$\omega(P_0) = \lambda^2 \sqrt{\frac{EJ}{\rho A}} \sqrt{1 + \frac{P_0 L^2}{\gamma EJ}}, \tag{3.40}$$

where the coefficients λL and γ depend on the boundary conditions. Table 3.2 collects their values for several boundary conditions. Note that λ coincides with λ_1, introduced in Table 3.1.

Equation 3.39 and hence Equation 3.40 can also be used for axially constrained beams if transverse oscillations can be considered small with respect to the beam height. This hypothesis, which is often reasonable for structural problems at the macroscale, is, in general, not valid for microstructures, such as those of MEMS resonators. In this case, the complete Equation 3.38 and the associated boundary conditions should be considered.

Numerical solutions for the transverse oscillation of the nonlinear beam can be obtained as shown in Comi (2009), using Hamilton's principle and searching for an approximate solution in the form of Equation 3.26. A convenient choice of $\psi(x)$ is the eigenfunction of the same problem but linearized, without *second-order* effects. Considering free oscillations ($p = 0$), the approximate equation of motion is

$$m\ddot{Y} + k_1 Y + k_3 Y^3 = 0 \tag{3.41}$$

with

$$\begin{aligned}
m &= \int_0^L \rho A \psi^2 \, \mathrm{d}x, \\
k_1 &= k_\mathrm{e} + k_\mathrm{G}, \quad k_\mathrm{e} = \int_0^L EJ(\psi'')^2 \, \mathrm{d}x, \quad k_\mathrm{G} = P_0 \int_0^L (\psi')^2 \, \mathrm{d}x, \\
k_3 &= \frac{1}{2} \frac{k_\mathrm{a} EA}{k_\mathrm{a} L + EA} \int_0^L (\psi')^2 \, \mathrm{d}x \int_0^L (\psi')^2 \, \mathrm{d}x.
\end{aligned} \tag{3.42}$$

where Equation 3.41 is known as the Duffing oscillator equation for a single-dof system, m is the equivalent mass, k_e is the equivalent linear elastic bending stiffness, k_G is the generalized geometric stiffness due to a constant initial axial load and k_3 is the cubic nonlinear constant. For the problem under consideration, of the nonlinear oscillations of a single-span beam, k_3, given in Equation 3.42 is always positive (*hard spring effect*). The

Table 3.2 Coefficients λL and γ in Equation 3.40 for single-span beams with different boundary conditions.

Constraints	λL	γ
Clamped–clamped	4.730	40.7
Clamped–free	1.875	2.66
Hinged–hinged	3.142	9.87
Sliding–sliding	3.142	9.87
Sliding–hinged	1.572	2.47

solution of Equation 3.41 can be obtained from a series of successive approximations, see Landau and Lifshitz (1969):

$$Y = Y^{(1)} + Y^{(2)} + Y^{(3)}, \tag{3.43}$$

where

$$Y^{(1)} = \overline{A}\cos\omega t, \quad Y^{(2)} = 0, \quad Y^{(3)} = -\frac{\overline{A}^3}{32}\frac{k_3}{k_1}\cos 3\omega t. \tag{3.44}$$

In Equation 3.44, $\overline{A}$ is the oscillation amplitude and ω is the actual value of the natural frequency, which differs from the reference value of the linear case $\omega_0 = \sqrt{k_1/m}$. The natural frequency of the nonlinear problem can be expressed as

$$\omega = \omega_0\left(1 + \frac{3}{8}\frac{k_3}{k_1}\overline{A}^2\right). \tag{3.45}$$

Note that, unlike a linear oscillator, the fundamental frequency ω of the Duffing oscillator does depend on the oscillation amplitude $\overline{A}$. Nonlinearity also affects the forced oscillations. Considering a harmonic driving term $p(t) = p_0\cos\omega_p t$, of frequency ω_p close to the natural frequency, neglecting damping, one obtains a relation between the amplitude p_0 and the frequency ω_p of the driving force and the amplitude A of the forced vibration of the beam (see e.g. Landau and Lifshitz (1969)):

$$\left(\frac{p_0}{k_1}\right)^2 = \left(2\left(1 - \frac{\omega_p}{\omega_0}\right)\overline{A} + \frac{3}{4}\frac{k_3}{k_1}\overline{A}^3\right)^2. \tag{3.46}$$

The solution of Equation 3.46 is shown in Figure 3.3 in terms of the dynamic amplification factor $\overline{A}/(p_0/k_1)$ versus the ratio ω_p/ω_0 between the frequency of the driving force ω_p and the reference natural frequency ω_0. For $k_3 > 0$, the resonance curve bends towards the right and, for sufficiently high frequencies, three solutions exist. Two solutions are stable while the third belongs to an unstable branch plotted with a dashed line

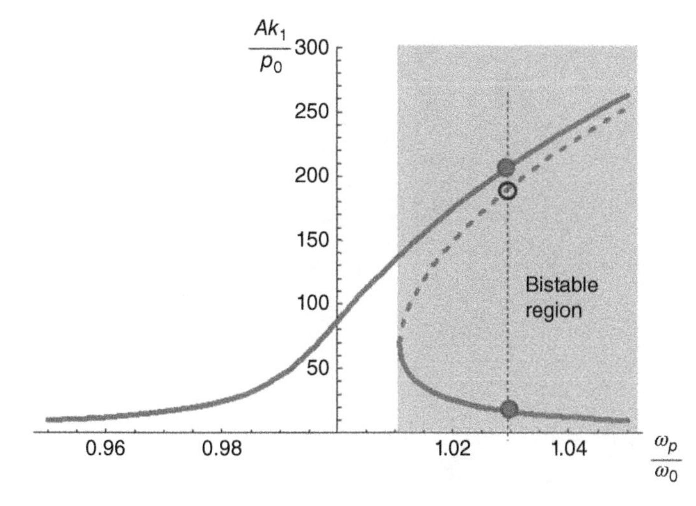

Figure 3.3 Forced frequency response of nonlinear resonator: normalized dynamic amplification factor versus normalized frequency.

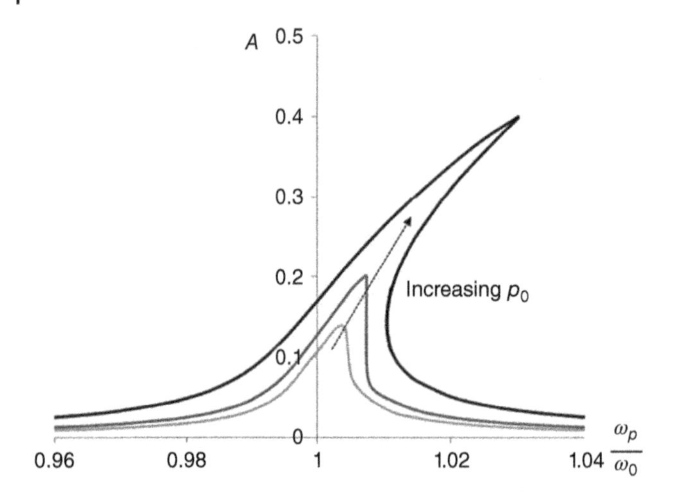

Figure 3.4 Forced frequency response of nonlinear resonator for increasing excitation load. Amplitude in [μm].

in Figure 3.3. In real resonators with damping, the resonance curve is asymmetric but, for low values of excitation, hysteresis phenomena are avoided, see Figure 3.4.

3.5 Example of Application: Bistable Elements

As a meaningful example of purely mechanical modelling implementing the concepts expounded in the previous sections, we address here the analysis of bistable beams and their application in the MEMS shock sensor (STMicroelectronics, 2013; Frangi *et al.*, 2015) depicted in Figure 3.5.

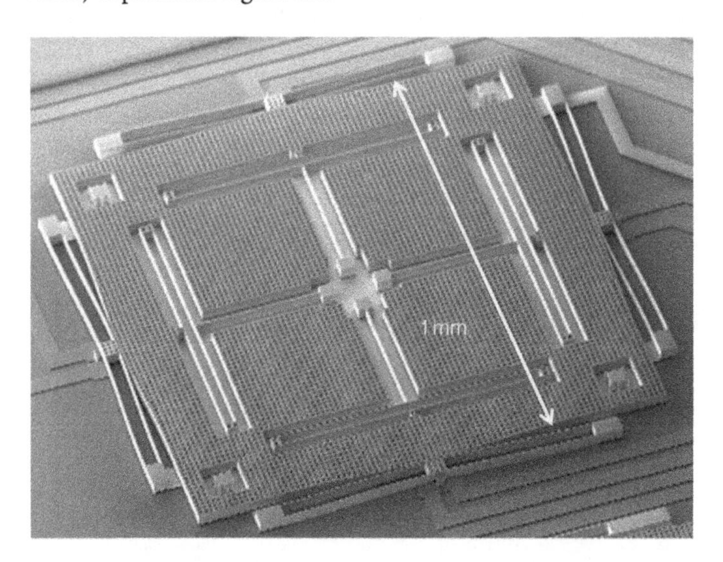

Figure 3.5 SEM of a MEMS shock sensor based on bistable beams. *Source*: Frangi *et al.* (2015), Figure 2. Reproduced with permission of IEEE. (*See color plate section for the color representation of this figure.*)

Multistable elements are often utilized for various applications (Howell, 2001), including energy harvesting, even at the microscale (Brake *et al.*, 2010). Multistability can be achieved in several ways, e.g. exploiting the magnetic field produced by permanent magnets or piezoelectric effects. Limiting our attention to the mechanical multistability of beams, the simplest possible multistable beam is a doubly clamped straight beam compressed above the Eulerian critical load. There are two stable equilibrium configurations that are perfectly symmetric with respect to the original beam axis. However, the production and control of such a device poses formidable challenges if it is to be fabricated using standard industrial surface micromachining processes. An alternative practical way of producing a bistable element is to couple two doubly clamped beams with a deformed initial shape, following an idea put forward by Qiu, Lang and Slocum (2001, 2004), as will be discussed next.

The sensor is made of polysilicon and we assume linear elastic isotropic constitutive behaviour with Young's modulus $E = 150$ GPa and mass density $\rho = 2330$ kg/m^3. The out-of-plane thickness of the device is fixed by the fabrication process at $H = 22$ μm. A suspended inertial shuttle of mass $m = 1.95 \times 10^{-8}$ kg, is attached to the substrate via four flexible springs; each of these springs is made of five thin elements of thickness $t_s = 2.6$ μm and length $L_s = 245$ μm.

The shuttle in its rest condition is positioned very close (with a gap of approximately 1 μm) to four bistable elements, one of which is schematically represented in Figure 3.6. Each bistable element is made of two doubly clamped beams that have the initially curved shape described analytically by:

$$v_0 = \delta\psi_1(x) \quad \text{with}$$
$$\psi_1 = \frac{1}{2}\left(1 - \cos\frac{2\pi x}{L}\right), \tag{3.47}$$

with $\delta = 10$ μm, $L = 600$ μm. The two doubly clamped beams are rigidly connected to each other in the central portion by a clamp in order to prevent rotation. As is well known in the literature (Qiu, Lang and Slocum, 2001, 2004), a single doubly clamped beam with initial configuration given by Equation 3.47 is not bistable. Indeed, if forced to the second equilibrium configuration (bottom of Figure 3.7), it would snap back to the first configuration (top of Figure 3.7) via an antisymmetric mode that is prevented here by means of the central clamp. Therefore, only symmetric deflections of the bistable elements will be considered in the following. Moreover a bistable mechanism can be obtained only if the curved beams are characterized by a sufficiently high geometrical coefficient $\Delta = \delta/t$, where t is the in-plane (vibrating) thickness. The ratio Δ has to respect the inequality $\Delta > 4/\sqrt{3}$, which sets an upper bound on t. In the current design of the device $t = 2$ μm and hence $\Delta = 5$.

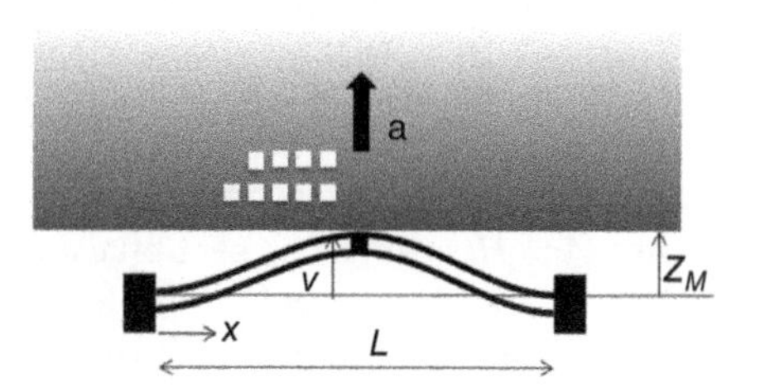

Figure 3.6 Simplified scheme: rigid shuttle and bistable element. The coordinate system is attached to the anchors; z_M denotes the shuttle position. *Source*: Frangi *et al.* (2015), Figure 4. Reproduced with permission of IEEE.

When an external acceleration reaches a given threshold (say 1000g for the device analysed, where g is the acceleration due to gravity), the contact force transmitted by the mass to the bistable element induces the transition to a second stable configuration.

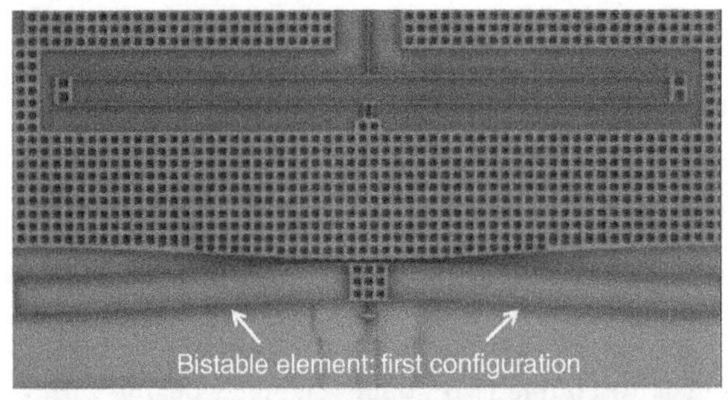

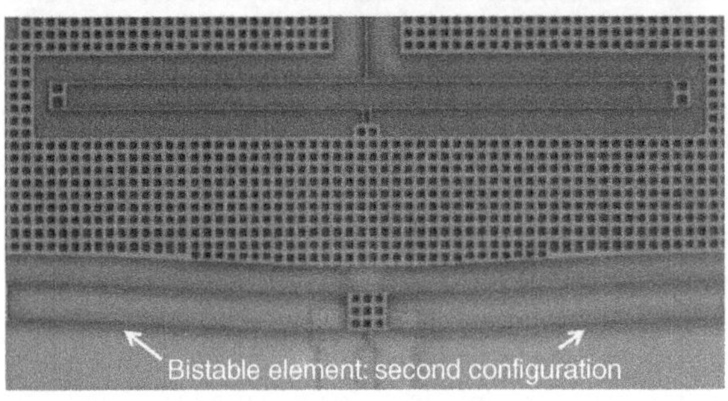

Figure 3.7 Close-up of bistable element in two stable configurations before and after actuation. *Source*: Frangi *et al*. (2015), Figure 3. Reproduced with permission of IEEE.

During the dynamic transition, the beam hits a weak link that breaks and opens a resistive path.

Modelling Strategy The beam dynamics can be described via the principle of virtual power (Section 3.2.1) for slender beams and moderately large displacements (Nayfeh, Kreider and Anderson, 1995; Maurini, Pouget and Vidoli, 2009; Giannopoulos, Monreal and Vantomme, 2007), with the addition of the inertia forces. For each of the two beams in the bistable element, the problem is formulated as follows.

Find the deflection $v(x) \in C^1(0)$ such that, for $\forall \delta v \in C^1(0)$:

$$2\int_0^L \left(\rho A(\ddot{v}+a)\delta v + EJ(v''-v''_0)\delta v'' - N[v]v'\delta v'\right)\,\mathrm{d}x = -F\delta v(L/2). \qquad (3.48)$$

The space $C^1(0)$ is here the space of functions w with continuous first derivative and $v = v' = 0$ in $x = 0, L$ and

$$N[v] = \frac{EA}{2L}\int_0^L \left((v_0')^2 - (v')^2\right)\mathrm{d}x = -EA\frac{\Delta L}{L}$$

is the compressive axial force assumed independent of the position along the beam (ΔL is the elongation of the beam axis). The force F is exerted by the shuttle on the central clamp when contact develops and is assumed to be equally distributed between the two beams. $A = Ht$ is the cross-sectional area of each beam; $J = (1/12)Ht^3$ is the momentum of inertia; ρA is the mass per unit length; EJ is the flexural stiffness. It is worth emphasizing that the term $EJ(v'' - v_0'')$ represents the bending moment in the beam and that the rotational inertia of the cross-sections has been neglected.

To derive a simple formula for design purposes, let us start by assuming that the bistable element is loaded quasi-statically with a force F applied at midspan and that its displacement is initially proportional to the first buckling mode, i.e. $v = \alpha\psi_1(x)$. Neglecting inertia forces and setting $y = \alpha/\delta$ and $\Delta = \delta/t$, the force–displacement relationship from Equation 3.48 is

$$F(y) = 2\frac{\pi^4 EI}{L^3}(\delta - \alpha)\left(2 + \frac{3}{2t^2}\alpha(\delta + \alpha)\right)$$
$$= 2\frac{\pi^4 EI\delta}{L^3}(1 - y)\left(2 + \frac{3}{2}\Delta^2 y(1 + y)\right). \quad (3.49)$$

The force that induces the snap-through is a parameter of paramount importance. As discussed by Qiu, Lang and Slocum (2004), since $\Delta^2 > 16/3$, snap-through occurs when the stiffness of the third buckling mode vanishes, i.e. when

$$y_T^2 = 1 - \frac{16}{3\Delta^2} \quad (3.50)$$

and the snap-through force F_T follows from Equation 3.49 as $F_T = F(y_T)$.

However, to reproduce more accurately the snap-through of the beams, a more sophisticated dynamic model is required, including contact with the shuttle. With the aim of simplifying the underlying equations as much as possible, we assume that the substrate is subjected to the external in-plane acceleration a aligned with the device, as depicted in Figure 3.6, and that the movement occurs along the same direction.

Contact will develop only with one bistable element. The stiffness of the spring system of the shuttle has been computed with standard procedures as $k \simeq 11$ N/m. The Couette flow between the mass and the substrate generates a dissipative force $b\dot{z}_M$. Using the techniques described in Chapter 15, a quality factor $Q = 40$ has been estimated and b fixed accordingly. If contact is activated between the mass and the bistable element, the mass will receive the force F. Globally, in a frame rigidly connected with the anchors, the shuttle dynamics are governed by the 1D model

$$m(\ddot{z}_M + a) + b\dot{z}_M + k(z_M - \delta) = F \quad (3.51)$$

where z_M denotes the mass coordinate.

Equations 3.51 and 3.48 must be complemented with the conditions

$$z_M - v(L/2) \geq 0, \quad F > 0 \quad (z_M - v(L/2))F = 0 \quad (3.52)$$

governing *perfectly hard* contact. The first condition of Equation 3.52 prevents penetration between the beam and the shuttle; the second guarantees that the force exchanged is repulsive and the third simply states that the force arises only when the gap is closed and vanishes otherwise.

Numerical Results and Experimental Validation To solve Equation 3.48 numerically, the simplest possible solution strategy involves assuming a 2D discretization space, such that the displacement v is expressed as a linear combination of the first and third buckling modes (see Section 3.3):

$$v(x,t) = \alpha(t)\psi_1(x) + \beta(t)\psi_3(x),$$

where ψ_1 is defined in Equation 3.47 and

$$\psi_3 = \frac{1}{2}\left(1 - \cos\frac{4\pi x}{L}\right) \quad (3.53)$$

Initial conditions impose $v = v_0$ and hence $\alpha(0) = \delta$ and $\beta(0) = 0$. The functions ψ_1 and ψ_3 are chosen as the buckling modes of a doubly clamped straight beam. The second antisymmetric mode,

$$\psi_2 = 1 - \frac{2x}{L} - \cos\left(N_1 \frac{x}{L}\right) + \frac{2}{N_1} \sin\left(N_1 \frac{x}{L}\right) \tag{3.54}$$

(with $N_1 \simeq 2.86\pi$), is not considered, owing to the presence of the central clamp.

Setting first $\delta v = \psi_1$ and then $\delta v = \psi_3$ in Equation 3.48, one obtains two second-order differential equations for α and β:

$$\rho AL \left(\frac{3\ddot{\alpha}}{8} + \frac{a}{2} + \frac{\ddot{\beta}}{4} \right) + EJ\frac{2\pi^4}{L^3}(\alpha - \delta) - \frac{\pi^2}{2L} N[\alpha, \beta]\alpha = -\frac{F}{2},$$

$$\rho AL \left(\frac{\ddot{\alpha}}{4} + \frac{a}{2} + \frac{3\ddot{\beta}}{8} \right) + EJ\frac{32\pi^4}{L^3}\beta - \frac{2\pi^2}{L} N[\alpha, \beta]\beta = 0,$$

with

$$N[\alpha, \beta] = EA\frac{\pi^2}{2L^2} \left(\frac{\delta^2}{2} - \frac{\alpha^2}{2} - 2\beta^2 \right). \tag{3.55}$$

A penalty approach is implemented to simulate contact. This involves replacing perfectly hard contact with a contact force that increases linearly with compenetration:

$$F = \lambda(\alpha - z_M)H(\alpha - z_M),$$

where H is the Heaviside function and λ is a positive large penalty coefficient. This set of equations is integrated with an explicit central difference scheme and adaptive step control.

Figure 3.8 presents the simulated history of the midspan deflection $\alpha = v(L/2)$ computed by applying an external acceleration of the form:

$$a(t) = \frac{\overline{N}g}{2}(1 - \cos(2\pi f_0 t)), \quad t < \frac{1}{f_0}, \tag{3.56}$$

with $f_0 = 1500$ Hz. The simulations have been run for increasing values of maximum acceleration $\overline{N}g$ in order to identify the threshold acceleration a_T inducing the transition to the second stable configuration. At each frequency, the response for two values of $\overline{N}g$ just below and above a_T is presented. In the former case, the beam goes back to the initial configuration; in contrast, in the latter situation, contact is lost at a certain time and the beam makes the transition to the second stable equilibrium configuration.

Numerical simulations compare well with the experiments reported in Frangi *et al.* (2015). MEMS have been attached to a frame undergoing controlled acceleration histories chosen according to the MIL.STD-202G protocol, which is typically adopted to check the suitability of electronic components when subjected to improper handling, transportation and in general operations that have as consequence a mechanical shock.

The tests were conducted as follows. All the chips were subjected to a series of shock tests with increasing maximum acceleration and the number of ruptures was recorded. The results are summarized in Figure 3.9, where the cumulative probability of rupture (the sum of the two orientations) is plotted versus the maximum $\overline{N}$ value of the test.

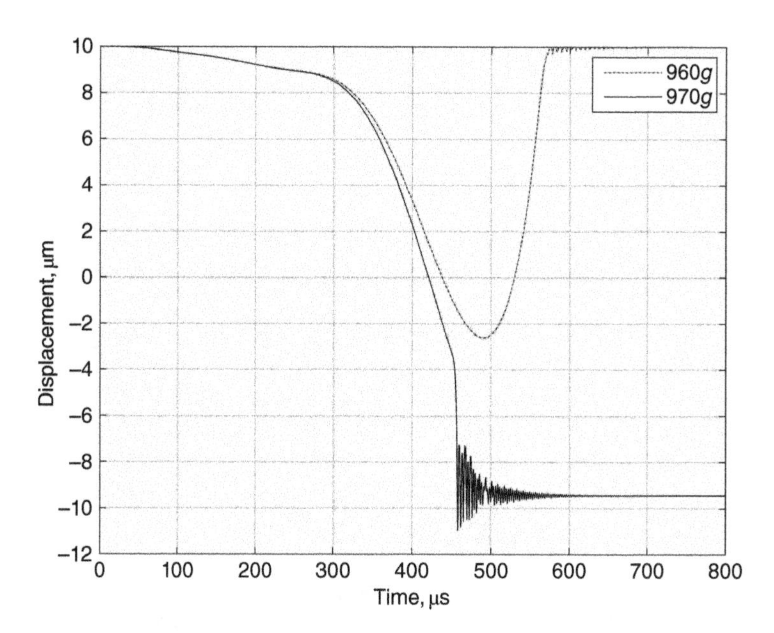

Figure 3.8 Simulated midspan deflection for two different values of $\overline{N}g$ (see Equation 3.56) just below and above the snap-through threshold. Design input frequency 1500 Hz. *Source*: Frangi *et al.* (2015), Figure 6. Reproduced with permission of IEEE.

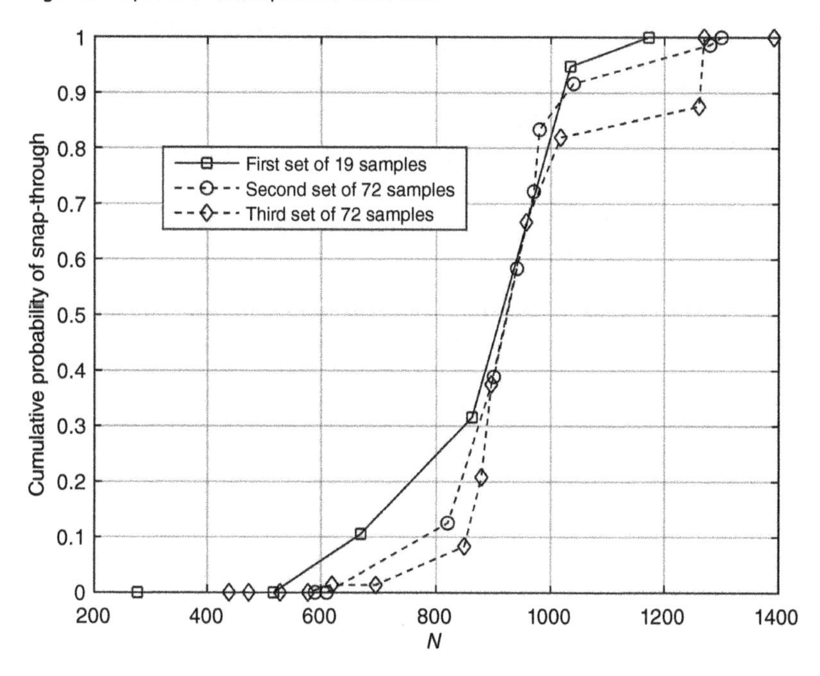

Figure 3.9 Cumulative probability of rupture versus maximum acceleration. *Source*: Frangi *et al.* (2015), Figure 11. Reproduced with permission of IEEE.

References

Bechtold, T., Schrag, G. and Feng, L. (eds) (2013) *System-Level Modeling of MEMS*, Wiley-VCH Verlag GmbH.

Bonnet, M., Frangi, A. and Rey, C. (2014) *The Finite Element Method in Solid Mechanics*, McGraw-Hill.

Brake, M.R., Baker, M.S., Moore, N.W. *et al.* (2010) Modeling and measurement of a bistable beam in a microelectromechanical system. *Journal of Microelectromechanical Systems*, **19**, 1503–1514.

Comi, C. (2009) On geometrical effects in micro-resonators. *Latin American Journal of Solids and Structures*, **6**, 73–87.

Frangi, A., De Masi, B., Confalonieri, F. and Zerbini, S. (2015) Threshold shock sensor based on a bistable mechanism: design, modeling, and measurements. *Journal of Microelectromechanical Systems*, **24**, 2019–2026.

Giannopoulos, G., Monreal, J. and Vantomme, J. (2007) Snap-through buckling behavior of piezoelectric bimorph beams: I. Analytical and numerical modeling. *Smart Materials and Structures*, **16**, 1148.

Howell, L. (2001) *Compliant Mechanisms*, John Wiley & Sons, Inc.

Korvink, J. and Paul, O. (eds) (2006) *MEMS: A Practical Guide to Design, Analysis and Applications*, Springer.

Landau, L.D. and Lifshitz, E.M. (1969) *Mechanics*, Pergamon Press.

Marsden, J.E. and Hughes, T.J. (1993) *Mathematical Foundations of Elasticity*, Dover.

Maurini, C., Pouget, J. and Vidoli, S. (2009) Bistable buckled beam: modelling and piezoelectric actuation. *Advances in Science and Technology*, **54**, 281–286.

Nayfeh, A., Kreider, W. and Anderson, T.J. (1995) Investigation of natural frequencies and mode shapes of buckled beams. *AIAA Journal*, **33**, 1121–1126.

Qiu, J., Lang, J.H. and Slocum, A.H. (2001) *A Centrally-Clamped Parallel-Beam Bistable MEMS Mechanism*, MEMS 2001. The 14th IEEE International Conference on Micro Electro Mechanical Systems, January 25, 2001, Interlaken, Switzerland, IEEE.

Qiu, J., Lang, J.H. and Slocum, A.H. (2004) A curved-beam bistable mechanism. *Journal of Microelectromechanical Systems*, **13**, 137–146.

STMicroelectronics (2013) Shock sensor with bistable mechanism and method of shock detection. US Patent US9316550 B2, filed Aug. 1, 2013 and issued Apr. 19 2016.

Part II

Devices

4

Accelerometers

4.1 Introduction

Accelerometers belong to the class of inertial sensors, i.e. sensors designed to transduce the inertia forces acting on a system into a measurable signal. MEMS accelerometers play an important role in the field of sensors with applications in various contexts, including automotive systems, vibration monitoring and portable electronics. The majority of microaccelerometers proposed in the literature and currently present on the market can be grouped into three main classes, on the basis of the sensing principle: capacitive, resonant and piezoresistive. In this chapter, we will describe, in some detail, capacitive and resonant MEMS accelerometers made using surface micromachining techniques. In general, these devices comprise a mobile part, denoted the *proof mass*, suspended with respect to the substrate by means of elastic elements, simply called *springs*, and fixed parts named *stators*, which cannot move with respect to the accelerometer package.

The accelerometer feels the external acceleration by means of the proof mass, which moves with respect to the stator when an external acceleration occurs.

The movement of the proof mass is equivalent to the movement of a person in an elevator: when the elevator goes up, the person feels a force that pushes towards the elevator's floor, when the elevator goes down, the person feels a force that pulls towards the elevator's roof. Therefore, the person in an elevator *feels* the external acceleration and acts as an *accelerometer*.

Another example of how accelerometers work can be taken from the movements of structures due to an earthquake: the earthquake causes accelerations of the soil, which, in turn, is responsible for acceleration delivered to the structure foundation. As a consequence, the whole portion of the structure outside the soil can feel inertia forces, which will be maximum where most of the mass is concentrated. Every floor of a building therefore acts as a proof mass in an accelerometer. This comparison also explains why, in many cases, the proof mass of an accelerometer is also called the *seismic mass*.

Capacitive accelerometers are discussed in Section 4.2, while resonant accelerometers are described in Section 4.3. Some examples are shown in Section 4.4, while Section 4.5 contains remarks on design and reliability issues.

Mechanics of Microsystems, First Edition. Alberto Corigliano, Raffaele Ardito, Claudia Comi, Attilio Frangi, Aldo Ghisi, and Stefano Mariani.

Companion website: www.wiley.com/go/corigliano/mechanics

4.2 Capacitive Accelerometers

4.2.1 In-Plane Sensing

The most widely diffused microaccelerometers make use of capacitive sensing. A simple scheme of a 1D capacitive accelerometer is shown in Figure 4.1. A proof mass m is attached to a fixed frame by means of an elastic spring of stiffness k and can translate in one direction (x in the figure). Its movement can be detected by the variation in capacitance of the parallel plates represented by the electrode attached to the mass and the electrodes attached to the fixed MEMS frame.

The motion of mass m can be studied in the noninertial reference frame system attached to the MEMS box.

The equation governing relative dynamics and presented in Chapter 2, (2.17), projected in the x-direction, reads:

$$m\left\{\mathbf{a}_{\mathrm{r}}+\left[\mathbf{a}_0+\frac{\mathrm{d}\boldsymbol{\omega}}{\mathrm{d}t}\wedge(P-O)+\boldsymbol{\omega}\wedge(\boldsymbol{\omega}\wedge(P-O))\right]+2\boldsymbol{\omega}\wedge\mathbf{v}_{\mathrm{r}}\right\}_x=F_x. \tag{4.1}$$

Considering that for the 1D case $(P-O)=x\mathbf{i}$, one obtains the following equation governing the motion in direction x:

$$ma_{\mathrm{r}x}=F_x-ma_{0x}+mx(\omega_y^2+\omega_z^2). \tag{4.2}$$

The force F_x is usually given by three contributions: an elastic term due to the spring, which is proportional to the displacement u, a damping term, assumed here as viscous, which is linearly dependent on the velocity $v=\dot{u}$, and two electrostatic terms (see Figure 4.1):

$$F_x=-ku-c\dot{u}+F_{\mathrm{right}}^{\mathrm{elec}}-F_{\mathrm{left}}^{\mathrm{elec}}. \tag{4.3}$$

In real accelerometers, the mass is a movable *shuttle* mass attached to the substrate by slender beams, possibly folded, as shown in Figure 4.2. It is usual to assume that only the springs can deform, while the shuttle is rigid. Using the solution for a doubly clamped beam presented in Section 2.3.2, the equivalent mechanical stiffness is expressed as

$$k=\frac{12}{n_{\mathrm{folds}}}\frac{EJ}{l^3}, \tag{4.4}$$

where n_{folds} is the number of folds of the beam, for example three in Figure 4.2.

The electrostatic forces depend on the gap variation induced by the motion, on the geometry of the fixed and moving electrodes, and on the potential applied to them. Considering the case of parallel plates of Figure 4.3 and assuming that the proof mass is kept

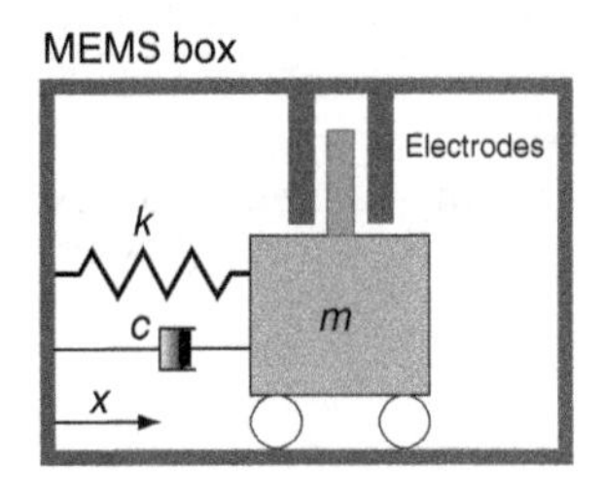

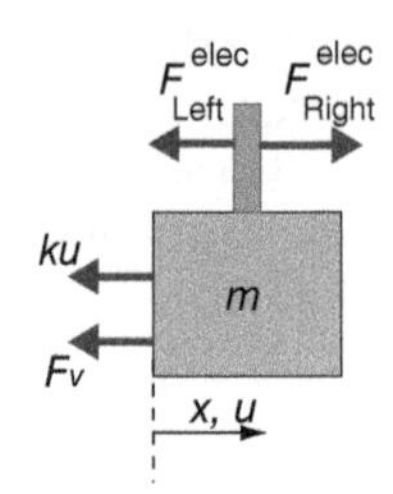

Figure 4.1 1D capacitive accelerometer.

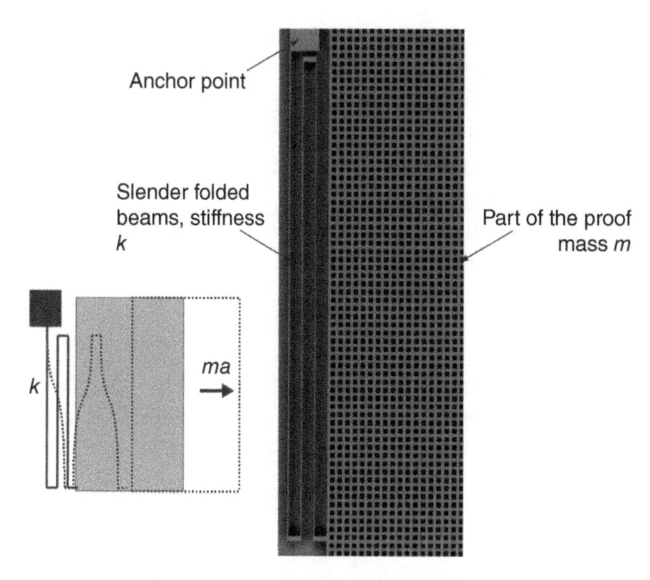

Figure 4.2 Folded spring of in-plane accelerometer. *Source:* Reproduced with permission of STMicroelectronics.

at a fixed potential φ_p, while the fixed electrodes are at ground, one has

$$F_{\text{right}}^{\text{elec}} - F_{\text{left}}^{\text{elec}} = \frac{1}{2}\epsilon_r\epsilon_0 S\varphi_p^2\left[\frac{1}{(d-u)^2} - \frac{1}{(d+u)^2}\right] \cong \frac{2}{d^3}\epsilon_r\epsilon_0 S\varphi_p^2 u, \tag{4.5}$$

where the approximation in this equation has been obtained by developing the function of u up to the linear term around $u = 0$.

With this approximation, electrostatic forces result in a term proportional to u, which can be combined with the elastic force, thus giving an effect of stiffness reduction due to an equivalent electrostatic stiffness k^{elec}

$$k^{\text{elec}} \equiv \frac{2}{d^3}\epsilon_r\epsilon_0 S\varphi_p^2. \tag{4.6}$$

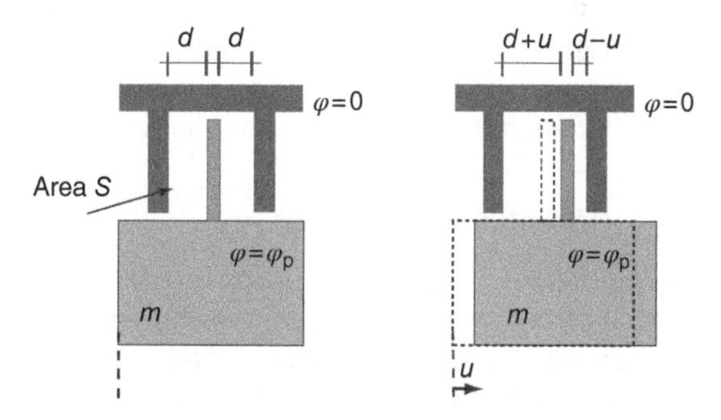

Figure 4.3 Parallel-plate sensing.

Accelerometers are usually designed such that the eigenfrequency of the proof mass, suspended by linear elastic springs, is much greater than the frequency of the external signal that should be measured. In this case, the effect of the external acceleration a_{0x} can be viewed as quasi-static and one can assume $a_{rx} = \ddot{u} \cong 0$ and $v = \dot{u} \cong 0$. By neglecting, in addition, the effect of external angular velocity, one obtains a relation between the external acceleration and the displacement:

$$(k - k^{\text{elec}})u = -ma_{0x}. \tag{4.7}$$

The external acceleration turns out to be proportional to the displacement of the mass and sensing can be achieved by measuring the displacement via the variation in capacitance. To this purpose, both comb fingers and parallel plates can be used. The latter enable high values of capacitance variation to be reached, with small dimension, but have the disadvantage of having a highly nonlinear behaviour and of showing the pull-in phenomenon, as discussed in Chapter 2, Section 2.5.2.

Let us consider Figure 4.3: when a parallel plate moves, as depicted in the figure, with a displacement u, the capacitance of the system varies as

$$\Delta C = \Delta C_{\rm l} - \Delta C_{\rm r} = \epsilon_{\rm r}\epsilon_0 S \left\{ \left[\frac{1}{d+u} - \frac{1}{d} \right] - \left[\frac{1}{d-u} - \frac{1}{d} \right] \right\}, \tag{4.8}$$

where $\Delta C_{\rm l}$ and $\Delta C_{\rm r}$ denote the capacitance variations in the left and right capacitors, respectively. By developing the expression (4.8) up to first order around the initial configuration $u = 0$, it is possible to approximate the capacitance variation ΔC as

$$\Delta C \cong -2\frac{\epsilon_{\rm r}\epsilon_0 S u}{d^2}, \tag{4.9}$$

By solving Equation 4.9 with respect to the displacement u, and combining this result with Equation 4.7, it is possible to compute the *sensitivity* of the accelerometer, given as the ratio of the measured variation of capacitance ΔC over the external acceleration a_{0x}:

$$\frac{\Delta C}{a_{0x}} = 2\frac{m}{k - k^{\rm elec}}\frac{\epsilon_{\rm r}\epsilon_0 S}{d^2}. \tag{4.10}$$

From this equation, it can be observed that the sensitivity of the capacitive accelerometer can be increased by means of material, structural and geometrical features of the chosen design: the material choice enters through the values of the mass density, permittivity constant and Young's modulus; the structural design enters through the value of the elastic suspension spring stiffness; the geometry enters through the superposed area of the capacitors, the gap and the volume of the proof mass. For fixed material choice, one can increase the sensitivity by increasing the mass m and the surface area S and reducing the stiffness $(k - k^{\rm elec})$ and the gap d.

4.2.2 Out-of-Plane Sensing

In MEMS fabricated through micromachining, the out-of-plane sensing of the acceleration usually requires a tilting proof mass. The scheme is similar to the one shown in Figure 4.1, but the mass has a rotational degree of freedom ϑ and is attached to the substrate by torsional springs. These latter are usually fabricated as slender beams. Figure 4.4 shows a close-up view of the spring in an out-of-plane accelerometer.

The dynamics of the tilting proof mass can be described by an equivalent one-dof system, in which displacement, velocity and acceleration are respectively substituted by rotation, angular velocity and angular acceleration; the mass is substituted by rotational inertia, the damping coefficient is suitably determined (see Chapter 15) and the stiffness is computed starting from the torsional behaviour of the suspension springs, as described next.

The torsional equivalent stiffness, relating the torsional moment $M_{\rm t}$ to the rotation angle ϑ, for slender beams with rectangular cross-section can be expressed as (see

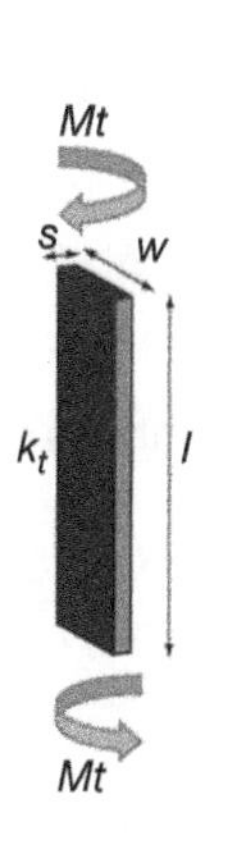

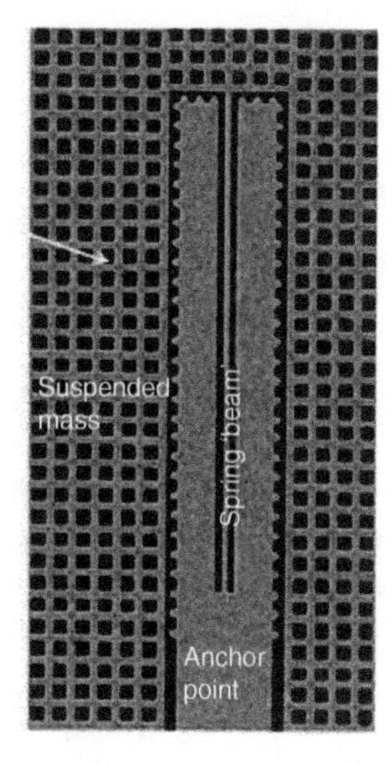

Figure 4.4 Torsional spring of an out-of-plane accelerometer. *Source:* Reproduced with permission of STMicroelectronics.

Chapter 2, Section 2.3.2)

$$k_{\mathrm{t}} = \frac{GJ_{\mathrm{t}}}{l}, \quad \text{with } J_{\mathrm{t}} = \frac{ws^3}{3}. \tag{4.11}$$

In this case, the external acceleration causes a rotation of the proof mass, which can be sensed by the electrodes on the substrate as a capacitance variation. By considering the scheme in Figure 4.5, a simplified analytical formula gives the differential

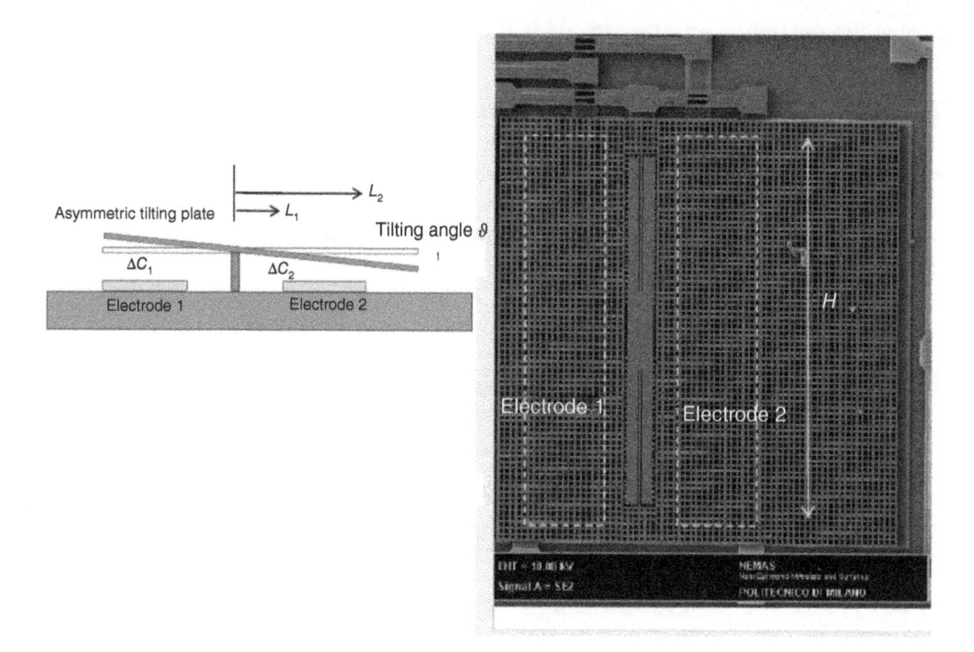

Figure 4.5 Capacitive sensing in an out-of-plane accelerometer. *Source:* Reproduced with permission of STMicroelectronics.

capacitance variation:

$$\Delta C_1 - \Delta C_2 = \epsilon_r \epsilon_0 H \left[\int_{L_2}^{L_1} \frac{dx}{d + \vartheta x} - \int_{L_2}^{L_1} \frac{dx}{d - \vartheta x} \right]. \tag{4.12}$$

The application of this equation allows for the determination of the rotation angle and, consequently, of the acceleration.

4.3 Resonant Accelerometers

In resonant accelerometers, the external acceleration produces a recordable shift of the resonance frequency of the structure, or of some part thereof. Resonant sensing, with respect to other sensing principles, has the advantage of direct frequency output, high potential sensitivity and large dynamic range. Sensitivity of resonant accelerometers is generally defined as the frequency shift produced by an external acceleration of $1g$.

Resonating accelerometers can be divided into two categories: the first type is composed of a resonating proof mass and a system for excitation and sensing, while the second type is composed of a moving proof mass, a resonator and a system for excitation and sensing.

Figure 4.6 shows, schematically, the two kinds of resonant accelerometer. Note that in Figure 4.6, and in the rest of this chapter, the dashpot representing the damping of the system is omitted for the sake of clarity.

4.3.1 Resonating Proof Mass

The basic principle of these accelerometers is to detect the variation of resonant frequency of the oscillating proof mass due to the external acceleration. The proof mass is usually kept at resonance by electrostatic actuation, as shown schematically in Figure 4.7. Let $f_0 = \frac{1}{2\pi}\sqrt{k/m}$ denote its eigenfrequency. If the mass is electrostatically actuated, the stiffness k should be understood as the actual stiffness, i.e. the difference between the mechanical stiffness k_m and the electrostatic stiffness k^{elec}, as discussed in Section 4.2.

By considering a parallel-plate actuation as in Figure 4.7 and by denoting by φ the difference between the voltage of the mass and of the electrode, one has

$$k^{\text{elec}} = -\frac{1}{d^3}\epsilon_r \epsilon_0 S \varphi^2.$$

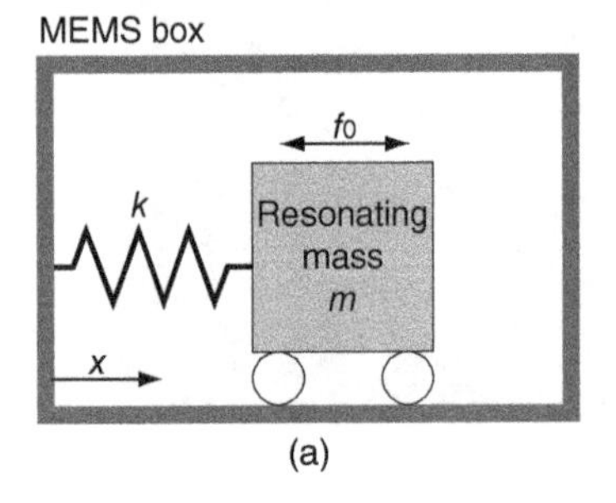

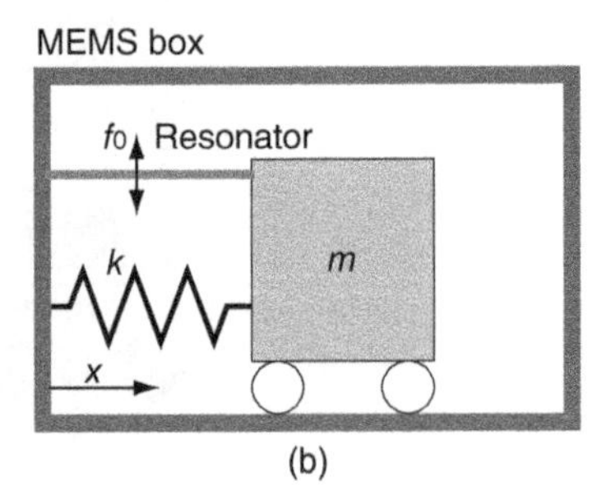

Figure 4.6 (a) Accelerometer with resonating proof mass. (b) Accelerometer with resonating beam device coupled to proof mass.

When an external acceleration is applied, the inertia force acting on the mass produces a displacement of the mass. By considering the mass as a one-dof oscillator, one has

$$m\ddot{u} + c\dot{u} + ku = ma_{0x}. \tag{4.13}$$

If the frequency of the external acceleration a_{0x} is much smaller than the resonance frequency f_0, the displacement of the mass due to external acceleration can be expressed as

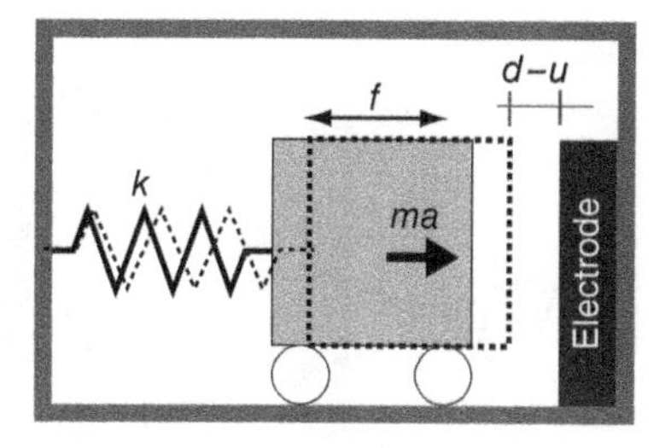

Figure 4.7 Function of resonant accelerometer.

$$u \cong \frac{m}{k}a_{0x}, \tag{4.14}$$

hence there is a gap variation between the mass and the fixed electrode. The change of the gap produces a variation of the electrostatic stiffness k^{elec} and a consequent variation of the oscillation frequency, as explained in Chapter 3:

$$f = \frac{1}{2\pi}\sqrt{\frac{k_{\mathrm{m}} - k^{\mathrm{elec}}}{m}} \tag{4.15}$$

This displacement modifies the mean gap between the mass and the electrode; the corresponding electrostatic stiffness variation, similarly to Equation 4.6, is given by

$$k^{\mathrm{elec}} = -\frac{1}{(d-u)^3}\epsilon_{\mathrm{r}}\epsilon_0 S\varphi^2. \tag{4.16}$$

Substituting Equation 4.16 into Equation 4.15, linearized for $k^{\mathrm{elec}}/k \ll 1$, one obtains:

$$\Delta f = f - f_0 \approx f_0\left(\frac{1}{(d-u)^3} - \frac{1}{d^3}\right)\frac{\epsilon_0 S\varphi^2}{2k} \approx f_0\frac{\epsilon_0 S\varphi^2}{2k}\frac{3}{d^4}u. \tag{4.17}$$

The sensitivity of the accelerometer can then be approximated by

$$\frac{\Delta f}{a_{0x}} \approx f_0\frac{3\epsilon_0 S\varphi^2}{2k^2 d^4}m. \tag{4.18}$$

The sensitivity thus depends on the geometry of the device, through its mass and stiffness, and increases with the square of the actuation voltage.

An accelerometer based on this principle has been proposed, e.g. by Sung, Lee and Kang (2003) for out-of-plane acceleration sensing.

4.3.2 Resonating Elements Coupled to the Proof Mass

In resonant accelerometers, more often the resonant device is a small oscillator coupled to the proof mass, as sketched in Figure 4.6b. The input acceleration is detected in terms of a shift in the resonant frequency of the oscillator. This shift can be induced by

- The electrostatic stiffness variation, as for the accelerometers discussed in Section 4.3.1,
- The variation of the momentum of inertia,
- The geometrical stiffness variation.

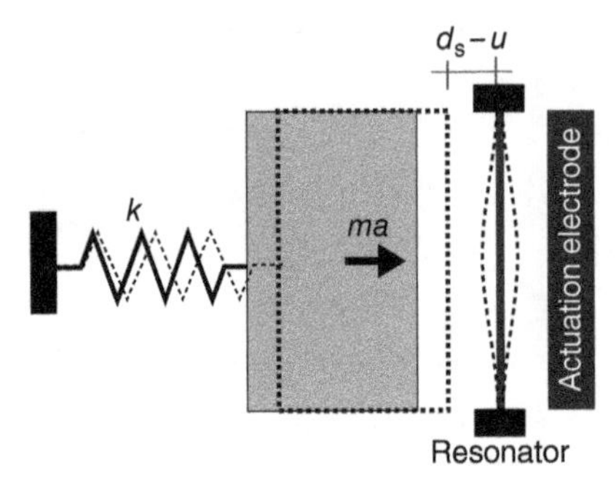

Figure 4.8 Uniaxial accelerometer with resonant beam.

The first case is shown in Figure 4.8, in which the resonator is represented by a vertical oscillating beam. When a horizontal acceleration is applied, the proof mass moves while the resonator is vibrating and the gap d_s between the resonator and the sensing electrode attached to the proof mass varies. The electrode therefore senses the resonant frequency variation due to the electrostatic stiffness variation. Seok, Kim and Chun (2004) have proposed a MEMS accelerometer for in-plane measurements exploiting this idea. In their proposal, the resonating part is constituted by a double-ended tuning-fork resonator, and two proof masses are present for differential sensing. Other types of resonator can also be coupled to the proof mass, see e.g. Kim *et al.* (2005). An example of an out-of-plane accelerometer with torsional resonators will be discussed in detail in Section 4.4.

The second scheme, based on the variation of the momentum of inertia, is used less often in existing accelerometers. The simplified scheme of Figure 4.9 illustrates the idea. The resonating part is a beam, oscillating in bending, constituted by two parts connected at their ends (parallel beams resonator). One of these parts is attached to the substrate at its central point, while the other is attached to the proof mass. The external acceleration produces a displacement of the proof mass and hence a variation of the central cross-section of the resonator. The corresponding variation of the momentum of inertia induces a change in the natural frequency, which can be measured and which provides the measure of the external acceleration. An example of a biaxial accelerometer based on this principle is proposed by Tabata and Yamamoto (1999).

In the third kind of resonant accelerometer, the sensing device is a resonating beam, which undergoes frequency variations due to the geometrical stiffness variation. The corresponding scheme is represented in Figure 4.10, where a resonating beam, shown horizontally, is the sensing device. The operating principle is based on the dependence

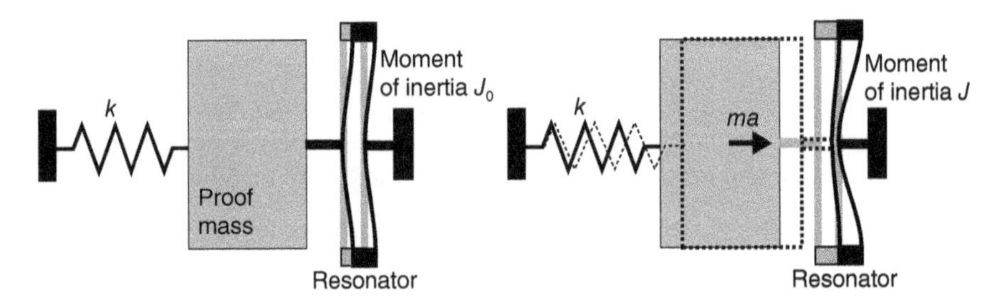

Figure 4.9 Resonant accelerometer with a change of momentum of inertia.

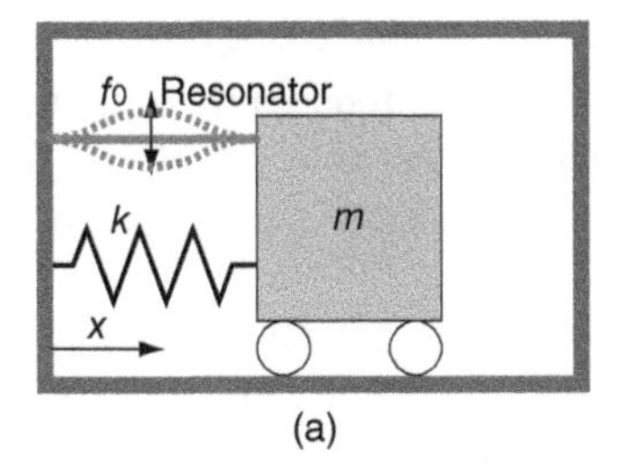

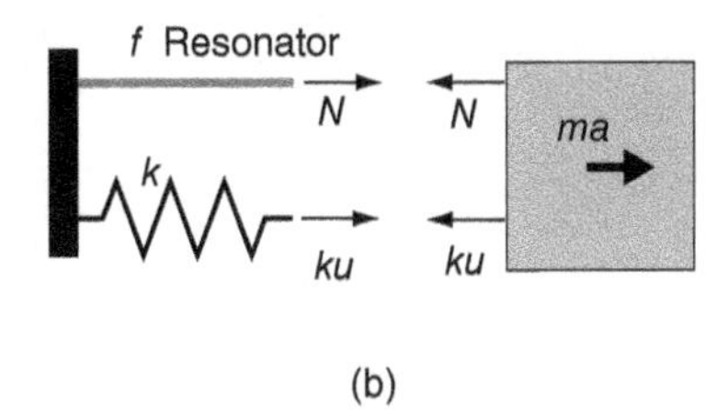

Figure 4.10 (a) Resonant accelerometer with a resonating beam; (b) forces acting when an external acceleration is applied.

of the resonant frequency on the axial force acting on the resonator. The external acceleration produces a force, $F = ma_{0x}$ on the inertial mass m. This force produces, in turn, an axial force N in the resonating beam. For a single-span beam, the frequency increases in the case of a tensile load and decreases in the case of a compressive load.

As explained in detail in Section 3.4.2, denoting by f_0 the fundamental frequency of the beam resonating without axial load, the resonant frequency f of the axially loaded beam with $N = P_0$ can be expressed as (cf. Equation 3.40)

$$f\left(P_0\right) = \frac{\omega\left(P_0\right)}{2\pi} = f_0\sqrt{1 + \frac{P_0 L^2}{\gamma EJ}}, \tag{4.19}$$

where L and J are the length and the momentum of inertia of the resonating beam and γ is the coefficient depending on the end constraints given in Table 3.2. The ratio between the axial force in the resonator and the inertia force, also called the amplification factor, depends on the peculiar geometry of the accelerometer and determines the sensitivity of the device. In the uniaxial accelerometer described by Aikele *et al.* (2001), the proof mass can rotate in-plane by means of a hinge-like connection and induces an axial force in the resonating beam amplified by a lever effect, as shown in Figure 4.11. Another proposal, endowed with high sensitivity and small dimensions, will be presented in detail in Section 4.4.

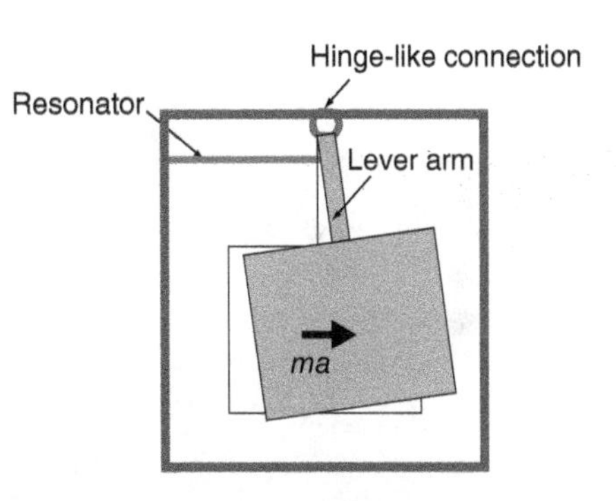

Figure 4.11 Resonant accelerometer proposed by Aikele *et al.* (2001). Reproduced with permission of Elsevier.

4.4 Examples

4.4.1 Three-Axis Capacitive Accelerometer

As an example of a MEMS accelerometer based on capacitive detection techniques, we discuss in detail, using the equations developed in Section 4.2, the three-axis accelerometer sensor proposed by Frangi, De Masi and Simoni (2011), which is able to detect components of linear acceleration acting along the three axes (x, y and z).

The micromechanical structure of polysilicon is obtained through surface micromachining and includes a single inertial mass, which has a main extension in a plane and is attached to the substrate by flexural and torsional springs that enable two in-plane translations and an out-of-plane rotation. Figure 4.12 shows a scheme of the accelerometer, while Figure 4.13 displays an SEM of the device. The proof mass of the microstructure comprises an external frame, attached to the substrate by slender beams, which allow for x- and y-translation and an internal mass attached to the frame

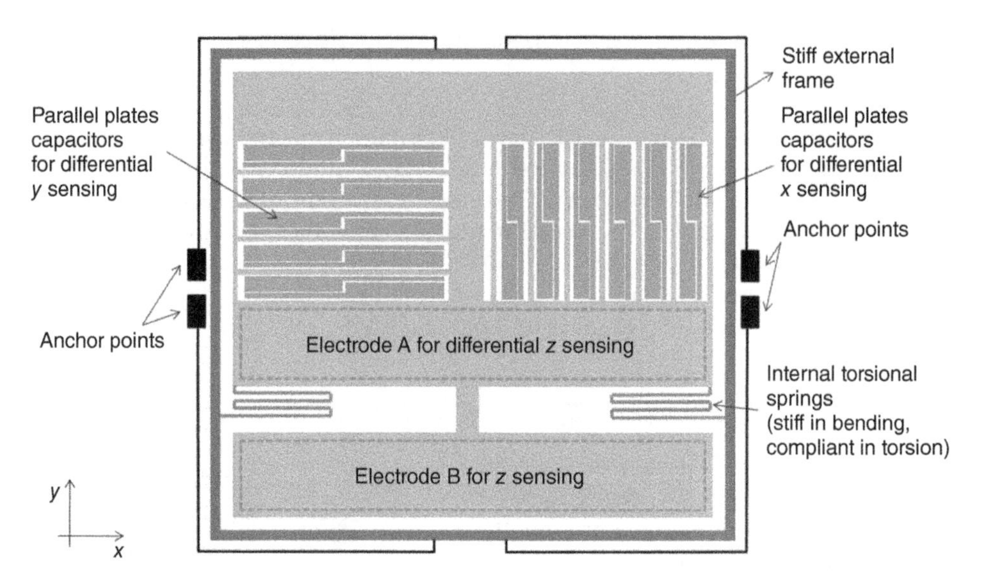

Figure 4.12 Three-axis capacitive accelerometer.

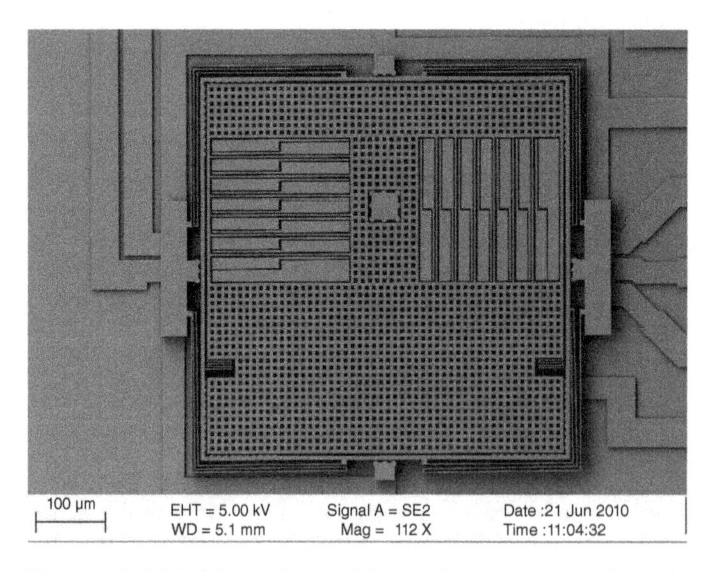

Figure 4.13 SEM of three-axis capacitive accelerometer. *Source:* Reproduced with permission of STMicroelectronics.

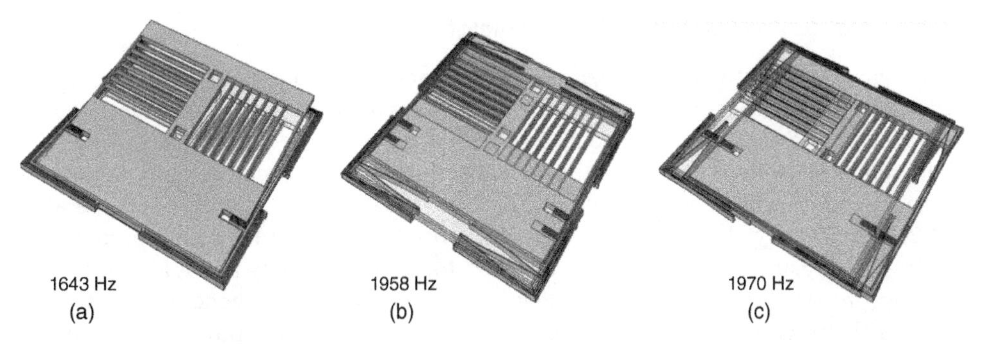

Figure 4.14 Eigenmodes and eigenfrequencies: (a) torsional mode; (b, c) translational modes.

by torsional springs, allowing for out-of-plane tilting. The springs are designed to decouple the three movements. There are parallel-plate capacitors for the differential x- and y-acceleration detection; other electrodes below the mass enable a differential reading scheme of the z-acceleration. Figure 4.14 shows the first three eigenmodes of the proof mass: an out-of-plane rotation and two in-plane translations. The presence of a single inertial mass, for detecting the external accelerations acting along the three axes, enables a considerable reduction of the dimensions of the micromechanical structure and of the corresponding sensor; the device, fabricated through the ThELMA (thick epitaxial layer for microactuators and accelerometers) process of STMicroelectronics, has a maximum size in the xy plane of about 600 × 600 μm, an out-of-plane thickness of 15 μm and a sensitivity of 5 fF/g (for a_{0z}) and 7 fF/g (for a_{0x} and a_{0y}). The use of a differential-detection scheme for the three detection axes allows for the achievement of an excellent linearity in the response to the external accelerations and a reduction of the thermal drift. Figure 4.15 shows a plot of the capacitive variation (expressed in femtofarads) along each of the detection axes, as a function of the corresponding input acceleration (expressed in g): a good linearity of operation is observed in the considered range of input values of acceleration ($\pm 10g$).

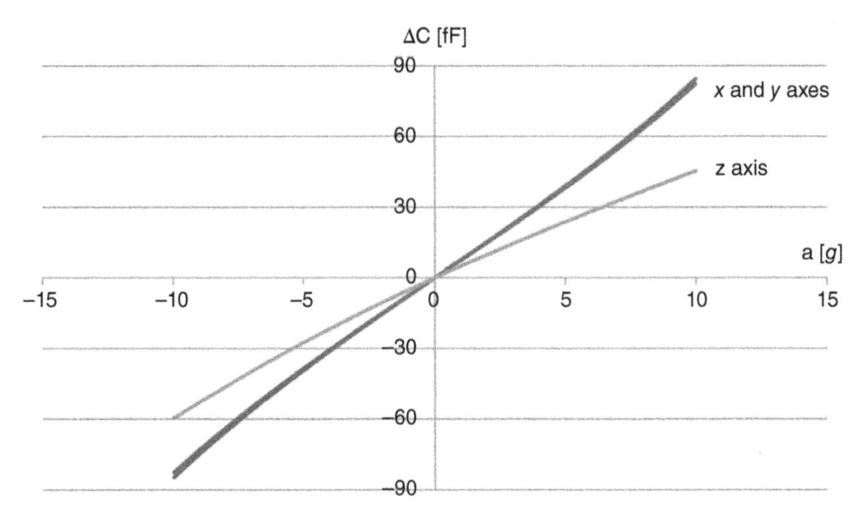

Figure 4.15 Capacitance variation for different external acceleration components.

4.4.2 Out-of-Plane Resonant Accelerometer

A recent example of a microaccelerometer with torsional resonators is given by Caspani *et al.* (2014a). This out-of-plane accelerometer makes use of the gap sensitive variation of electrostatic stiffness of the torsional resonators and allows for differential sensing. The differential sensitivity, defined as the shift in resonance frequencies corresponding to a 1g acceleration, obtained by this device, packaged at a pressure of 1 mbar and operated at 2.5 V is 14 Hz/g. The accelerometer is composed of a suspended planar proof mass, with out-of-plane thickness 22 μm, attached to the substrate by two folded torsional springs and two torsional resonators. Figure 4.16a shows a SEM of the accelerometer: the torsional resonators A and B consisting of a mass of in-plane dimensions $L \times 2b$ and two folded torsional springs attached to the proof mass are visible. Driving and sensing of the resonators is achieved by two parallel electrodes attached to the substrate. The torsional elements are kept in resonance according to their torsional natural mode. When in the rest position, the mass is at distance d_0 from both electrodes and the torsional resonator has the nominal frequency

$$f_0 = \frac{1}{2\pi}\sqrt{\frac{K_m - K^{elec}}{\rho J_{Tmass}}}, \quad K_m = \frac{2GJ_T}{l}, \quad K^{elec} = \frac{2\epsilon_0 L}{3d_0^3}\varphi_p^2(b^3 - c^3), \tag{4.20}$$

where J_T is the torsional momentum of inertia of the springs, ρJ_{Tmass} is the centroidal mass moment of inertia of the rigid mass, ρ is the mass density, l is the total length of one of the folded torsional springs, φ_p is the polarization voltage applied to the mass and $2c$ is the distance between the sensing and driving electrodes. As shown in Figure 4.16b, when an external out-of-plane acceleration a_z is applied, the proof mass rotates around the axis a–a and the gap between the resonators and the electrodes changes; this induces a variation in the electrostatic stiffness and, hence, of the frequency. Combining the

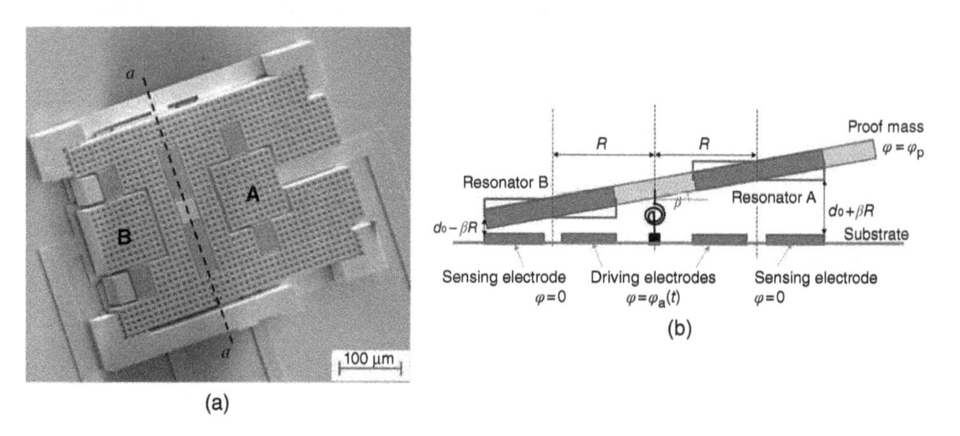

Figure 4.16 (a) SEM of z-axis accelerometer with two torsional resonators A and B; (b) side view of electrostatically actuated accelerometer, inclined due to external acceleration. *Source:* Caspani *et al.* (2014a), Figure 1. Reproduced with permission of Elsevier. (*See color plate section for the color representation of this figure.*)

readings of the two torsional resonators, the following expression is obtained (Caspani *et al.*, 2014a), allowing for the measure of the external acceleration a_z:

$$f_A - f_B \approx \frac{3}{2} f_0 \frac{K^{\text{elec}}}{K_{\text{m}} - K^{\text{elec}}} \frac{R R_G \bar{l}}{G \bar{J}_{\text{T}} d_0} m a_z, \tag{4.21}$$

where R_G is the distance between the a–a axis of rotation of a proof mass and the centre of gravity of the same mass, $\bar{l}$ is the total length of one folded torsional spring attaching the mass to the substrate and $\bar{J}_{\text{T}}$ its torsional momentum of inertia.

4.4.3 In-Plane Resonant Accelerometer

In this section, we describe the silicon uniaxial resonant accelerometer proposed by Comi *et al.* (2010), produced through the 15 μm thick surface micromachining ThELMA process of STMicroelectronics. The device is based on the frequency variation of axially loaded beams, as discussed in Section 4.3.2. The inertial sensing element is a square proof mass (400 μm × 400 μm), suspended by springs and linked to two beams, which constitute the resonating elements, as shown in Figure 4.17.

The proof mass is attached to the substrate by means of springs of length d, which restrain its movement to be a uniaxial translation. The resonators are very thin beams attached to the substrate at one end and to the springs at the other end, at distance d_1 from the anchor point. Electrostatic driving and sensing of each resonator is achieved by means of two parallel electrodes attached to the substrate.

When the reference frame is subject to an external acceleration a in the direction indicated by the arrow in Figure 4.17, the inertial mass m translates and one resonator (say resonator 1) is subject to a tension and the other (say resonator 2) to a compression of the same magnitude, $N_1 = -N_2$, as shown in Figure 4.17b. The axial forces produce a change in the resonance frequency of the two beams, which provides differential sensing of the external acceleration. The position of the resonating beams with respect to the

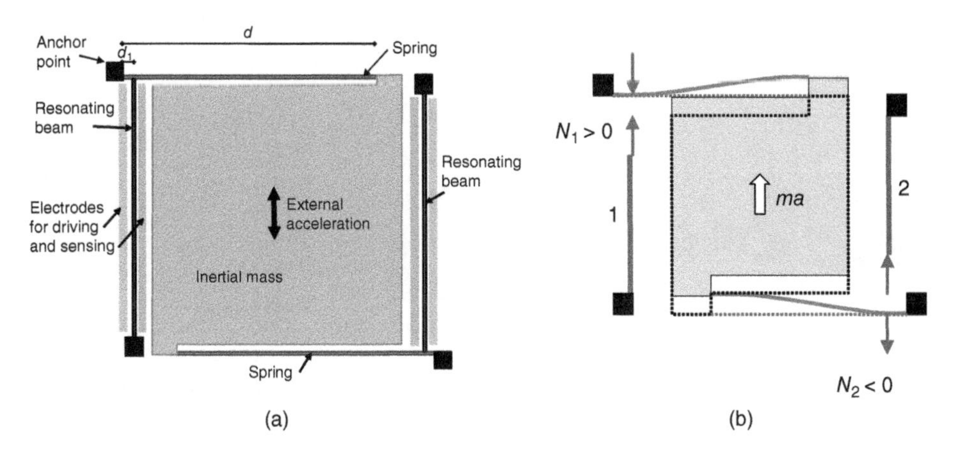

Figure 4.17 (a) Resonant accelerometer. (b) Effect of external acceleration *a*. *Source:* Comi *et al.* (2010), Figure 1. Reproduced with permission of IEEE.

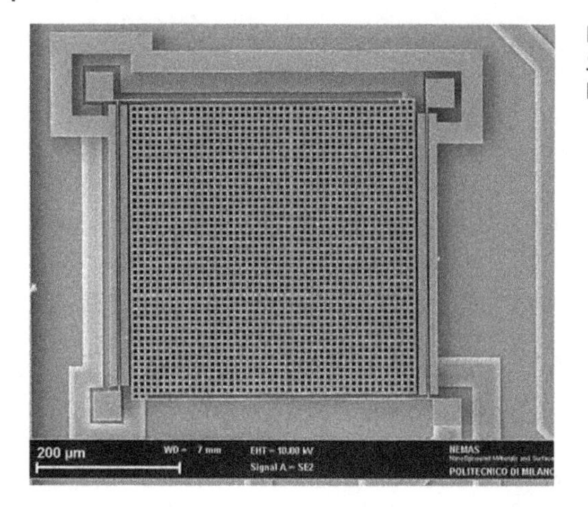

Figure 4.18 SEM of resonant accelerometer. *Source:* Comi *et al.* (2010), Figure 6. Reproduced with permission of IEEE.

anchor point of the spring was optimized to maximize the amplification factor for the axial force. The optimal position for the resonator is very close to the anchor point of the spring at about one sixtieth of its length. Figure 4.18 shows an SEM of the fabricated device. The very thin resonating beams are visible on the right and left of the proof mass in the vertical direction between the sensing and excitation electrodes. The square inertial mass has holes to allow for complete oxide removal beneath it. The experimental differential sensitivity obtained with this accelerometer is 455 Hz/g. Figure 4.19a, from Comi *et al.* (2010), shows the different power spectra of the circuit output signal for a single resonating beam when the accelerometer is subjected to external accelerations

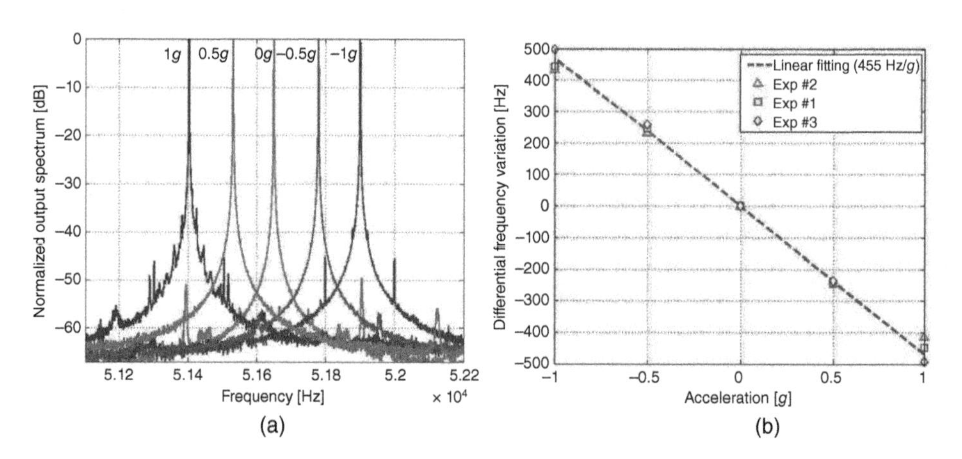

Figure 4.19 (a) Normalized output spectrum of the oscillating circuit for a single resonating beam evaluated for four different applied accelerations, 0g, ±0.5g, ±1g. (b) Variation of the peak frequency difference between the resonators $\Delta f - \Delta f_0$ as a function of the external acceleration in the range of ±1g for three different devices. Δf_0 corresponds to the peak frequency difference at 0g. *Source:* Comi *et al.* (2010), Figure 15. Reproduced with permission of IEEE.

in the range of $\pm 1g$. Good linearity is observed in the range of operation, as shown in Figure 4.19b.

4.5 Design Problems and Reliability Issues

As described, the main parameter of interest of a MEMS accelerometer is sensitivity. In general, a reduction in the device's dimensions induces a reduction in sensitivity, hence increasingly strict requirements of miniaturization render the mechanical design particularly challenging.

In capacitive accelerometers, to maintain high sensitivity with reduced dimensions, the mechanical stiffness should be reduced, but this entails the risk of pull-in instability, see Chapter 2, Section 2.5.1.

Resonant sensing can solve this problem. However, miniaturization of resonators can reduce their linear operating regime. The electrostatic stiffness becomes important when compared with a low mechanical stiffness and nonlinear electrostatic effects can no longer be neglected, see e.g. Caspani *et al.* (2014). Furthermore, mechanical nonlinearities can be relevant for very thin structures and particular provisions should be taken to enhance the linearity, as discussed in Tocchio *et al.* (2011).

Another important issue for inertial sensor performance is that of bias stability. This parameter gives an estimate of the achievable long-term noise performance, which is typically limited by offset drifts rather than white noise. In the case of frequency-modulated devices, such as the frequency-modulated accelerometers of Section 4.3, a critical issue is given by the temperature coefficient of frequency of the differential modes of the two torsional resonators.

References

Aikele, M., Bauer, K., Ficker, W. *et al.* (2001) Resonant accelerometer with self-test. *Sensors and Actuators*, **A 92**, 161–167.

Caspani, A., Comi, C., Corigliano, A. *et al.* (2014a) A differential resonant micro accelerometer for out-of-plane measurements. *Procedia Engineering*, **87**, 640–643.

Caspani, A., Comi, C., Corigliano, A. *et al.* (2014) Dynamic nonlinear behavior of torsional resonators in MEMS. *Journal of Micromechanics and Microengineering*, **24**, 095025.

Comi, C., Corigliano, A., Langfelder, G. *et al.* (2010) A resonant micro-accelerometer with high sensitivity operating in an oscillating circuit. *Journal of Microelectromechanical Systems*, **19**(5), 1140–1152.

Frangi, A., De Masi, B. and Simoni, B. (2011) Microelectromechanical three-axis capacitive accelerometer, US Patent US8863575 B2, filed Jun.15 2011 and issued Oct. 21 2014.

Kim, H., Seok, S., Kim, I. *et al.* (2005) *Inertial-Grade Out-of-Plane and In-Plane Differential Resonant Silicon Accelerometers (DRXLs)*, TRANSDUCERS '05. The 13th International Conference on Solid-State Sensors, Actuators and Microsystems, June 5–9, 2005, Seoul, South Korea, IEEE.

Seok, S., Kim, H. and Chun, K. (2004) *An Inertial-Grade Laterally-Driven MEMS Differential Resonant Accelerometer*, 2004 IEEE SENSORS, October 24–27, 2004, Vienna, Austria, IEEE.
Sung, S., Lee, J. and Kang, T. (2003) Development and test of MEMS accelerometer with self-sustatined [sic] oscillation loop. *Sensors and Actuators A: Physical*, **109**, 1–8.
Tabata, A. and Yamamoto, T. (1999) Two-axis detection resonant accelerometer based on rigidity change. *Sensors and Actuators A: Physical*, **75**, 53–59.
Tocchio, A., Comi, C., Langfelder, G. *et al.* (2011) Enhancing the linear range of MEMS resonators for sensing applications. *IEEE Sensor Journal*, **11**, 3202–10.

5

Coriolis-Based Gyroscopes

5.1 Introduction

Microelectromechanical gyroscopes, or MEMS gyros, are devices that measure angular velocity. The angular velocity or speed of rotation is expressed in degrees per second (°/s) or in revolutions per second. The three components of angular velocity are called *roll*, *pitch* and *yaw* and are defined as shown in Figure 5.1. Micromechanical gyroscopes, an alternative to classical rate gyroscopes, play an important role in inertial navigation and control systems of flight vehicles, and may have applications in automobile design, defence, consumer electronics and biomedical engineering. The benefits of micromechanical gyroscopes over classical gyroscope are robustness, low power consumption, potential for miniaturization and low cost.

Vibrating structure gyroscopes, or Coriolis vibratory gyroscopes, constitute a wide group of MEMS gyroscopes. The underlying physical principle is that a vibrating object tends to continue to vibrate in the same plane as its support rotates. In gyroscopes, generally, an inertial mass is oscillating at the natural resonance frequency and the effect of the Coriolis force that originates on one or more detection elements in the presence of an external angular velocity is measured. A variety of structures can be used to implement MEMS Coriolis vibratory gyroscopes. Broadly, these are grouped into lumped proof mass–spring structures, which involve one or more rigid masses translating or rotating, and ring, disc or shell structures, which involve coupled flexural deformation modes of the whole structure.

In this chapter, we will first illustrate the basic working principle of a yaw gyroscope and then we will discuss the mechanical features of several different Coriolis vibratory MEMS gyroscopes. A variety of other solutions are proposed in the literature; the interested reader can refer, for example, to Acar and Shkel (2009).

5.2 Basic Working Principle

To illustrate the working principle of a Coriolis-based MEMS gyroscope, we recall first the governing dynamic equations presented in Chapter 2. Consider the system shown in Figure 5.2: x'–y'–z' is an inertial system and x–y–z is the noninertial system attached to the MEMS, which is moving with respect to the inertial frame.

Mechanics of Microsystems, First Edition. Alberto Corigliano, Raffaele Ardito, Claudia Comi, Attilio Frangi, Aldo Ghisi, and Stefano Mariani.

Companion website: www.wiley.com/go/corigliano/mechanics

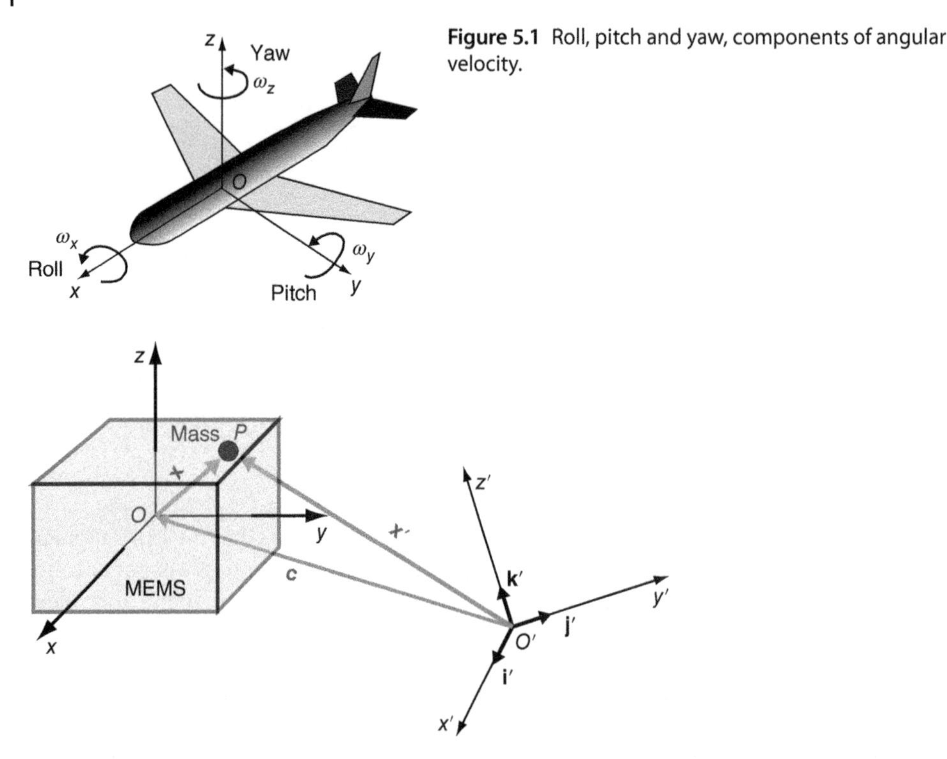

Figure 5.1 Roll, pitch and yaw, components of angular velocity.

Figure 5.2 Reference frame *x–y–z* attached to MEMS containing a proof mass and inertial reference frame *x′–y′–z′*.

The purpose of a MEMS gyroscope is to measure the angular velocity of the noninertial reference system. Analogously to microaccelerometers, inside the MEMS, at point P, is located a proof mass, which can move with respect to the MEMS box. The vibration of the mass m, considered as a material point, can be studied in the noninertial reference frame system attached to the MEMS box.

We consider here a yaw surface micromachined gyroscope, as depicted in Figure 5.3, which schematically shows the mechanical structure composed of a proof mass attached to the substrate by four springs, which can vibrate in the x- and y-directions. The governing dynamic equation (see Chapter 2) reads:

$$m\left\{\mathbf{a}_{\mathrm{r}}+\left[\mathbf{a}_0+\frac{\mathrm{d}\boldsymbol{\omega}}{\mathrm{d}t}\wedge(P-O)+\boldsymbol{\omega}\wedge(\boldsymbol{\omega}\wedge(P-O))\right]+2\boldsymbol{\omega}\wedge\mathbf{v}_{\mathrm{r}}\right\}=\mathbf{F}. \tag{5.1}$$

If, at rest position, the mass is located at the origin O of the MEMS reference frame, the position vector at time t coincides with the displacement vector, i.e.:

$$(P-O)^{\mathrm{T}}=\left\{u_x\ u_y\ 0\right\}. \tag{5.2}$$

Projecting Equation 5.1 in the x- and y-directions, and assuming that the frame has a constant angular velocity ω_z around the z-axis only, one obtains

$$ma_{\mathrm{r}x}=F_x-ma_{0x}+mu_x\omega_z^2+2m\omega_z v_{\mathrm{r}y}, \tag{5.3}$$

$$ma_{\mathrm{r}y}=F_y-ma_{0y}+mu_y\omega_z^2-2m\omega_z v_{\mathrm{r}x}. \tag{5.4}$$

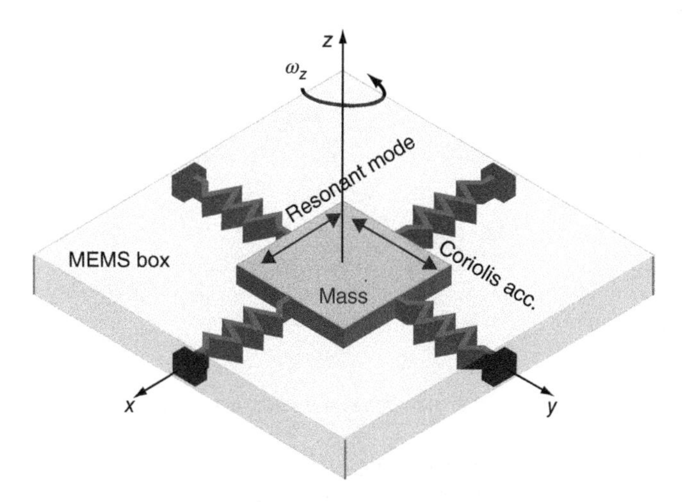

Figure 5.3 Coriolis vibratory yaw rate gyroscope.

As in the case of accelerometers (see Chapter 4), the forces F_x and F_y have three contributions: an elastic term due to the springs, which is proportional to the displacements u_x and u_y, respectively; a damping term, assumed here as viscous, which is linearly dependent on the velocities v_{rx} and v_{ry}, and an electrostatic force

$$F_x = -k_x u_x - c_x v_{rx} + F_x^{\text{elec}}, \tag{5.5}$$

$$F_y = -k_y u_y - c_y v_{ry} + F_y^{\text{elec}}. \tag{5.6}$$

The basic idea that is exploited in Coriolis vibratory gyroscopes is to control the motion in one direction (the *driving direction*) and to measure ω_z from the movement in the orthogonal direction (the *sensing direction*). Let us assume that the motion in direction x is driven at resonance by an electrostatic force. The mass then has a relative velocity v_{rx}. The motion in the y-direction is governed by Equation 5.4. Denoting by $\overline{\omega}_y$ the angular eigenfrequency in the y-direction, one obtains

$$a_{ry} - \frac{c_y}{m} v_{ry} + (\overline{\omega}_y^2 - \omega_z^2) u_y = \frac{F_y^{\text{elec}}}{m} - a_{0y} - 2\omega_z v_{rx}. \tag{5.7}$$

The last term in Equation 5.7 represents the Coriolis' acceleration that is produced along the sense axis (y-axis), and that is perpendicular to both drive and rotation axes. The apparent force related with this acceleration is the Coriolis' force,

$$F_c = 2m\omega_z v_{rx}. \tag{5.8}$$

Usually, it is assumed that $\omega_z^2 \ll \overline{\omega}_y^2$ and that the external acceleration a_{0y} is null, so that from Equation 5.7 one obtains the explicit dependence of the response of the proof mass on the external angular velocity:

$$a_{ry} - \frac{c_y}{m} v_{ry} + \overline{\omega}_y^2 u_y = \frac{F_y^{\text{elec}}}{m} - 2\omega_z v_{rx}. \tag{5.9}$$

Thus, one can determine the rotation rate through the detection of the vibration in the sensing direction.

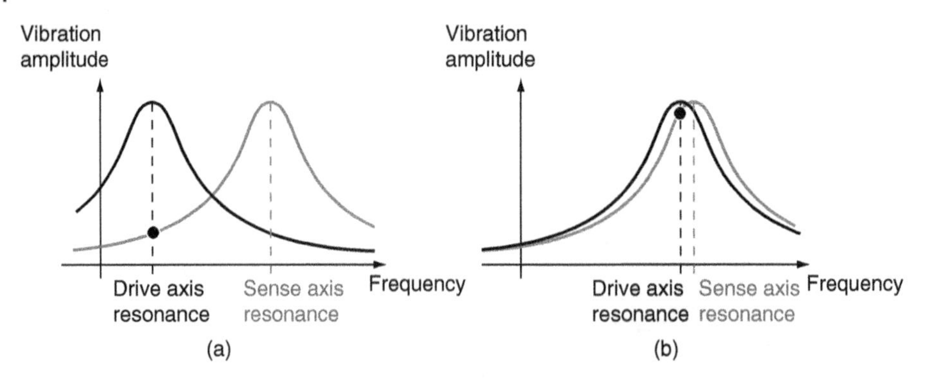

Figure 5.4 Driving and sensing mode resonance frequencies: (a) separated; (b) matched.

To take advantage of the large mechanical gain at resonance, the driven axis is usually forced at its resonant frequency, and the design is conceived in such a way that this resonant frequency matches that of the sensing axis. If the gyro is not *mode-matched,* as illustrated in the left panel of Figure 5.4, the displacement along the sensing axis will be small. If, however, the device is mode-matched, as illustrated on the right panel of Figure 5.4, the displacement along the sensing axis is amplified by the mechanical gain at resonance.

Ideally, the two axes of the device would be coupled only through the Coriolis force; however, manufacturing defects introduce coupling from one axis to another through both stiffness and damping. This effect, referred to as *quadrature error,* limits the performances of the gyroscope and must be carefully studied and partially reduced.

MEMS gyros are generally driven electrostatically, through comb drives or parallel-plate electrodes. Often, the detection is also made by means of the capacitive technique, see, e.g., Willig and Mörbe (2003), Alper and Akin (2004), Prandi *et al.* (2011), Trusov *et al.* (2011) and Nitzan *et al.* (2015), while there exist a few examples of microgyroscopes with resonant detection, see, e.g., Seshia, Howe and Montague (2002), Li *et al.* (2010) and Zega *et al.* (2014).

5.2.1 Sensitivity of Coriolis Vibratory Gyroscopes

There are two different ways to measure the effect induced by the Coriolis force in a Coriolis vibratory gyroscope: the first is called the *open-loop* method and consists of direct detection of the displacement in the sensing direction as a measure of the angular rate, while the second is called the *closed-loop* method and involves dynamically driving to zero the motion of the proof mass in the sensing direction and computing the angular rate through the force needed to nullify the motion. In open-loop amplitude-modulated gyroscopes, the external angular rate is obtained by measuring the amplitude of the sense signal, often through capacitance variation measurements.

The sensitivity of the gyroscope can, hence, be defined as the displacement (or as the corresponding capacitance variation) induced by a unit angular rate, i.e., for a yaw gyroscope:

$$\text{sensitivity} = \frac{U_y}{\omega_z}. \tag{5.10}$$

Usually the proof mass is driven to resonance by an external electrostatic force $F_x^{\text{elec}} = F\sin(\omega t)$, with $\omega \approx \overline{\omega}_x = \sqrt{k_x/m}$. Assuming that the external angular rate is much less than the operating frequency ($\omega_z \ll \overline{\omega}_x$) and noting that the sense axis response is much smaller in amplitude than the response of the drive mode, the drive oscillation turns out to be governed by

$$ma_{rx} + k_x u_x + c_x v_{rx} = F \sin(\omega t). \tag{5.11}$$

As discussed in Chapter 2, the solution of this equation is $u_x = U_x \sin(\omega t + \varphi_x)$, with

$$U_x = \frac{F}{k_x \sqrt{\left[1 - \left(\frac{\omega}{\overline{\omega}_x}\right)^2\right]^2 + \left[\frac{\omega}{Q_x \overline{\omega}_x}\right]^2}}, \tag{5.12}$$

where $Q_x = m\overline{\omega}_x / c_x$ is the quality factor. If the gyroscope works exactly at the natural frequency of the driving mode, the amplitude of the drive motion becomes

$$U_x = Q_x \frac{F}{k_x} \tag{5.13}$$

and the relative velocity reads

$$v_{rx} = \overline{\omega}_x U_x \cos(\overline{\omega}_x t + \varphi_x). \tag{5.14}$$

Substituting this relation into Equation 5.9 with zero electrostatic force, one obtains the equation of motion in the sensing direction,

$$ma_{ry} + c_y v_{ry} + k_y u_y = -2m\omega_z \overline{\omega}_x U_x \cos(\overline{\omega}_x t + \varphi_x). \tag{5.15}$$

The solution is again an oscillating function with amplitude U_y

$$U_y = \omega_z \frac{2mU_x \overline{\omega}_x}{k_y \sqrt{\left[1 - \left(\frac{\overline{\omega}_x}{\overline{\omega}_y}\right)^2\right]^2 + \left[\frac{\overline{\omega}_x}{Q_y \overline{\omega}_y}\right]^2}}, \tag{5.16}$$

which provides an explicit expression for the sensitivity. The factor that multiplies the angular velocity is also called the *scale factor* of the device. The sensing signal, and hence the sensitivity, is maximized if the device is mode-matched, i.e. if $\overline{\omega}_x = \overline{\omega}_y$. Furthermore, the sensitivity increases with the driving amplitude and with the quality factor of the device.

5.3 Lumped-Mass Gyroscopes

This class of MEMS gyroscope is characterized by the presence of proof masses, which can be considered as rigid bodies, attached to the substrate by suitable systems of springs. Their fundamental resonant mode is usually an in-plane translation, while the sensing modes can be translations or out-of-plane rotations. In the following, we will describe some representative arrangements.

5.3.1 Symmetric and Decoupled Gyroscope

As already pointed out, to increase the sensitivity of MEMS gyroscopes, it is essential to use matched resonance frequencies for the driving and sensing mode vibrations. This leads to the design of symmetric suspensions for the driving and sensing modes, which is also important in keeping the temperature-dependent drift small. The disadvantage of symmetric suspensions is that they usually yield undesired mechanical coupling between driving and sensing modes. Several proposals have been developed

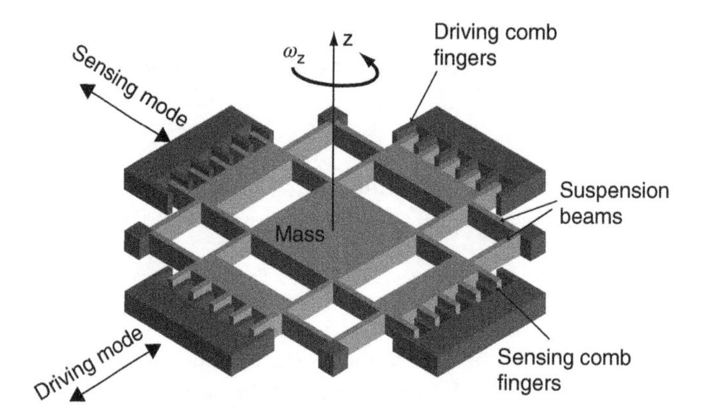

Figure 5.5 Symmetric, decoupled yaw gyroscope, as proposed by Alper and Akin (2004). Reproduced with the permission of Elsevier.

to achieve a good decoupling of modes. Figure 5.5 shows a possible arrangement of a symmetric and decoupled yaw gyroscope structure as proposed by Alper and Akin (2004). The drive mode (x-axis) of the gyroscope is electrostatically excited to resonance by comb fingers. When an angular yaw rate input is applied about the z-axis, the sense mode also starts vibrating, owing to the Coriolis' coupling from the drive mode, and this vibration is sensed by the comb fingers in the y-direction. The oscillation modes are decoupled by means of a system of suspension beams connecting the mass with the moving comb fingers.

5.3.2 Tuning-Fork Gyroscope

A single-mass gyroscope, such as the one shown in Figure 5.3, has the disadvantage that the effect on the sensing motion of an external acceleration a_{0y} cannot be distinguished from the effect, through the Coriolis' acceleration, of the angular rate ω_z (see Equation 5.7). To cancel the effect of the external acceleration, dual-mass (Willig and Mörbe, 2003) and quad-mass gyroscopes (Trusov *et al.*, 2011) have been proposed.

Figure 5.6 shows a dual-mass tuning-fork yaw gyroscope. The proof masses translate in opposite directions; therefore, when the angular rate ω_z is applied, the Coriolis' forces have opposite sign, while the inertia forces due to a linear acceleration have the

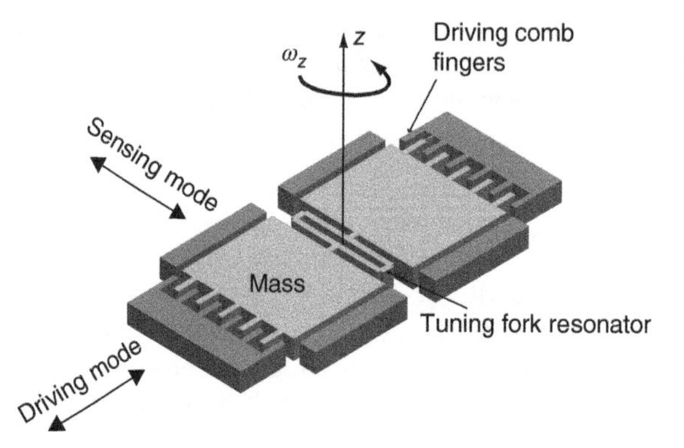

Figure 5.6 Dual-mass tuning-fork yaw gyroscope.

same sign; a differential sensing thus enables the effect of external acceleration to be cancelled. Trusov *et al.* (2011) use four proof masses translating in opposite directions. The quadruple-mass gyroscope preserves and expands the structural advantages of the dual-mass design. The quadruple-mass architecture provides true mechanical rejection of external vibrations and mechanical shocks along both the driving and sensing axes. The structural symmetry of the device suggests improved robustness to fabrication imperfections and temperature-induced frequency drifts.

5.3.3 Three-Axis Gyroscope

A three-axis angular rate gyroscope can be obtained using a single structure resonating at the primary mode only and using secondary modes for sensing rotations. An example developed by STMicroelectronics and described by Prandi *et al.* (2011), is shown in Figure 5.7. The proof mass consists of four trapezoidal portions connected by springs to the outer corners and by additional springs to a central cross-beam. The vibration is driven using comb-finger actuators on two opposite sides. Each trapezoid moves in and out from the centre in the primary mode, but the trapezoids all move in a coordinate manner, so the whole unit alternately expands and contracts, giving it the name *beating heart* (see Figure 5.7b). Sensing is done by measuring capacitance at key places as a result of the secondary vibration modes. There is a secondary mode, whereby each of the opposing trapezoids vibrates out-of-plane in opposite directions. This is used to detect roll and pitch (see Figure 5.7a). Another mode, characterized by the translation of two opposite masses in the y-direction, perpendicular to their drive movement, is detected by parallel plates, sensing yaw rotation. The symmetry and the differential approach adopted in the structure design assure a high level of rejection of linear acceleration that may act on the sensor due to shocks or vibrations.

5.3.4 Gyroscopes with Resonant Sensing

The microgyroscopes proposed by Seshia, Howe and Montague (2002) and Li *et al.* (2010) are endowed with resonant detection: the Coriolis force generates axial stresses in resonator elements, which modify their resonance frequency accordingly, enabling

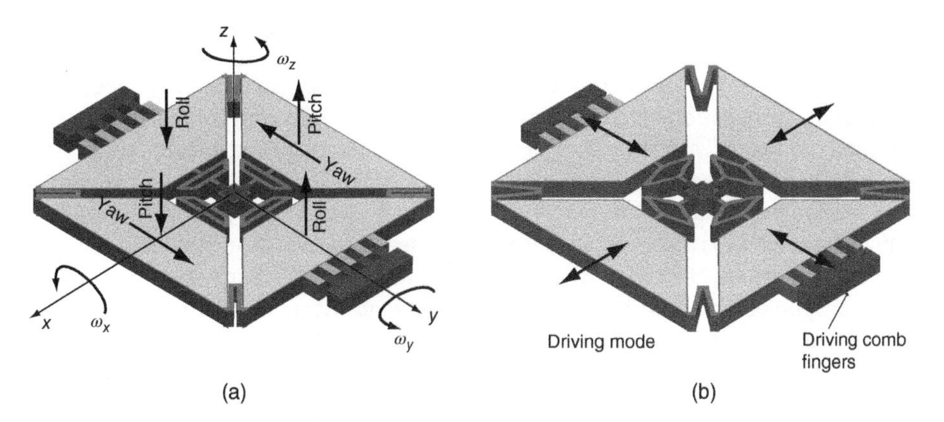

Figure 5.7 Three-axis heart-beating gyroscope: (a) scheme of sensing modes for roll, pitch and yaw as a consequence of an *expansion*; (b) drive mode.

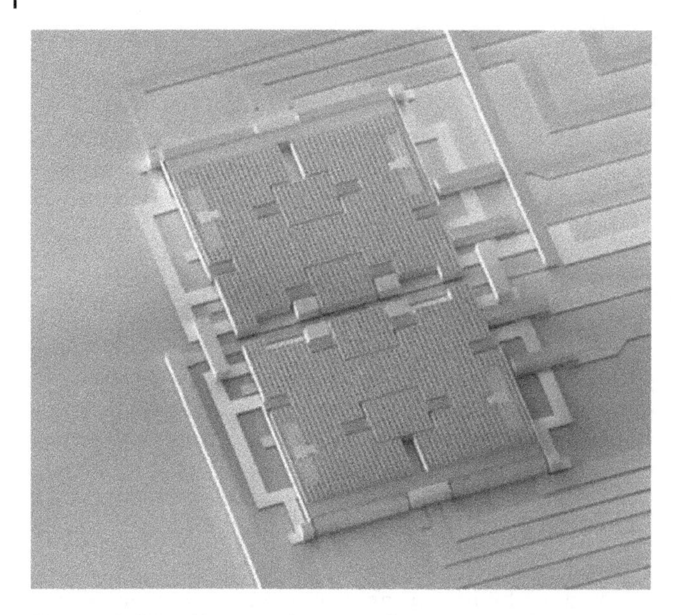

Figure 5.8 SEM of integrated structure for resonant microgyroscope and accelerometer. *Source*: Zega *et al.* (2014), Figure 5. Published under the Open Journal Systems 2.4.8.1, freely distributed by the Public Knowledge Project under the GNU General Public License. http://www.gruppofrattura.it/pdf/rivista/numero29/numero_29_art_29.pdf. CC-BY 4.0.

detection of the angular velocity. Comi, Corigliano and Baldasarre (2004) and Zega *et al.* (2014) proposed an integrated dual-mass structure for detection of acceleration and angular (yaw and pitch) velocity with resonant sensing. An SEM of the device, fabricated in polysilicon through the ThELMA micromachining process of STMicroelectronics, is shown in Figure 5.8. Figure 5.9 shows, schematically, a planar view of the structure, which is composed of two proof masses coupled with four slender-beam bending resonators, labelled I, II, III and IV, and four torsional resonators, labelled 1, 2, 3 and 4. The bending resonators are kept in resonance by electrostatic actuation, through the driving plates shown in Figure 5.9, while the torsional resonators are kept in resonance through the electrodes located on the substrate beneath them. The driving mode is a translation of the two masses in the x-direction.

As shown in Figure 5.10, the external yaw angular velocity ω_z and the linear acceleration a_{0y} induce a translation of the masses in the y-direction, through the Coriolis forces and the inertia forces, respectively. The proof masses translate out-of-phase for yaw (Figure 5.10a) and in-phase for external acceleration (Figure 5.10b). These movements cause an axial force in the bending resonators that changes their frequency, as explained in detail in Chapter 4 for resonant accelerometers (see Section 4.3). In particular, two of the resonators are subject to tension and therefore their frequency increases, while the other two are subject to compression and their frequency decreases. By properly combining the frequency variations of the four resonators one can sense the yaw and acceleration simultaneously. The device provides a differential measure of both quantities.

The pitch angular velocity ω_y and the linear out-of-plane acceleration a_{0z} cause a tilt of the proof masses around the a–a axis (Figure 5.9). As shown in Figure 5.11, this induces

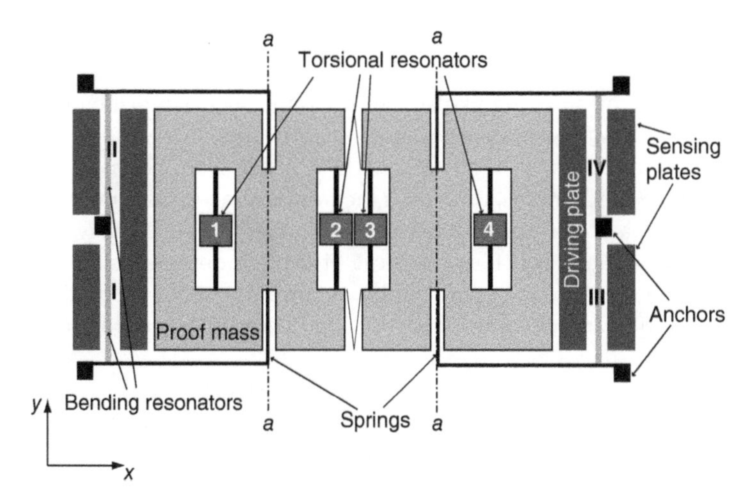

Figure 5.9 Plan view of the structure of Figure 5.8, for detection of acceleration and angular velocity.

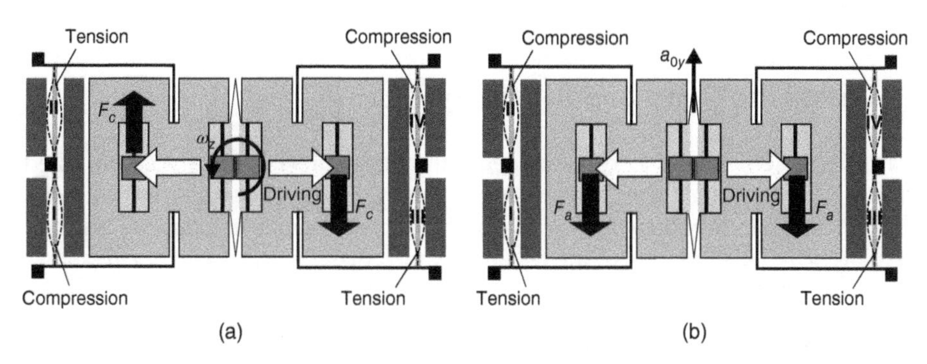

Figure 5.10 (a) Detection of yaw angular velocity ω_z; (b) detection of linear acceleration a_{0y}.

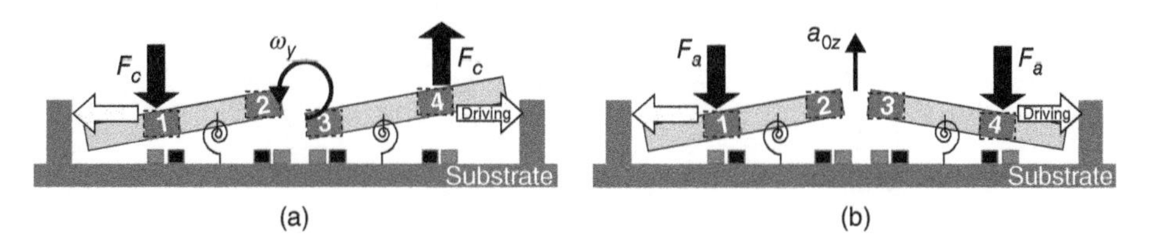

Figure 5.11 (a) Detection of pitch angular velocity ω_y; (b) detection of linear acceleration a_{0z}.

a variation of the gap between the torsional resonators 1, 2, 3 and 4 and the electrodes located on the substrate beneath them. The related change in electrostatic stiffness provokes a change in the frequency of the torsional resonators which can be sensed, as explained in detail in Section 4.4.2 for the out-of-plane resonant accelerometer. By combining the frequency variations of the four resonators, one obtains a simultaneous differential measure of both the pitch angular velocity and the out-of-plane linear acceleration.

5.4 Disc and Ring Gyroscopes

Like lumped-mass gyroscopes, disc or ring gyroscopes operate using two orthogonal vibration modes. Rather than a lumped mass translating linearly, these gyroscopes rely on two flexural radial vibration modes of the structure. These mode shapes have deformation proportional to $\cos(n\vartheta)$ and $\sin(n\vartheta)$, where n is the mode number. Generally the 2ϑ or 3ϑ modes, which are separated from each other by 45° and 30°, respectively, are used for gyroscope operation (Figure 5.12). The earliest MEMS gyroscopes using the 2ϑ modes were ring gyroscopes. Since ring gyros employ symmetric oscillation modes, they exhibit reduced acceleration and vibration sensitivity, and are more robust to shock (Yoon, Lee and Najafi, 2011).

Disc resonator gyroscopes take advantage of the symmetric structure and central anchor of a ring, but the mass is increased by adding concentric internal rings, as shown in Figure 5.13, which is taken from Nitzan *et al.* (2015). The device is realized by a single-crystal silicon slotted disc supported by a central cylindrical anchor and surrounded by capacitive electrodes used to force and sense vibration; the disk has orthogonal elliptical modes, separated by 45° (Figure 5.13b).

5.5 Design Problems and Reliability Issues

We have seen that high sensitivity is obtained with a high-Q device in the mode-matched condition. This results, however, in a narrow bandwidth of the device; furthermore, small fluctuations in the frequency split between the driving and sensing axes result in large changes in the amplitude of the response, which decreases the *scale-factor* stability. Many commercial MEMS gyroscopes operate in the non-mode-matched condition to obtain good stability at the price of lower sensitivity. Alternatively, the devices can be electrostatically tuned with a dedicated set of electrodes to follow the mode-matched conditions in operation. In an ideal gyroscope, the driving and sensing axes would be coupled only through the Coriolis force; however, manufacturing defects introduce coupling from one axis to another through both stiffness and damping. This effect introduces additional terms into the forces in Equations 5.5 and 5.6 that become

$$F_x = -k_x u_x - k_{xy} u_y - c_x v_{rx} - c_{xy} v_{ry} + F_x^{\text{elec}}, \tag{5.17}$$

$$F_y = -k_{yx} u_x - k_y u_y - c_{yx} v_{rx} - c_y v_{ry} + F_y^{\text{elec}}, \tag{5.18}$$

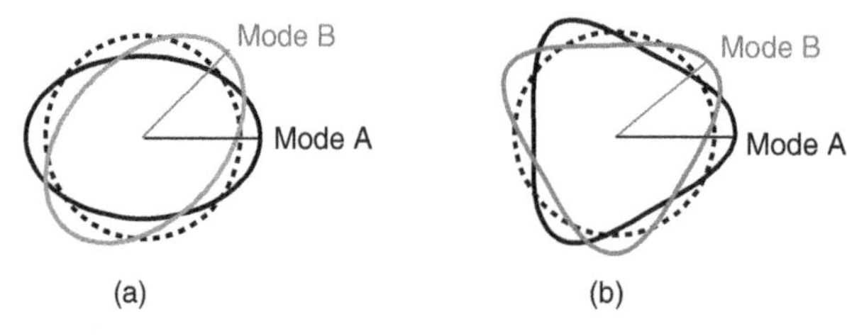

Figure 5.12 Ring resonator gyroscope: (a) shape of two orthogonal modes, with 45° separation; (b) shape of two orthogonal modes, with 30° separation.

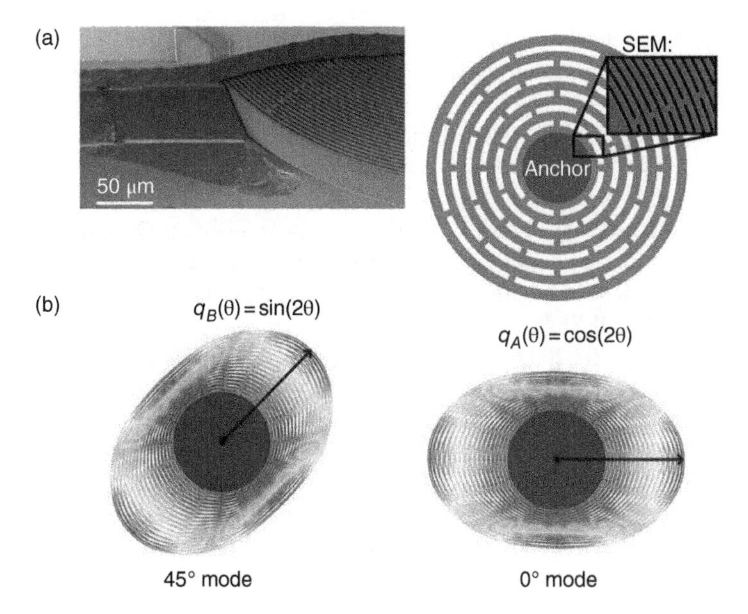

Figure 5.13 Disc resonator gyroscope: (a) SEM of disc resonator gyroscope and drawing of disc resonator gyroscope shape, with inset SEM of rings; (b) orthogonal elliptical mode shapes, with contours of displacement. *Source*: Nitzan *et al*. (2015), Figure 1. Licensed under a Creative Commons Attribution 4.0 International License http://creativecommons.org/licenses/by/4.0/.

where $k_{xy} = k_{yx}$, and c_{xy} and c_{yx} are the stiffness and damping coupling, respectively, between the two modes. The stiffness coupling due to k_{yx} in Equation 5.18 is proportional to the displacement u_x of the driven axis, and is thus shifted through 90° relative to the Coriolis force, which is proportional to the velocity of the driven axis v_{rx}. Accordingly, the stiffness coupling is referred to as quadrature. Quadrature errors can be cancelled out with a dedicated set of tuning electrodes, which, however, increase the size and energy consumption of the device. Because the damping force is proportional to the velocity, like the Coriolis force, the damping coupling introduces an offset term in the output that cannot be distinguished from the effect of an external angular velocity. If this offset is constant, it can be subtracted, but variation of the offset during operation introduces drift in the output signal. Therefore, mechanical designs try to minimize both stiffness and damping couplings.

References

Acar, C. and Shkel, A. (2009) *MEMS Vibratory Gyroscopes*, Springer.

Alper, S.E. and Akin, T. (2004) Symmetrical and decoupled nickel microgyroscope on insulating substrate. *Sensors and Actuators A*, **115**, 336–350.

Comi, C., Corigliano, A. and Baldasarre, L. (2004) Integrated resonant structure for the detection of acceleration and angular velocity, and related MEMS sensor device. US Patent US7104129 B2, filed Feb. 2, 2004 and issued Sept. 12, 2006.

Li, J., Fang, J., Dong, H. and Tao, Y. (2010) Structure design and fabrication of a novel dual-mass resonant output micromechanical gyroscope. *Microsystems Technology*, **16**, 543–552.

Nitzan, S., Zega, V., Li, M. *et al.* (2015) Self-induced parametric amplification arising from nonlinear elastic coupling in a micro-mechanical resonating disk gyroscope. *Scientific Reports*, **5**, 9036.

Prandi, L., Caminada, C., Coronato, L. *et al.* (2011) *A Low-Power 3-Axis Digital-Output MEMS Gyroscope with Single Drive and Multiplexed Angular Rate Readout*, IEEE International Solid-State Circuits Conference Digest of Technical Papers, February 20–24, 2011, San Francisco, CA, IEEE.

Seshia, A., Howe, R. and Montague, S. (2002) *An Integrated Microelectromechanical Resonant Output Gyroscope*, Fifteenth IEEE International Conference on Micro Electro Mechanical Systems, January 24, 2002, Las Vegas, NV, IEEE.

Trusov, A., Prikhodko, I., Zotov, S. and Shkel, A. (2011) Low-dissipation silicon tuning fork gyroscopes for rate and whole angle measurements. *IEEE Sensors Journal*, **11**, 2763–2770.

Willig, R. and Mörbe, M. (2003) New generation of inertial sensor cluster for ESP and future vehicle stabilizing systems in automotive applications. *SAE Technical Paper* 2003-01-0199.

Yoon, S., Lee, S. and Najafi, K. (2011) Vibration sensitivity analysis of MEMS vibratory ring gyroscopes. *Sensors and Actuators A: Physical*, **171**, 163–177.

Zega, V., Comi, C., Corigliano, A. and Valzasina, C. (2014) Integrated structure for a resonant micro-gyroscope and accelerometer. *Frattura ed Integritá Strutturale*, **29**, 334–342.

6

Resonators

6.1 Introduction

MEMS resonators are small electromechanical structures that vibrate at high frequencies. They are used in many fields for timing references, signal filtering in radio and mobile communication systems, mass sensing, biological sensing and motion sensing. MEMS resonators with frequencies less than 100 MHz are called low-frequency oscillators and are used in low-precision consumer electronics. High-frequency oscillators (between 100 MHz and 900 MHz) are still being developed in a research environment, while for frequencies greater than 900 MHz a number of filters, including film bulk acoustic resonators, have many applications.

A classification of MEMS resonators can be based on the mode of vibration. One can distinguish four basic categories: (i) flexural mode resonators, (ii) contour mode or Lamb-wave resonators, (iii) thickness-extensional mode resonators and (iv) shear mode resonators; see Figure 6.1.

Flexural resonators usually consist of slender beams with different boundary conditions (e.g. clamped–clamped, as in Figure 6.1(i)). The mode shape can be described as a combination of sinusoidal and hyperbolic functions (see Section 3.4.1). The frequency can easily be changed by selecting the proper length and thickness, but the frequency that can be attained is quite low, less than 10 MHz.

Contour mode or Lamb-wave resonators have their frequency set primarily by lithographic processes (i.e. by the lateral, in-plane, dimensions of the structure). These devices have high mechanical stiffness and are suitable for use at high frequencies. Figure 6.1(ii) shows, schematically, the resonator developed by Piazza *et al.* (2005), which is composed of an aluminium nitride film sandwiched between two metal electrodes. When an AC signal is applied across the thickness of the device, a contour–extensional mode of vibration is excited through the equivalent d_{31} piezoelectric coefficient of aluminium nitride (mode 3-1, see Figure 2.33).

Thickness-extensional mode resonators, or thin-film bulk acoustic resonators oscillate out of the plane by converting electrical energy into mechanical energy and vice versa. A thin-film bulk acoustic resonator consists of a piezoelectric thin film sandwiched

Mechanics of Microsystems, First Edition. Alberto Corigliano, Raffaele Ardito, Claudia Comi, Attilio Frangi, Aldo Ghisi, and Stefano Mariani.

Companion website: www.wiley.com/go/corigliano/mechanics

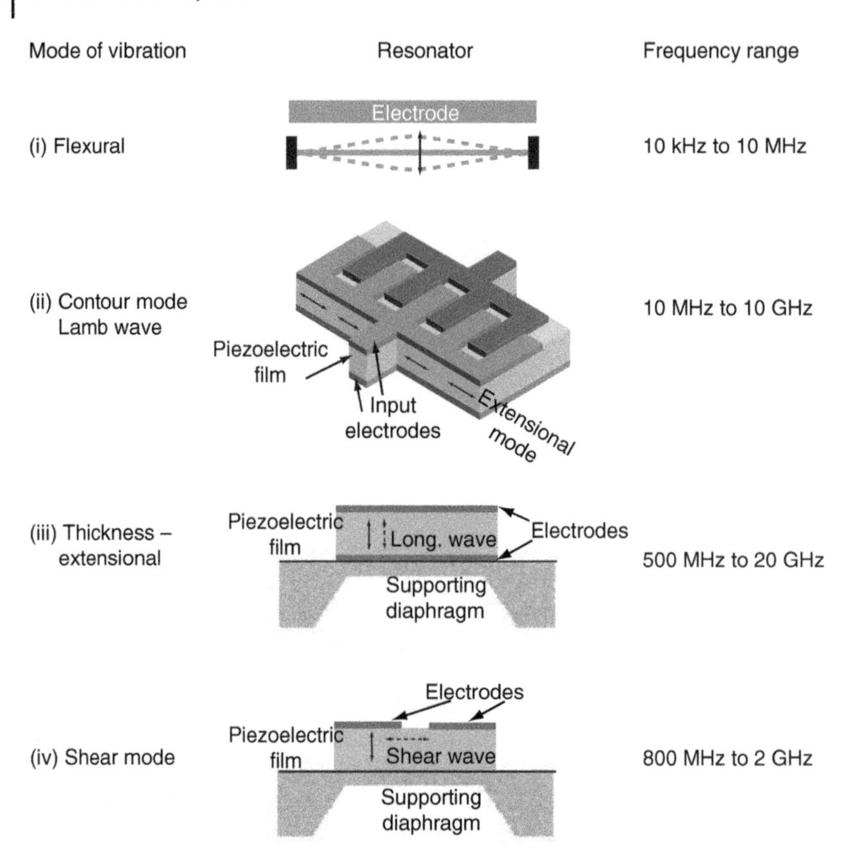

Figure 6.1 Classification of MEMS resonators based on mode of vibration: (i) flexural mode resonators, (ii) contour mode or Lamb-wave resonators, (iii) thickness-extensional mode resonators and (iv) shear mode resonators.

between two metal layers, as shown in Figure 6.1(iii). The fundamental resonant frequency is inversely proportional to the thickness of the piezoelectric material used.

In shear mode resonators, the wave generated by the electrodes induces shear across the film thickness. These devices have a high stiffness and can operate at high frequencies. Since the mode of oscillation does not produce volumetric strain, there is no thermoelastic damping and high quality factors Q can be obtained. As with thin-film bulk acoustic resonators, the frequency is set by the film thickness; therefore, only one frequency can be obtained per wafer.

Several transduction mechanisms have been explored and used at the MEMS level. The dominant ones are electrostatic (mainly used in the production of low-frequency oscillators) and piezoelectric (mainly used in the production of high-frequency filters and oscillators) transduction, while electrostrictive and thermal transduction are less-often used.

6.2 Electrostatically Actuated Resonators

In the case of electrostatic actuation, the electrodes do not contact the vibrating body and are placed in such a way as to maximize coupling into the desired mode of vibration. This type of actuation has been used to excite any mode of vibration. The simplest scheme is that of a flexural oscillating beam with different boundary conditions. Considering surface-micromachining technology, one can consider an in-plane oscillation or an out-of-plane oscillation, see Figure 6.2

Out-of-plane doubly clamped microresonators were proposed e.g. by Bannon, Clark and Nguyen (2000). Micromechanical in-plane flexural resonators constitute the basic vibratory unit in a number of MEMS applications, which include inertia sensors, such as the resonant accelerometer proposed by Comi *et al.* (2010). The frequency strongly depends on the geometry and can be easily adjusted by changing the length and the in-plane dimensions of the beam's cross-section. As discussed in Section 3.4.1, for a doubly clamped beam the frequency reads

$$\omega_0 = \frac{4.73^2}{L^2}\sqrt{\frac{EJ}{\rho A}}. \tag{6.1}$$

Since the resonator is electrostatically actuated through parallel plates, the frequency also depends on the electrostatic stiffness, as discussed in Section 2.5.2 for the one-dof mechanical oscillator. For a slender-beam flexural resonator, the expression of the electrostatic stiffness can be computed in a similar way, by considering that the transverse displacement is nonuniform $v = v(x,t)$. When the resonator is biased at voltage φ_p and is actuated by an AC voltage $\varphi_a(t)$, as shown in Figure 6.2a, the electrostatic force per unit length applied to the resonating beam is expressed by Equation 3.22, here rewritten for clarity:

$$p(v) = \frac{1}{2}\frac{dC_a}{dv}(\varphi_p - \varphi_a)^2 + \frac{1}{2}\frac{dC_s}{dv}\varphi_p^2, \tag{6.2}$$

where C_a and C_s are the parallel-plate capacitances per unit length formed by the driving and sensing electrodes with the resonator. Denoting the out-of-plane thickness of the

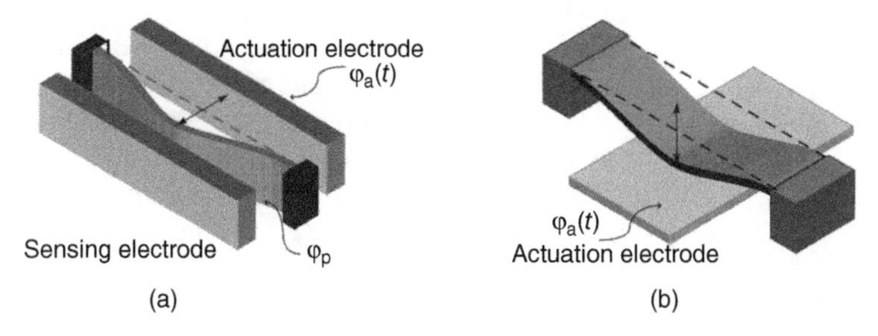

Figure 6.2 Flexural mode MEMS resonators: (a) clamped–clamped in-plane resonator; (b) clamped–clamped out-of-plane resonator.

resonator by s, and the initial gaps by g_0, the capacitances read

$$C_a = \frac{\epsilon_0 s}{g_0 + v}, \quad C_s = \frac{\epsilon_0 s}{g_0 - v}. \tag{6.3}$$

If $\varphi_a \ll \varphi_p$ and for small values of v it is possible to linearize the expression of p (6.2), in the form

$$p(v) \approx \overline{p} + \overline{k}_{\mathrm{elec}} v, \quad \text{with} \quad \overline{k}_{\mathrm{elec}} = 2\frac{\epsilon_0 s}{g_0^3}\varphi_p^2. \tag{6.4}$$

By substituting this force into the dynamic equilibrium equation of the beam, considering the free oscillation ($\overline{p} = 0$) and following the same procedure explained in Section 3.4, one obtains the expression of the frequency of the clamped–clamped resonator, actuated by parallel plates:

$$\omega_0 = \frac{4.73^2}{L^2}\sqrt{\frac{EJ - \frac{L^4}{4.73^4}\frac{2\epsilon_0 s}{g_0^3}\varphi_p^2}{\rho A}}. \tag{6.5}$$

The frequency, hence, turns out to depend on the polarization potential. It should be noted that the electrostatic actuation through parallel plates is a source of electrostatic nonlinearity, which is non-negligible if v is not sufficiently small with respect to the gap g_0. This nonlinear effect, as the nonlinear mechanical effect arising for large displacements, can prevent the proper functioning of the device.

Besides frequency, the most important attribute of a resonator is its quality factor Q, which influences the oscillator performance. In doubly clamped flexural resonators operating in near-vacuum conditions, the main sources of dissipation are the thermoelastic dissipation, see, e.g., Ardito *et al.* (2008), and the anchor losses, as described in Chapter 15. Doubly clamped resonators have a frequency range limited to tens of megahertz, since the anchor losses increase with frequency and hence at high frequencies Q becomes too low. To limit anchor losses, the resonator should be attached to the substrate at the nodal points of its oscillation shape. An example is the free–free beam resonator proposed by Wang, Wong and Nguyen (2000), which oscillates at 100 MHz and is endowed with $Q = 11500$ in vacuum.

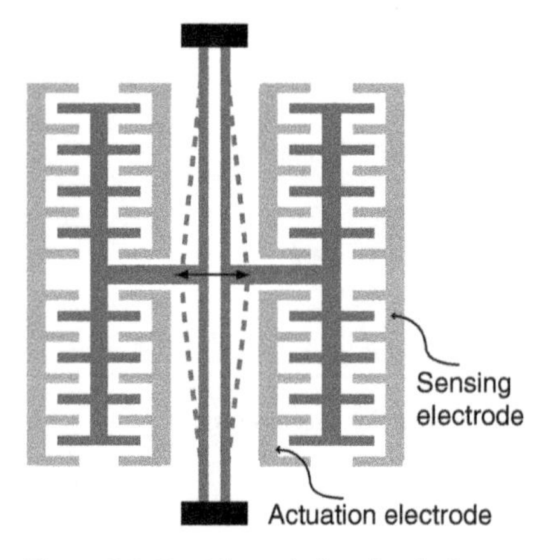

Figure 6.3 Double-ended tuning-fork resonator.

The electrostatic actuation can also be implemented by comb-finger electrodes, as in the case of double-ended tuning-fork resonators, see Figure 6.3. This scheme of actuation ensures a good electrostatic linearity. Several MEMS, such as resonant accelerometers, contain a double-ended tuning-fork resonator as the sensing element, see, e.g., Seshia, Howe and Montague. (2002).

Another class of frequently used resonators that are capacitively transduced is that of extensional bulk acoustic wave resonators. Figure 6.4 shows the geometry and the resonant mode of the square extensional bulk acoustic wave resonator proposed by Kaajakari *et al.* (2003), which is constituted by a square

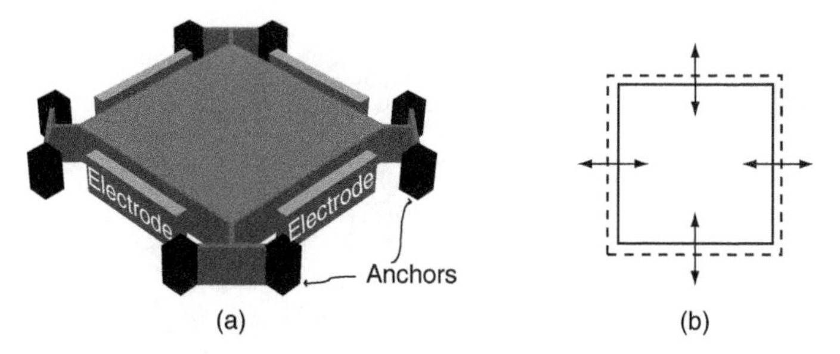

Figure 6.4 Extensional bulk acoustic wave resonator: (a) geometry; (b) resonant mode.

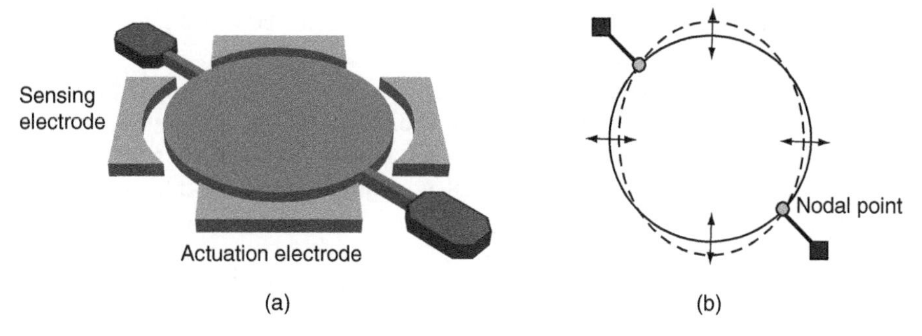

Figure 6.5 Wine glass disc resonator: (a) geometry; (b) resonant mode.

plate, supported at the corners by springs, electrostatically driven by the four electrodes in its extensional mode. The frequency is proportional to the inverse of the square of the side length and can therefore be increased by scaling. This represents a major advantage of this capacitively transduced resonator over the class of piezoelectric resonators, whose resonance frequencies are governed mainly by film thickness, for which a new fabrication run is required to achieve the different frequencies. In addition, capacitive transduction also offers voltage-controlled frequency tunability.

A different geometry that allows for high frequency and high quality factor is that studied by Lin *et al.* (2004) of the radial contour mode disc, which achieved $Q > 10\,000$ at a frequency of 1.5 GHz. The device, schematically shown in Figure 6.5, consists of a disc, supported by two beams attached to the disc at quasi-nodal points. The disc is driven to resonance by two opposite electrodes of the four that surround it and it expands along one axis and contracts along the orthogonal axis.

6.3 Piezoelectric Resonators

Piezoelectric-film resonators, such as aluminium nitride resonators, are widely used as radio-frequency filters in mobile communication systems because of their miniaturization and high electric performance. The piezoelectric transduction technique, commonly used in quartz resonators, is slow to develop in thin-film form mainly because of difficulties with material deposition. Piezoelectric transduction requires metal electrodes directly on the thin-film layer to apply electric field. Usually, in MEMS the piezoelectric material has the polarization axis (three-axis) orthogonal to the

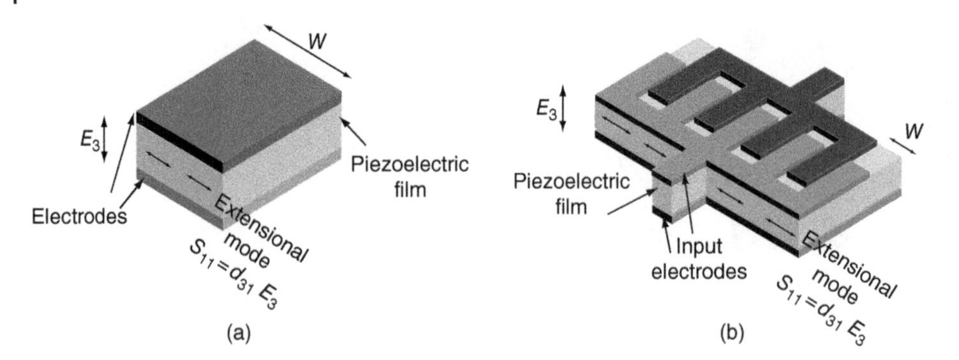

Figure 6.6 Contour mode resonators: (a) simple scheme; (b) multiscaled subresonators.

film plane. If the film is sandwiched between the two electrodes, the applied electric field is in the 3-direction and the piezoelectric coefficients d_{31} and d_{33} allow for the activation of the desired mode of vibration in the plane or through the thickness. In contour mode resonators, when an AC signal is applied across the film thickness, a contour-extensional mode of vibration is excited through the d_{31} piezoelectric coefficient, as shown in Figure 6.6a.

In contour mode resonators, the frequency is set by the width W of the laterally vibrating part through

$$\omega_0 = \frac{1}{2W}\sqrt{\frac{E_{\mathrm{eq}}}{\rho_{\mathrm{eq}}}}, \tag{6.6}$$

where E_{eq} and ρ_{eq} are the equivalent Young's modulus and the equivalent mass density of the material stack that forms the resonator. To increase the frequency, without reducing the device capacitance, one can mechanically couple a large number of subresonators with reduced width W, as shown in Figure 6.6b. Rinaldi *et al.* (2009) demonstrated this kind of resonator with a frequency greater than 3 GHz.

Anchor losses are the dominant energy loss mechanism in aluminium nitride contour mode resonators at low temperature or at room temperature and low frequency (200 MHz) for most anchor designs, while at high temperature an additional dissipation mechanism, probably related to the piezoelectric material viscosity, plays a more important role in setting Q as shown in Segovia-Fernandez *et al.* (2015).

In thickness-extensional resonators, the vibration mode, exploiting the d_{33} coefficient of the piezoelectric material, is in the normal direction to the film plane. Therefore, the resonant frequency is inversely proportional to the thickness of the film.

Another common configuration of thin-film piezoelectric resonators is characterized by both electrodes deposited above the film. In that case, the electric field is in the 1 direction and the piezoelectric coefficient d_{15} is exploited.

6.4 Nonlinearity Issues

In general, the correct functioning of filters, oscillators and resonant inertial sensors requires a linear response of the resonators. However, the continuous miniaturization process may entail a limited linear range and the appearance of nonlinear effects that should be studied. In the simplest case of a flexural beam resonator that can be described as a one-dof oscillator, we have seen in Section 3.4.2 that the large deflection can induce a nonlinear behaviour, and the resonator is described as a Duffing resonator with a mechanical *hard spring* effect (Figure 3.4). If the resonator is electrostatically actuated through parallel plates, the electrostatic stiffness also presents some nonlinearities. Under the assumption $\varphi_a \ll \varphi_p$, the electrostatic force (6.2) can be approximated through its series expansion up to the third order, obtaining the following expression:

$$p(v) \approx \overline{p} + \overline{k}_{\text{elec}} v + \overline{k}_{3\text{elec}} v^3, \quad \text{with} \quad \overline{k}_{3\text{elec}} = 4\frac{\epsilon_0 S}{g_0^5}\varphi_p^2. \tag{6.7}$$

As discussed by Tocchio *et al.* (2011), the presence of the third-order nonlinear term, $\overline{k}_{3\text{elec}}$, in the electrical stiffness mitigates the hard spring effect generally associated with the third-order term of the elastic stiffness.

For sufficiently high values of the biasing voltage, the electrical nonlinear terms can overcome the mechanical counterparts, causing a 'soft-spring' effect, as shown in Figure 6.7. The electrostatic nonlinearities also intervene in torsional resonators, such as those studied by Caspani *et al.* (2014) and Comi *et al.* (2016). In this case, the mechanical behaviour is linear and hence a soft-spring effect is always observed. Figure 6.8a shows the SEM of a torsional resonator fabricated through the ThELMA process. The folded springs allow the rotation of the square mass around the axis a–a, they are subject to torsion and ensure a linear elastic mechanical behaviour. The electrostatic actuation is realized by an electrode placed on the substrate below the resonating mass. Figure 6.8b shows the experimental frequency responses obtained by varying the actuation voltage, $|\varphi_a|$, for $\varphi_p = 4V$, plotted together with the numerical spectra computed by solving the full dynamic equations of motion. For low values of $|\varphi_a|$, the usual linear frequency response is obtained. By increasing $|\varphi_a|$, first a hysteretic behaviour (see jumps in the backward frequency sweep) is observed, then

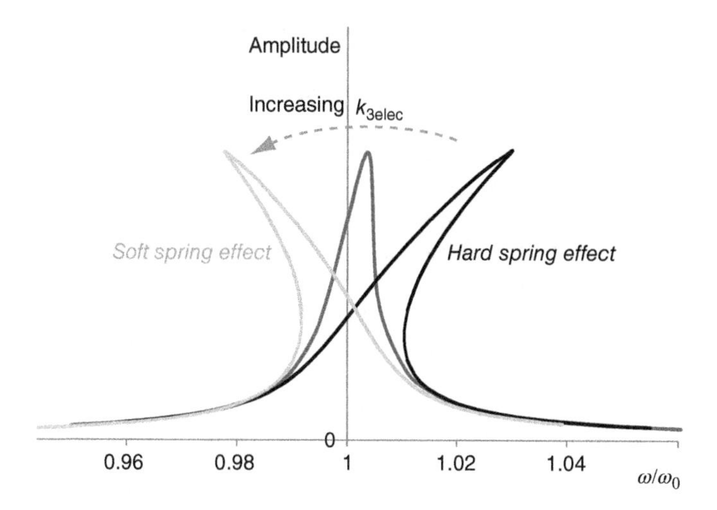

Figure 6.7 Frequency response of nonlinear resonator for different values of electrostatic nonlinear stiffness $\overline{k}_{3\text{elec}}$.

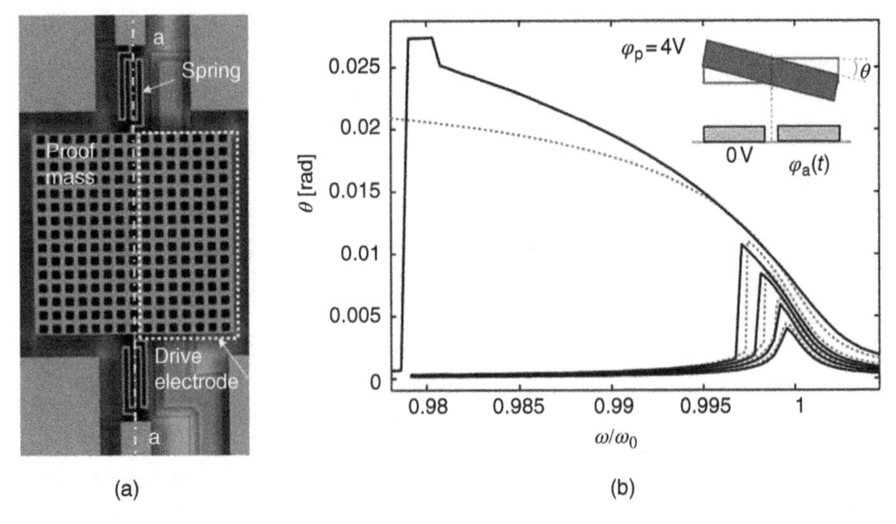

(a) (b)

Figure 6.8 Torsional resonator: (a) SEM of polysilicon resonator fabricated through ThELMA surface micromachining process developed by STMicroelectronics; (b) comparison between experimental data (continuous lines) and numerical predictions (dotted lines) for $\varphi_p = 4V$ and $|\varphi_a| = 45, 65, 85, 105$ and 150 mV – only the backward frequency sweeps are reported for sake of clarity. *Source:* Comi *et al.* (2016), Figures 1 and 3. Open access article published by Elsevier under the CC BY-NC-ND license (http://creativecommons.org/licenses/by-nc-nd/4.0/). (*See color plate section for the color representation of this figure.*)

dynamic pull-in occurs: the numerical curve reaches a horizontal asymptote while, experimentally, a jump to the maximum physical admissible amplitude is experienced.

References

Ardito, R., Comi, C., Corigliano, A. and Frangi, A. (2008) Solid damping in micro electro mechanical systems. *Meccanica*, **43**, 419–428.

Bannon, F., Clark, J. and Nguyen, C.C. (2000) High-*Q* HF microelectromechanical filters. *IEEE Journal of Solid-State Circuits*, **35**, 512–526.

Caspani, A., Comi, C., Corigliano, A. *et al.* (2014) Dynamic nonlinear behavior of torsional resonators in MEMS. *Journal of Micromechanics and Microengineering*, **24**, 095025.

Comi, C., Corigliano, A., Doti, M. *et al.* (2016) Torsional microresonator in the nonlinear regime: experimental, numerical and analytical characterization. *Procedia Engineering*, **168**, 933–936.

Comi, C., Corigliano, A., Langfelder, G. *et al.* (2010) A resonant micro-accelerometer with high sensitivity operating in an oscillating circuit. *Journal of Micro-electromechanical Systems*, **19**, 1140–1152.

Kaajakari, V., Mattila, T., Oja, A. *et al.* (2003) *Square Extensional Mode Single-Crystal Silicon Micromechanical RF resonator*, 12th International Conference on TRANSDUCERS, Solid-State Sensors, Actuators and Microsystems, June 8–12, 2003, Boston, MA, IEEE.

Lin, Y.W., Lee, S., Li, S.S. *et al.* (2004) Series-resonant VHF micromechanical resonator reference oscillators. *IEEE Journal of Solid-State Circuits*, **39**, 2477–2491.

Piazza, G., Stephanou, P., Wijesundara, M. and Pisano, A. (2005) *Single-Chip Multiple-Frequency Filters Based on Contour-Mode Aluminium Nitride Piezoelectric Micromechanical Resonators*, TRANSDUCERS '05: The 13th International Conference on Solid-State Sensors, Actuators and Microsystems, June 5–9, 2005, Seoul, South Korea, IEEE.

Rinaldi, M., Zuniga, C., Chengjie, Z. and Piazza, G. (2009) *AlN Contour-Mode Resonators for Narrow-Band Filters Above 3 GHz*, IEEE International Frequency Control Symposium, Joint with the 22nd European Frequency and Time Forum, April 20–24, 2009, Besançon, France, IEEE.

Segovia-Fernandez, J., Cremonesi, M., Cassella, C. *et al.* (2015) Anchor losses in AlN contour mode resonators. *Journal of Microelectromechanical Systems*, **24** (2), 265–275.

Seshia, A., Howe, R. and Montague., S. (2002) *An Integrated Microelectromechanical Resonant Output Gyroscope*, The Fifteenth IEEE International Conference on Micro Electro Mechanical Systems, January 24, 2002, Las Vegas, NV, IEEE.

Tocchio, A., Comi, C., Langfelder, G. *et al.* (2011) Enhancing the linear range of MEMS resonators for sensing applications. *IEEE Sensors Journal*, **11**, 3202–3210.

Wang, K., Wong, A.C. and Nguyen, C.T.C. (2000) VHF free–free beam high-*Q* micromechanical resonators. *Journal of Microelectromechanical Systems*, **9**, 347–360.

7

Micromirrors and Parametric Resonance

7.1 Introduction

Micromirrors are emerging as a very successful component (Lab4MEMS II, 2014) within the family of optical MEMS. In general, these are MEMS merged with micro-optics and involve sensing or manipulating optical signals on a very small scale, using integrated mechanical, optical and electrical systems.

Micromirrors based on optical MEMS come in a variety of forms, ranging from single mirrors with one or two degrees of freedom, to large mirror matrices with millions of individually addressable mirrors. Devices that consist of a single mirror element and that are typically designed to scan a laser beam are usually called microscanners.

Among the most successful applications of microscanners, one can cite laser scanners for 3D imaging and miniaturized image projectors, or pico-projectors. Their technology has progressed rapidly in the last few years, with the first products already on the market. Depending on the applications, there are different types of operating mode. In the *point-to-point mode*, also called the quasi-static mode, both axes are controlled without resonance. Therefore, the mirror can either hold a static position or move at a steady velocity for raster scanning and vector graphics. Point-to-point mode is preferable to other operations in that it is simple to synchronize the laser illumination to the mirror movement. Moreover, it is straightforward to ensure that each point is exposed by an equal amount, which makes it simpler to achieve a uniform picture. It is, however, very challenging to produce a single mirror with both large deflection and high-frequency scanning in two directions. For this reason, two other modes have been applied in microscanners for image projection.

In the resonant mode, both axes use the resonance of the mechanical structure to obtain a large optical deflection amplitude. The main drawback is the limitation of a very narrow-band frequency range and sinusoidal trajectories with a phase lag with respect to the applied voltage. Resulting 2D motions are therefore often limited to special shapes like circles or ellipses.

The last mode basically combines the previous ones. One axis is driven in resonance very quickly (e.g. at a few kilohertz) to create, e.g., horizontal lines for pico-projection, while the second axis is driven quasi-statically with a sawtooth-like waveform to create a raster pattern that covers a rectangular display or imaging area in a controlled fashion.

Mechanics of Microsystems, First Edition. Alberto Corigliano, Raffaele Ardito, Claudia Comi, Attilio Frangi, Aldo Ghisi, and Stefano Mariani.

Companion website: www.wiley.com/go/corigliano/mechanics

At present, this is the preferred mode of operation of a microscanner for image projection applications.

Several physical principles and designs have been reported in the literature, having different forms of actuation: electrostatics, electromagnetic, electrothermal and piezoelectric. Electrostatically actuated devices have been widely investigated since the early contributions by Fan and Wu (1998) and Lee (2004). They offer high speed, but provide limited mechanical scanning ranges at nonresonance (typically 2–3°) and up to 20° at resonance. Moreover this actuation mode often involves a large actuator footprint, and relatively high voltages, above 40 V, are required. One advantage of electrostatic actuators is that standard robust surface micromachining processes can be applied, to reduce the number of failure modes of scanners based on this actuation principle.

Electrothermal mirrors have also been extensively investigated (Jain *et al.*, 2004; Singh *et al.*, 2005). Both 1D and 2D scanning at large scanning angles (up to ±30°) and low driving voltages have been demonstrated. Critical issues are the high driving current, the requirement for thermal isolation of the actuator and a rather slow dynamic response.

Electromagnetic (Cho and Ahn, 2003; Yalcinkaya *et al.*, 2006) scanners use a lower voltage to generate relatively large forces, enabling the scanning mirror to have a large deflection angle and reliable operation. The main drawbacks of this actuation mode remain its power consumption and the complex structures and bulky sizes for 2D actuation.

Although piezoelectric actuation for optical MEMS modes has not been investigated as much as other alternatives and no commercial products are yet available, it presents several potential advantages, such as large deflection angles using bimorph structures with low voltage and low power consumption (Schroth *et al.*, 1999; Filhol *et al.*, 2005). It is expected that piezoelectric devices will eventually outperform their electrostatic counterparts in many applications.

The remainder of this chapter contains a discussion concerning an electrostatically actuated single-axis resonant micromirror. This choice is motivated by the commercial interest of its already available industrial applications, but also by the fact that their dynamic response is based on the characteristic phenomenon of parametric resonance, as explained in the following sections.

7.2 Electrostatic Resonant Micromirror

As a representative example, we focus here on the resonant 1D micromirror depicted in Figure 7.1. The response of the micromirror is based on parametric resonance, which is a known phenomenon in the nonlinear dynamics of structures.

Parametric resonance has been well understood in many areas of science (Nayfeh and Mook, 1979; Nayfeh, 1981; Younis, 2011; Lacarbonara, 2013), including the stability of ships, the forced motion of a swing and Faraday surface wave patterns on water.

Microsystems have provided several examples of parametric resonance, starting from the pioneering work of Rugar and Grütter (1991). Applications include mass sensing, parametric amplification (Zalalutdinov *et al.*, 2001; Carr *et al.*, 2000; Zhang, Baskaran and Turner, 2002; Rhoads *et al.*, 2005; Thompson and Horsley, 2011; Shmulevich, Grinberg and Elata, 2015). In particular, Rhoads *et al.* (2006) focused on the design of a highly sensitive mass sensor and used a nonlinear generalized Mathieu equation to

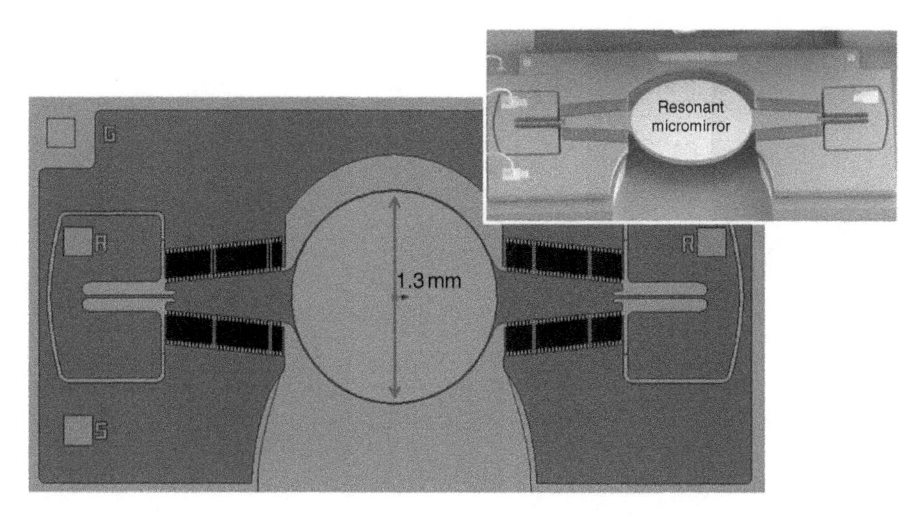

Figure 7.1 Resonant micromirror: SEM of device and layout. *Source*: Frangi *et al.* (2017), Figure 1. Reproduced with permission of IEEE.

model the problem. Analytical results show that nonlinearity significantly changes the stability characteristics of parametric resonance.

In the application of interest herein, the elastic stiffness is linear and all the nonlinearities are due to electrostatic forcing. However, these cannot be approximated with linear and cubic terms, and a very accurate numerical procedure is mandatory.

The 2D layout and a SEM of the micromirror are depicted in Figure 7.1. The device was fabricated by STMicroelectronics with dedicated silicon-on-insulator technology.

The central circular reflecting surface is attached to the substrate via two coaxial beams acting as torsional springs. Four sets of 33 fingers each are anchored to the trapezoidal regions directly attached to the mirror. These plates, interdigitated with their stator counterparts, form a comb-drive structure providing the electrostatic actuation mechanism. Sensing of the opening angle is performed, during operation, via the same comb-drive electrodes (Figure 7.2).

Following the discussion published by Frangi *et al.* (2017), the mirror is treated as a rigid body hinged in its centre and connected to the substrate via the two elastically deformable springs. The torsional response is hence governed by the simple 1D model:

$$I\ddot{\psi} + b\dot{\psi} + k\psi = \varepsilon_0 M(\psi)\Phi^2(t), \tag{7.1}$$

where $I = 2.375 \times 10^{10}$ ng µm^2 is the inertia around the torsional axis; $k = 2.704 \times 10^7$ µN µm is the torsional stiffness of the springs; $b = 1.448 \times 10^6$ µN µm µs is a damping coefficient and $\varepsilon_0 M$ is the electrostatic torque due to a unit voltage bias. The voltage Φ is given in volts, and the time in Equation 7.1 is measured in microseconds. Since $\varepsilon_0 = 8.85 \times 10^{-6}$ pF/µm, M is measured in micrometres and is a purely geometrical feature. It is worth stressing that the dissipation term is still an open issue, which deserves further attention in future investigations. Preliminary results have been obtained on a similar structure (Mirzazadeh, Mariani and De Fazio, 2014); however, in the work described, the value of b was guessed, starting from the known dissipation mechanisms in infinitesimal transformations (Frangi, Spinola and Vigna, 2006). The torsional eigenfrequency of the mirror is $f_0 = 5370$ Hz.

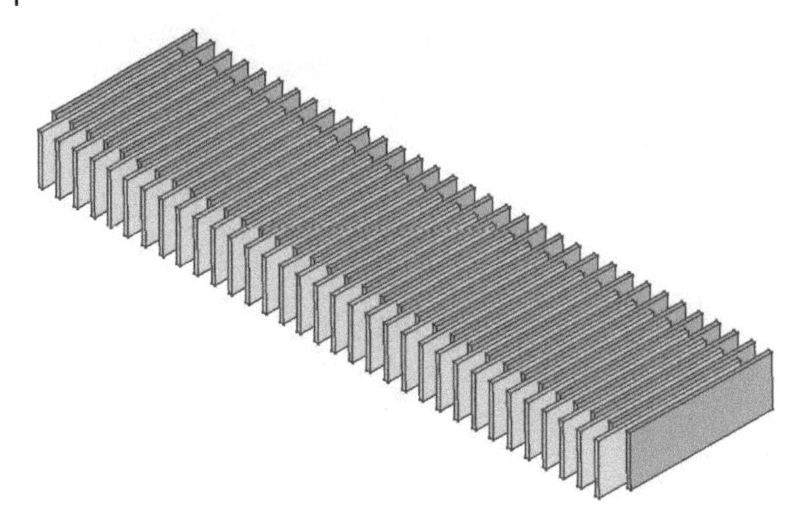

Figure 7.2 Geometrical model of one complete series of interdigitated fingers. *Source*: Frangi *et al.* (2017), Figure 2. Reproduced with permission of IEEE.

Owing to symmetry, the electrostatic torque around the torsional axis vanishes for $\psi = 0$, i.e. in the rest configuration, $\forall \Phi(t)$. This trivial solution becomes, however, unstable for some combinations of the input voltage and frequency and triggers the mirror rotation. Indeed, Frangi *et al.* (2017) show that Equation 7.1 reduces, in the case of small rotation and negligible dissipation, to the well-known Mathieu equation (Nayfeh and Mook, 1979), written in terms of a nondimensional time τ:

$$\psi'' + (\delta + \epsilon \cos 2\tau)\psi = 0, \tag{7.2}$$

which admits a nontrivial response only in specific regions of the δ–ϵ plane, called instability tongues, that emanate from δ axis. Theory predicts that instability tongues are associated with drive frequencies f near $2f_0/n$, where f_0 is the system's natural frequency and n is any positive integer. However, in macroscopic systems only the first and largest instability tongue can typically be observed, because of damping and of the exponential narrowing in the resonance regions with increasing n. In this case, the system is excited by a periodic forcing function at frequency $2f_0$, which is totally different from classical phenomena of harmonic resonance. An exception is described by Turner *et al.* (1998), where the torsional oscillations in a single-crystal silicon MEMS have been analysed. Five instability regions have been measured, owing to the low damping, stability and precise frequency control achievable in that system.

In standard working conditions, the mirror is excited near the resonance peak and the phase between the driving signal and mirror rotation is kept constant by a closed loop. The mirror is driven with a nonideal square wave with finite ramp-up and ramp-down speeds, typically at 150 V with a duration of <25% of the duty cycle. Figure 7.3 presents the square wave for a chosen frequency of 5300 Hz, close to the natural torsional frequency of the mirror. It is worth stressing, for future discussion, that the fast Fourier transform of the square wave contains almost all the multiples of the fundamental frequency, as evidenced in Figure 7.4, which illustrates the power spectrum.

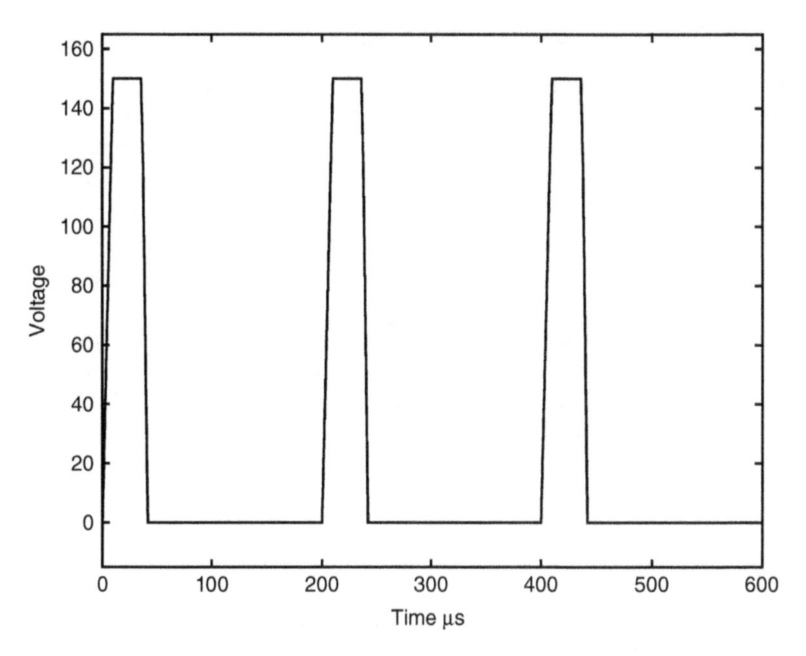

Figure 7.3 Input voltage: square wave at 5300 Hz, 150 V. *Source*: Frangi *et al*. (2017), Figure 5. Reproduced with permission of IEEE.

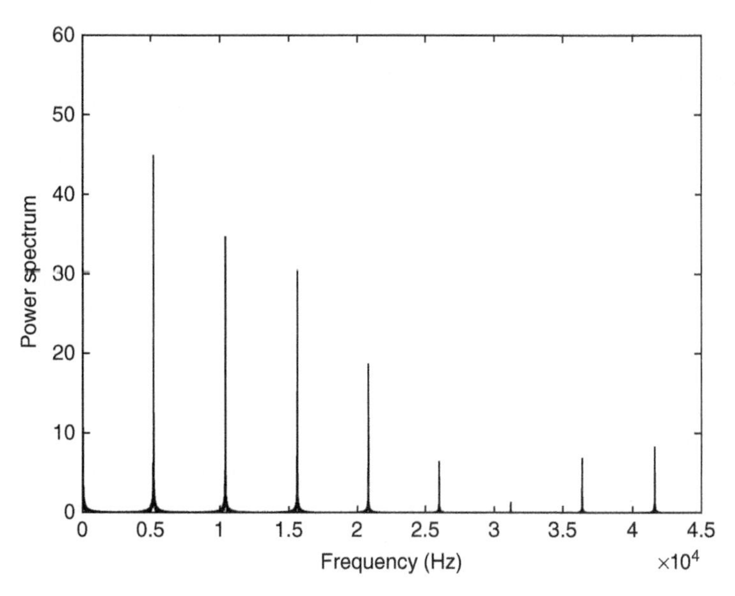

Figure 7.4 Power spectrum for the square wave of Figure 7.3. *Source*: Frangi *et al*. (2017), Figure 6. Reproduced with permission of IEEE.

7.2.1 Numerical Simulations with a Continuation Approach

The numerical simulation of Equation 7.1 can be performed with several techniques. The brute force approach, which closely reproduces the actual operation of the mirror, performs a sweep over the frequencies of interest and, for each frequency, simulates a sufficient number of cycles by direct integration in time to reach a steady state; the amplitude is then recorded and the next frequency is addressed, using the final amplitude and phase of the previous analysis as the initial conditions. This is a very robust technique, which, however, enables simulation of only the stable branches of the amplitude versus frequency response.

On the contrary, the continuation approach (Wagg and Neild, 2015; Doedel and Kernévez, 1986) with arc length control is more versatile and is adopted herein. The model in Equation 7.1 is rewritten as a first-order nonautonomous differential system of equations in terms of the fictitious time $\tau = t/T$, with $\tau \in [0, 1]$:

$$\begin{aligned} y_1' &= Ty_2, \\ y_2' &= -\frac{BT}{I}y_2 - \frac{KT}{I}y_1 - \frac{T}{I}\varepsilon_0 M(y_1)\Phi^2(2\pi\beta\tau). \end{aligned} \tag{7.3}$$

In Equation 7.3, the prime denotes differentiation with respect to τ. We limit our attention to periodic forcing functions with $\Phi(0) = \Phi(2\pi\beta)$ and to periodic solutions, so that our problem can be rewritten in condensed form as

$$\mathbf{y}'(\tau) = T\mathbf{A}\mathbf{y}(\tau) + T\mathbf{f}(y_1, \tau), \quad \tau \in [0, 1], \quad \mathbf{y}(0) = \mathbf{y}(1), \tag{7.4}$$

where $\mathbf{A}$ is a constant matrix and $\mathbf{f}$ contains the forcing function. It is worth stressing that, according to our assumptions:

i) In the interval T of analysis, $\mathbf{f}$ and $\mathbf{y}$ might contain $n \geq 1$ and $m \geq 1$ cycles, respectively, with $n \neq m$ in general. This is essential for the simulation of different instability tongues.
ii) The system could be generalized and transformed into an autonomous one by adding a nonlinear oscillator generating the desired periodic forcing, as done for instance in the software AUTO (Doedel and Kernévez, 1986).

Let us suppose now that $\mathbf{y}_n$, T_n is a known solution of the system. The simplest choice consists of taking T as a continuation parameter: we fix $T_{n+1} = T_n + \Delta T$ and solve Equation 7.4 for $\mathbf{y}_{n+1}$ through an iterative Newton–Raphson procedure using a suitable initial guess.

However, this classical parameter continuation fails in the presence of an unstable branch, as is clear from Figure 7.5. Indeed, by imposing an increment $\Delta T > 0$ (and hence $\Delta f < 0$) at the peak, the solution would jump to the $\psi = 0$ stable solution, omitting completely the unstable dashed branches. For this reason, it is customary to introduce an arc length control, in which ΔT is part of the unknowns, the abscissa s along the solution branch is taken as the continuation parameter, and a new constraint $\mathcal{F}(\Delta\mathbf{y}, \Delta T) = 0$ is added. A typical choice is

$$\mathcal{F}(\Delta\mathbf{y}, \Delta T) = \Delta\mathbf{y} \cdot \Delta\mathbf{y} + (\Delta T)^2 - (\Delta s)^2 = 0.$$

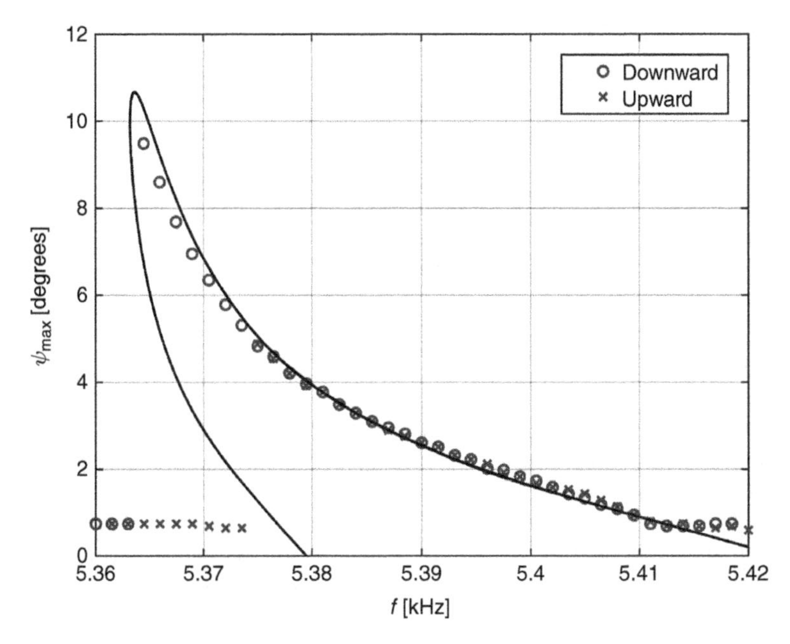

Figure 7.5 Sinusoidal excitation at 5 kHz, $\Phi_0 = 55$ V; experimental upward and downward sweeps (discrete symbols) and numerical continuation (continuous line). *Source*: Frangi *et al.* (2017), Figure 11. Reproduced with permission of IEEE.

An alternative is the Keller's pseudo-arc length method (Doedel and Kernévez, 1986), in which the increment $\Delta\mathbf{y}, \Delta T$ is sought such that its projection along a specific direction (typically the tangent to the $\mathbf{y}, T$ manifold) equals a fixed arc length Δs.

In our simple implementation, we enforce Equation 7.4 in a weak manner, fixing a suitable space C of vector test functions $\tilde{\mathbf{y}}$:

$$
\begin{aligned}
&\text{Find } \mathbf{y}_{n+1}, T_{n+1} \text{ such that } \mathbf{y}_{n+1}(0) = \mathbf{y}_{n+1}(1) \text{ and}\\
&\mathcal{R}(\mathbf{y}_{n+1}, T_{n+1}; \mathbf{y}) = \int_0^1 \left(\mathbf{y}'_{n+1} - T_{n+1}\mathbf{A}\mathbf{y}_{n+1} - T_{n+1}\mathbf{f}(y_1, \tau)\right) \cdot \tilde{\mathbf{y}} \, \mathrm{d}\tau = 0, \quad \forall \tilde{\mathbf{y}} \in C\\
&\mathcal{F}(\Delta\mathbf{y}, \Delta T) = 0.
\end{aligned} \tag{7.5}
$$

The procedure is then repeated for every n of interest.

For the numerical solution of Equation 7.5, the segment [0–1] is partitioned into N equal elements and the unknown function $\mathbf{y}$ is discretized over each element with quadratic Lagrangian shape functions. By contrast, the space of test functions is selected as the space of piecewise Legendre orthogonal polynomials P_2. If Equation 7.5 is integrated numerically over each element with a two-point Gauss–Legendre quadrature rule, this is equivalent to the method of orthogonal collocation (Doedel and Kernévez, 1986), in which Equation 7.4 is collocated at the two zeros of P_2 in each element.

Equation 7.5 is solved iteratively by means of a Newton–Raphson procedure. In the generic iteration, given an estimate for $\mathbf{y}_{n+1}^{[k]}, T_{n+1}^{[k]}$ (and hence for $\Delta\mathbf{y}^{[k]}, \Delta T^{[k]}$), a small correction $\delta\mathbf{y}, \delta T$ is sought such that

$$\mathbf{y}_{n+1}^{[k+1]} = \mathbf{y}_{n+1}^{[k]} + \delta\mathbf{y}, \qquad T_{n+1}^{[k+1]} = T_{n+1}^{[k]} + \delta T \tag{7.6}$$

is a solution of the linearized system:

Find $\delta\mathbf{y}, \delta T$ such that $\delta\mathbf{y}(0) = \delta\mathbf{y}(1)$ and

$$\int_0^1 \left(\delta\mathbf{y}' - \delta T \mathbf{A}\mathbf{y}_{n+1}^{[k]} - T_{n+1}^{[k]}\mathbf{A}\delta\mathbf{y} - \delta T \mathbf{f}\left(\mathbf{y}_{n+1}^{[k]}, \tau\right) - T_{n+1}^{[k]} \frac{\partial \mathbf{f}}{\partial y_1}\left(\mathbf{y}_{n+1}^{[k]}, \tau\right)\delta y_1\right) \cdot \tilde{\mathbf{y}}\, \mathrm{d}\tau$$

$$= -\mathcal{R}\left(\mathbf{y}_{n+1}^{[k]}, T_{n+1}^{[k]}; \mathbf{y}\right), \quad \forall \tilde{\mathbf{y}}(\tau) \in C_t$$

$$\frac{\partial \mathcal{F}}{\partial \mathbf{y}}\left(\Delta\mathbf{y}^{[k]}, \Delta T^{[k]}\right) \cdot \delta\mathbf{y} + \frac{\partial \mathcal{F}}{\partial T}\left(\Delta\mathbf{y}^{[k]}, \Delta T^{[k]}\right)\delta T = -\mathcal{F}\left(\Delta\mathbf{y}^{[k]}, \Delta T^{[k]}\right) \tag{7.7}$$

The implementation of Equation 7.7 requires the computation of the derivative of the electrostatic torque $M(y_1)$ with respect to the rotation angle $y_1 = \psi$. A commonly adopted strategy is to compute M numerically using some dedicated software and then differentiate it using standard finite differences. However this approach often lacks the required accuracy for the iterative procedure to converge. In the following section, we detail an alternative direct formulation, based on the notion of material derivative.

7.2.1.1 Computation of the Electrostatic Torque and its Derivative via Direct Finite Element Method

In this work, a finite element method has been selected to compute the electrostatic torque directly by post-processing the potential field of the classical electrostatic problem. Although the procedure is presented here as a specific application to the micromirror problem, the approach is quite general.

The rigid fingers of the mirror are collected into two groups: the 'stator' Ω_T and the 'shuttle' Ω_H, the former being fixed and the latter being movable. The stator and the shuttle are immersed in the infinite space Ω truncated at S_∞, where homogeneous Neumann conditions are enforced.

Let us express the solution as $\varphi(\mathbf{x})\Phi$, where Φ is a constant scaling factor. We will assume, without any loss of generality, that the unknown function φ is governed by the differential problem $\nabla^2\varphi = 0$ in Ω, subjected to the following boundary conditions: $\varphi = 0$ on $\partial\Omega_\mathrm{T}$, $\varphi = 1$ on $\partial\Omega_\mathrm{H}$, $\mathrm{grad}\varphi \cdot \mathbf{n} = 0$ on S_∞.

If $C(\overline{\varphi})$ denotes the space of sufficiently continuous functions respecting Dirichlet boundary conditions on φ, the weak form of electrostatic problem is:

Find $\varphi \in C(\overline{\varphi})$ such that:

$$\int_\Omega \mathrm{grad}\tilde{\varphi} \cdot \mathrm{grad}\varphi \, \mathrm{d}\Omega = 0 \qquad \forall \tilde{\varphi} \in C(0). \tag{7.8}$$

Direct Computation of Torque A straightforward method of computing forces and couples by a simple post-processing of φ flows from the notion of material derivative applied to the total potential energy (Bonnet, 1995; Ren, Qu and Xu, 2012; Ardito, Baldasarre and Corigliano, 2009).

Let us assume that a change of configuration $\Phi(\mathbf{x}, p)$ is generated by the variation of a given parameter p (e.g. the rotation angle ψ of the shuttle) and is associated with the fictitious velocity $\boldsymbol{\theta} = (\partial\Phi/\partial p)\dot{p}$. Following classical texts of optimization (Bonnet,

1995), the material derivative (also called the total or particle derivative) of a scalar f, of a gradient ∇f and of a volume element $\mathrm{d}\Omega$ due to $\boldsymbol{\theta}$ are, respectively:

$$\overset{\star}{f} = \frac{\partial f}{\partial p}\dot{p} + \mathrm{grad}f\cdot\boldsymbol{\theta}, \qquad \overset{\star}{(\mathrm{grad}f)} = \mathrm{grad}\overset{\star}{f} - \mathrm{grad}f\cdot\mathrm{grad}\boldsymbol{\theta}, \qquad \overset{\star}{\mathrm{d}\Omega} = \mathrm{div}\boldsymbol{\theta}\,\mathrm{d}\Omega. \tag{7.9}$$

The restriction to $\partial\Omega_{\mathrm{H}}$ of any admissible $\boldsymbol{\theta}$ must be a rigid-body motion $\mathbf{U}(\mathbf{x}) = \mathbf{V} + \boldsymbol{\omega}\wedge\mathbf{x}$, while $\boldsymbol{\theta}$ must vanish on $\partial\Omega_{\mathrm{T}}$. The following expression for the virtual velocity will be adopted:

$$\boldsymbol{\theta}(\mathbf{x}) = \varphi(\mathbf{x})\,(\mathbf{V} + \boldsymbol{\omega}\wedge\mathbf{x}) = \varphi(\mathbf{x})\mathbf{U}(\mathbf{x}), \tag{7.10}$$

where φ is the solution of problem Equation 7.8. Now,

$$\mathrm{grad}\boldsymbol{\theta} = \mathbf{U}\otimes\mathrm{grad}\varphi + \varphi\boldsymbol{\omega}^{\wedge}, \qquad \mathrm{div}\boldsymbol{\theta} = \mathbf{U}\cdot\mathrm{grad}\varphi, \tag{7.11}$$

where $\boldsymbol{\omega}^{\wedge}$ is such that $\boldsymbol{\omega}^{\wedge}\cdot\mathbf{b} = \boldsymbol{\omega}\wedge\mathbf{b}$, $\forall\mathbf{b}$. Let $\overset{\star}{W}$ denote the material derivative of the total potential energy W. If $\varepsilon_0\mathbf{M}\Phi^2$ is the torque exerted on $\partial\Omega_{\mathrm{H}}$,

$$\overset{\star}{W} = \overset{\star}{W}_{\mathrm{elec}} - \varepsilon_0\mathbf{M}\cdot\boldsymbol{\omega}, \quad \text{with} \quad W_{\mathrm{elec}} = \frac{\varepsilon_0}{2}\int_{\Omega}\mathrm{grad}\varphi\cdot\mathrm{grad}\varphi\,\mathrm{d}\Omega \tag{7.12}$$

where W_{elec} is the electrostatic energy for a unit voltage bias. Applying Equation 7.9,

$$\overset{\star}{W}_{\mathrm{elec}} = \varepsilon_0\int_{\Omega}\mathrm{grad}\overset{\star}{\varphi}\cdot\mathrm{grad}\varphi - \mathrm{grad}\varphi\cdot\mathrm{grad}\boldsymbol{\theta}\cdot\mathrm{grad}\varphi + \frac{1}{2}\mathrm{grad}\varphi\cdot\mathrm{grad}\varphi\,\mathrm{div}\boldsymbol{\theta}\,\mathrm{d}\Omega. \tag{7.13}$$

However, the first term vanishes, since

$$\int_{\Omega}\mathrm{grad}\overset{\star}{\varphi}\cdot\mathrm{grad}\varphi\,\mathrm{d}\Omega = 0, \tag{7.14}$$

owing to Equation 7.8, written with $\tilde{\varphi} = \overset{\star}{\varphi}$. Indeed, $\overset{\star}{\varphi}$ is an admissible test function that vanishes on both $\partial\Omega_{\mathrm{H}}$ and $\partial\Omega_{\mathrm{T}}$. At equilibrium, $\overset{\star}{W} = 0$, hence

$$\mathbf{M}\cdot\boldsymbol{\omega} = \int_{\Omega} -\mathrm{grad}\varphi\cdot\mathrm{grad}\boldsymbol{\theta}\cdot\mathrm{grad}\varphi + \frac{1}{2}\mathrm{grad}\varphi\cdot\mathrm{grad}\varphi\,\mathrm{div}\boldsymbol{\theta}\,\mathrm{d}\Omega, \tag{7.15}$$

yielding the final general formula

$$\mathbf{M}\cdot\boldsymbol{\omega} = -\frac{1}{2}\int_{\Omega}\|\mathrm{grad}\varphi\|^2(\mathrm{grad}\varphi\cdot\mathbf{U})\,\mathrm{d}\Omega, \tag{7.16}$$

since $\mathrm{grad}\varphi\cdot\boldsymbol{\omega}^{\wedge}\cdot\mathrm{grad}\varphi = 0$. The torque is obtained by setting $\mathbf{V} = \mathbf{0}$

$$\mathbf{M} = -\frac{1}{2}\int_{\Omega}\|\mathrm{grad}\varphi\|^2\mathbf{a}\,\mathrm{d}\Omega, \tag{7.17}$$

where $\mathbf{a} = \mathbf{x}\wedge\mathrm{grad}\varphi$. It is worth stressing that Equation 7.17 for the torque does not involve the material derivative $\overset{\star}{\varphi}$ of the potential. This is, however, required to differentiate $\mathbf{M}$.

Material Derivative of the Potential Since the material derivative of the weak form of Equation 7.8 vanishes identically,

$$\int_{\Omega} -\operatorname{grad}\tilde{\varphi}\cdot\operatorname{grad}\boldsymbol{\theta}\cdot\operatorname{grad}\varphi + \operatorname{grad}\tilde{\varphi}\cdot\left(\operatorname{grad}\overset{\star}{\varphi} - \operatorname{grad}\varphi\cdot\operatorname{grad}\boldsymbol{\theta}\right)$$
$$+ \operatorname{grad}\tilde{\varphi}\cdot\operatorname{grad}\varphi\ \operatorname{div}\boldsymbol{\theta}\ \mathrm{d}\Omega = 0, \tag{7.18}$$

we have

$$\int_{\Omega}\operatorname{grad}\tilde{\varphi}\cdot\operatorname{grad}\overset{\star}{\varphi}\ \mathrm{d}\Omega = \int_{\Omega}\|\operatorname{grad}\varphi\|^2(\operatorname{grad}\tilde{\varphi}\cdot\mathbf{U})\ \mathrm{d}\Omega \tag{7.19}$$
$$= \left(\int_{\Omega}(\mathbf{x}\wedge\operatorname{grad}\tilde{\varphi})\|\operatorname{grad}\varphi\|^2\ \mathrm{d}\Omega\right)\cdot\boldsymbol{\omega}$$

It is worth stressing that the discretization of Equation 7.19 leads to the same matrix as the primary problem, which can be efficiently factorized only once and solved for different right-hand-side vectors.

Let $\overset{\star}{\varphi}_{\omega}$ denote a vector such that

$$\overset{\star}{\varphi} = \overset{\star}{\varphi}_{\omega}\cdot\boldsymbol{\omega}. \tag{7.20}$$

The components of $\overset{\star}{\varphi}_{\omega}$ can be computed from Equation 7.19 by fixing three independent choices for $\boldsymbol{\omega}$.

Differentiation of Torque The procedure can be iterated by applying material derivatives to Equation 7.17. It turns out that $\overset{\star}{\mathbf{C}} = \overset{\star}{\mathbf{C}}_{\omega}\cdot\boldsymbol{\omega}$, with

$$\overset{\star}{\mathbf{C}}_{\omega} = \int_{\Omega} -\mathbf{a}\otimes(\operatorname{grad}\overset{\star}{\varphi}_{\omega}\cdot\operatorname{grad}\varphi) - \frac{1}{2}\|\operatorname{grad}\varphi\|^2\mathbf{x}^{\wedge}\cdot\nabla^{\mathrm{T}}\overset{\star}{\varphi}_{\omega} + \|\operatorname{grad}\varphi\|^2\mathbf{a}\otimes\mathbf{a}$$
$$- \frac{1}{2}\varphi\|\operatorname{grad}\varphi\|^2(\mathbf{x}\otimes\operatorname{grad}\varphi - \operatorname{grad}\varphi\otimes\mathbf{x})\ \mathrm{d}\Omega. \tag{7.21}$$

While Equation 7.21 is general and could be applied to any rigid-body rotation of the structure, for the case of interest herein only $\partial M(\psi)/\partial\psi$ is required (Figure 7.6). The virtual velocity is a rotation around the torsional axis, of unit vector $\mathbf{e}$; hence $\mathbf{V} = \mathbf{0}$, $\boldsymbol{\omega} = \mathbf{e}$ and the material derivative of the potential is $\overset{\star}{\varphi}_e = \mathbf{e}\cdot\overset{\star}{\varphi}_{\omega}$. Moreover $\partial M(\psi)/\partial\psi = \overset{\star}{\mathbf{C}}_{\omega}\cdot\mathbf{e}$. Eventually, setting $\overset{\star}{\mathbf{a}} = \mathbf{x}\wedge\operatorname{grad}\overset{\star}{\varphi}_e$:

$$\frac{\partial M(\psi)}{\partial\psi} = \int_{\Omega} -(\mathbf{a}\cdot\mathbf{e})(\operatorname{grad}\overset{\star}{\varphi}_e\cdot\operatorname{grad}\varphi) - \frac{1}{2}\|\operatorname{grad}\varphi\|^2(\overset{\star}{\mathbf{a}}\cdot\mathbf{e}) + \|\operatorname{grad}\varphi\|^2(\mathbf{a}\cdot\mathbf{e})^2\ \mathrm{d}\Omega. \tag{7.22}$$

7.2.2 Experimental Set-Up

The opening angle of the electrostatic micromirror was monitored through experimental measurement of the deflection of a laser beam incident on the device. The set-up scheme for this test is reported in Figure 7.7. In particular, the activation signal was produced by the function generator 'Agilent 33521 A'. A He-Ne laser was preferred in

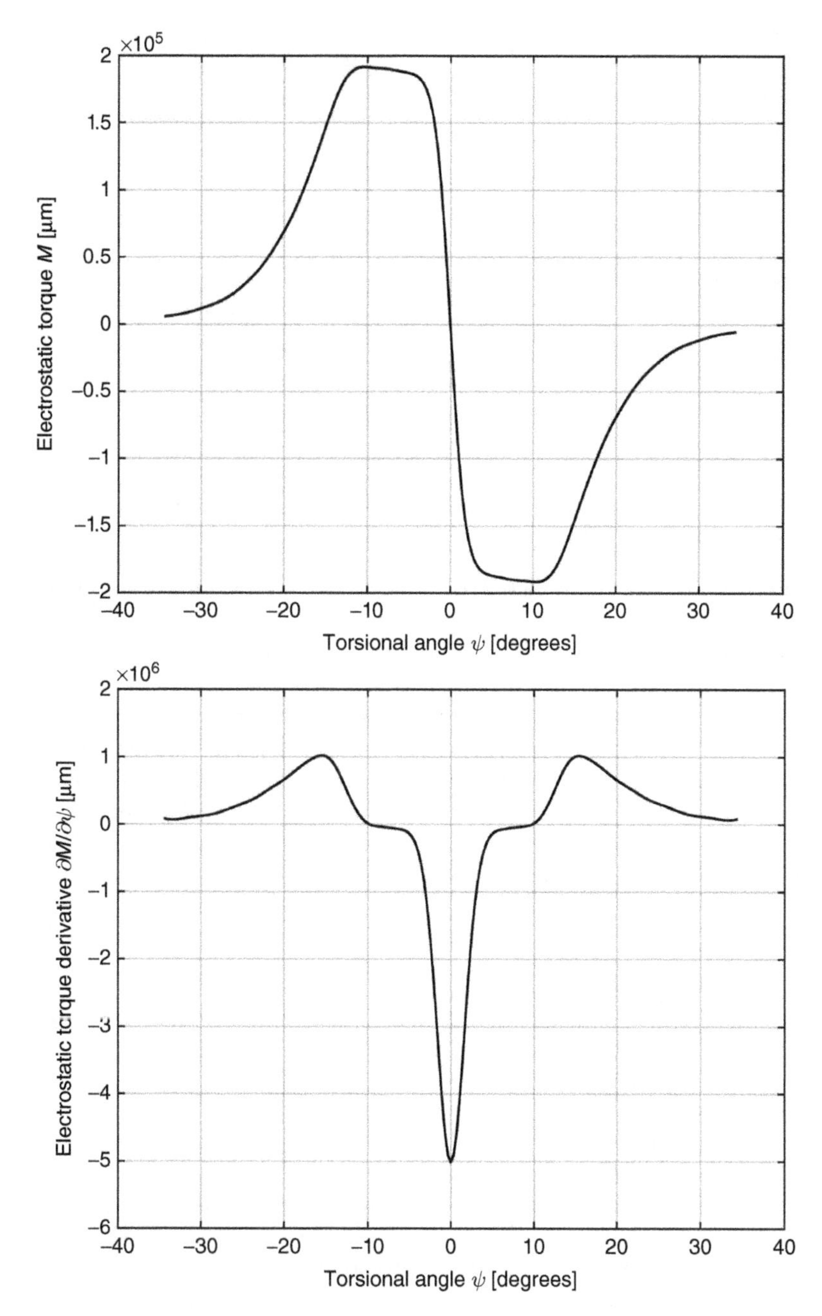

Figure 7.6 Electrostatic torque and derivative of electrostatic torque for one set of comb fingers with respect to torsional angle ψ.

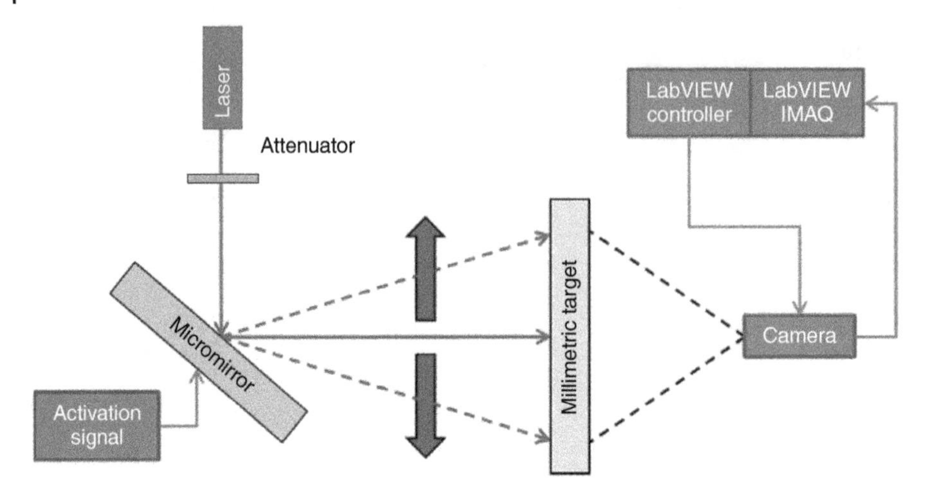

Figure 7.7 Set-up for dynamic characterization of the mirror. *Source*: Frangi *et al.* (2017), part of Figure 8. Reproduced with permission of IEEE.

order to have a spot size smaller than the mirror diameter, without using other optical instrumentation for beam collimation.

An attenuator was positioned in front of the laser to reduce the laser intensity and avoid saturating the camera. The mirror was mounted on an optical alignment bench to provide a 45° laser beam incidence and reflection angle and a perpendicular projection of the reflected beam on the millimetric target. The resonant movement of the mirror spans a laser segment, which is acquired by a camera and processed through a dedicated LabVIEW script. Accurate calibration of the camera was performed to obtain the relation between camera pixels and physical lengths. The mechanical opening angle is calculated through the following relation:

$$\psi_{\text{mech}} = \frac{1}{2}\psi_{\text{opt}} = \frac{1}{2}\arctan\left(\frac{L}{2d}\right) \tag{7.23}$$

where L is the distance between the target and the mirror and d is the laser segment length. The control software gives the possibility to set the scanning frequency, range, and step and direction of sweep (up or down).

7.2.2.1 Sinusoidal Excitation

In the case of pure tone excitation, the mirror was actuated with sinusoidal voltages $\Phi(t) = \Phi_0 \cos \omega t$ in the range of frequencies around 5 kHz.

The driving term Φ^2 contains the term $\cos 2\omega t$ and the theory recalled in the previous section predicts that, in the absence of dissipation, a nontrivial response is obtained in regions of the space ω, Φ (instability tongues) emanating from the axis $\Phi = 0$ at $\omega = \omega_0/n, n > 0$. In the 5 kHz case, corresponding to $n = 1$, the mirror shows classical nontrivial softening behaviour. Figure 7.5 collects experimental results and numerical simulations for $\Phi_0 = 55$ V, for both the upward and downward sweep. Considering that only stable branches could be reproduced in the experiments, the agreement with numerical data is, in general, impressive. It is worth recalling that the model is almost exact, the only limitation being the simplified dissipative term $b\dot{\psi}$. From the comparison with experimental data, it is apparent that b should be, in general, a (weak) function of ψ.

Tongues with $n > 1$ are very narrow and difficult to measure experimentally or even simulate numerically. Moreover, in the presence of important damping, as for the micromirror, tongues with $n > 1$ only exist for large Φ.

7.2.2.2 Square Wave Excitation

If the square wave excitation is utilized, peculiar phenomena occur that could not be explained in the context of the classical harmonic resonance. When the input frequency is in the range of 2 kHz, no Fourier component for Φ^2 equals the torsional eigenfrequency, but the mirror is actually parametrically driven by the (fifth) Fourier component at 11 kHz, i.e. $2\omega_0$, which is indeed rather small in amplitude. This explains why the maximum opening angles are limited in Figure 7.8, relevant to $\Phi_0 = 110$ V. Moreover, the hysteresis loop is extremely narrow, which makes the continuation tricky. It is worth stressing that the continuation approach only works if the forcing and mirror response are periodic, which means that two periods of the square wave and five periods of the mirror response must be simulated as one 'global' period T (see Equation 7.3). On increasing the frequency of the input square wave, the torsional mode of the mirror is activated whenever there exists an integer n such that $n\omega \simeq 2\omega_0$. For instance, at 5 kHz, one has $n = 2$ and at 10 kHz, $n = 1$. These two cases have been investigated numerically and experimentally, the former being the normal operating mode of the mirror. Figure 7.9 refers to the 5 kHz case.

Finally, Figure 7.10 presents the comparison when the mirror is actuated with a square wave in a frequency range around 10 kHz. In this case, in the power spectrum, the first high energy Fourier component is at $2\omega_0$ and hence energetically activates the mirror

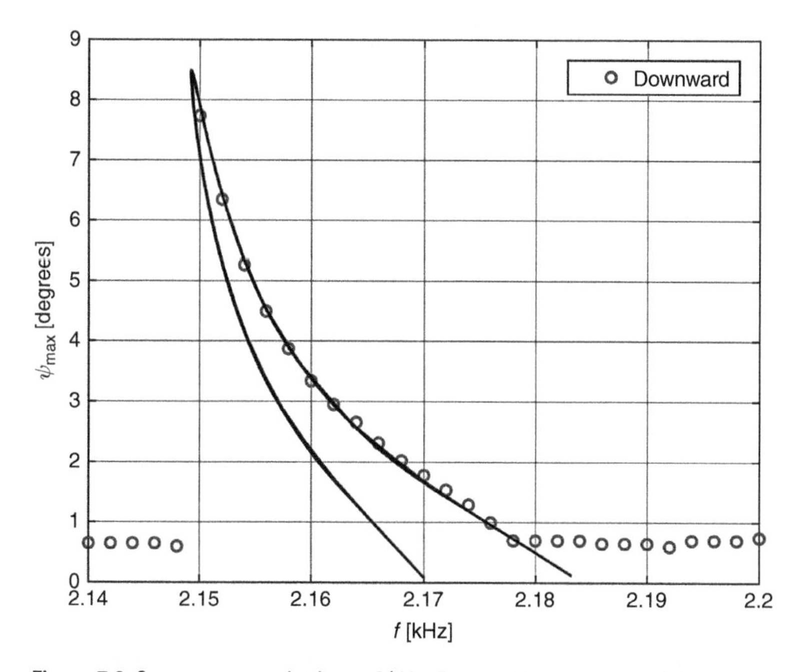

Figure 7.8 Square wave excitation at 2 kHz, $\Phi_0 = 110$ V; experimental downward sweep (discrete symbols) and numerical continuation (continuous line). *Source*: Frangi *et al.* (2017), Figure 14. Reproduced with permission of IEEE.

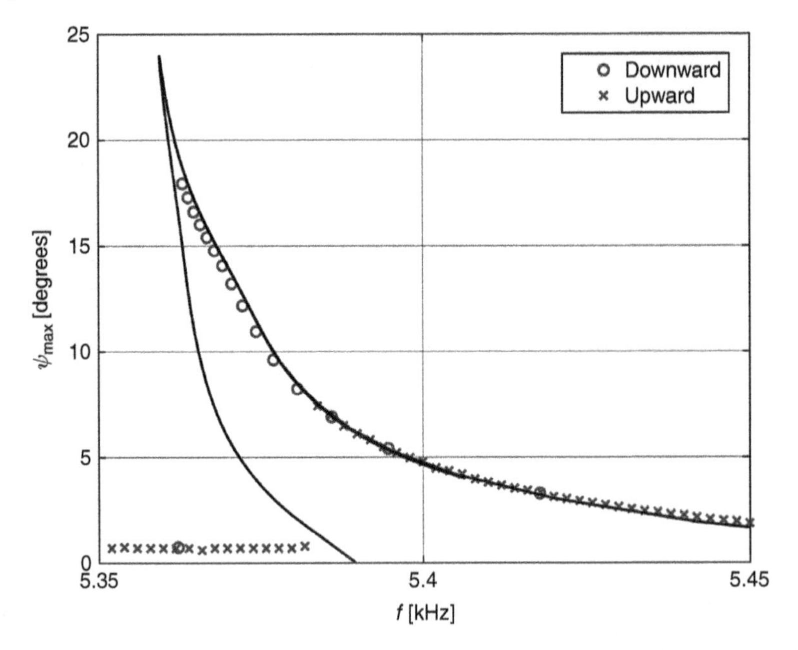

Figure 7.9 Square wave excitation at 5 kHz, $\Phi_0 = 110$ V; experimental upward and downward sweeps (discrete symbols) and numerical continuation (continuous line). *Source*: Frangi *et al.* (2017), Figure 17. Reproduced with permission of IEEE.

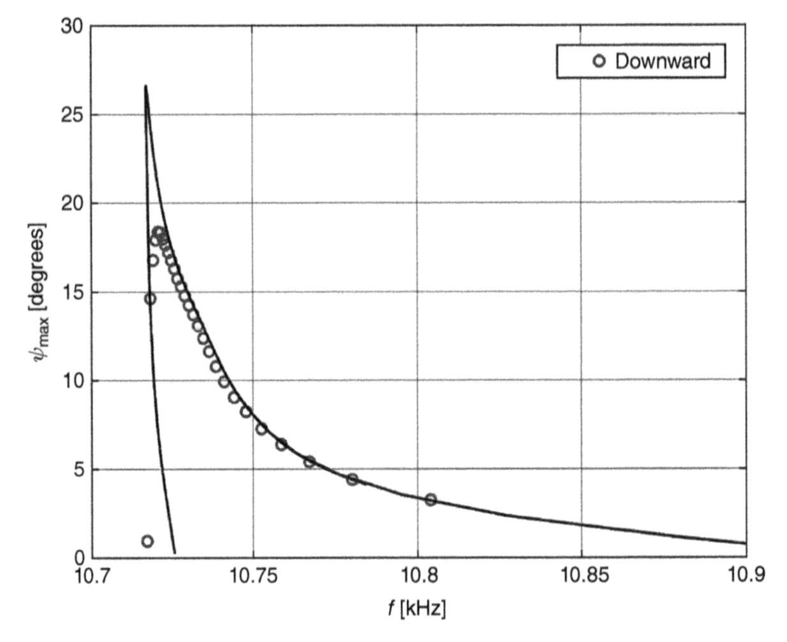

Figure 7.10 Square wave excitation at 10 kHz, $\Phi_0 = 90$ V; experimental downward sweep (discrete symbols) and numerical continuation (continuous line). *Source*: Frangi *et al.* (2017), Figure 19. Reproduced with permission of IEEE.

response. Even though the Φ_0 voltage utilized in this case is rather low, the mirror experiences very large rotations, inducing partial failure of the experimental set-up, which limits the validity of experimental data at the peak.

References

Ardito, R., Baldasarre, L. and Corigliano, A. (2009) *On the Numerical Evaluation of Capacitance and Electrostatic Forces in MEMS*, EuroSimE 2009 – 10th International Conference on Thermal, Mechanical and Multi-Physics Simulation and Experiments in Microelectronics and Microsystems, April 26–29, 2009, Delft, Netherlands, IEEE.

Bonnet, M. (1995) Regularized BIE formulations for first- and second-order shape sensitivity of elastic fields. *Computers & Structures*, **56**, 799–811.

Carr, D.W., Evoy, S., Sekaric, L. *et al.* (2000) Parametric amplification in a torsional microresonator. *Applied Physics Letters*, **77**, 1545–1547.

Cho, H.J. and Ahn, C.H. (2003) Magnetically-driven bi-directional optical microscanner. *Journal of Micromechanics and Microengineering*, **13**, 383.

Doedel, E.J. and Kernévez, J.P. (1986) AUTO: Software for continuation problems in ordinary differential equations with applications, *Tech. Rep.*, Applied Mathematics, California Institute of Technology.

Fan, L. and Wu, M.C. (1998) *Two-Dimensional Optical Scanner with Large Angular Rotation Realized by Self-Assembled Micro-elevator*, IEEE/LEOS Summer Topical Meeting on Broadband Optical Networks and Technologies: An Emerging Reality/Optical MEMS/Smart Pixels/Organic Optics and Optoelectronics, July 20–24, 1998, Monterey, CA, IEEE.

Filhol, F., Defay, E., Divoux, C. *et al.* (2005) Resonant micro-mirror excited by a thin-film piezoelectric actuator for fast optical beam scanning. *Sensors and Actuators A: Physical*, **123–124**, 483–489.

Frangi, A., Guerrieri, A., Carminati, R. and Mendicino, G. (2017) Parametric resonance in electrostatically actuated micromirrors. *IEEE Transactions on Industrial Electronics*, **64**, 1544–1551.

Frangi, A., Spinola, G. and Vigna, B. (2006) On the evaluation of damping in MEMS in the slip-flow regime. *International Journal for Numerical Methods in Engineering*, **68**, 1031–1051.

Jain, A., Qu, H., Todd, S. *et al.* (2004) *Electrothermal SCS Micromirror with Large-Vertical-Displacement Actuation*, Solid-State Sensor and Actuator Workshop, June 6–10, 2004, Hilton Head Island, SC, IEEE.

Lab4MEMS II (2014) *ENIAC Project, Micro-Optical MEMS, Micro-mirrors and Pico-projectors*, http://www.lab4mems2.ite.waw.pl/ (accessed 5 July 2017).

Lacarbonara, W. (2013) *Nonlinear Structural Mechanics*, Springer.

Lee, C. (2004) Design and fabrication of epitaxial silicon micromirror devices. *Sensors and Actuators A: Physical*, **115**, 581–590.

Mirzazadeh, R., Mariani, S. and De Fazio, M. (2014) Modeling of Fluid Damping in Resonant Micro-Mirrors with Out-of-Plane Comb-Drive Actuation, International Electronic Conference on Sensors and Applications, June 1–16, 2014, Sciforum.

Nayfeh, A.H. (1981) *Introduction to Perturbation Techniques*, John Wiley & Sons, Inc.

Nayfeh, A.H. and Mook, D.T. (1979) *Nonlinear Oscillations*, John Wiley & Sons, Inc.

Ren, Z., Qu, H. and Xu, X. (2012) Computation of second order capacitance sensitivity using adjoint method in finite element modeling. *IEEE Transactions on Magnetics*, **48**, 231–234.

Rhoads, J., Shaw, S., L. Turner, K., and Baskaran, R. (2005) Tunable microelectromechanical filters that exploit parametric resonance. *Journal of Sound and Acoustics*, **127**, 423–430.

Rhoads, J.F., Shaw, S.W., Turner, K.L. *et al.* (2006) Generalized parametric resonance in electrostatically actuated microelectromechanical oscillators. *Journal of Sound and Vibration*, **296**, 797–829.

Rugar, D. and Grütter, P. (1991) Mechanical parametric amplification and thermomechanical noise squeezing. *Physical Review Letters*, **67**, 699–702.

Schroth, A., Lee, C., Matsumoto, S. and Maeda, R. (1999) Application of sol-gel deposited thin PZT film for actuation of 1D and 2D scanners. *Sensors and Actuators A: Physical*, **73**, 144–152.

Shmulevich, S., Grinberg, I.H. and Elata, D. (2015) A MEMS implementation of a classic parametric resonator. *Journal of Microelectromechanical Systems*, **24**, 1285–1292.

Singh, J., Gan, T., Agarwal, A. *et al.* (2005) 3D free space thermally actuated micromirror device. *Sensors and Actuators A: Physical*, **123–124**, 468–475.

Thompson, M.J. and Horsley, D.A. (2011) Parametrically amplified Z-axis Lorentz force magnetometer. *Journal of Microelectromechanical Systems*, **20**, 702–710.

Turner, K.L., Miller, S.A., Hartwell, P.G. *et al.* (1998) Five parametric resonances in a microelectromechanical system. *Nature*, **396**, 149–152.

Wagg, D. and Neild, S. (2015) *Nonlinear Vibration with Control*, Springer.

Yalcinkaya, A.D., Urey, H., Brown, D. *et al.* (2006) Two-axis electromagnetic microscanner for high resolution displays. *Journal of Microelectromechanical Systems*, **15**, 786–794.

Younis, M.I. (2011) *MEMS Linear and Nonlinear Statics and Dynamics*, Springer.

Zalalutdinov, M., Olkhovets, A., Zehnder, A. *et al.* (2001) Optically pumped parametric amplification for micromechanical oscillators. *Applied Physics Letters*, **78**, 3142–3144.

Zhang, W., Baskaran, R. and Turner, K.L. (2002) Effect of cubic nonlinearity on auto-parametrically amplified resonant MEMS mass sensor. *Sensors and Actuators A: Physical*, **102**, 139–150.

8

Vibrating Lorentz Force Magnetometers

8.1 Introduction

The purpose of this chapter is to discuss recent advancements in the design of vibrating Lorentz force magnetometers.

A magnetometer is a sensor designed to measure the magnetic field surrounding the instrument, typically the Earth's magnetic field, as in the case of a compass. Magnetic field is a vectorial quantity, therefore, like accelerometers and gyroscopes, magnetometers can be designed as single-axis or multiaxis sensors, where each axis measures one component of the vectorial magnetic field projected onto the reference frame attached to the MEMS box.

Inertial measurement units based on multiaxis MEMS devices integrated in the same chip have several applications, including civil and military aviation, space satellites, trains, ships, unmanned or remote operated vehicles, stabilization systems and consumer electronics (Robin and Perlmutter, 2011). In particular, the integration of a three-axis accelerometer, a three-axis gyroscope, a three-axis magnetometer and a pressure sensor, all based on the same MEMS process, can result in a '10-dof' high-resolution, low-cost, low-power miniaturized system for position, motion, heading and altitude monitoring. The state of the art is represented by a six-axis MEMS unit for acceleration and angular rate sensing (STMicroelectronics, 2011; InvenSense, 2012) while magnetometers of inertial measurement units are still based on non-MEMS technologies. For instance, the six-axis system-in-package for acceleration and magnetic field sensing described by STMicroelectronics (2009) includes magnetometers based on anisotropic magnetoresistance (Afzal, Renaudin and Lachapelle, 2011).

However, as the anisotropic magnetoresistance magnetometer for the z-axis (orthogonal to the plane of the surface mounting of the board hosting the inertial measurement unit) is exactly the same as for the x- and y-axes, but twisted by 90°, a large volume is wasted in the height of the final packaging owing to the required out-of-plane assembly. Therefore, most research has concentrated on the development of a z-axis MEMS magnetometer (Emmerich and Schofthaler, 2000). In an interesting intermediate step, a nine-dof inertial measurement unit can be realized with a two-axis anisotropic magnetoresistance magnetometer and a seven-dof MEMS die.

Mechanics of Microsystems, First Edition. Alberto Corigliano, Raffaele Ardito, Claudia Comi, Attilio Frangi, Aldo Ghisi, and Stefano Mariani.

Companion website: www.wiley.com/go/corigliano/mechanics

In this chapter, we focus on the design of z-axis Lorentz-force-based capacitive MEMS magnetometers operating at resonance in a typical industrial packaging, for consumer applications (digital compass, heading, dead reckoning, map rotation). These generally require a resolution of around a few microtesla over a bandwidth of approximately 50 Hz, with a full-scale linear range less than 1 mT. Several constraints are imposed by the industrial process and packaging (Bahreyni and Shafai, 2007): the maximum device dimension should fit into the typical MEMS die (with a side length of 2 mm, including dead areas); the required vacuum level should cope with industrial packages (with a typical pressure in the range of 1 mbar); the resonance frequency should lie outside the acoustic bandwidth (>20 kHz); power dissipation should be competitive with respect to anisotropic magnetoresistance technology (<1.5 mW).

As in other chapters of Part Two of this book concerning devices, we discuss herein examples taken from the authors' experience; these are meant as meaningful, practical applications of general mechanical and multiphysics principles applied to microsystems and should not be interpreted as a complete review of what was recently proposed in the literature concerning the various devices discussed.

8.2 Vibrating Lorentz Force Magnetometers

8.2.1 Classical Devices

In the scientific literature, there are several examples of MEMS magnetometers based on the Lorentz force principle. In all these devices, in the presence of a magnetic field of intensity B and direction $\mathbf{e}$, a driving current i flows in a suspended structure and generates a force orthogonal to both $\mathbf{e}$ and the current direction.

Though the induced motion can be sensed in different ways (e.g. through a change of resistance in piezoresistors (Ghorba *et al.*, 2007), or through a change in the resonance frequency of suitably designed resonators (Bahreyni and Shafai, 2007)), the most frequently used technique is the capacitive read-out. Figure 8.1 shows a typical z-axis parallel-plate MEMS magnetometer. Two springs of length L constitute the suspending element of a central frame, which forms, through a set of fixed stators, two differential parallel-plate capacitors C_1 and C_2 for sensing.

The AC current $i(t) = i_0 \sin(2\pi f_0 t)$ flows through the springs at the device resonant frequency f_0.

A 1D model of the device can easily be built, assuming: (i) the central shuttle is rigid and its horizontal displacement is denoted by $Y(t)$; (ii) each of the four half-springs deforms according to the classical cubic function $\psi(x)$, which is defined to be compliant with the boundary conditions $\psi(0) = 0$, $\psi'(0) = \psi'(L) = 0$ and $\psi(L/2) = Y$, where $x = 0$ at the anchor and $x = L/2$ at the shuttle.

As explained in Chapter 3, this gives, for the whole device, an equivalent stiffness k and an equivalent mass m:

$$k = 4EJ \int_0^{L/2} (\psi''(x))^2 \, dx = 384 \frac{EJ}{L^3},$$

$$m = m_S + 4\rho A \int_0^{L/2} \psi^2(x) \, dx = m_S + \frac{26}{35} \rho A L,$$

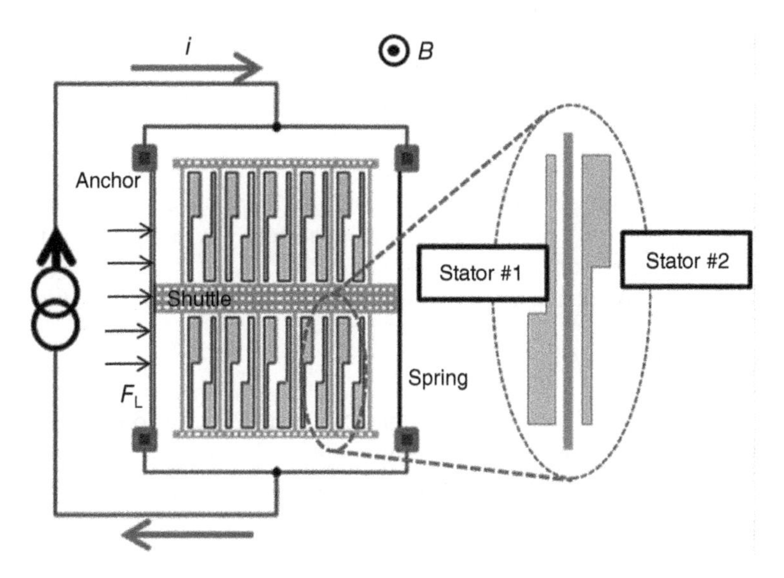

Figure 8.1 *z*-axis parallel-plate MEMS magnetometer. The suspended mass is subject to a Lorentz force in the presence of a magnetic field; the corresponding displacement can be sensed through the pair of differential capacitors. *Source*: Langfelder *et al.* (2013), Figure 1. Reproduced with permission of IEEE.

where m_S is the mass of the central shuttle and of the attached plates, EJ is the bending stiffness of the springs and ρA is the mass per unit length of the springs. The equivalent Lorentz force F_L acting on a half spring is

$$F_L(t) = i(t)B \int_0^{L/2} \psi(x)\,\mathrm{d}x = i(t)BL/4$$

and the 1D model finally reads:

$$m\ddot{Y} + b\dot{Y} + kY = 4F_L.$$

In this formula, b is the damping coefficient. Even if the MEMS of interest in this investigation has relatively low package pressures, in the range of 1 mbar, the dominant loss mechanism is still associated with the interaction of the rarefied gas with the structure. It is worth stressing that, for these devices, the Knudsen number, defined as $\mathrm{Kn} = \lambda/\ell$, where ℓ is a characteristic length of the flow (e.g. the gap between parallel plates) and λ the mean free path of gas molecules, is typically well above 10. Working conditions hence fall within the free-molecule regime, in which collisions between molecules can be neglected (see Chapter 15). For frequencies in the range of 20 kHz, the following quasi-static assumption holds:

$$\frac{f}{\sqrt{2\mathcal{R}T}} Y_0 \ll 1,$$

where f is the vibration frequency, $\mathcal{R}$ is the universal gas constant divided by the molar mass, T is the working temperature and Y_0 is the maximum amplitude of the oscillation. As a consequence, the damping coefficient has the form:

$$b = \xi A_\xi \rho_g \sqrt{2\mathcal{R}T} = A_\xi b_A, \tag{8.1}$$

where ρ_g is the gas density and A_ξ is the overall facing area of the parallel plates, that is the product of the process height H, the length of each parallel plate L_C, and the number of differential sensing cells N (i.e. the number of shuttle plates). The coefficient b_A is the damping per unit facing area.

The nondimensional constant ξ depends only on the problem geometry and is computed using the integral approach described in Chapter 15. In our case, we have $\xi = 23.5$. It is clear from Equation 8.1 that, at constant temperature, the viscous force is proportional, through the density, to the working pressure p.

Globally, the shuttle displacement can be expressed as

$$Y(t) = \frac{i(t)BLQ}{k}, \qquad Q = 2\pi f_0 \frac{m}{b},$$

where Q is the quality factor of the device (see Chapter 2, Section 2.2.3). The differential capacitance variation for a displacement $Y(t)$, in the assumption that displacements are much smaller than the nominal air gap between parallel plates g and neglecting fringe effects as a first approximation, can be written as

$$\Delta C(t) = 2C_0 \frac{Y(t)}{g} = 2\epsilon_0 A_\xi \frac{Y(t)}{g^2} = 2\epsilon_0 L_C NH \frac{Y(t)}{g^2},$$

where ϵ_0 is the electrical permittivity inside the package (assumed as that of vacuum) and A_ξ is the overall facing area of the parallel plates. The mechanical sensitivity $\Delta C/(\Delta B)$ of the magnetometer, defined as the capacitance variation per variation of magnetic field, finally becomes

$$\frac{\Delta C}{\Delta B} = 2\epsilon_0 L_C NH \frac{i(t)LQ}{g^2 k} = \epsilon_0 L_C NH \frac{i(t)L\sqrt{mk}}{g^2 kb} = \epsilon_0 L_C NH \frac{i(t)L}{g^2 2\pi f_0 b}.$$

It thus turns out that the sensitivity is almost independent of the number of parallel-plate cells:

$$\frac{\Delta C}{\Delta B} = \frac{\epsilon_0 i(t)L}{4\pi g^2 f_0 b_A}. \tag{8.2}$$

Looking at Equation 8.2, some considerations can be made about possible optimization of the design:

1) The minimum air gap is set by the technology $g = g_0$. Choosing the minimum gap means increasing the process variance and decreasing the repeatability from part to part, necessitating self-calibration required.
2) The resonance frequency f_0 is set in the range 20–30 kHz to avoid acoustic interference, as commonly done for gyroscopes.
3) The damping coefficient per unit area b_A is constrained by the packaging pressure and by the open-loop bandwidth of the device, $BW = f_0/2Q$. As an example, to give a bandwidth of 50 Hz, the quality factor should not exceed around 200–300.
4) The spring length L can be increased to increase the sensitivity (adjusting the spring width accordingly, to cope with point 2) but 1 mm is typically an upper bound with the current technology.
5) The driving current i can be increased to boost the sensitivity at the cost of an augmented power dissipation by the Joule effect in the spring resistance.

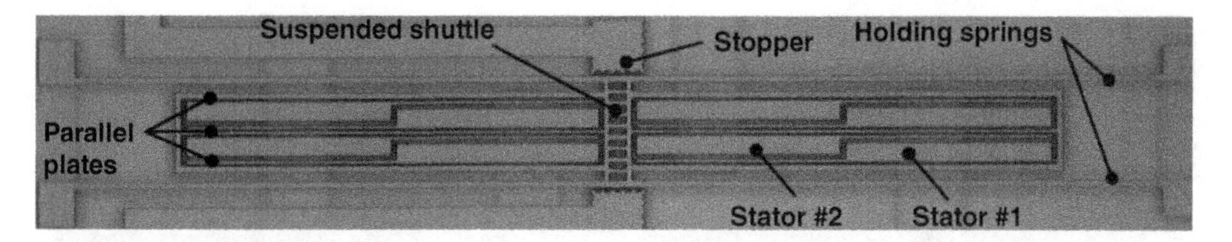

Figure 8.2 Simple *z*-axis magnetometer. Each spring is 980 μm long, while the dimension in the orthogonal direction is 89 μm. *Source*: Langfelder *et al.* (2012), Figure 2. Reproduced with permission of IEEE.

Table 8.1 Parameters of the designed magnetometers (theoretical resonance frequency $f_0 = 28$ kHz).

Number of plates	m [nkg]	k [N/m]	Q	$\frac{\Delta C}{i\Delta B}$ [aF/(μT mA)]
4	0.56	17.85	481	0.841
8	0.78	24.50	335	0.843
16	0.90	28.35	258	0.844
32	1.80	57.05	223	0.845

The mentioned independence of the sensitivity on the number of sensing cells is relevant as it suggests the design of an almost 1D z-axis MEMS magnetometer (see, for instance, the example shown in Figure 8.2), which takes up a small area and fits in a lateral side of the same package of the gyroscope and which shares with the magnetometer the same pressure requirement. This is a rigorous result, which contrasts with several examples of parallel-plate magnetometers in the literature showing a large number of parallel plates (Emmerich and Schofthaler, 2000).

Some experimental data taken from Langfelder *et al.* (2013) and collected in Table 8.1 further validate this conclusion. In this device, $H = 15$ μm and the gap is chosen to be the minimum, $g = 2$ μm, so that the sensitivity is maximized. Finally, the sensitivity does not depend on the length of the parallel plates L_C, yet the open-loop bandwidth $f_0/2Q$ is a function of the damping coefficient and, in turn, of the length L_C. Therefore, we chose to have an overall parallel-plate length of 330 μm, which, from simulations, guarantees a maximum Q of around 450 at the reference pressure.

8.2.2 Improved Design

From the discussion presented in the previous section it is clear that, in order to improve the sensitivity of this family of magnetometers, an innovative mechanical design is required.

In Frangi *et al.* (2015a), a very simple device has been proposed, where an AC current flows through a doubly clamped beam at its first resonating frequency. The movement of the beam is sensed via two capacitors (stators) having different configurations. In the first option, the beam is placed within standard continuous parallel plates; in the second, suitable interruptions have been etched in the stators (*block-shaped stators*); in the third, the beam is designed within two *comb-shaped stators*, as shown in Figure 8.3.

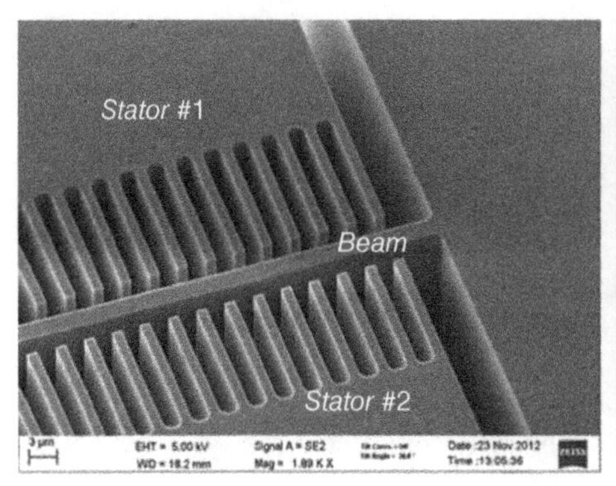

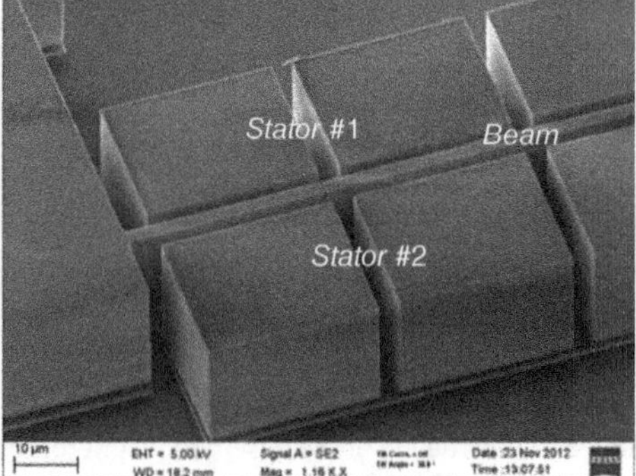

Figure 8.3 Magnetometers featuring a current flowing through the clamped–clamped beam, with motion sensed via optimized stators having a comb-shaped configuration (left) or a block-shaped configuration (right). *Source*: Frangi *et al.* (2015a), Figure 1. Reproduced with permission of IEEE.

Devices were built using the ThELMA process from STMicroelectronics. The beam length and the process thickness are 987.6 μm and 22 μm, respectively. The beam width is nominally 3.4 μm, with a ±0.15 μm tolerance. The nominal air gap between the beam and the stators is 2.4 μm for the parallel plates and block-shaped devices and 2 μm for the comb-shaped devices, respectively. The possibility of using a smaller gap for the comb-shaped structure is associated with the production technology utilized. When the gap between the beam and the stators is too narrow, the dry hydrofluoric acid attack is not effective in etching the oxide beneath the polysilicon and releasing the suspended beam, as desired. This sets a limit on the smallest admissible gap. In the case of the comb-shaped structure, the length of the regions where the gap is small is very limited and comparable to the gap size itself. This enables the gap to be reduced without issues in the release process, as demonstrated by the correct operation of these devices. The gap of 2 μm for the comb-shaped design is used to compensate the lower sensitivity caused by its smaller facing area.

Each block-shaped stator is formed of 30 blocks (29.3 μm long and 23.7 μm deep) separated by a 3.5 μm air gap. The electrical connection of the blocks is guaranteed by a high-conductivity polysilicon route placed beneath the structure. Comb-shaped stators are composed of 306 comb fingers (1.2 μm thick and 14.3 μm deep) laterally separated by a 2 μm air gap.

To estimate the mechanical parameters, in the framework of the finite element method, equilibrium is enforced in a weak form via the principle of virtual work for slender beams (see Chapter 3, Section 3.2.1). Letting x be a coordinate running along the axis of the beam, the problem is formulated as follows. Find a kinematically compatible displacement field $v(x,t) \in C$, such that, $\forall \delta v \in C(0)$:

$$\int_0^L \rho A \ddot{v}(x,t)\delta v(x)\,\mathrm{d}x + \int_0^L EJ v''(x,t)\delta v''(x)\,\mathrm{d}x = \int_S f(\mathbf{x},t)\delta v(x)\mathrm{d}S, \tag{8.3}$$

where δv is a virtual kinematic field; $C(0)$ is a suitable space of sufficiently continuous functions v, such that $v(0) = v(L) = 0$, $v'(0) = v'(L) = 0$; ρA is the mass per unit length; EJ is the bending stiffness and $f(\mathbf{x},t)$ is the component of the forces orthogonal to the beam surface.

It is worth stressing that the last integral in Equation 8.3 is extended to the whole lateral surface S of the beam since $f(\mathbf{x})$ depends, a priori, on all three Cartesian coordinates. A simplified, but accurate, simulation technique involves choosing C as a 1D space with

$$v(x,t) = \psi(x)Y(t), \qquad \delta v(x) = \psi(x)\delta Y,$$

where $\psi(x)$ is the first modal shape for a doubly clamped beam (see Section 3.4):

$$\psi(x) = B_1(\cos(\lambda x) - \cosh(\lambda x)) + C_1(\sin(\lambda x) - \sinh(\lambda x)),$$

with $\lambda = 4.73/L$ and

$$\frac{C_1}{B_1} = \frac{\cos(\lambda L) - \cosh(\lambda L)}{\sin(\lambda L) - \sinh(\lambda L)},$$

normalized such that $\psi(L/2) = 1$.

Inserting the assumed interpolation in Equation 8.3, one gets the standard 1D model:

$$m\ddot{Y}(t) + kY(t) = F(t),$$

where

$$m = \int_0^L \rho A\psi^2(x)\,\mathrm{d}x = 0.396\ \rho AL, \tag{8.4}$$

$$k = \int_0^L EJ(\psi'')^2(x)\,\mathrm{d}x = 198.46\ EJ/L^3, \tag{8.5}$$

$$F(t) = \int_S f(\mathbf{x},t)\psi(x)\mathrm{d}S. \tag{8.6}$$

Two different contributions to the force F are analysed here. One is the Lorentz generalized force F_L due to the magnetic field B and current $i(t)$:

$$F_\mathrm{L}(t) = Bi(t)\int_0^L \psi(x)\,\mathrm{d}x = 0.523\ i(t)BL. \tag{8.7}$$

The other is the force due to the interaction of the vibrating beam with the surrounding gas, which generates the dominant dissipative contribution, $-b\dot{x}(t)$.

The viscous damping coefficient c can be expressed (see the previous paragraph and Chapter 15) as

$$b = \xi hL\rho\sqrt{2\mathcal{R}T}. \tag{8.8}$$

Table 8.2 reports values of ξ computed for the three different configurations described here (parallel plates, block-shaped devices and comb-shaped devices). It is worth stressing that the viscous damping coefficient b and, hence, the quality factor Q can be computed for any value of pressure p. For convenience, in the last column of Table 8.2 the quality factors are listed only at the nominal working pressure of 1 mbar, for each device. Considering that the quality factor is proportional to $1/p$ in the free-molecule regime, the value of Q at any pressure can be easily computed starting from the third column of Table 8.2. Other sources of dissipation have been neglected in the 1D model.

Geometrical dimensions and extracted mechanical parameters of the presented devices are summarized in Table 8.3. The first resonance mode of the beam, assuming the nominal width, can be calculated to be at 28.82 kHz.

Table 8.2 Values of ξ for the different stator configurations.

Device	ξ constant	Q at 1 mbar
Parallel plates	9.75	120
Block-shaped	7.75	150
Comb-shaped	3.85	303

Table 8.3 Device parameters.

Parameter	Value	
	Block-shaped	Comb-shaped
Gap (x_0)	2.4 μm	2.0 μm
Beam length (L)	987.6 μm	
Beam width (w)	3.4 μm	
Thickness (h)	22 μm	
Young's modulus (E)	150 GPa	
Polysilicon density (ρ)	2320 kg/m^3	
Beam stiffness (k)	2.23 N/m	
Effective mass (m)	0.068 nkg	
Resonance frequency (f_0)	28.82 kHz	
Overetch spread	±0.15 μm	
Nominal package pressure (p)	1 mbar	

The electrostatic sensitivities ($S_{el} = \Delta C/\Delta Y$, with Y the amplitude of the oscillation at the centre of the beam) are predicted using the finite element method for both devices. The domain boundaries of the finite element method simulations resemble the fabrication process as closely as possible, taking into account the substrate lying beneath the structures and the oxide underneath the stators. Simulations are performed in three dimensions, by fixing Dirichlet boundary conditions for the substrate and the stators (grounded) and for the beam (whose voltage is set to an arbitrary DC value). Neumann conditions are set at the boundaries of the air volume surrounding the device. The finite element method software solves for the electrostatics equations and gives as an output the differential capacitance variation as a function of the centre displacement. For the block-shaped configuration, one has $\Delta C/\Delta Y = 35.7$ aF/nm, while for the comb-shaped configuration $\Delta C/\Delta Y = 45.7$ aF/nm. For comparison, the response of a standard parallel-plate configuration with a 2.4 μm gap was also simulated, obtaining $\Delta C/\Delta Y = 37.3$ aF/nm.

A rough analytical estimate of the absolute value of the differential capacitance variation for a beam within two continuous stators (parallel plates) can be obtained from:

$$S_{el} = \frac{\Delta C}{\Delta Y} = 2\varepsilon_0 \frac{H}{g_0^2} \int_0^L \psi(x)\, dx = 35.0 \text{ aF/nm}.$$

This estimate indeed neglects all fringe effects but can be fruitfully employed for comparison with the finite element method simulations.

Looking at the simulation results, one can note that the sensitivity of the block-shaped stator device is slightly lower (≈95%) than the parallel-plate one, while that of the comb-shaped stator device is even higher; this gain is obtained thanks to the smaller gap allowed by this particular shape (see Table 8.3). For the block-shaped device, the 5% loss in capacitance variation is due to the slight reduction of the facing area of the plates. For the comb-shaped device, the higher loss in facing area is compensated partially by the reduction of the gap and partially by increased fringing effects. The result is a capacitance variation that is 1.23 times greater than that achieved for the parallel-plate configuration. The purpose of changing the geometry (from the standard parallel-plate configuration) without worsening the electrostatic sensitivity $\Delta C/\Delta Y$ is thus fulfilled.

The differential capacitance variation (the two-sided difference $|\Delta C_1 - \Delta C_2|$ is assumed here) per unit input force, at resonance, can be written as

$$\frac{\Delta C}{\Delta F} = \frac{\Delta C}{\Delta Y}\frac{Y_0}{\Delta F} = \frac{\Delta C}{\Delta Y}\frac{Q}{k}. \tag{8.9}$$

Given an external magnetic field variation ΔB in the out-of-plane direction and a peak driving current i_0, the corresponding peak Lorentz force variation can be written as $\Delta F = \Delta B i_0 L$. The theoretical differential capacitance variation per unit magnetic field is thus

$$\frac{\Delta C}{\Delta B} = \frac{\Delta C}{\Delta Y}\frac{Q}{k} i_0 L. \tag{8.10}$$

Theoretical predictions, normalized to the driving current amplitude, are 1.15 aF/(μT mA) and 3.01 aF/(μT mA) for the block-shaped and the comb-shaped devices, respectively. The same parameter for the parallel-plate device turns out to be 0.97 aF/(μT mA).

These results can be compared with those obtained for a typical multi-parallel-plate configuration built with the same fabrication process, whose value at resonance is 0.84 aF/(μT mA), as reported in Table 8.1.

8.2.3 Further Improvements

Frangi *et al.* (2015b) have further pushed these ideas to the limit. Test devices, shown in the left panel of Figure 8.4, are formed by 3.6 μm-wide, 1050 μm-long clamped–clamped beams featuring a central suspended mass to set the frequency in the 15–20 kHz range, typical of MEMS gyroscopes. The devices are in the same package to guarantee identical pressure and similar etching conditions.

Four geometries are used for the sensing electrodes: standard parallel plates (Figure 8.4b), a holed stator (Figure 8.4c), a comb-shaped stator (Figure 8.4d) and an arrow-shaped stator (Figure 8.4e). All the electrodes are at a nominal distance of 1.8 μm from the suspended beam. The geometries are a trade-off between quality factor and sensitivity. The differential capacitance variation $\Delta C/\Delta B$ per unit magnetic field predicted numerically for the arrow-shaped stator is 9.01 aF/(μT mA), which represents a 3× improvement over the devices of the previous section and a 10× improvement over the classical devices analysed in Section 8.2.1.

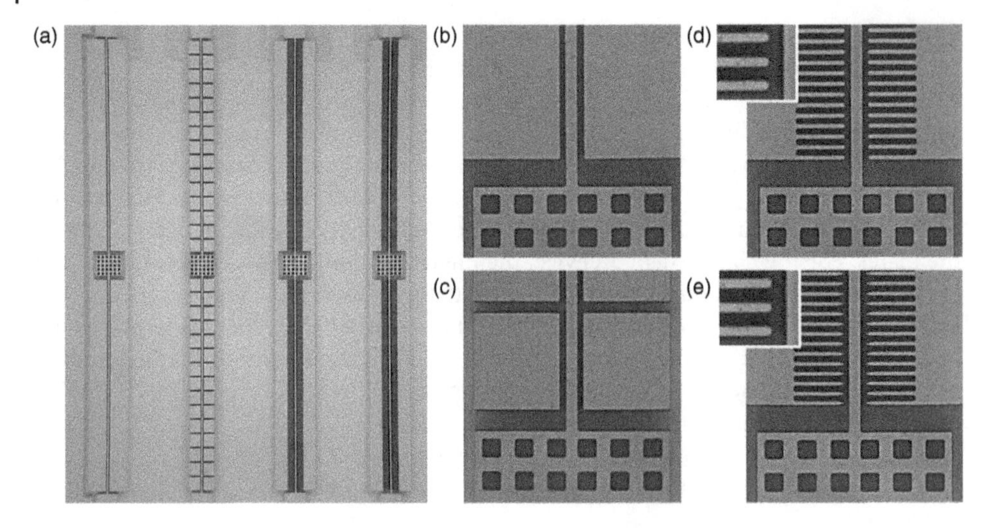

Figure 8.4 Optimized magnetometers made using a doubly clamped resonant shuttle beam and four different types of stator. *Source*: Frangi 2015, figure 1. Reproduced with permission of Elsevier.

8.3 Topology or Geometry Optimization

Starting from a slightly different perspective, Bagherinia *et al.* (2014) optimized the geometry of the resonant part of a z-axis Lorentz force magnetometer so as to maximize its sensitivity to the magnetic field and minimize its power consumption.

An SEM of the relevant vibrating structure is shown in Figure 8.5; since the whole device was composed of four equal beams resonating at the same frequency, the details reported in what follows refer to a single beam only. The flexible structure is a clamped–clamped beam with uniform rectangular cross-section, resonating in the plane parallel to the substrate surface due to an out-of-plane magnetic field B and a

Figure 8.5 SEM of the resonating structure of a z-axis Lorentz force magnetometer. *Source*: Bagherinia *et al.* (2014), Figure 1. Reproduced with permission of Elsevier.

current $i(t)$ flowing in its longitudinal direction. Sensing is achieved with two parallel plates attached to the midspan cross-section of the beam. Assuming that, at resonance, beam deflection $v(x,t)$ can be multiplicatively decomposed as:

$$v(x,t) = \frac{1}{2}\,(1 - \cos(2\pi x/L))\,Y(t), \tag{8.11}$$

where x is the lateral displacement at midspan, the equation of motion of the structure reads:

$$m\ddot{Y}(t) + b\dot{Y}(t) + k_1 Y(t) + k_3 Y^3(t) = F_{\mathrm{L}}(t). \tag{8.12}$$

In Equation 8.12, owing to the constraints and to the assumed deformation field, the effective values of terms are:

$$\begin{aligned}
m &= \frac{3}{8}\rho AL + 2\rho A^{\star} L,\\
c &= \frac{\mu h^3 L}{8g^3},\\
k_1 &= 2\pi^4 \frac{EJ}{L^3} - \frac{\pi^2}{3L}\alpha EA\Delta T_{\mathrm{m}} - 2\epsilon_0 V_0^2 \frac{hL}{g^3},\\
k_3 &= \frac{\pi^4}{8L^3} EA - 4\epsilon_0 V_0^2 \frac{hL}{g^5},\\
F_{\mathrm{L}} &= \frac{1}{2} iBL.
\end{aligned}$$

Besides the variables already introduced in the previous sections, in these definitions, $A^{\star}$ is the cross-sectional area of each attached plate, which is not necessarily the same as that of the beam; μ is the viscosity coefficient of the squeezed fluid; α is the coefficient of thermal expansion of the beam in the longitudinal direction (in the case of thermal anisotropy); ΔT_{m} is the value at midspan of the temperature raise caused by the Joule effect and V_0 is the bias voltage.

The Duffing equation (8.12), see Section 3.4.2, has the nonlinear term k_3, which is governed by the relaxation of the axial constraint while the beam is stretched, and by a higher-order electrostatic term. An analytical formulation for the absolute value of the maximum lateral displacement $|x_{\mathrm{max}}|$ was next adopted to drive the sensitivity maximization, see Bagherinia *et al.* (2014).

As far as the overall power consumption is instead concerned, results turn out to be proportional to

$$\text{power} = \frac{L}{h}. \tag{8.13}$$

The optimization strategy adopted by (Bagherinia *et al.*, (2014) was to minimize the following objective function,

$$\varphi = -\beta_x \frac{|x_{\mathrm{max}}|}{|x_{\mathrm{max}}|_r} + \beta_p \frac{\text{power}}{\text{power}_r}, \tag{8.14}$$

under given upper and lower bounds on the resonance frequency, handling the length L and width w of the beam as the variables to be optimized. In Equation 8.14, β_x and β_p are dimensionless weighting factors, to be selected in the design phase (imposing e.g. $\beta_x + \beta_p = 1$) on the basis of the relative importance of sensitivity and power consumption for the specific device; the negative sign before the sensitivity term means that sensitivity itself has to be maximized when function φ is minimized; terms $|x_{\max}|_r$ and power$_r$ at the denominators were introduced to scale functions $|x_{\max}|$ and power, which may take values in intervals differing by orders of magnitude. The constraints on the resonance frequency were introduced to ensure that it lies above the acoustic range, and to prevent a too slender geometry of the beam, which would be prone to buckling.

Two examples of the objective function are reported in Figure 8.6, as attained for different values of β_x and β_p. An example of the path followed by the solution, in terms of L, w, as provided by the adopted CONLIN minimizer (Fleury, 1989), is shown in Figure 8.7; to assess the robustness of the method, solutions were sought even moving from an initial guess that did not satisfy the constraints on the resonance frequency.

The same optimization strategy was further adopted by Bagherinia *et al.* (2015) for a resonating double-ended tuning-fork structure, see Figure 8.8. In that case, the optimal

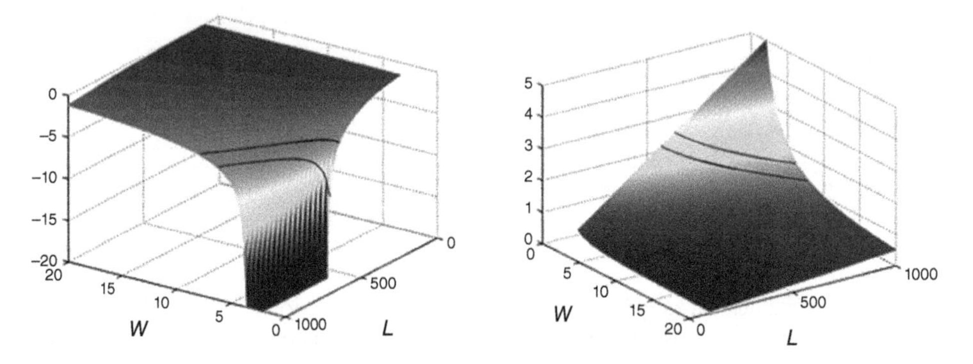

Figure 8.6 Objective function and lines representing its intersections with the bounds on the resonance frequency: (left) $\beta_x = 1$ and $\beta_p = 0$; (right) $\beta_x = 0$ and $\beta_p = 1$. *Source*: adapted from Bagherinia *et al.* (2014), Figures 2 and 5. Reproduced with permission of Elsevier.

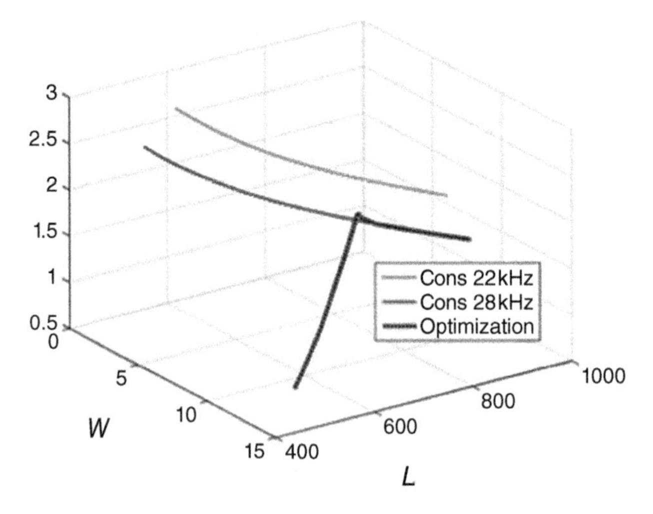

Figure 8.7 Example of the path followed by the solution provided by the CONLIN minimizer ($\beta_x = 0$ and $\beta_p = 1$). *Source*: adapted from Bagherinia *et al.* (2014), Figure 7. Reproduced with permission of Elsevier.

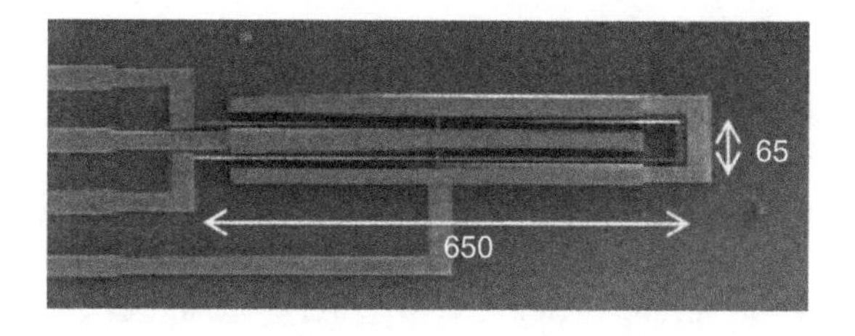

Figure 8.8 SEM of resonating double-ended tuning-fork structure of a *z*-axis Lorentz force magnetometer, with reference dimensions (in micrometres). *Source*: Bagherinia *et al.* (2015), Figure 1. Reproduced with permission of IEEE.

solution (in terms of beam length L and width w, and of sensing plate length and width) was sought to maximize sensitivity and bandwidth simultaneously, and to minimize power consumption and resolution. Depending on the assumed values of the weight factors coupling all the performance indices, as in Equation 8.14, different optimal geometries were obtained. With a rather compact design, a mechanical sensitivity of 0.43 aF/(μT mA), a bandwidth of 134 Hz and a resolution of 25 nTmA/$Hz^{1/2}$ were attained. It was also reported that, owing to the somewhat coupled multiphysics formulation, issues linked to multiple (by far different) optimal solutions might arise.

References

Afzal, M.H., Renaudin, V. and Lachapelle, G. (2011) *Magnetic Field Based Heading Estimation for Pedestrian Navigation Environments*, 2011 International Conference on Indoor Positioning and Indoor Navigation, September 21–23, 2011, Guimaraes, Portugal, IEEE.

Bagherinia, M., Bruggi, M., Corigliano, A. *et al.* (2014) Geometry optimization of a Lorentz force, resonating MEMS magnetometer. *Microelectronics Reliability*, **54**, 1192–1199.

Bagherinia, M., Bruggi, M., Corigliano, A. *et al.* (2015) An efficient earth magnetic field MEMS sensor: modeling, experimental results, and optimization. *Journal of Microelectromechanical Systems*, **24** (4), 887–895.

Bahreyni, B. and Shafai, C. (2007) A resonant micromachined magnetic field sensor. *IEEE Sensors Journal*, **7** (9), 1326–1334.

Emmerich, H. and Schofthaler, M. (2000) Magnetic field measurements with a novel surface micromachined magnetic-field sensor. *IEEE Transactions on Electron Devices*, **47** (5), 972–977.

Fleury, C. (1989) CONLIN: an efficient dual optimizer based on convex approximation concepts. *Structural Optimization*, **1** (2), 81–89.

Frangi, A., Laghi, G., Langfelder, G. *et al.* (2015a) Optimization of sensing stators in capacitive MEMS operating at resonance. *Journal of Microelectromechanical Systems*, **24** (4), 1077–1084.

Frangi, A., Laghi, G., Minotti, P. and Langfelder, G. (2015b) Effect of stators geometry on the resonance sensitivity of capacitive MEMS. *Procedia Engineering*, **120**, 294–297.

Ghorba, M.E., Andre, N., Sobieski, S. and Raskin, J.P. (2007) *CMOS Compatible Out-of-Plane In-Plane Magnetometers*, TRANSDUCERS 2007 – International Solid-State Sensors, Actuators and Microsystems Conference, June 10–14, 2007, Lyons, France, IEEE.

InvenSense (2012) MPU-6100 and MPU-6150 Product Specification, Revision 1.1, Sunnyvale, CA.

Langfelder, G., Buffa, C., Frangi, A. *et al.* (2013) *z*-axis magnetometers for MEMS inertial measurement units using an industrial process. *IEEE Transactions on Industrial Electronics*, **60** (9), 3983–3990.

Langfelder, G., Buffa, C., Tocchio, A. *et al.* (2012) *Design criteria for MEMS magnetometers resonating in free-molecule flow and out of the acoustic bandwidth*, International Frequency Control Symposium, May 21–24, 2012, Baltimore, MD, IEEE.

Robin, L. and Perlmutter, M. (2011) IMU and High Performance Inertial MEMS, *Tech. Rep.*, Yole Développement, France.

STMicroelectronics (2009) LSM303DLH: 3-Axis Accelerometer and 3-Axis Magnetometer Module, Geneva, Switzerland.

STMicroelectronics (2011) LSM330DLC: 3D Accelerometer + 3D Gyroscope Technical Datasheet, Geneva, Switzerland.

9

Mechanical Energy Harvesters

9.1 Introduction

In 1999, Ashton (2009) introduced the definition of the *Internet of Things*:

> If we had computers that knew everything there was to know about things – using data they gathered without any help from us – we would be able to track and count everything, and greatly reduce waste, loss and cost. We would know when things needed replacing, repairing or recalling, and whether they were fresh or past their best.

The concept of the Internet of Things is rapidly spreading in several areas of interest, leading to the extended idea of the *Internet of Everything*. Connecting everything by means of smart wireless sensors and actuators is a fascinating and ambitious objective that poses new and complex problems. One of the main issues in wireless electronics consists of efficiently powering remote devices. A few years ago, batteries and wiring were the only available solutions to provide the energy supply to electronic devices, but today there is a noticeable trend towards reduction in power consumption of sensors. This fact paves the way for the development of on-chip energy-harvesting solutions, with the final aim of achieving battery-less autonomous sensors and network systems.

An energy harvester, or scavenger, is a device that transforms and, possibly, stores the free energy present in the environment (solar power, thermal energy, wind energy or kinetic energy) into electric energy, in order to power small devices, such as wearable electronics or wireless sensor networks. If the energy harvester is realized at the MEMS scale, the harvested power might be on the order of milliwatts or even less (Kim, Ptiya and Kanno, 2012); nevertheless, that tiny amount of energy can be sufficient to feed modern sensors, whose technology is increasingly oriented to the use of ultra-low amount of power.

Piezoelectric materials (see Section 2.6) represent a spontaneous option for energy harvesters, because of the natural electromechanical coupling that transforms strain derived from a mechanical action (motor or traffic-induced vibrations in buildings, human movement) into electrical current (direct piezoelectric effect). It is worth noting that piezoelectric materials, e.g. lead zirconate titanate (PZT) or aluminium

Mechanics of Microsystems, First Edition. Alberto Corigliano, Raffaele Ardito, Claudia Comi, Attilio Frangi, Aldo Ghisi, and Stefano Mariani.

Companion website: www.wiley.com/go/corigliano/mechanics

nitride, can be embedded in the microfabrication process in the form of a thin film, which can be realized over a silicon substrate by means of different techniques: pulsed laser deposition (Horwitz *et al.*, 1991), sputtering (Jacobsen *et al.*, 2010), or the sol-gel procedure (Schroth *et al.*, 1999). It is common practice (Trolier-McKinstry and Muralt, 2004) to derive the *effective properties* for piezoelectric thin films, on the basis of bulk parameters, by imposing suitable conditions on the stress and strain components, dictated by the thin-film geometry.

In the design of a MEMS energy harvester, two criteria should be taken into account. First, the application and the amount of energy required for the wireless device (the target is about 100 μW) and, second, the necessity to have small dimensions to integrate the device with the electronics. The big issue is that the energy generated by the harvester decreases rapidly when the size of the device reduces; a trade-off between size and energy scavenged has to be found. In this perspective, piezoelectric materials are advantageous because they provide high energy density and they are not too much affected by the size scaling.

Piezoelectric energy harvesters consist of a vibrating spring–mass system coupled with a piezoelectric element, which is designed to convert elastic energy into electric energy (Roundy, Wright and Rabaey, 2004). Clearly, the best solution is a resonant system, which is the case of most of the works proposed in the literature. Unfortunately, resonant piezoelectric energy harvesters suffer from a series of drawbacks. (i) In common applications (e.g. building or human body energy harvesting), the external vibrations have very low frequencies, while the smaller the device the higher its natural frequencies. (ii) Even if the energy harvester matches the source frequency, the displacements induced to the system are very high and might induce a brittle mechanical failure of the device. (iii) The external vibrations are usually random and the performance of the energy harvesters decreases rapidly out of the resonance regime.

This chapter contains a detailed description of the mechanical modelling for an inertial energy harvester, in its simplest form (namely, a piezoelectric cantilever with a tip mass), along with a short survey of the possible provisions that could be adopted to overcome the frequency mismatch. The special case of energy harvesting in the presence of fluid–structure interaction is treated separately, in view of the peculiar features and of the increasing importance of that kind of devices.

9.2 Inertial Energy Harvesters

9.2.1 Classification of Resonant Energy Harvesters

Resonant harvesters consist of a mass–spring system clamped to a moving support that oscillates according to an external acceleration. Such a dynamic system is essentially described by its frequency response function, which is a frequency-dependent measure that links the input signal and the magnitude and phase of the output signal as a function of frequency.

The concept of resonant harvesters has been introduced by Williams and Yates (1996), in terms of an undefined device, which generates electricity when excited by a sinusoidal motion of the reference. In spite of the fact that they considered an electromagnetic generation mechanism, their model can be applied to every kind of transduction, provided that the induced electrical damping is suitably updated. Even their conclusions, related to the maximization of energy harvester performances, are of broad interest: a large mass is needed; the displacement of the mass should be as large as possible; the first natural frequency should match the excitation frequency; the quality factor of the device should be very high. Another example of an electromagnetic harvester has been proposed at the University of Southampton (Glynne-Jones *et al.*, 2004): the device consists of a series of magnets coupled to a coil that is attached to a vibrating cantilever. In recent times, the same group has proposed a miniaturized version of the same device ($0.15\,cm^3$), suitable for low ambient vibrations (Beeby *et al.*, 2007). The final device, which is characterized by a resonant frequency of 52 Hz, is able to produce 46 μW with a resistive load of 4 kΩ, if the maximum external acceleration is $0.59\,ms^{-2}$. Von Büren and Tröster (2007) have shown the design and optimization of a linear electromagnetic generator, aimed at supplying power to wearable sensors. The design is based on a tubular layout and a flexible translator, with an overall volume of $0.25\,cm^3$. If the energy harvester is installed on the human body during walking, the output power attains a value of 2–25 μW, depending on its position on the body.

The application of electromagnetic generation in the field of MEMS is hindered by the fact that magnetic induction is not efficiently scalable at the submillimeter scale. Conversely, electrostatic mechanisms of energy conversion are suited to the case of MEMS, which commonly include electrostatic sensing and actuation. A group at MIT (Meninger *et al.*, 2001) has studied an electrostatic generator based on the variation of in-plane overlap for a comb-driven structure, thus achieving 8 μW in the case of harmonic motion at 2.5 kHz. However, electrostatic harvesters do not assure the power density required for usual applications and they furthermore need an initial polarizing voltage or charge, which results in consumption of energy.

Piezoelectric transduction seems the best option to implement energy harvesting in MEMS. Pioneering work in this field have been carried out at Berkeley University (Roundy, Wright and Rabaey, 2004) and MIT (Du Toit, Wardle and Kim, 2005). The former research group focused on considering the correct piezoelectric transduction mechanism for a device connected to an external load resistance. Two structural layouts have been optimized within an overall space limit of $1\,cm^3$. The experimental value of harvested power is about $375\,\mu Wcm^{-3}$ for excitation with maximum acceleration equal to $2.5\,ms^{-2}$ at 120 Hz. The MIT group proposed an electromechanical model of a piezolaminated microscale cantilever beam, with experimental validation against a MEMS-scale nonoptimized device. The main result is represented by the identification of two optimal frequencies for power extraction, corresponding to the resonance and antiresonance frequencies of the device. The same group (Jeon *et al.*, 2005) proposed a piezoelectric energy harvester represented by a laminated beam with PZT deposited over a structural layer (Figure 9.1). The innovation of this work has been represented by the use of interdigitated electrodes on top of the piezoelectric layer with the purpose of harnessing the piezoelectric stretching mode (the so-called 33-mode) in lieu of the conventional transverse mode (the 31-mode). The proposed device, with an overall size of 170 μm × 260 μm, was able to scavenge up to 1 μW of power when excited at the resonance frequency of 13.9 kHz and connected to an external circuit with a 5.2 MΩ resistive load.

An improved piezoelectric energy harvester has been proposed by Muralt *et al.* (2009), still by adopting interdigitated electrodes. The cross-section of the beam was constituted of a 2 μm thick PZT layer over a 5 μm thick silicon layer. Experimental measurements

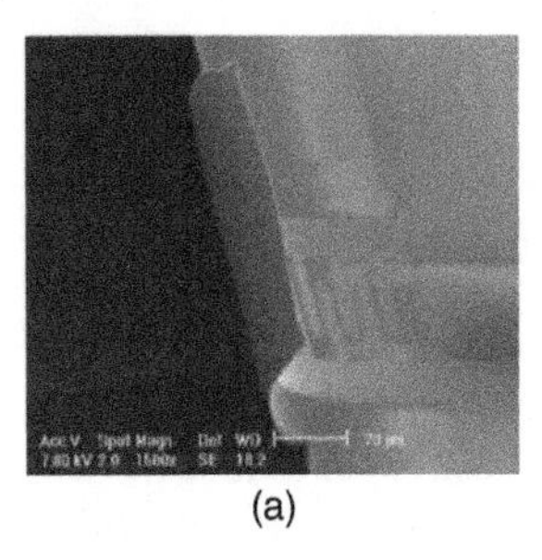
(a)

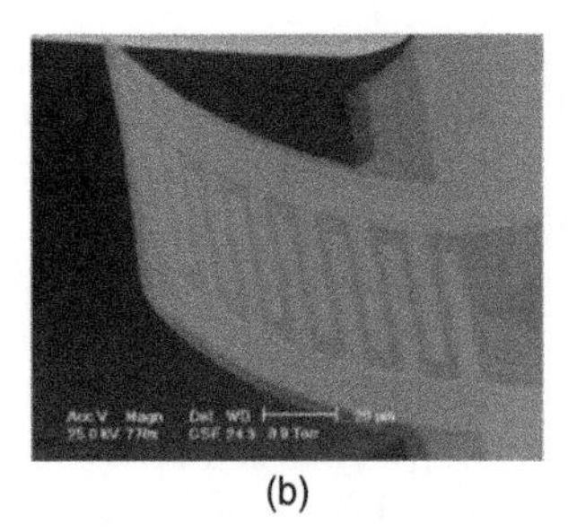
(b)

(c)

Figure 9.1 SEM of piezoelectric generator proposed by Jeon *et al.* (2005). The comb-finger electrodes are viewable on the top of three layered piezoelectric cantilevers with different materials used to control the initial curvature. *Source*: Jeon *et al.* (2005), Figure 5. Reproduced with permission of Elsevier.

provided encouraging results: an optimal output voltage of about 1.6 V, a harvested power of 1.4 μW, for a 0.8 mm × 1.2 mm cantilever with an active area of 0.8 mm × 0.4 mm and a mass in the remaining space. The energy source was simulated by applying a 2.0*g* acceleration at the resonant frequency of 870 Hz.

Elfrink *et al.* (2009) have realized a piezoelectric energy harvester that exploits the piezoelectric behaviour of aluminium nitride. Aluminium nitride represents an interesting material for several reasons: first, the figure of merit for energy transduction is better than that of PZT; second, the well-known sputter deposition process can be adopted; third, it is a lead-free material. The devices studied by Elfrink *et al.* (2009) have resonance frequencies that range between 200 Hz and 1200 Hz; the maximum power output for the unpackaged device is 60 μW for an external acceleration equal to 2.0*g* at 572 Hz.

In most cases, resonant energy harvesters work in the linear regime and this will be the main focus of the this chapter. Nevertheless, several attempts have been made in the past to exploit various nonlinear behaviours with the main purpose of enlarging the frequency bandwidth. For instance, Hajati and Kim (2011) proposed a piezoelectric energy harvester based on nonlinear geometric behaviour due to stretching: a doubly clamped piezoelectric beam, shown in Figure 9.2, is subject to transverse acceleration so that a nonlinear Duffing-type oscillator is obtained (see Section 3.4.2). The proposed device has been modelled in some detail by Gafforelli *et al.* (2013); subsequently, a successful comparison with experimental data has been obtained by Gafforelli *et al.* (2014). The main drawback of such devices is that the intrinsically hardening behaviour of the nonlinear oscillator entails the widest bandwidth only in the direction towards high frequencies. Considering that the resonance frequency of MEMS is usually high, one would desire exactly the opposite behaviour. Conversely, this approach is promising for MEMS-scale integration, since it allows for micromanufacturing processes already available and does not require additional materials or external interventions to induce nonlinearity. Indeed, the Duffing oscillator can also be achieved by exploiting

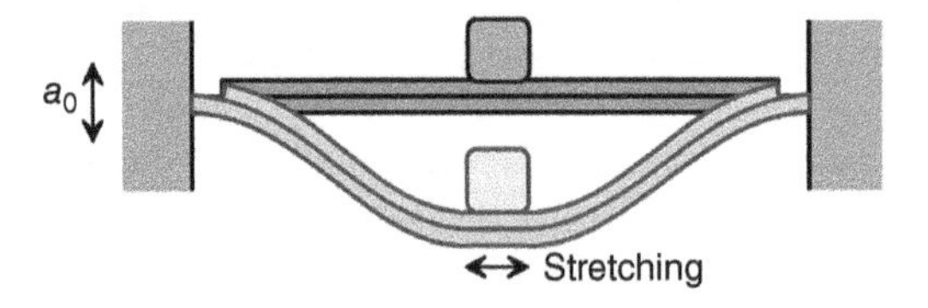

Figure 9.2 Stretching nonlinear harvester, as proposed by Hajati and Kim (2011), Gafforelli *et al.* (2013) and Gafforelli *et al.* (2014).

the interaction between magnets. Challa *et al.* (2008) proposed a device based on a piezoelectric cantilever beam: two permanent magnets were attached at its free end, and two further magnets were fixed on the top and bottom surfaces, vertically aligned with the magnets on the beam. The magnets induce attractive and repulsive magnetic forces on each side of the beam, so that the apparent stiffness of the structure is modified. By changing the sign of the magnetic force, the natural frequency of the beam can be tuned to higher and lower frequencies with respect to the initial resonance frequency of the piezoelectric beam.

Among the recent proposals of nonlinear devices, it is worth citing the possible exploitation of parametric resonance, which has been explained in Chapter 7, with reference to micromirror dynamics. The theoretical issues of parametric resonance of energy harvesters have been considered by Daqaq *et al.* (2008) and Abdelkefi, Nayfeh and Hajj (2011), with the main objective of boosting the resonant mechanism and, consequently, the harvested power. Jia *et al.* (2013) developed a MEMS piezoelectric energy harvester prototype which was able to access five orders of parametric resonance, showing both power increase and frequency broadening.

9.2.2 Mechanical Model of a Simple Piezoelectric Harvester

This section shows how to formulate a mechanical model of the simplest case of a piezoelectric energy harvester, namely a cantilever beam, characterized by a layered cross-section, with a tip mass (Figure 9.3). No external forces will be considered: the excitation will be represented by an external acceleration in the transverse direction, a_0. In what follows, we shall borrow many concepts illustrated in Chapters 2 and 3: first, we shall consider the specific modification of the piezoelectric constitutive law in the case of 1D elements (i.e. beams in space); then we shall establish the principle of virtual power for a layered beam with piezoelectric elements; finally, we shall consider a simplification of the model by means of the Rayleigh–Ritz procedure and we shall show the solution in the frequency domain.

9.2.2.1 Piezoelectric Constitutive Law for Beams

The piezoelectric constitutive law, which has been described in Section 2.6.2, is now revised to take into account the specific restrictions of the beam model. Indeed, in the presence of almost 1D bodies (i.e. long and narrow beams), it is quite natural to introduce the uniaxial stress hypothesis, namely:

$$T_{22} = 0, \quad T_{33} = 0. \tag{9.1}$$

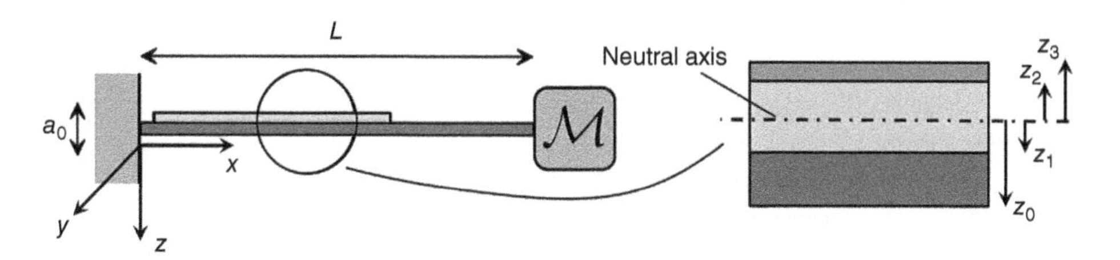

Figure 9.3 Cantilever beam used as energy harvester: the tip mass is shown, along with a magnified view of the layered cross-section.

The validity of such hypotheses has been thoroughly discussed by Ardito, Corigliano and Gafforelli (2015) and Gafforelli, Ardito and Corigliano (2015), who have successfully introduced a modification of the constitutive law in order to obtain accurate results for narrow beams (for which the uniaxial stress hypothesis is valid), wide beams (which can be studied by assuming that the transverse strain is null) and any intermediate situation. The proposed model, whose description is beyond the scope of this book, has been validated with respect to fully 3D simulations, with satisfactory results in all cases. The model has also been used to assess the performance of a realistic piezoelectric energy harvester (Ardito *et al.*, 2016). For the sake of simplicity, this chapter is focused on narrow beams, so that Equation 9.1 holds.

If the piezoelectric layer is very thin and is polarized in the vertical direction, so that it works in the 31-mode, one can reasonably assume that the only nonzero component of the electric field is E_3, i.e. along the vertical axis. On the basis of such an assumption, the piezoelectric law for isotropic elastic materials can be written in the following form, as shear stress and strain are not relevant for Euler–Bernoulli beams:

$$
\begin{aligned}
T_{11} &= \frac{E}{(1+\nu)(1-2\nu)}[(1-\nu)S_{11} + \nu S_{22} + \nu S_{33}] - e_{31}E_3, \\
0 &= \frac{E}{(1+\nu)(1-2\nu)}[\nu S_{11} + (1-\nu)S_{22} + \nu S_{33}] - e_{32}E_3, \\
0 &= \frac{E}{(1+\nu)(1-2\nu)}[\nu S_{11} + \nu S_{22} + (1-\nu)S_{33}] - e_{33}E_3.
\end{aligned} \tag{9.2}
$$

By employing the second and the third equations in Equation 9.2, one can eliminate S_{22} and S_{33} and the following constitutive relationship is obtained:

$$T_{11} = E\ S_{11} - (e_{31} - \nu e_{32} - \nu e_{33})E_3 \equiv E\ S_{11} - e_{\text{eff}}E_3. \tag{9.3}$$

Similarly, the electric displacement can be expressed in terms of the axial strain S_{11} and the transverse electric field:

$$
\begin{aligned}
D_3 &= (e_{31} - \nu e_{32} - \nu e_{33})\ S_{11} - \left[\frac{1-\nu^2}{E}\left(e_{32}^2 - 2\frac{\nu}{1-\nu}e_{32}\ e_{33} + e_{33}^2\right) + \epsilon_{33}^S\right] E_3 \\
&\equiv e_{\text{eff}}\ S_{11} + \epsilon_{\text{eff}}E_3.
\end{aligned} \tag{9.4}
$$

An example of the numerical values of the effective properties is reported in Table 9.1, which lists electrical and mechanical properties of the piezoelectric material. The case of a transversally isotropic material is considered herein, so that $e_{31} = e_{32}$. It is worth noting that the effective piezoelectric coefficient e_{eff} defined herein is different from the coefficient that is usually defined for thin films, see Trolier-McKinstry and Muralt (2004). In that case, the in-plane strain components are assumed to be zero, which is a reasonable hypothesis for a thin film with indefinite in-plane extension.

Table 9.1 Example of standard and effective piezoelectric properties for uniaxial stress ($\epsilon_0 = 8.854\ 10^{-12}$ F/m).

E [GPa]	ν [-]	e_{31}, e_{32} [N/mV]	e_{33} [N/mV]	ϵ_{33}^S [ϵ_0]	e_{eff} [N/mV]	ϵ_{eff} [ϵ_0]
100	0.3	−10.0	15.0	1200	−16.5	2107

9.2.2.2 The Principle of Virtual Power for a Piezoelectric Cantilever Beam

The principle of virtual power for the case of beams has been described in Section 3.2.4; however, it will be reconsidered here in view of the specific features of the problem at hand.

First, we have to consider that the beam is endowed with a layered cross-section: each layer has different mechanical properties and one or, in some cases, two layers will represent the active piezoelectric elements. We denote here each layer by means of the subscript k: for instance, the thickness of each layer is t_k and its Young's modulus is E_k. To decouple the bending problem from the axial problem, as shown in Section 2.3.2, it is important to set the origin of the reference system not at the *centre of mass* of the cross-section but at its *centre of stiffness*, which belongs to the *neutral axis* in the case of pure bending. By setting a temporary axis $\tilde{z}$, which has its origin on the upper surface of the beam and points downwards, one finds that the position of the centre of stiffness is obtained by the following formula:

$$\tilde{z}_{\text{CS}} = \frac{\sum_k E_k t_k \tilde{z}_k}{\sum_k E_k t_k}, \tag{9.5}$$

where $\tilde{z}_k$ indicates the position of the centre of mass of each layer.

For the specific application, only the bending problem is considered, so that the kinematic field is represented by the axial strain S_{11} and the displacement components u_x and u_z. The principle of virtual power for the dynamic case reads:

$$\int_{\Omega} T_{11} \delta \dot{S}_{11} \mathrm{d}\Omega + \int_{\Omega} \rho \left[\ddot{u}_x \delta \dot{u}_x + \ddot{u}_z \delta \dot{u}_z \right] \mathrm{d}\Omega = 0. \tag{9.6}$$

For the Euler–Bernoulli beam, the kinematic model reads:

$$u_x = -z \frac{\mathrm{d}w}{\mathrm{d}x}, \tag{9.7}$$

$$u_z = w, \tag{9.8}$$

whereas, if the external acceleration is denoted a_0, the transverse acceleration including the drag contribution is

$$\ddot{u}_z = a_0 + \ddot{w}. \tag{9.9}$$

By considering the integral along the whole beam, it is possible to obtain the following form of the principle of virtual power:

$$\int_0^l B \sum_k \int_{z_{k-1}}^{z_k} -T_{11} z \, \mathrm{d}z \frac{\mathrm{d}^2 \delta \dot{w}}{\mathrm{d}x^2} \mathrm{d}x + \int_0^l B \sum_k \rho_k \int_{z_{k-1}}^{z_k} z^2 \, \mathrm{d}z \frac{\mathrm{d}\ddot{w}}{\mathrm{d}x} \frac{\mathrm{d}\delta \dot{w}}{\mathrm{d}x} \mathrm{d}x$$
$$+ \int_0^l B \sum_k \rho_k t_k (a_0 + \ddot{w}) \delta \dot{w} \, \mathrm{d}x + \delta P_{\mathcal{M}} = 0, \tag{9.10}$$

where B is the beam width (along the y-axis) and $\delta P_{\mathcal{M}}$ is the virtual power due to the inertia of the mass tip. The latter contribution can be obtained if one considers that the displacement of the centre of the tip mass and its rotation are expressed in terms of the displacement of the beam tip, namely $w_L = w(x = l)$ and

$$\phi_L = \left. \frac{\mathrm{d}w}{\mathrm{d}x} \right|_{x=l},$$

as:

$$\begin{aligned}\ddot{u}_{\mathcal{M}x} &= -\ddot{\phi}_L \Delta z_{\mathcal{M}},\\ \ddot{u}_{\mathcal{M}z} &= \ddot{w}_L + \ddot{\phi}_L \Delta x_{\mathcal{M}} + a_0,\\ \ddot{\phi}_{\mathcal{M}} &= \ddot{\phi}_L.\end{aligned} \tag{9.11}$$

In Equation 9.11, $\Delta x_{\mathcal{M}}$ and $\Delta z_{\mathcal{M}}$ represent the relative coordinates of the centre of the tip mass with respect to the centre of stiffness of the beam tip. After some algebra, denoting by $\mathcal{M}$ the tip mass and by $I_{\mathcal{M}}$ its moment of inertia, one finds that

$$\begin{aligned}\delta P_{\mathcal{M}} = {}& \mathcal{M}(\ddot{w}_L + \Delta x_{\mathcal{M}}\ddot{\phi}_L + a_0)\delta\dot{w}_L +\\ &+ \left[\mathcal{M}\Delta x_{\mathcal{M}}\ddot{w}_L + (\mathcal{M}\Delta x^2_{\mathcal{M}} + \mathcal{M}\Delta z^2_{\mathcal{M}} + I_{\mathcal{M}})\ddot{\phi}_L + \mathcal{M}\Delta x_{\mathcal{M}}a_0\right]\delta\dot{\phi}_L.\end{aligned} \tag{9.12}$$

So far, we have just considered the mechanical equilibrium expressed through the principle of virtual power. For the problem at hand, it is also necessary to provide an equivalent statement for the balance of the electric flux. If one defines the virtual electric potential $\delta\varphi$ and the corresponding virtual electric field $\delta\mathbf{E} = -\nabla(\delta\varphi)$, the electric flux conservation reads:

$$\int_\Omega \mathbf{D}\cdot\delta\mathbf{E}\,\mathrm{d}\Omega - \int_\Omega q_\Omega\delta\varphi\,\mathrm{d}\Omega - \int_\Gamma q_\Gamma\delta\varphi\,\mathrm{d}\Gamma = 0. \tag{9.13}$$

For this specific case, only the E_3 component is different from zero. Moreover, to induce the 31-mode in the piezoelectric layer, a couple of equipotential electrodes are introduced on its upper and lower surfaces, so that the electric potential (and, consequently, the E_3 field) does not depend on the x and y coordinates. Finally, we consider only the presence of a distributed charge on the upper electrode, so that the governing equation becomes

$$\int_0^l B\sum_k \int_{z_{k-1}}^{z_k} D_3\ \delta E_3 \mathrm{d}z\,\mathrm{d}x - \int_0^l B\sum_k q_k\delta\varphi_k\,\mathrm{d}x = 0. \tag{9.14}$$

Of course, in this equation ,the sum should include only the piezoelectric layers: for the sake of conciseness, but without any loss of generality, from now on we shall consider the presence of a single piezoelectric layer.

9.2.2.3 Governing Equations via the Weighted Residuals Approach

The governing integral equations can be reduced to a set of algebraic equations if some suitable approximations are introduced. First, the electric potential φ is assumed to be linear across the thickness of the pth piezoelectric layer:

$$\varphi \cong V(t)\frac{z_p}{t_p} \quad \Longrightarrow \quad E_3 \cong -\frac{V(t)}{t_p}. \tag{9.15}$$

In this way, the electric field is governed by a scalar function of time, namely the difference of electric potential V between the upper and the lower electrode, which we assume to be grounded. The approximate nature of the relationship in Equation 9.15 is quite obvious if one considers the piezoelectric constitutive law (9.4) and the beam kinematics (9.7), which entails a linear variation of the strain component S_1 along the z coordinate. As a matter of fact, the electric potential should be a nonlinear function. Nevertheless, since the thickness of the piezoelectric layer is generally very small in comparison with the overall thickness of the beam, such a nonlinear function can be approximated by means of its first-order expansion, thus achieving a satisfactory degree of accuracy.

The mechanical field, namely the transverse displacement w, can be expressed, in the framework of the weighted residuals approach (see Section 3.3), in terms of a shape function $\psi_w(x)$, which guarantees that the kinematic boundary conditions are fulfilled. For a cantilever beam, this means:

$$w(x=0)=0, \qquad \left.\frac{\mathrm{d}w}{\mathrm{d}x}\right|_{x=0}=0. \tag{9.16}$$

A cubic shape function can be chosen, according to:

$$\psi_w(x)=\frac{3}{2}\left(\frac{x}{l}\right)^2-\frac{1}{2}\left(\frac{x}{l}\right)^3, \tag{9.17}$$

so that:

$$w\cong\psi_w W(t), \quad w_L\cong W, \quad \phi_L\cong\frac{3}{2}\frac{W}{l}. \tag{9.18}$$

By combining Equations 9.3, 9.4, 9.10, 9.15 and 9.18, one finds the following expression of the principle of virtual power:

$$\begin{aligned}\delta\dot{W}\Bigg\{&\int_0^l B\sum_k\int_{z_{k-1}}^{z_k}E_kz^2\,\mathrm{d}z\left(\frac{\mathrm{d}^2\psi_w}{\mathrm{d}x^2}\right)^2\mathrm{d}x\,W-\int_0^l B\int_{z_{p-1}}^{z_p}e_{\mathrm{eff}}\frac{z}{t_p}\mathrm{d}z\frac{\mathrm{d}^2\psi_w}{\mathrm{d}x^2}\mathrm{d}x\,V\\&+\int_0^l B\sum_k\rho_k\left[\int_{z_{k-1}}^{z_k}z^2\,\mathrm{d}z\left(\frac{\mathrm{d}\psi_w}{\mathrm{d}x}\right)^2+t_k\psi_w^2\right]\mathrm{d}x\,\ddot{W}+\int_0^l B\sum_k\rho_kt_k\psi_w\,\mathrm{d}x\,a_0\\&+\mathcal{M}\left(1+3\frac{\Delta x_{\mathcal{M}}}{l}+\frac{9}{4}\frac{\Delta x_{\mathcal{M}}^2+\Delta z_{\mathcal{M}}^2+\frac{I_{\mathcal{M}}}{\mathcal{M}}}{l^2}\right)\ddot{W}+\mathcal{M}\left(1+\frac{3}{2}\frac{\Delta x_{\mathcal{M}}}{l}\right)a_0\Bigg\}=0.\end{aligned} \tag{9.19}$$

We can now define the generalized stiffness of the layered beam, whose dimension is force over length, as

$$k\equiv\int_0^l B\left(\sum_k\int_{z_{k-1}}^{z_k}E_kz^2\,\mathrm{d}z\right)\left(\frac{\mathrm{d}^2\psi_w}{\mathrm{d}x^2}\right)^2\mathrm{d}x, \tag{9.20}$$

the generalized piezoelectric coefficient, whose dimension is charge over length, as

$$\Theta\equiv\int_0^l B\left(\int_{z_{p-1}}^{z_p}e_{\mathrm{eff}}\frac{z}{t_p}\mathrm{d}z\right)\frac{\mathrm{d}^2\psi_w}{\mathrm{d}x^2}\mathrm{d}x, \tag{9.21}$$

the generalized mass as

$$\begin{aligned}m\equiv&\int_0^l B\sum_k\rho_k\left[\int_{z_{k-1}}^{z_k}z^2\,\mathrm{d}z\left(\frac{\mathrm{d}\psi_w}{dx}\right)^2+t_k\psi_w^2\right]\mathrm{d}x\\&+\mathcal{M}\left(1+3\frac{\Delta x_{\mathcal{M}}}{l}+\frac{9}{4}\frac{\Delta x_{\mathcal{M}}^2+\Delta z_{\mathcal{M}}^2+\frac{I_{\mathcal{M}}}{\mathcal{M}}}{l^2}\right)\end{aligned} \tag{9.22}$$

and the generalized inertia connected to the drag acceleration as

$$m_a \equiv \int_0^l B \sum_k \rho_k t_k \psi_w \,\mathrm{d}x + \mathcal{M}\left(1 + \frac{3}{2}\frac{\Delta x_{\mathcal{M}}}{l}\right). \tag{9.23}$$

It is worth noting that the two inertial terms are different from one another. The two parameters tend to the same value only if the inertial contributions of the rotational field are neglected. In this special case one finds:

$$m = m_a = \int_0^l B \sum_k \rho_k t_k \psi_w \,\mathrm{d}x + \mathcal{M}. \tag{9.24}$$

In general, the thickness of the beam is considerably small if compared with its length, so that the rotational inertia of the cross-section can be neglected without any loss of accuracy. Conversely, the rotational inertia of the tip mass can provide significant contribution, unless very compact masses are used. For this reason, the use of Equation 9.24 should be limited to cases of cantilever beams without tip mass.

By considering the electrical flux conservation, Equation 9.14, along with Equations 9.4, 9.15 and 9.18, one finds

$$\delta V \left\{ W \int_0^l B \left(\int_{z_{p-1}}^{z_p} e_{\mathrm{eff}} \frac{z}{t_p} \mathrm{d}z \right) \frac{\mathrm{d}^2 \psi_w}{\mathrm{d}x^2} \mathrm{d}x + V \int_0^l B \frac{\epsilon_{\mathrm{eff}}}{t_p} \mathrm{d}x - \int_0^l B q_p \,\mathrm{d}x \right\} = 0. \tag{9.25}$$

The first term contains the generalized piezoelectric coefficient, which has already been defined with reference to the equilibrium equation. We can now define the generalized capacitance of the piezoelectric layer, whose dimension is charge squared over force times length, as

$$C_{\mathrm{E}} \equiv \int_0^l B \frac{\epsilon_{\mathrm{eff}}}{t_p} \mathrm{d}x \tag{9.26}$$

and the overall charge on the upper electrode as

$$q \equiv \int_0^l B q_p \,\mathrm{d}x. \tag{9.27}$$

Wrapping up the relationships that have been examined in this section, one finds the set of governing equations for the coupled piezoelectric problem, reduced to a two-dof problem:

$$m\ddot{W} + c_{\mathrm{M}} \dot{W} + kW - \Theta V = -m_a\, a_0, \tag{9.28}$$

$$C_{\mathrm{E}}\, V + \Theta\, W - q = 0. \tag{9.29}$$

The equilibrium equation now contains the mechanical damping, which has been introduced by means of the generalized damping coefficient c_{M}, which is inversely proportional to the mechanical quality factor Q_{M}.

The electric charge q, collected by the upper electrode, is managed by an external circuit, which provides the power supply for the self-powered electronic device. Different circuit schemes have been investigated by Guyomar *et al.* (2005). The harvester provides an AC voltage and the simplest solution is the coupling with an external load resistance:

$$\dot{q} = -R^{-1}V. \tag{9.30}$$

The final system of equations reads:

$$m\ddot{W} + c_{\mathrm{M}}\dot{W} + kW - \Theta V = -m_a\, a_0, \tag{9.31}$$

$$C_{\mathrm{E}}\dot{V} + \Theta\dot{W} + R^{-1}\, V = 0. \tag{9.32}$$

9.2.2.4 Solution in the Frequency Domain

As already stated, the response of a resonant piezoelectric energy harvester can be successfully represented by the frequency response function. The governing equations (9.31)–(9.32) should be modified, taking account of the fact that the external excitation is harmonic ($-m_a a_0 = F\mathrm{e}^{\mathrm{i}\omega t}$), as well as the two degrees of freedom ($W = \widetilde{W}\mathrm{e}^{\mathrm{i}\omega t}$, $V = \widetilde{V}\mathrm{e}^{\mathrm{i}\omega t}$). One finds the following system of algebraic equations:

$$\begin{aligned} -\frac{\omega^2}{\omega_{\mathrm{r}}^2}\widetilde{W} + 2\mathrm{i}\xi_{\mathrm{M}}\frac{\omega}{\omega_{\mathrm{r}}}\widetilde{W} + \widetilde{W} - \frac{\Theta}{k}\widetilde{V} &= \frac{F}{k}, \\ \mathrm{i}\omega\widetilde{V} + \mathrm{i}\omega\frac{\Theta}{C_{\mathrm{E}}}\widetilde{W} + \frac{1}{RC_{\mathrm{E}}}\widetilde{V} &= 0, \end{aligned} \tag{9.33}$$

where $\omega_{\mathrm{r}} = m^{-1}k$ is the resonance frequency of the device and $\xi_{\mathrm{M}} = (2m\omega_{\mathrm{r}})^{-1}c_{\mathrm{M}}$ is the mechanical damping ratio. The product RC_{E} is the time constant of the electrical circuit, and is a measure of the time required to charge the capacitor from 0 to 63.2%, in the absence of piezoelectric coupling. Such a time constant can be also expressed in terms of the cut-off frequency, given by

$$\omega_{RC} = \frac{1}{RC_{\mathrm{E}}}. \tag{9.34}$$

The second equation in Equation 9.33 can easily be solved to achieve the voltage parameter:

$$\widetilde{V} = -\frac{\mathrm{i}\omega}{\mathrm{i}\omega + \omega_{RC}}\frac{\Theta}{C_{\mathrm{E}}}\widetilde{W}. \tag{9.35}$$

The application of well-known formulae for the ratio of complex numbers yields the following equation:

$$\widetilde{V} = -\left(\frac{\omega^2}{\omega^2 + \omega_{RC}^2} + \mathrm{i}\frac{\omega\omega_{RC}}{\omega^2 + \omega_{RC}^2}\right)\frac{\Theta}{C_{\mathrm{E}}}\widetilde{W}. \tag{9.36}$$

It is now possible to introduce the so-called global coupling coefficient κ_{E}^2. It is a dimensionless parameter that describes, for resonating structures, the efficiency of piezoelectric transduction, taking into account the piezoelectric coefficient, the effective capacitance and the global stiffness of the structure:

$$\kappa_{\mathrm{E}}^2 = \frac{\Theta^2}{k\, C_{\mathrm{E}}}. \tag{9.37}$$

The final equation can be written in a simpler form if one introduces the normalized excitation frequency and the normalized cut-off frequency, in both cases with respect to the resonance frequency ω_{r}:

$$\Omega_{\mathrm{M}} = \frac{\omega}{\omega_{\mathrm{r}}}, \quad \Omega_{\mathrm{E}} = \frac{\omega_{RC}}{\omega_{\mathrm{r}}}. \tag{9.38}$$

By introducing Equation 9.36 into the first of Equation 9.33, one can easily obtain the normalized frequency response function for the displacement H_W, i.e. the ratio between the parameter $\widetilde{W}$ and the static displacement F/k:

$$H_W = \frac{1}{\left(1 - \Omega_{\mathrm{M}}^2 + \kappa_{\mathrm{E}}^2 \frac{\Omega_{\mathrm{M}}^2}{\Omega_{\mathrm{M}}^2+\Omega_{\mathrm{E}}^2}\right) + \mathrm{i}\left(2\,\xi_M\,\Omega_{\mathrm{M}} + \kappa_{\mathrm{E}}^2 \frac{\Omega_{\mathrm{M}}\Omega_{\mathrm{E}}}{\Omega_{\mathrm{M}}^2+\Omega_{\mathrm{E}}^2}\right)}. \tag{9.39}$$

Similarly, going back to Equation 9.36, one can compute the frequency response function of the voltage H_V, normalized with respect to the ratio $-F/\Theta$:

$$H_V = \frac{\kappa_{\mathrm{E}}^2}{\Omega_{\mathrm{M}}^2 + \Omega_{\mathrm{E}}^2} \frac{\Omega_{\mathrm{M}}^2 + \mathrm{i}\Omega_{\mathrm{M}}\Omega_{\mathrm{E}}}{\left(1 - \Omega_{\mathrm{M}}^2 + \kappa_{\mathrm{E}}^2 \frac{\Omega_{\mathrm{M}}^2}{\Omega_{\mathrm{M}}^2+\Omega_{\mathrm{E}}^2}\right) + \mathrm{i}\left(2\,\xi_{\mathrm{M}}\Omega_{\mathrm{M}} + \kappa_{\mathrm{E}}^2 \frac{\Omega_{\mathrm{M}}\Omega_{\mathrm{E}}}{\Omega_{\mathrm{M}}^2+\Omega_{\mathrm{E}}^2}\right)}. \tag{9.40}$$

Let us first examine the frequency response function of the displacement (9.39). It has the same form of the standard frequency response function of a single-dof damped oscillator, with two important differences. In the real part, there is an additional stiffness contribution that depends on the coupling coefficient and the cut-off frequency of the external circuit: one finds a stiffening effect due to the presence of the active piezoelectric layer, so that the equivalent stiffness of the piezoelectric resonator can be written as

$$\tilde{k} = k\left(1 + \kappa_{\mathrm{E}}^2 \frac{\Omega_{\mathrm{M}}^2}{\Omega_{\mathrm{M}}^2 + \Omega_{\mathrm{E}}^2}\right). \tag{9.41}$$

In the imaginary part of Equation 9.39 it is possible to recognize an additional source of damping due to the piezoelectric effect; the equivalent damping ratio can be defined as

$$\tilde{\xi} = \xi_{\mathrm{M}} + \frac{1}{2}\kappa_{\mathrm{E}}^2 \frac{\Omega_{\mathrm{E}}}{\Omega_{\mathrm{M}}^2 + \Omega_{\mathrm{E}}^2}. \tag{9.42}$$

The additional damping (so-called electric damping) is connected to the conversion between mechanical and electrical energy and the subsequent dissipation due to the Joule effect in the resistor: this is the key parameter for evaluation of the harvested energy.

The magnitudes of the frequency response functions are plotted in Figure 9.4 with respect to the normalized cut-off frequency and the excitation frequency. It is worth noting that the frequency response function of displacement shows two asymptotic values for $\Omega_{\mathrm{E}} \to \infty$ and for $\Omega_{\mathrm{E}} \to 0$. The former case corresponds to the short-circuit condition, i.e. no electric resistance in the electric circuit. In this case, the piezoelectric effects on stiffness and damping are null and the resonance frequency corresponds to that of the purely mechanical system. Another consequence of the absence of resistance is that the voltage tends to zero. Conversely, the case $\Omega_{\mathrm{E}} \to 0$ corresponds to the open-circuit conditions, which means that the electric resistance tends to infinity. In this case, the resonance phenomenon happens for a slightly different frequency with respect to the mechanical system, namely

$$\Omega_{\mathrm{M}} = \sqrt{1 + \kappa_{\mathrm{E}}^2}, \tag{9.43}$$

which is commonly referred to as the antiresonance frequency. The open-circuit condition corresponds to the asymptotic value of the voltage. In both cases, i.e. short circuit

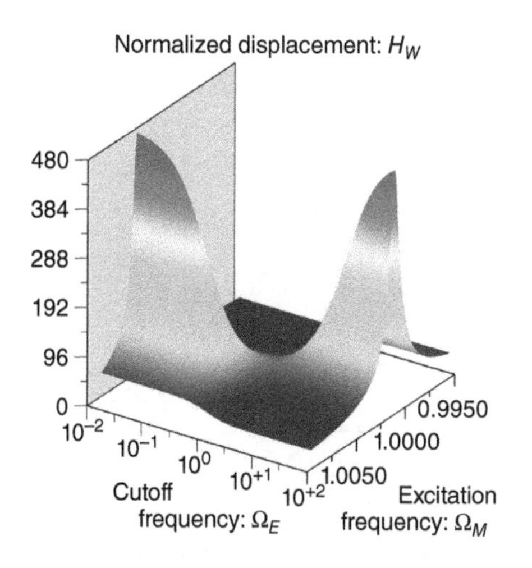

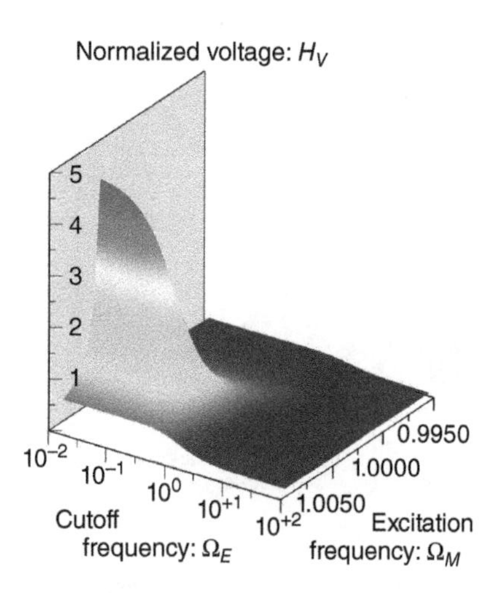

Figure 9.4 Normalized frequency response function for displacement and voltage for a cantilever harvester with $\kappa_E = 0.1$ and $Q_M = 500$. (*See color plate section for the color representation of this figure.*)

and open circuit, the electric damping defined in Equation 9.42 tends to zero, which means that there is no harvested power (as one can easily deduce on the basis of the features of the electric circuit). This fact can be investigated by computing the frequency response function of the harvested power, according to the well-known formula of resistive circuits:

$$P = \frac{|V|^2}{R}. \tag{9.44}$$

The amplitude of power generation is normalized with respect to the power at resonance, defined as

$$P_0 = \frac{1}{2}\frac{F^2}{\Theta^2}\frac{1}{R\,\Omega_E}, \tag{9.45}$$

so that one obtains

$$H_P = \frac{2\kappa_E^4}{\Omega_M^2 + \Omega_E^2}\frac{\Omega_M^2\Omega_E}{\left(1 - \Omega_M^2 + \kappa_E^2\frac{\Omega_M^2}{\Omega_M^2+\Omega_E^2}\right)^2 + \left(2\xi_M\Omega_M + \kappa_E^2\frac{\Omega_M\Omega_E}{\Omega_M^2+\Omega_E^2}\right)^2}. \tag{9.46}$$

As expected, the power amplitude tends to zero in the short-circuit situation ($\Omega_E \to \infty$) and in the open-circuit situation ($\Omega_E \to 0$). The maximum harvested power is attained for a specific value of the normalized cut-off frequency, which can be easily obtained by applying the stationarity condition to the power amplitude. One finds

$$\Omega_E^{\text{opt}} = \Omega_M\sqrt{\frac{\left(1 - \Omega_M^2 + \kappa_E^2\right)^2 + 4\xi_M^2\Omega_M^2}{\left(1 - \Omega_M^2\right)^2 + 4\xi_M^2\Omega_M^2}}. \tag{9.47}$$

The optimal cut-off frequency corresponds to an optimal value of the resistance of the electric circuit. Such a value obviously depends on the excitation frequency. It is possible to show that the maximum value of the optimal cut-off frequency corresponds to the

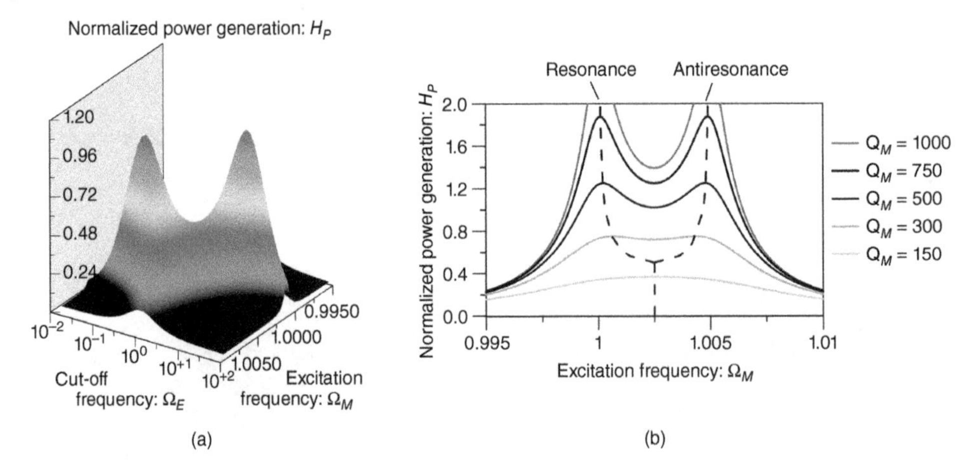

Figure 9.5 (a) Normalized frequency response function for harvested power for a cantilever beam with $\kappa_E = 0.1$ and $Q_M = 500$. (b) Optimal normalized power versus excitation frequency for different levels of mechanical damping ($\kappa_E = 0.1$). (*See color plate section for the color representation of this figure.*)

mechanical resonance, whereas at the antiresonance the optimal value tends to zero (it would be exactly zero in the ideal case of null mechanical damping). One can compute, for each excitation frequency, the harvested power corresponding to the optimal value of the cut-off frequency. Figure 9.5 shows a plot of the power amplitude along with a plot of the optimal power generation. The latter graph also shows the variation of the optimal power with respect to the mechanical damping. Two peaks are shown, corresponding to resonance and antiresonance. As expected, the higher the quality factor, the higher the harvested power.

9.3 Frequency Upconversion and Bistability

One serious drawback of piezoelectric energy harvesters based on linear oscillators is that they work well only in a narrow band of excitation. The efficiency of such devices is dramatically reduced if the ambient vibrational energy is distributed over a wide spectrum, if the spectral density of the input signal changes over time, or if the spectrum is dominated by low-frequency components. For these reasons, many researchers have addressed their attention to techniques for extending the energy harvester efficiency over a wide frequency bandwidth. A possible solution can be found by exploiting an array of multifrequency piezoelectric converters, as proposed by Ferrari *et al.* (2008). Alternatively, one could apply active systems to tune the resonance frequency of the device to match the external frequency, which varies with time. However, whatever the employed technique, it costs active energy to be implemented, resulting in a lower effective power generation (Leland and Wright, 2006; Eichhorn, Goldschmidtboeing and Woias, 2009). Finally, a nonlinear oscillator, briefly addressed in Section 9.2.1, can be used. The spring stiffening effect for large deformations was successfully employed by Marzencki, Defosseux and Basrour (2009), who proposed this approach to achieve a completely passive tuning of the resonance frequency. The device consists of a seismic mass suspended by two piezolaminated beams that stretches when the mass oscillates. As the beam stretches, the stiffness increases and the natural frequency is modified.

The main problem of MEMS devices is that the natural frequency is usually high, whereas the excitation frequency can be very low. Consequently, it is not sufficient to enlarge the bandwidth around the resonance frequency, but it is also mandatory to introduce a sort of interface in order to establish the interaction between the input signal and the vibrating device. To this purpose, the concept of frequency upconversion can be applied. A classical approach to frequency upconversion is referred to the adoption of bistable systems, which have been studied in Section 3.5. Harne and Wang (2013) proposed a detailed and extensive review of bistable harvesters, which also includes frequency upconversion harvesters. The governing equations of classical single-dof bistable oscillators have been extended to include electromechanical coupling from a general point of view. The resulting equations were specialized according to the considered transduction mechanism. The benefits and disadvantages of bistable devices due to stochastic excitation have not yet been conclusively determined: there is still an open issue on the potentiality of vibration energy harvesting in random excitation environments (Ferrari *et al.*, 2010; Daqaq, 2012; Cottone *et al.*, 2012).

Frequency upconversion could be based on magnetic interactions between distinct components of the device. A device proposed by Kulah and Najafi (2008) consisted of two resonators, a low-frequency resonator provided with a large permanent magnet and an array of high resonance frequency beams with a magnetic tip that can be attracted by the large permanent magnet. As the large magnet resonates in response to the external vibration, it moves closer to the cantilever beam located beneath it. The distance between the two is adjusted such that the magnet catches the cantilever at a certain point of its movement, pulls it up and releases it at another point. Thus, the cantilever beams start to freely oscillate at their own resonant frequency and generate power. The same concept was extensively developed by the group at University of Michigan (Galchev, Aktakka and Najafi, 2012), employing both electromagnetic and piezoelectric transduction. They fabricated a frequency increased generator, with a volume of 1.2 cm^3, which generates a peak power of 100 μW and an average power of 3.25 μW from an input acceleration of 9.8 m/s^2 at 10 Hz. The device operates over a frequency range of 24 Hz. In any case, no conclusive design has been obtained for a feasible application in MEMS or NEMS.

In recent times, Procopio *et al.* (2014) have proposed a magnetic frequency upconversion device that is fully compatible with MEMS or NEMS, since it adopts piezoelectric transduction and avoids impact or contacts between the magnets, with great improvements in reliability. The basic principle is shown in Figure 9.6. A central low-frequency resonator is endowed with a certain number of permanent magnets, which can interact, in a contactless manner, with the magnets located on the tip of the surrounding cantilevers. The central resonator induces, via magnetic interaction, the deformation of the cantilever beams, which are suddenly released when the relative distance between the magnets becomes larger than a certain threshold. The threshold can be computed by considering the force versus distance law of the magnetic interaction: when the magnetic force becomes smaller than the elastic recovery force, the beams are released and vibrate at their natural frequency. Frequency upconversion is therefore achieved in the framework of a compete compatibility with microfabrication constraints: the object is basically flat, the different materials can be introduced via successive depositions and

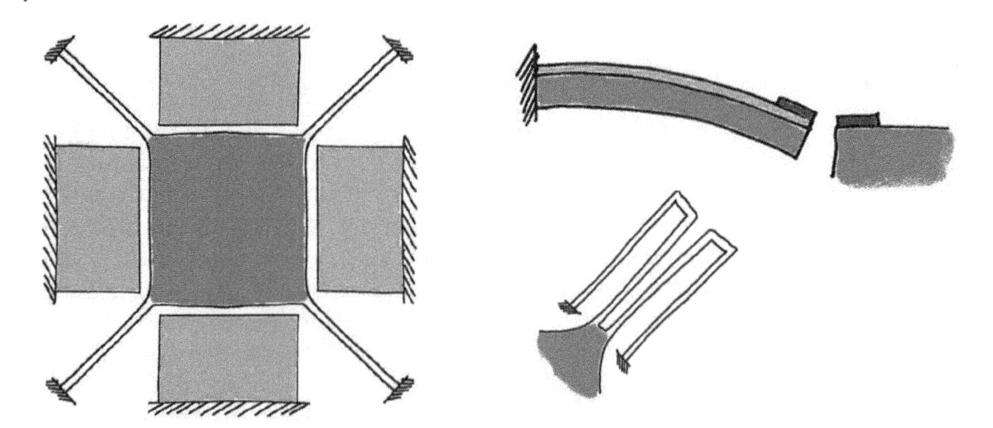

Figure 9.6 Operating principle of the magnetic frequency upconversion device proposed by Procopio *et al.* (2014). Left: plan view showing central low-frequency oscillator and four high-frequency piezoelectric cantilevers. Top right: cross-section; magnetic interaction allows the cantilever to bend. Bottom right: alternative scheme for the supporting springs.

there is no contact between parts. The key design parameters of the proposed devices are as follows:

1) The tip displacement of the beams, owing to magnetic interaction, should be maximized; to this purpose, the length of the beams should be carefully selected.
2) An impulsive effect on the beams should be achieved, in the sense that the cantilevers should not follow the large mass in a quasi-static fashion; this condition is used to fix the dimension of the magnets with respect to the beam stiffness.
3) The mass displacement, induced by the external acceleration, should exceed the magnetic attraction, so that the low-frequency resonator is not stopped by the presence of the surrounding magnets; the mass dimension plays a paramount role in this sense.

It can be shown that such a frequency upconversion device may yield a mean power of about 10 μW from an input acceleration of 50g at 20 Hz, with an overall volume of the device of about 0.1 cm^3.

9.4 Fluid–Structure Interaction Energy Harvesters

A current frontier in research on energy harvesting is represented by the synergic application of inertial devices and fluid flow effects. Some interesting proposals for fluid energy harvesters can be found in the literature, although none of them is related to MEMS devices. Gao, Shih and Shih (2013) have proposed an energy harvester device based on a piezoelectric cantilever beam with a cylinder at the top. The flow-induced vibrations on the cylinder deflect the cantilever and, consequently, the piezoelectric material converts mechanical energy into electrical energy. The vibration is caused both by turbulent motion generated by the flow and by vortex shedding. Similarly, Hobbs and Hu (2012) have devised a system characterized by four cylinders in series, aligned along the flow direction. Abdelkefi, Hajj and Nayfeh (2012) exploited vortex-induced vibrations of a cylinder, simply supported at both ends, using piezoelectric elements.

De Marqui Jr., Erturk and Inman (2010) studied a different aeroelastic phenomenon from that studied by the previous authors. They focused on flutter instability, to understand the piezo-aeroelastic behaviour of a thin piezolaminated plate.

These solutions, however, are referred to mesoscale structures. When dealing with the problem at the MEMS scale, some limitations should be considered. Only some specific types of structures can be adopted, since circular cylinders are extremely complicated to realize; therefore, in the case of a flow energy harvester, the vibrating objects will present sections with straight edges. Moreover, an important outcome of the literature analysis is that a significant level of harvested power can be obtained only if a synergic approach is adopted, namely by combining inertial effect and fluid–structure interaction.

9.4.1 Synopsis of Aeroelastic Phenomena

The flow around an immersed body produces a surface pressure on it. If the body moves or deforms appreciably under the forces, the new configuration, by changing the boundary conditions of the flow, will affect the fluid behaviour. Aeroelasticity is the study of the phenomena wherein aerodynamic forces and structural motions interact significantly. The aeroelastic phenomena considered herein are vortex-induced vibration and flutter instability.

9.4.1.1 Vortex-Induced Vibration

Vortex-induced vibration occurs when vortex shedding interacts with the elastic behaviour of the immersed body. The vortex shedding frequency f_s is related to the Strouhal number St, the fluid velocity U and a characteristic dimension D, through

$$f_s = \frac{St\,U}{D}, \tag{9.48}$$

where D is assumed equal to the thickness of the beam. The Strouhal number is affected by the Reynolds number (Re) and by the aspect ratio of the rectangular cross-sectional area of the beam (Okajima, 1982). In the framework of a simplified approach, the results proposed in the literature on square prisms are adopted; for instance, the empirical formula suggested by Sohankar, Norberg and Davidson (1999) for laminar flow reads:

$$St = 0.18 - \frac{3.7}{Re}. \tag{9.49}$$

According to the characteristics of the vortex shedding phenomenon, the lift force can be assumed to be like a sinusoidal action of this type:

$$f_L = \frac{1}{2}\rho_f D C_L U^2 \sin(\omega_s t), \tag{9.50}$$

in which ρ_f is the density of the fluid, ω_s is the circular frequency of the vortex shedding and C_L is the lift coefficient, which usually depends on the shape of the cross-section. In the framework of the weighted residuals approach, discussed in Section 9.2.2, the force component is obtained as

$$F_L = \int_0^l f_L \psi_w \, dx. \tag{9.51}$$

For the purpose of energy harvesting, the synchronization (lock-in) condition is required: the fluid velocity is sought so that the vortex shedding frequency is close to the vibration frequency of the cantilever. The width of the lock-in domain and the amplitude of oscillation are strictly affected by the mass and the damping ratio, m and ξ (Williamson and Govardhan, 2004). The effects of these two parameters are synthesized by the Scruton number:

$$Sc = \frac{4\pi m\xi}{\rho_f D^2}, \tag{9.52}$$

where m is the mass per unit length of the beam. To amplify the response and the harvested power for a large range of fluid velocities, it is necessary to work with small values of Sc.

9.4.1.2 Flutter Instability

In this section, the so-called *classical flutter* is analysed. This aeroelastic phenomenon is characterized by two mechanical degrees of freedom, the rotation and the vertical translation of the beam, coupled in a flow-driven, unstable oscillation. According to the characteristics of the instability mechanisms, the motion of the structure will either decay or diverge according to whether the energy of motion extracted from the flow is less than or exceeds the energy dissipated by the system through mechanical damping. What distinguishes these two conditions is then recognized as the critical flutter condition. In the present context, the formulation proposed by Scanlan and Simiu (1996) is employed. The equations of motion of the beam read:

$$m\ddot{W} + c_M \dot{W} + kW - \Theta V = L_w, \tag{9.53}$$

$$I_\theta \ddot{\theta} + c_\theta \dot{\theta} + k_\theta \theta = M_\theta, \tag{9.54}$$

$$C_E \dot{V} + \Theta \dot{W} + R^{-1} V = 0. \tag{9.55}$$

The inertial term I_θ, the damping parameter c_θ and the torsional stiffness k_θ can be obtained in the framework of the weighted residuals approach, taking into account a shape function for the torsional rotation ψ_θ. The generalized force L_w and torque M_θ are respectively given by

$$L_w = \int_0^l l_w \psi_w \mathrm{d}x, \quad M_\theta = \int_0^l m_\theta \psi_\theta \mathrm{d}x, \tag{9.56}$$

The self-excited aerodynamic lift and moment components are computed as follows:

$$l_w = \frac{1}{2}\rho_f U^2 D \left(KH_1^* \frac{\dot{W}}{U} + KH_2^* \frac{\dot{\theta} D}{U} + K^2 H_3^* \theta + K^2 H_4^* \frac{W}{D} \right), \tag{9.57}$$

$$m_\theta = \frac{1}{2}\rho_f U^2 D^2 \left(KA_1^* \frac{\dot{W}}{U} + KA_2^* \frac{\dot{\theta} D}{U} + K^2 A_3^* \theta + K^2 A_4^* \frac{W}{D} \right), \tag{9.58}$$

where $K = D\omega/U$ is the reduced circular frequency and H_i^* and A_i^* are the flutter derivatives, which are nondimensional functions of K. In a microstructure, the Reynolds number is very low; for this reason the flutter derivatives computed by Bruno and Fransos (2008) are adopted.

To evaluate the instability condition of the system and the correspondence critical velocity, the quadratic eigenvalue problem is analysed:

$$\left(\lambda^2 \overline{\mathbf{M}} + \lambda \overline{\mathbf{C}} + \overline{\mathbf{K}}\right) \mathbf{X}^* = \mathbf{0}, \quad \mathbf{X}^* = \left[W^* \quad \theta^* \quad V^*\right]^{\mathrm{T}}, \tag{9.59}$$

where $\overline{\mathbf{M}}$, $\overline{\mathbf{C}}$ and $\overline{\mathbf{K}}$ are the mass, viscous damping and stiffness matrices, respectively. The electromechanical coupling is introduced according to the type of electric external circuit adopted. For an RC circuit:

$$\overline{\mathbf{M}} = \begin{bmatrix} m & 0 & 0 \\ 0 & I_\vartheta & 0 \\ 0 & 0 & 0 \end{bmatrix}, \quad \overline{\mathbf{C}} = \begin{bmatrix} c_M - L_w^d & -L_{w\vartheta}^d & 0 \\ -M_{\vartheta w}^d & c_\vartheta - M_\vartheta^d & 0 \\ \Theta & 0 & C_E \end{bmatrix},$$

$$\overline{\mathbf{K}} = \begin{bmatrix} k - L_w^k & -L_{w\vartheta}^k & -\Theta \\ -M_{\vartheta w}^k & k_\vartheta - M_\vartheta^k & 0 \\ 0 & 0 & 1/R \end{bmatrix}. \tag{9.60}$$

The problem is solved by transforming the quadratic eigenvalue problem into an equivalent standard eigenvalue problem, as explained by Ardito *et al.* (2008). The real part of the eigenvalue λ stands for the damping component of the system; therefore, the flutter instability occurs when this term becomes negative. The corresponding velocity is the *critical flutter velocity.*

9.4.2 Energy Harvesting through Vortex-Induced Vibration

This section, as well as the subsequent one, shows the paradigmatic application for a simple cantilever beam, shown in Figure 9.7, with a silicon structural layer and a piezoelectric active layer. The frequency response function of the harvested power can be computed through the procedure described in Section 9.2.2, taking into account the specific form of the external force given by Equations 9.50 and 9.51. The results are plotted in Figure 9.8. On the horizontal axis, instead of frequency ratios, there is a dimensionless fluid velocity, defined as the ratio between the velocity (related to frequency via Equation 9.48) and the velocity that corresponds to vortex shedding at resonance, $U^* = U/U_{\text{res}}$. In this preliminary analysis, the lift coefficient, C_L, is assumed equal to 0.25 and two different values of damping ratio are considered; $\xi_1 = 0.01$, $\xi_2 = 0.001$.

The bifurcation phenomenon, connected to the presence of two peaks, depends on the damping ratio. It can be demonstrated that there is a bifurcation limit for the damping

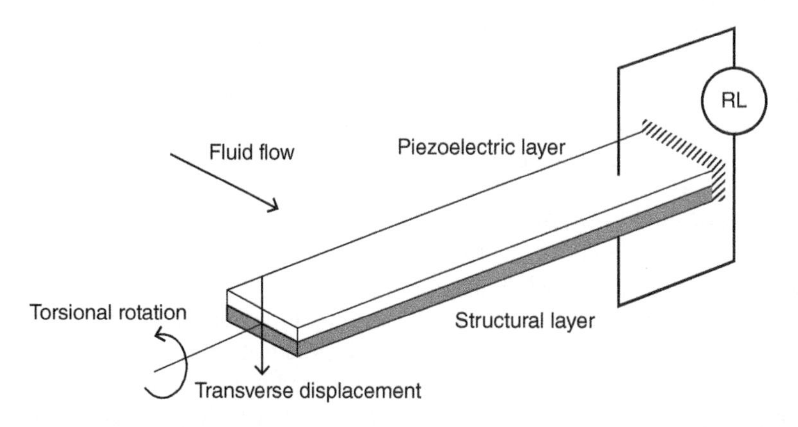

Figure 9.7 Piezoelectric beam subject to the effects of a fluid flow.

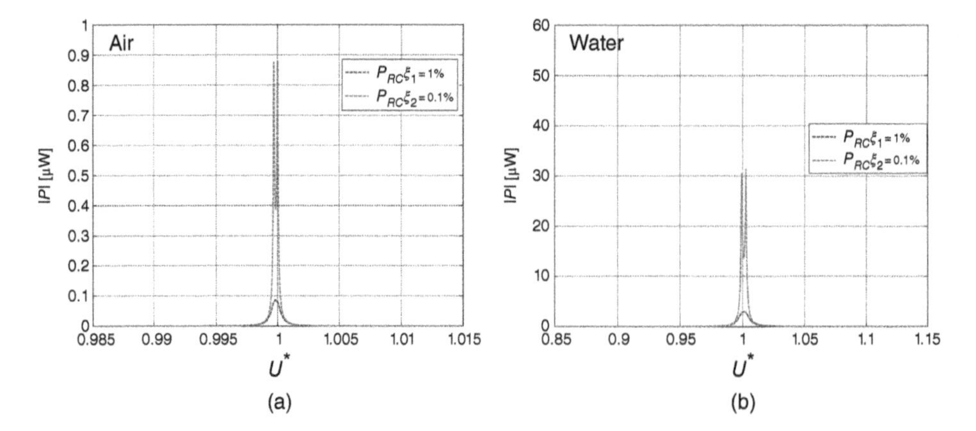

Figure 9.8 Frequency response function of harvested power due to vortex-induced vibration, corresponding to optimal electric resistance: (a) the fluid is air; (b) the fluid is water.

ratio, ξ_b, which is a function of the coupling coefficient (Renno, Daqaq and Inman, 2009). Herein, this quantity is equal to $\xi_b = 0.007$, which explains clearly the different trends shown in Figure 9.8. The magnitude of harvested power is much larger for the case of water: the density of the fluid plays a key role, since it affects the amplitude of the lift force, the value of the resonance velocity and the Scruton number. It is worth noting that the resonance velocity for air (80.7 m/s) is an order of magnitude larger than that of water (6.9 m/s). The low efficiency of air, as the fluid medium, is also emphasized by the large Scruton number.

9.4.3 Energy Harvesting through Flutter Instability

According to the flutter derivatives computed by Bruno and Fransos (2008) for small Reynolds numbers, the coefficient H_1^* gives the major contribution to the aeroelastic behaviour, so that the unstable phenomenon involves the transverse degree of freedom. The critical velocity obtained is equal to $U_{cr} = 3.25$ m/s. By introducing the electromechanical coupling, this value increases according to the increase in equivalent damping (Ardito and Musci, 2016). The critical flutter velocity depends on the features of the external circuit, as does the electrical damping: consequently, it is possible to obtain the optimal value of the electric resistance so that the critical velocity is maximized.

Flutter instability can be exploited for the energy-harvesting purpose, joining this aeroelastic phenomenon with another type of excitation, such as an inertial forcing, which can trigger the beam's vibration. For instance, one can consider the time-variant evolution for the cantilever subject to the fluid flow and to an initial tip displacement equal to the thickness of the beam. The aeroelastic effect is included by considering a fluid velocity that is slightly smaller than the critical value. In the presence of flutter, the tip displacement presents a much less damped response, with beneficial effects on the overall harvested energy. The reduction of damping, however, does not affect the peak amplitude of the harvested power, which remains practically the same. On the contrary, if one consider a different external circuit with an additional inductance, the dynamic response is largely affected by the aeroelastic effect, showing larger oscillation amplitude: the peak power is one order of magnitude larger than in the case of the RC circuit.

References

Abdelkefi, A., Hajj, M. and Nayfeh, A. (2012) Phenomena and modeling of piezoelectric energy harvesting from freely oscillating cylinders. *Nonlinear Dynamics*, **70** (2), 1377–1388.

Abdelkefi, A., Nayfeh, A.H. and Hajj, M.R. (2011) Global nonlinear distributed-parameter model of parametrically excited piezoelectric energy harvesters. *Nonlinear Dynamics*, **67**, 1147–1160.

Ardito, R., Comi, C., Corigliano, A. and Frangi, A. (2008) Solid damping in micro electro mechanical systems. *Meccanica*, **43** (4), 419–428.

Ardito, R., Corigliano, A. and Gafforelli, G. (2015) A highly efficient simulation technique for piezoelectric energy harvesters. *Journal of Physics: Conference Series*, **660** (1), 012141.

Ardito, R., Corigliano, A., Gafforelli, G. *et al.* (2016) Advanced model for fast assessment of piezoelectric micro energy harvesters. *Frontiers in Materials*, **3**, 17.1–17.9.

Ardito, R. and Musci, R. (2016) *Flutter Analysis of Piezoelectric Beams in MEMS*, First International Symposium on Flutter and its Application, JAXA Special Publication, May 15–17, 2016, Tokyo, Japan, Japan Aerospace Exploration Agency, **JAXA-SP-16-008E**, 121–130.

Ashton, K. (2009) That 'Internet of Things' thing. *RFID Journal*, http://www.rfidjournal.com/articles/view?4986 (accessed 6 July 2017).

Beeby, S.P., Torah, R.N., Tudor, M.J. *et al.* (2007) A micro electromagnetic generator for vibration energy harvesting. *Journal of Micromechanics and Microengineering*, **17**, 1257–1265.

Bruno, L. and Fransos, D. (2008) Evaluation of Reynolds number effects on flutter derivatives of a flat plate by means of a computational approach. *Journal of Fluids and Structures*, **24** (7), 1058–1076.

Challa, V.R., Prasad, M.G., Shi, Y. and Fisher, F.T. (2008) A vibration energy harvesting device with bidirectional resonance frequency tunability. *Smart Materials and Structures*, **17**, 015035.

Cottone, F., Gammaitoni, L., Vocca, H. *et al.* (2012) Piezoelectric buckled beams for random vibration energy harvesting. *Smart Materials and Structures*, **21** (3), 035021.

Daqaq, M. (2012) On intentional introduction of stiffness nonlinearities for energy harvesting under white Gaussian excitations. *Nonlinear Dynamics*, **69** (3), 1063–1079.

Daqaq, M.F., Stabler, C., Qaroush, Y. and Seuaciuc-Osorio, T. (2008) Investigation of power harvesting via parametric excitations. *Journal of Intelligent Material Systems and Structures*, **20**, 545–557.

De Marqui Jr., C., Erturk, A. and Inman, D. (2010) Piezoaeroelastic modeling and analysis of a generator wing with continuous and segmented electrodes. *Journal of Intelligent Material Systems and Structures*, **21** (10), 983–993.

Du Toit, N.E., Wardle, B.L. and Kim, S.G. (2005) Design considerations for MEMS-scale piezoelectric mechanical vibration energy harvesters. *Integrated Ferroelectrics*, **71** (1), 121–160.

Eichhorn, C., Goldschmidtboeing, F. and Woias, P. (2009) Bidirectional frequency tuning of a piezoelectric energy converter based on a cantilever beam. *Journal of Micromechanics and Microengineering*, **19** (9), 094006.

Elfrink, R., Kamel, T.M., Goedbloed, M. *et al.* (2009) Vibration energy harvesting with aluminum nitride-based piezoelectric devices. *Journal of Micromechanics and Microengineering*, **19**, 094005.

Ferrari, M., Ferrari, V., Guizzetti, M. *et al.* (2008) Piezoelectric multifrequency energy converter for power harvesting in autonomous microsystems. *Sensors and Actuators, A: Physical*, **142** (1), 329–335.

Ferrari, M., Ferrari, V., Guizzetti, M. *et al.* (2010) Improved energy harvesting from wideband vibrations by nonlinear piezoelectric converters. *Sensors and Actuators, A: Physical*, **162** (2), 425–431.

Gafforelli, G., Ardito, R. and Corigliano, A. (2015) Improved one-dimensional model of piezoelectric laminates for energy harvesters including three dimensional effects. *Composite Structures*, **127**, 369–381.

Gafforelli, G., Corigliano, A., Xu, R. and Kim, S.G. (2014) Experimental verification of a bridge-shaped, nonlinear vibration energy harvester. *Applied Physics Letters*, **105**, 203901.

Gafforelli, G., Xu, R., Corigliano, A. and Kim, S. (2013) Modelling of a bridge-shaped nonlinear piezoelectric energy harvester. *Journal of Physics: Conference Series*, **476**, 012100.

Galchev, T., Aktakka, E. and Najafi, K. (2012) A piezoelectric parametric frequency increased generator for harvesting low-frequency vibrations. *Journal of Microelectromechanical Systems*, **21** (6), 1311–1320.

Gao, X., Shih, W.H. and Shih, W. (2013) Flow energy harvesting using piezoelectric cantilevers with cylindrical extension. *IEEE Transactions on Industrial Electronics*, **60** (3), 1116–1118.

Glynne-Jones, P., Tudor, M.J., Beeby, S.P. and White, N.M. (2004) An electromagnetic, vibration-powered generator for intelligent sensor systems. *Sensors and Actuators A: Physical*, **110**, 344–349.

Guyomar, D., Badel, A., Lefeuvre, E. and Richard, C. (2005) Toward energy harvesting using active materials and conversion improvement by nonlinear processing. *IEEE Transactions on Ultrasonics, Ferroelectrics and Frequency Control*, **52**, 584–595.

Hajati, A. and Kim, S.G. (2011) Ultra-wide bandwidth piezoelectric energy harvesting. *Applied Physics Letters*, **99** (8), 083105.

Harne, R. and Wang, K. (2013) A review of the recent research on vibration energy harvesting via bistable systems. *Smart Materials and Structures*, **22** (2), 023001.

Hobbs, W. and Hu, D. (2012) Tree-inspired piezoelectric energy harvesting. *Journal of Fluids and Structures*, **28**, 103–114.

Horwitz, J., Grabowski, K., Chrisey, D. and Leuchtner, R. (1991) *In situ* deposition of epitaxial $PbZr_xTi_{(1-x)}O_3$ thin films by pulsed laser deposition. *Applied Physics Letters*, **59**, 1565–1567.

Jacobsen, H., Prume, K., Wagner, B. *et al.* (2010) High-rate sputtering of thick PZT thin films for MEMS. *Journal of Electroceramics*, **25**, 198–202.

Jeon, Y.B., Sood, R., Jeong, J.H. and Kim, S.G. (2005) MEMS power generator with transverse mode thin film PZT. *Sensors and Actuators A: Physical*, **122**, 16–22.

Jia, Y., Yan, J., Soga, K. and Seshia, A. (2013) Multi-frequency operation of a MEMS vibration energy harvester by accessing five orders of parametric resonance. *Journal of Physics: Conference Series*, **476**, 012126.

Kim, S.G., Ptiya, S. and Kanno, I. (2012) Piezoelectric MEMS for energy harvesting. *MRS Bulletin*, **37**, 1039–1050.

Kulah, H. and Najafi, K. (2008) Energy scavenging from low-frequency vibrations by using frequency up-conversion for wireless sensor applications. *IEEE Sensors Journal*, **8** (3), 261–268.

Leland, E. and Wright, P. (2006) Resonance tuning of piezoelectric vibration energy scavenging generators using compressive axial preload. *Smart Materials and Structures*, **15** (5), 1413–1420.

Marzencki, M., Defosseux, M. and Basrour, S. (2009) MEMS vibration energy harvesting devices with passive resonance frequency adaptation capability. *Journal of Microelectromechanical Systems*, **18** (6), 1444–1453.

Meninger, S., Mur-Miranda, J.O., Amirtharajah, R. *et al.* (2001) Vibration-to-electric energy conversion. *IEEE Transactions on Very Large Scale Integration (VLSI) Systems*, **9**, 64–76.

Muralt, P., Marzencki, M., Belgacem, B. *et al.* (2009) Vibration energy harvesting with PZT micro device. *Procedia Chemistry*, **1**, 1191–1194.

Okajima, A. (1982) Strouhal numbers of rectangular cylinders. *Journal of Fluid Mechanics*, **123**, 379–398.

Procopio, F., Valzasina, C., Corigliano, A. *et al.* (2014) Piezoelectric transducer for an energy-harvesting system. US Patent US20150035409 A1, filed Jul. 29, 2014 and issued Feb. 5, 2015.

Renno, J., Daqaq, M. and Inman, D. (2009) On the optimal energy harvesting from a vibration source. *Journal of Sound and Vibration*, **320** (1–2), 386–405.

Roundy, S.J., Wright, P.K. and Rabaey, J.M. (2004) *Energy Scavenging for Wireless Sensor Networks*, Kluwer Academic Publishers.

Scanlan, R.H. and Simiu, E. (1996) *Wind Effects on Structures*, John Wiley & Sons, Inc.

Schroth, A., Lee, C., Matsumoto, S. and Maeda, R. (1999) Application of sol-gel deposited thin PZT film for actuation of 1D and 2D scanners. *Sensors and Actuators, A: Physical*, **73**, 144–152.

Sohankar, A., Norberg, C. and Davidson, L. (1999) Simulation of three-dimensional flow around a square cylinder at moderate Reynolds numbers. *Physics of Fluids*, **11** (2),288–306.

Trolier-McKinstry, S. and Muralt, P. (2004) Thin film piezoelectrics for MEMS. *Journal of Electroceramics*, **12**, 7–17.

Von Büren, T. and Tröster, G. (2007) Design and optimization of a linear vibration-driven electromagnetic micro-power generator. *Sensors and Actuators A: Physical*, **135**, 765–775.

Williams, C.B. and Yates, R.B. (1996) Analysis of a micro-electric generator for microsystems. *Sensors and Actuators, A: Physical*, **52**, 8–11.

Williamson, C. and Govardhan, R. (2004) Vortex-induced vibrations. *Annual Review of Fluid Mechanics*, **36**, 413–455.

Color Plates

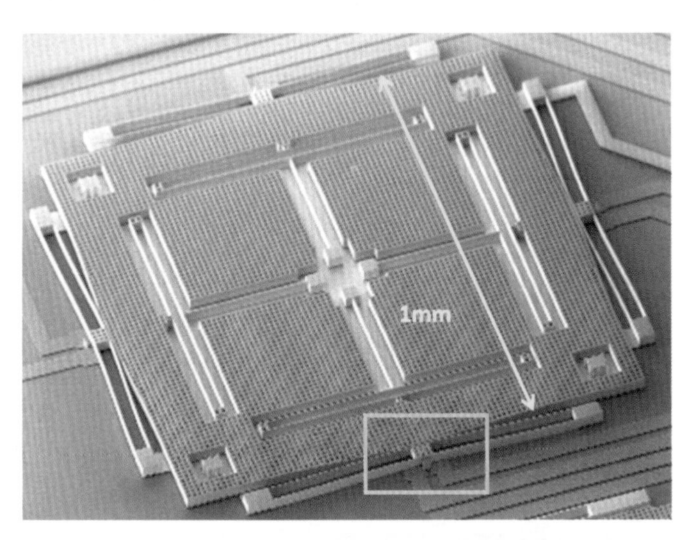

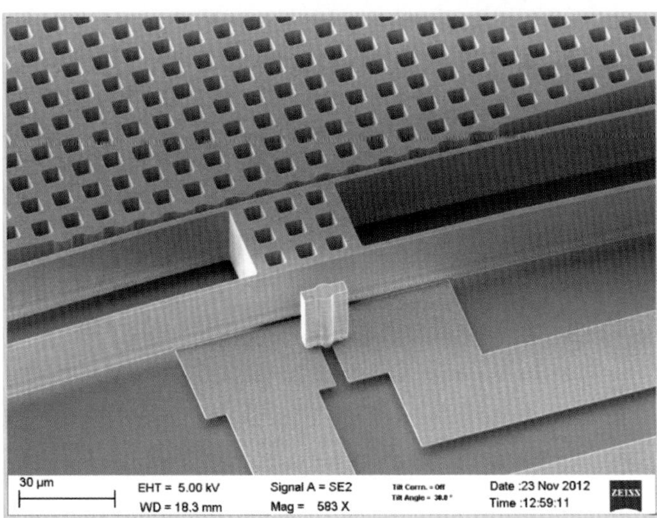

Figure 3.5 SEM of a MEMS shock sensor based on bistable beams and close-up view of the stopper. *Source*: Frangi *et al*. (2015), Figure 2. Reproduced with permission of IEEE.

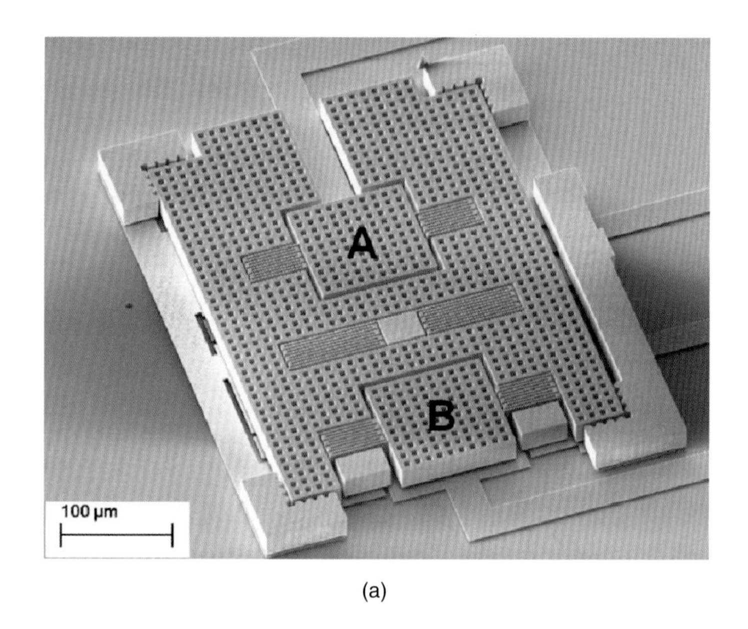

(a)

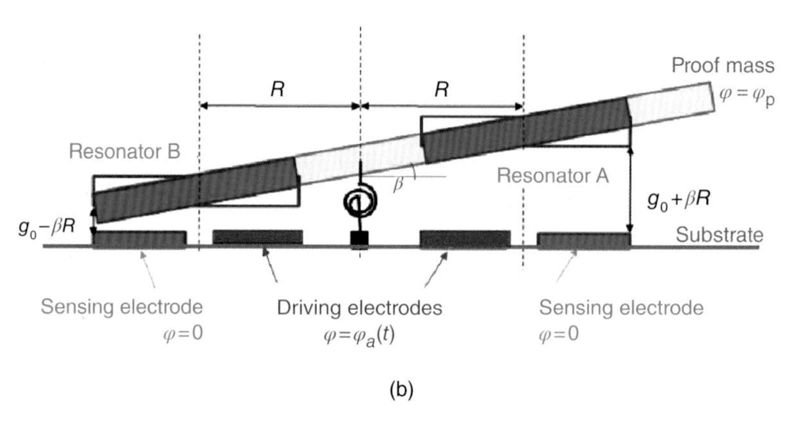

(b)

Figure 4.16 (a) SEM of z-axis accelerometer with two torsional resonators A and B; (b) side view of electrostatically actuated accelerometer, inclined due to external acceleration. *Source*: Caspani *et al.* (2014a), Figure 1. Reproduced with permission of Elsevier.

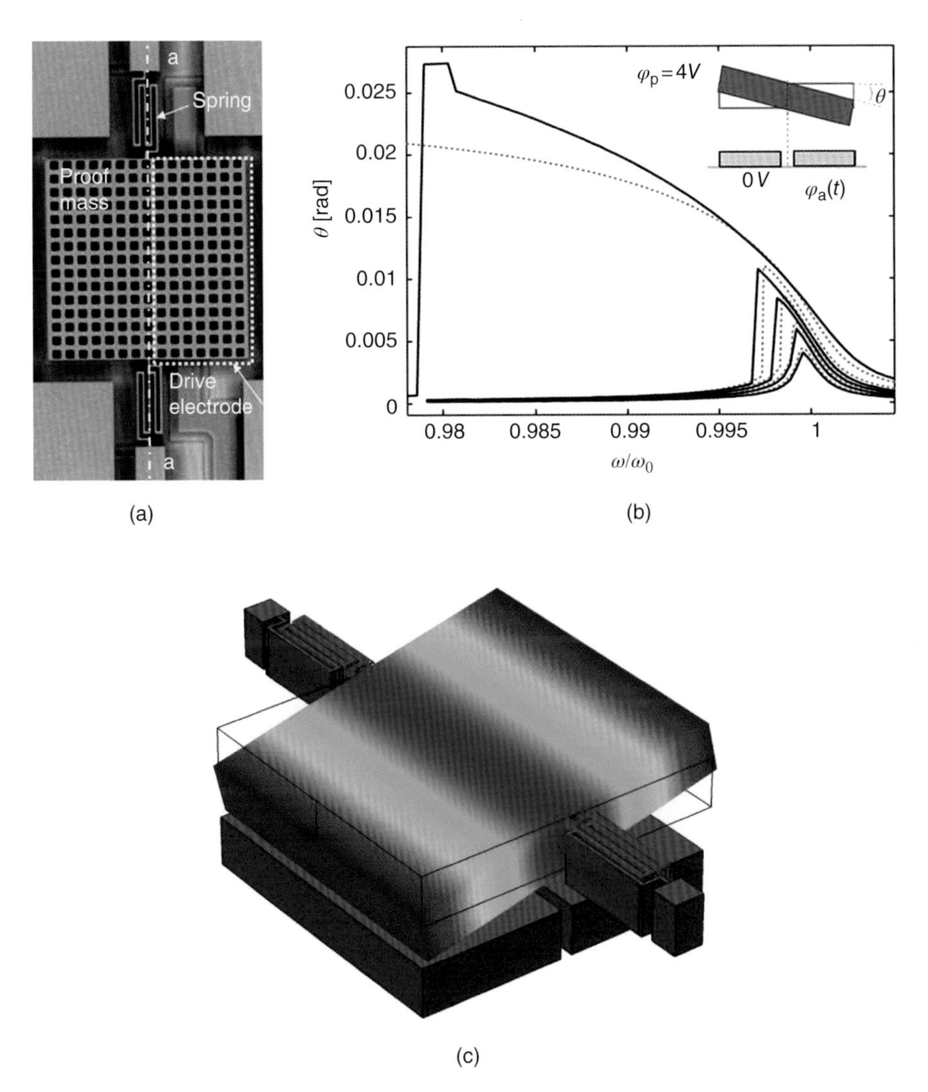

Figure 6.8 Torsional resonator: (a) SEM of polysilicon resonator fabricated through ThELMA surface micromachining process developed by STMicroelectronics; (b) comparison between experimental data (continuous lines) and numerical predictions (dotted lines) for $\varphi_p = 4V$ and $|\varphi_a| = 45, 65, 85, 105$ and 150 mV – only the backward frequency sweeps are reported for sake of clarity; (c) first resonating mode. *Source*: Comi *et al.* (2016), Figures 1 and 3. Open access article published by Elsevier under the CC BY-NC-ND license (http://creativecommons.org/licenses/by-nc-nd/4.0/).

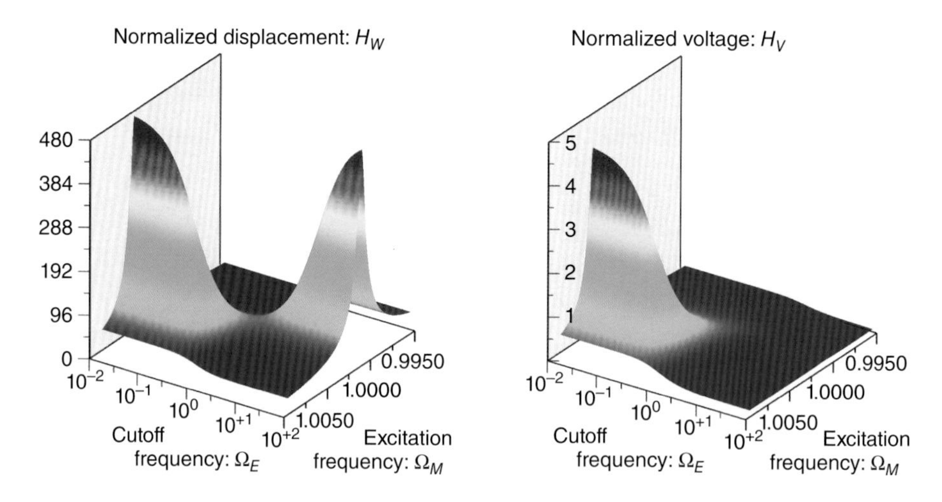

Figure 9.4 Normalized frequency response function for displacement and voltage for a cantilever harvester with $\kappa_E = 0.1$ and $Q_M = 500$.

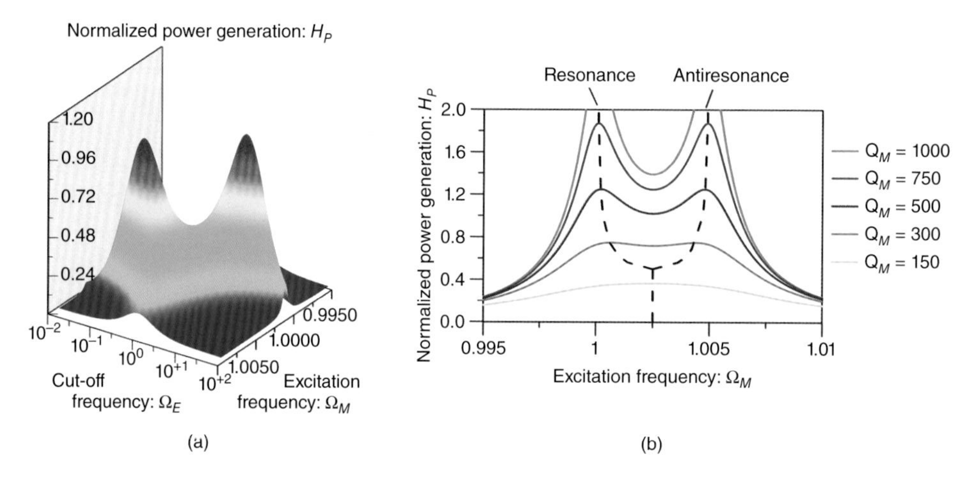

Figure 9.5 (a) Normalized frequency response function for harvested power for a cantilever beam with $\kappa_E = 0.1$ and $Q_M = 500$. (b) Optimal normalized power versus excitation frequency for different levels of mechanical damping ($\kappa_E = 0.1$).

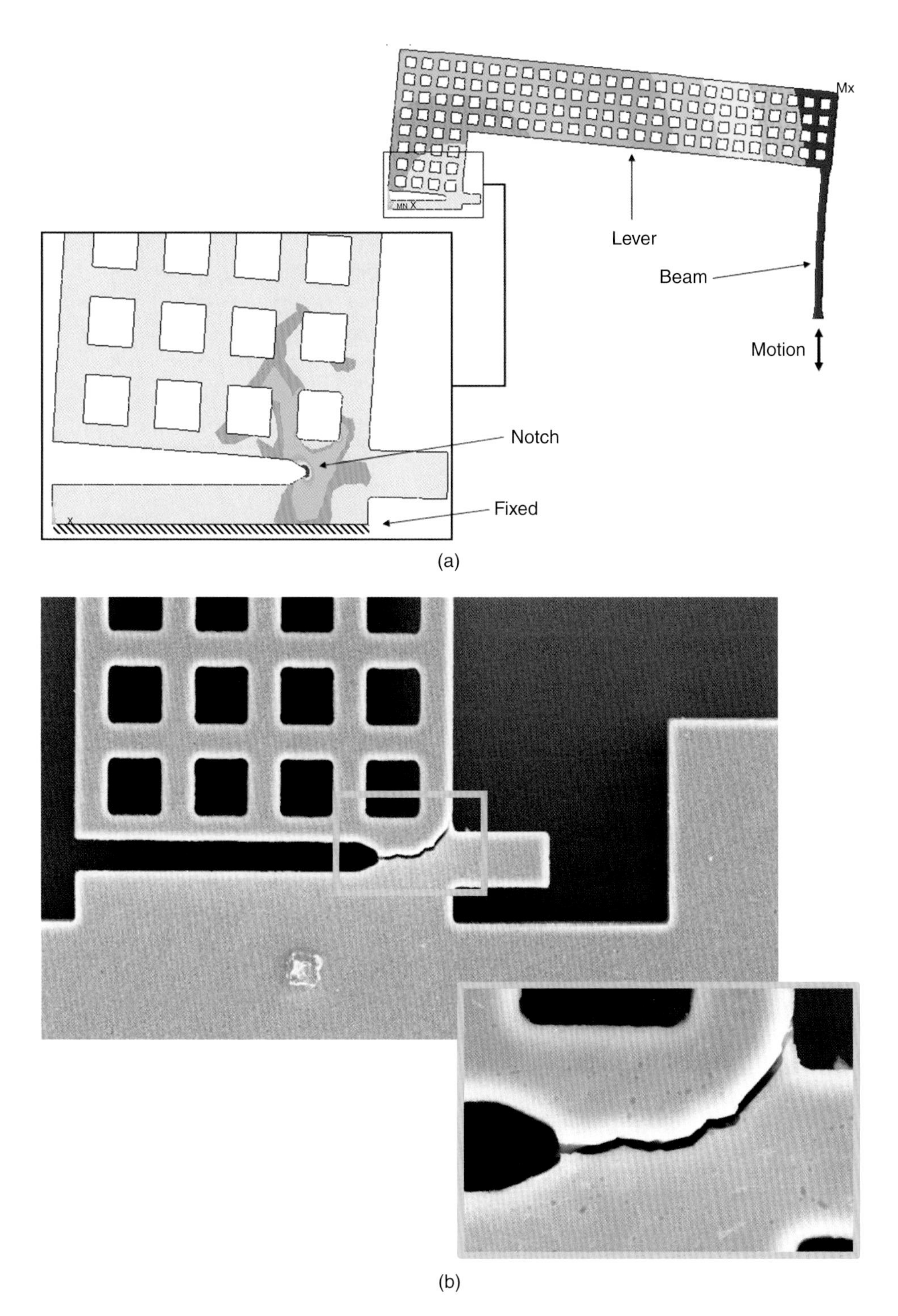

Figure 11.21 (a) Test structure for thick polysilicon: deformed shape of the specimen and contour plot of principal tensile stress of the notched zone. *Source*: Corigliano, Cacchione and Zerbini (2008), Figure 19. Reproduced with permission of Springer. (b) SEM of the fractured device and close-up view.

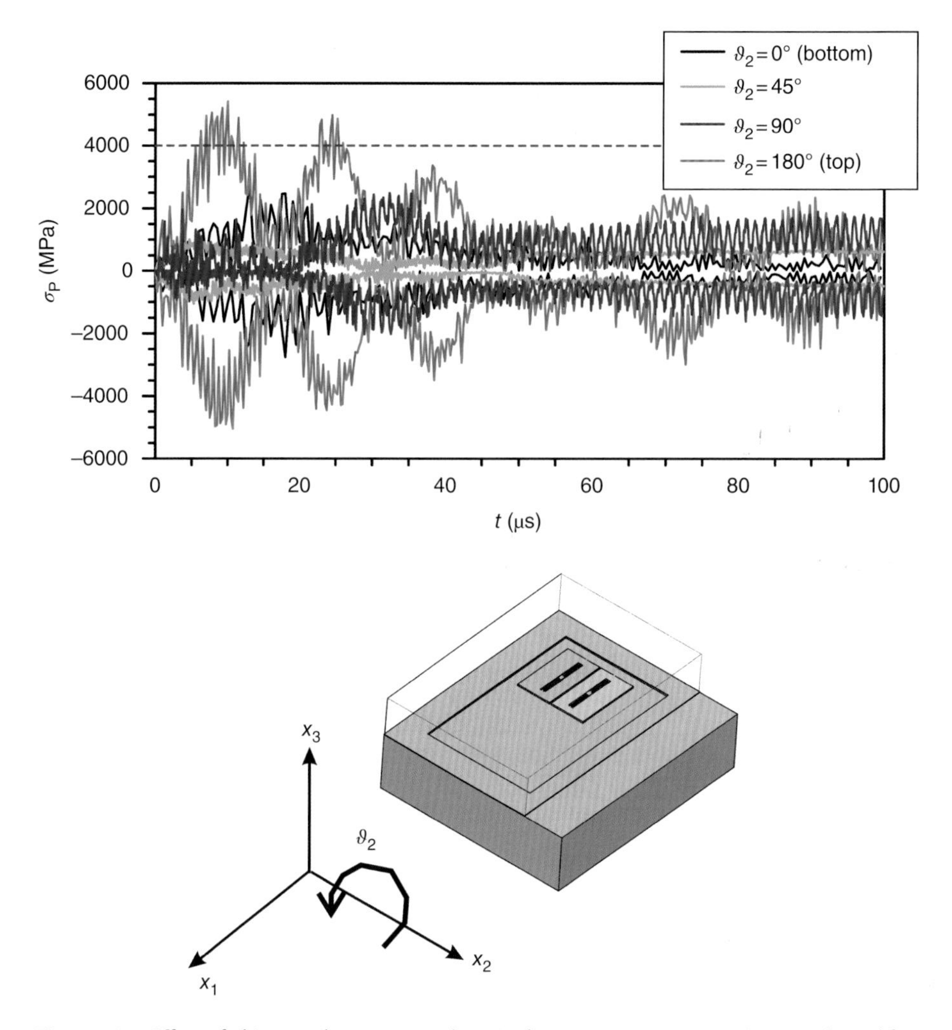

Figure 13.6 Effect of tilting on the stress envelope in the suspension springs. *Source*: adapted from Mariani *et al*. (2011a), Figure 14. Reproduced with permission of Nova Science Publishers.

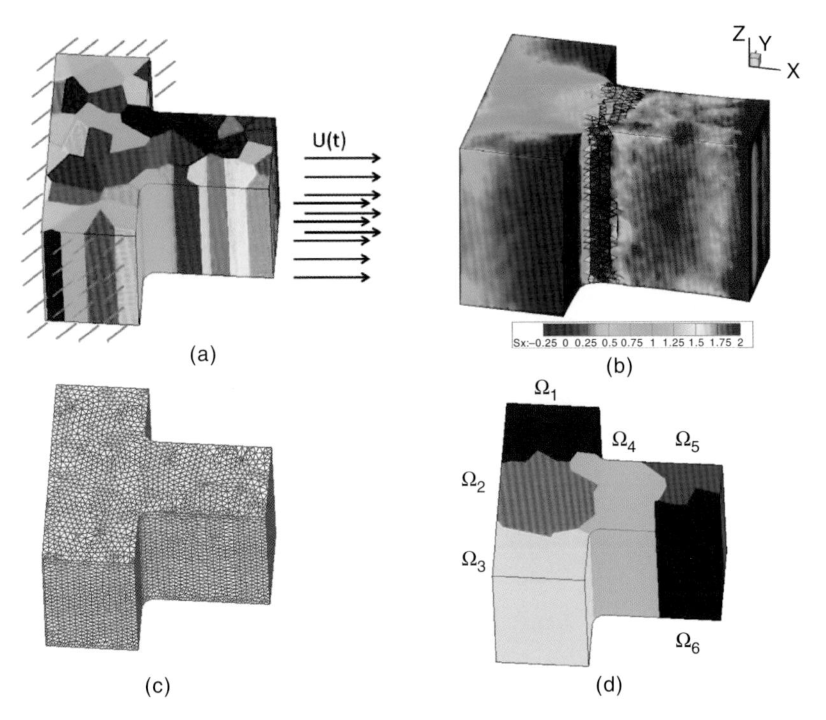

Figure 13.10 (a) 3D Voronoi tessellation of the most stressed polysilicon region; (b) longitudinal stress (in MegaPascals) and crack pattern under tensile loading directed along the *x*-direction; (c) example of discretization; (d) spatial decomposition in six domains.

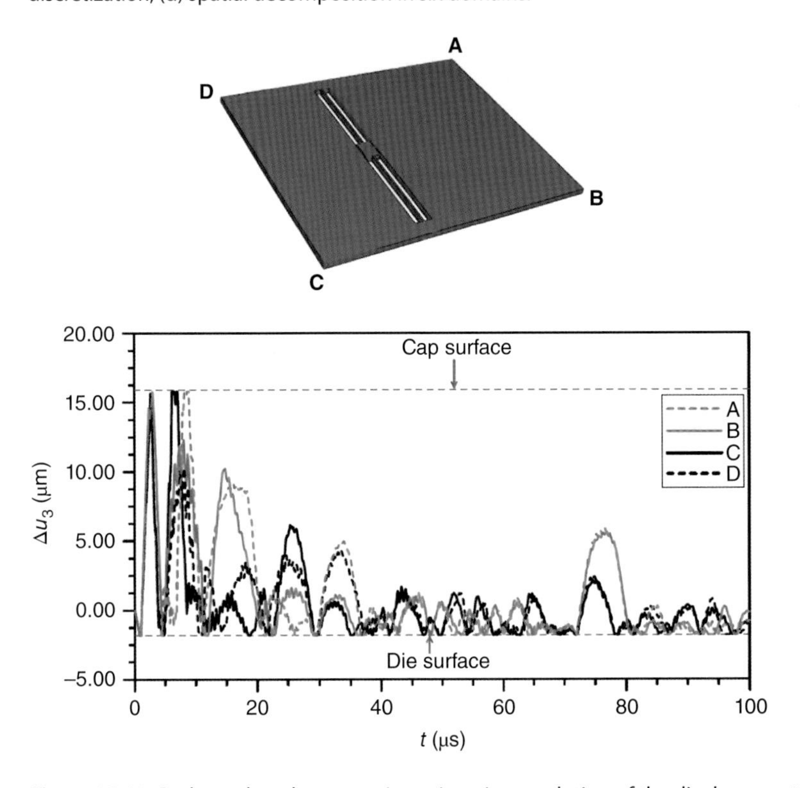

Figure 13.11 Package drop, bottom orientation: time evolution of the displacement at plate corners relative to die or cap surfaces. *Source*: adapted from Ghisi *et al*. (2009a), Figure 5. Reproduced with permission of Elsevier.

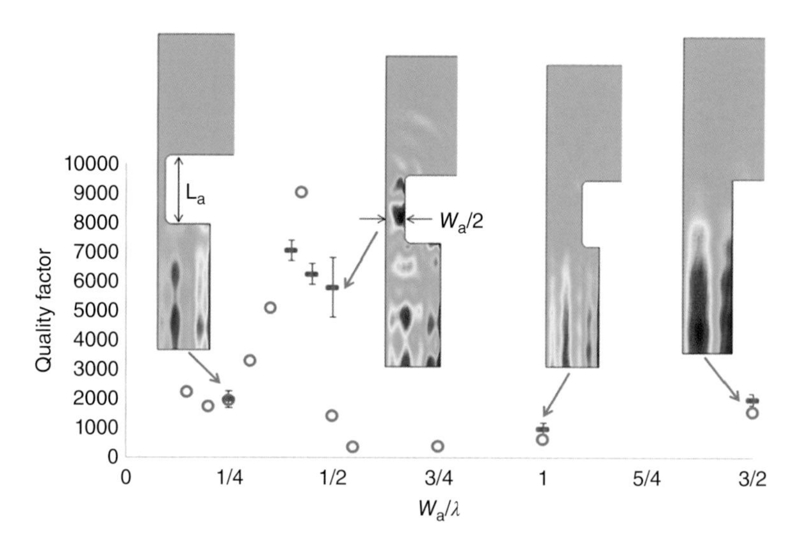

Figure 15.23 220 MHz devices, $L_a/\lambda = 1$. Numerically predicted Q_{anc} (circles) for varying anchor width versus experimental data. Bars denote average experimental data with superposed standard deviations. *Source*: Segovia-Fernandez *et al.* (2015), Figure 9. Reproduced with permission of IEEE.

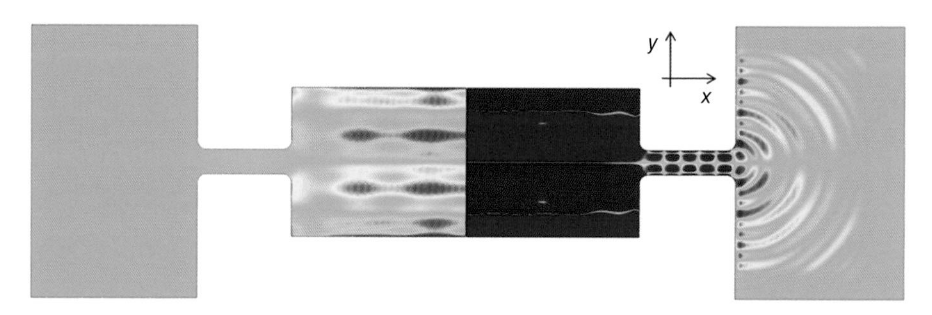

Figure 15.24 u_y for $W_a/\lambda = 1/4$ and $L_a/\lambda = 1$. *Source*: Segovia-Fernandez *et al.* (2015), Figure 11. Reproduced with permission of IEEE.

10

Micropumps

10.1 Introduction

A pump is a device that is able to promote the motion of a fluid, for instance in a network of pipes or channels. The integration of pumping mechanisms in MEMS technology has led to the development of several kinds of silicon micropump (Iverson and Garimella, 2008), that represent the *heart* of active microfluidic systems for achieving precise manipulation of fluids. Among the different applications of microfluidics, the most important are in the field of biomedical and life science diagnostics (Wolff *et al.*, 2003). In fact, microscale analytical machines, such as polymerase chain reaction devices, point-of-care testing systems, labs-on-a-chip and total microanalysis systems, present many advantages with respect to their macroscopic counterparts: the different steps of complex chemical reactions can be integrated in a unique device; the assays are by far more accurate than those performed by hand; smaller volumes of reagents are used. Moreover, it is worth remembering the importance of microfluidics in sorting, sampling, separating and treating cells and other biological materials (Zhang and Nagrath, 2013). Another application of micropumps in the field of life science is represented by drug delivery systems. It is well known that the therapeutic effects of drugs can be highly enhanced if an automatic delivery system is implemented. This is most important for chronic disease, such as diabetes. In fact, standard insulin injections do not mimic normal insulin–glucose kinetics, so peaks and troughs outside the normal range of glucose are unavoidable. Excess glucose is cumulatively toxic, finally yielding severe health complications. It is important to note that half of the social cost for diabetes in the USA (about $180 billions) is spent on complications (Taylor, 2005). Automated insulin delivery would be the best answer to such problems; micropumps are essential components in integrated drug delivery systems.

Microfluidics is important for many fields other than the biomedical one, for instance the thermal management of electronic devices. The increase in performance of microprocessors has been accompanied by a rapid rise of heat generation in microchips and of the heat fluxes that should be dissipated. Garimella, Singhal and Liu (2006) have proposed a fluid-based cooling system for microprocessors, with a

Mechanics of Microsystems, First Edition. Alberto Corigliano, Raffaele Ardito, Claudia Comi, Attilio Frangi, Aldo Ghisi, and Stefano Mariani.

Companion website: www.wiley.com/go/corigliano/mechanics

coolant liquid that circulates in microchannels directly realized on the chip surface. Such a solution could be advantageous for many reasons, among them a reduction in size of the cooling systems and a great efficiency of heat dissipation. Last, but not least, the fabrication process can be fully integrated in the standard process line of the electronic device. Clearly, the solution is feasible only if a suitable micropump is embedded in the microchannel system. One further, and technologically important, application of micropumps is represented by inkjet printheads for professional printers. The main purpose of a printhead is the formation of ink droplets, endowed with any desired volume, velocity and reliability (Wijshoff, 2010). The complete machining of microfluidics ink modules by MEMS technology entails several advantages: very high density of the nozzles; high frequency of droplet firing; dramatic decrease in droplet volume; high reliability; low manufacturing cost.

Micropumps are often classified in terms of the actuation system, which can be nonmechanical or mechanical (Nisar *et al.*, 2008; Woias, 2005). In the former case, so-called continuous flow micropumps aim to transform certain available nonmechanical energy into kinetic momentum so that the fluid can be driven in microchannels. Typically, the increase in pressure is achieved by exploiting characteristic phenomena, so that electro-osmotic, electrohydrodynamic and magnetohydrodynamic pumps have been proposed (Laser and Santiago, 2004). Moreover, other working principles have been recently reported, e.g. electrochemical, surface tension and capillary micropumps (Zhang, Xing and Li, 2007).

In spite of their technological interest, nonmechanical pumps are beyond the scope of this book. Attention is here focused on mechanical micropumps, which need a physical mechanism of actuation, e.g. electrostatic, piezoelectric, thermopneumatic or shape memory alloy. (Iverson and Garimella, 2008). More specifically, this chapter deals with diaphragm micropumps, with electrostatic or piezoelectric actuation, which entails the possibility of building a device by means of standard MEMS fabrication processes. After these introductory remarks, the specific modelling issues for the considered micropumps are described. Subsequent sections are devoted to the actuator modelling and to some reliability issues. Finally, an example of a complete multiphysics model is presented.

10.2 Modelling Issues for Diaphragm Micropumps

In electrostatic micropumps, a membrane is forced to deflect as an appropriate voltage is applied across opposing electrostatic plates, see Figure 10.1 (Zengerle *et al.*, 1995a,b; Jiang, Ng and Lam, 2000; Teymoori and Abbaspour-Sani, 2005; Bertarelli *et al.*, 2010). During this so-called *expansion stroke*, the decrease of pressure in the pumping chamber allows for fluid aspiration through the inlet valve into the pumping chamber. By releasing the applied voltage, the deflected membrane tends to return to its initial position, so that the increase of pressure in the pump chamber pushes the fluid through the outlet valve. Piezoelectric micropumps rely on a similar functioning principle, with the difference of adopting the piezoelectric effect in order to deflect the diaphragm of the pumping chamber (van Lintel, van De Pol and Bouwstra, 1988; Truong and Nguyen, 2004; Ardito *et al.*, 2013; Singh *et al.*, 2015). Again, the fluid in the reservoir is forced to flow in the

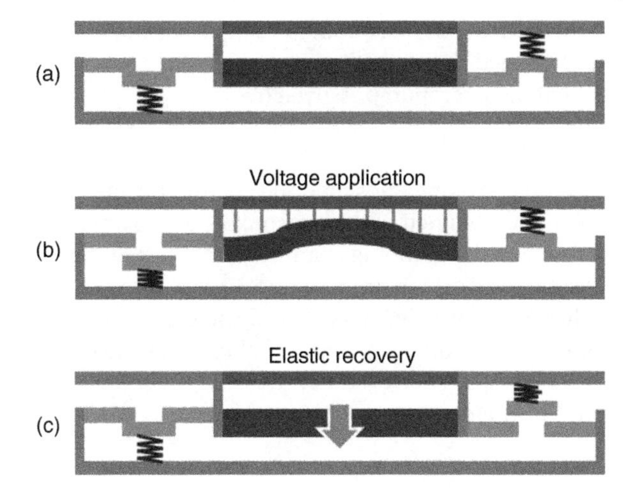

Figure 10.1 Device functioning over a pumping cycle: (a) rest position; (b) filling phase; (c) fluid discharge.

microchannels by the pressure difference induced by the membrane deflection in the pump chamber.

One of the main difficulties in designing a micropump is represented by the necessity of multiphysics simulations of a complex system, composed of the reservoir, the valves, the pumping chamber and the microchannels. A suitable model, coupling electric, structural and fluid dynamic phenomena (i.e. all the fields studied in Chapter 2), should be used as a design tool, not only to optimize the overall performance of the system but also to reduce both the time and cost of development. In the pioneering work of Voigt, Schrag and Wachutka (1998), a complete model of the electrofluidic microsystem previously realized by Zengerle *et al.* (1995a,b) is presented. The complexity of the final model is reduced by the application of a *lumped parameter* approach, including simple components for the pump chamber with an electrostatic actuated membrane, inlet and outlet valves and externally located connecting pipes. The behaviour of each component is described by a physically based compact model extracted from continuum theory or from experimental data. Although this approach is useful for a first characterization of the system, it is necessary to notice the number of multiphysics 3D finite element simulations, which have been introduced in recent times owing to the availability of huge computing power, even for personal desktop computers (Nisar *et al.*, 2008; Tsui and Lu, 2008; Ha *et al.*, 2009; Kang and Choi, 2011).

In many cases, some simple computations are still useful in the preliminary stages of research and development. For instance, the performance of the mechanical micropumps considered in this chapter can be measured by a simple quantity, the *stroke volume* Δv, i.e. the volume of fluid displaced during one actuation cycle. Such a quantity is of paramount importance in the design phase of the micropump and many authors have proposed simplified formulae for its expeditious computation. In the case of a diaphragm micropump where the fluid discharge is obtained by elastic recovery only, Cao, Mantell and Polla (2001) and Teymoori and Abbaspour-Sani (2005) have proposed the assumption that the diaphragm ideally bounces back to the undeformed configuration at each pumping cycle. Under this assumption, the pumping chamber variation in quasi-static conditions v_s represents a good approximation of the stroke volume, $\Delta v \approx v_s$. The stroke volume can easily be computed by carrying out a quasi-static analysis and by computing the volume between the deformed and the undeformed shapes of the diaphragm. For a given actuation frequency f_p, the ideal flow rate of the micropump is given by

$$q_{\text{ideal}} = v_s \cdot f_p. \tag{10.1}$$

In this way, one finds an upper bound for the actual flow rate. First of all, the stroke volume is obtained with a reasonable approximation only if a negligible pressure drop is found in the pumping chamber and through the inlet and outlet valves and channels. Second, the formula in Equation 10.1 is based on the assumption that the whole fluid volume v_s is pumped at each cycle, irrespective of the fluid inertia and of other effects, which depend on the actuation velocity (i.e. the pumping frequency). A better indicator of the pumping performance includes the so-called *pumping efficiency*, which depends on the actuation frequency f_p:

$$\mathcal{E}(f_p) = \frac{\Delta v(f_p)}{v_s} \cdot f_p. \tag{10.2}$$

Clearly, correct evaluation of the pumping efficiency requires fully coupled simulations in order to capture the fluid–structure interactions. Finally, for a given pumping frequency the actual flow is given by

$$q = v_s \cdot \mathcal{E}(f_p). \tag{10.3}$$

10.3 Modelling of Electrostatic Actuator

10.3.1 Simplified Electromechanical Model

This section is devoted to a quasi-static analysis of the pumping diaphragm subject to electrostatic actuation. It is worth remembering that, in many practical cases, the pumping chamber has a circular plan section and is actuated by means of circular electrodes: in such a case, which is schematically depicted in Figure 10.2, the problem is *axisymmetric* and the model is quite simple, since none of the mechanical fields depend on the coordinate ϑ. In the following developments, the actuation is assumed to be controlled

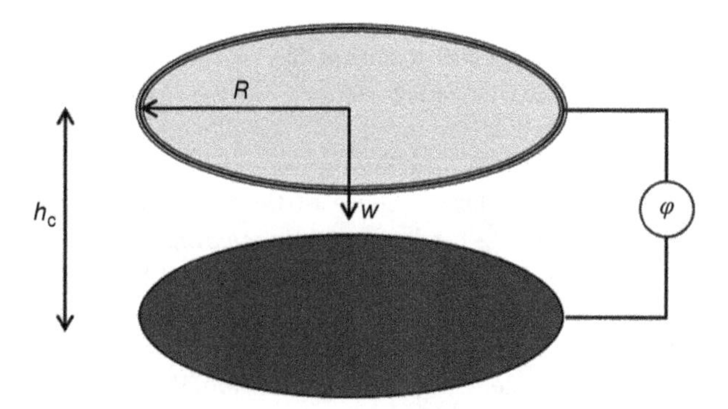

Figure 10.2 Actuator under analysis: the light grey circular plate is the diaphragm of the pumping chamber, which is actuated by the black electrode.

by imposing the applied voltage φ: some remarks will finally be introduced on other possible control strategies.

10.3.1.1 The Principle of Virtual Work for an Axisymmetric Plate

We consider that the pumping diaphragm is an elastic homogeneous plate, whose radius is denoted by R and thickness is denoted by h_{m}. The ratio $2R/h_{\mathrm{m}}$ is very large for typical MEMS devices, so that the diaphragm can be modelled by applying the Kirchhoff approximation for thin plates (Timoshenko and Woinosky-Krieger, 1959), which can be considered as an extension to the plate problems of the Euler–Bernoulli beam theory briefly discussed in Section 2.3.2. In this theory, the shear deformations can be neglected and the governing kinematic field is represented by the out-of-plane displacement $w(r)$. Taking into account the axisymmetric feature, the kinematic model for the plate bending problem in a cylindrical reference system reads:

$$\begin{aligned} u_r &= -z\frac{\mathrm{d}w}{\mathrm{d}r}, \\ u_\vartheta &= 0, \\ u_z &= w, \end{aligned} \tag{10.4}$$

where the hoop displacement u_ϑ is equal to zero because of axial symmetry.

The deformation field can be reduced to the radial strain ε_r and the hoop strain ε_ϑ:

$$\begin{aligned} \varepsilon_r &= -z\frac{\mathrm{d}^2w}{\mathrm{d}r^2} = z\chi_r, \\ \varepsilon_\vartheta &= -z\frac{\mathrm{d}w}{r\,\mathrm{d}r} = z\chi_\vartheta, \end{aligned} \tag{10.5}$$

with obvious definition of the radial and hoop curvatures, χ_r and χ_ϑ, respectively. Accordingly, it is possible to define the radial and hoop bending moments:

$$\begin{aligned} M_r &= \int_{-h_{\mathrm{m}}/2}^{h_{\mathrm{m}}/2} z\sigma_r\,\mathrm{d}z, \\ M_\vartheta &= \int_{-h_{\mathrm{m}}/2}^{h_{\mathrm{m}}/2} z\sigma_\vartheta\,\mathrm{d}z. \end{aligned} \tag{10.6}$$

On the basis of the definitions in Equations 10.5 and 10.6, the principle of virtual power (see Section 3.2.1) for the axisymmetric plate reads:

$$\int_0^R (M_r\delta\dot{\chi}_r + M_\vartheta\delta\dot{\chi}_\vartheta)\,r\,\mathrm{d}r - \int_0^R p\,\delta\dot{w}\,r\,\mathrm{d}r = 0. \tag{10.7}$$

The elastic constitutive law is written on the basis of the plane stress assumption, which is customary adopted for plate models:

$$\begin{aligned} \sigma_r &= \frac{E}{1-\nu^2}(\varepsilon_r + \nu\varepsilon_\vartheta) \quad \Rightarrow \quad M_r = D(\chi_r + \nu\chi_\vartheta), \\ \sigma_\vartheta &= \frac{E}{1-\nu^2}(\varepsilon_\vartheta + \nu\varepsilon_r) \quad \Rightarrow \quad M_\vartheta = D(\chi_\vartheta + \nu\chi_r), \end{aligned} \tag{10.8}$$

where the plate stiffness $D = (Eh_{\mathrm{m}}^3)/(1-\nu^2)$ has been introduced. After some algebraic developments, from the principle of virtual power, one can obtain the governing equation for the axisymmetric plate, namely:

$$\nabla^4 w = \frac{p}{D}, \tag{10.9}$$

where the double Laplace operator ∇^4 should be evaluated in cylindrical coordinates, taking account of the fact that w does not depend on ϑ.

10.3.1.2 Electrostatic Forces

The problem of electrostatic actuation should be analysed in the framework of the fully coupled electromechanical problem, as better explained in Section 2.5.3. In the presence of a deformable body, the solution is generally obtained by applying some numerical method, such as the finite element method. Nonetheless, for the specific problem treated in this section, a simplified model can be introduced so that a closed-form solution can be obtained with a reasonable degree of accuracy (Batra, Porfiri and Spinello, 2007; Bertarelli *et al.*, 2011a).

The electrostatic force is produced by a potential difference between a deformable electrode (the diaphragm) and a fixed electrode. It is possible to assume that the interaction is dominated by vertical forces, which are distributed on the surface of the diaphragm. The governing equation is Equation 10.9, provided that a suitable expression of the nonuniform pressure p due to electrostatic interaction is obtained. To this purpose, the so-called *proximity force approximation* can be adopted, as is commonly done in the computation of small-scale interaction forces (DelRio *et al.*, 2005; Ardito, Corigliano and Frangi, 2013). Each point is endowed with a portion of area and interacts with a corresponding point on the opposite surface. The electrostatic force on the considered area is obtained by assuming that the surfaces around the interacting points are flat and parallel to one another. The standard formula for a parallel-plate capacitor (see Section 2.5.1) can be applied, with the important point that the initial gap h_c is modified, at each point, by the transverse displacement w of the deformable diaphragm. One finally obtains

$$p = \frac{\epsilon_r \epsilon_0}{2} \frac{\varphi^2}{(h_c - w)^2}. \tag{10.10}$$

The simplified electromechanical model is thus governed by a nonlinear differential equation, in terms of the transverse displacement w:

$$\nabla^4 w - \frac{\epsilon_r \epsilon_0}{2\,D} \frac{\varphi^2}{(h_c - w)^2} = 0. \tag{10.11}$$

The validity of the simplified approach is strictly related to the accuracy of the proximity force approximation, which provides good results if the surfaces are separated by small gaps. A numerical confirmation can be obtained by carrying out a critical comparison with the results of fully coupled numerical analyses.

10.3.1.3 Governing Equations via the Weighted Residuals Approach

The analytical solution of the nonlinear differential equation (10.11) is quite complicated. Alternatively, an algebraic formulation can be obtained by exploiting the virtual power principle and the weighted residuals approach. The transverse displacement is approximated by means of a suitable shape function, which is able to fulfil the kinematic boundary conditions. In the case of a fully clamped circular plate, one should enforce that:

$$w(r = R) = 0, \quad \left.\frac{\mathrm{d}w}{\mathrm{d}r}\right|_{r=R} = 0. \tag{10.12}$$

The simplest choice for the shape function is the analytical solution for deflection under uniform pressure, so that:

$$w \cong W\psi_w = W\left(\frac{r^2}{R^2} - 1\right)^2. \tag{10.13}$$

The governing parameter W represents the maximum displacement, in the centre of the plate.

By computing the approximate strain field by means of Equation 10.5 and by enforcing the principle of virtual power (10.7), the governing relation can be reduced to a one-dof equation:

$$kW - \frac{\epsilon_\mathrm{r}\epsilon_0\varphi^2R^2}{2\ h_\mathrm{c}^2}f(W) = 0, \tag{10.14}$$

where the generalized stiffness k is given by

$$k = \int_0^R D\left[\left(\frac{\mathrm{d}^2\psi_w}{\mathrm{d}r^2}\right)^2 + 2\nu\frac{\mathrm{d}^2\psi_w}{\mathrm{d}r^2}\frac{\mathrm{d}\psi_w}{r\ \mathrm{d}r} + \left(\frac{\mathrm{d}\psi_w}{r\ \mathrm{d}r}\right)^2\right] r\ \mathrm{d}r \tag{10.15}$$

and the nonlinear force amplitude $f(W)$ is given by

$$f(W) = \frac{1}{R^2}\int_0^R \frac{\psi_w r\ \mathrm{d}r}{\left(1 - \frac{W}{h_\mathrm{c}}\psi_w\right)^2}. \tag{10.16}$$

After some algebra, the load–displacement relationship can be obtained in a closed form. To this purpose, the nondimensional displacement $\mathcal{W}$ and nondimensional load parameter λ are adopted:

$$\mathcal{W} = \frac{W}{h_\mathrm{c}}, \quad \lambda = \frac{\epsilon_\mathrm{r}\epsilon_0\varphi^2R^4}{2Dh_\mathrm{c}^3}. \tag{10.17}$$

The solution reads:

$$\lambda = \frac{128}{3}\frac{\sqrt{\mathcal{W}^5}}{\frac{\sqrt{\mathcal{W}}}{\mathcal{W}-1} + \mathrm{arctanh}(\sqrt{\mathcal{W}})}. \tag{10.18}$$

The load–displacement curve is plotted in Figure 10.3, along with the solutions achieved using the refined numerical technique. More specifically, the figure contains the results obtained using: (i) the fully coupled electromechanical finite element model; (ii) the finite element method with displacement-dependent boundary conditions, which means that the proximity force approximation is used to compute the pressure on each node of the finite element mesh; (iii) the finite difference method with displacement-dependent boundary conditions. The maximum relative error is limited to 2.6%: the validity of the approximate procedure is therefore confirmed.

The closed-form solution for the load–displacement relationship can be used to carry out parametric analyses, which are very useful, in the design phase, for investigating

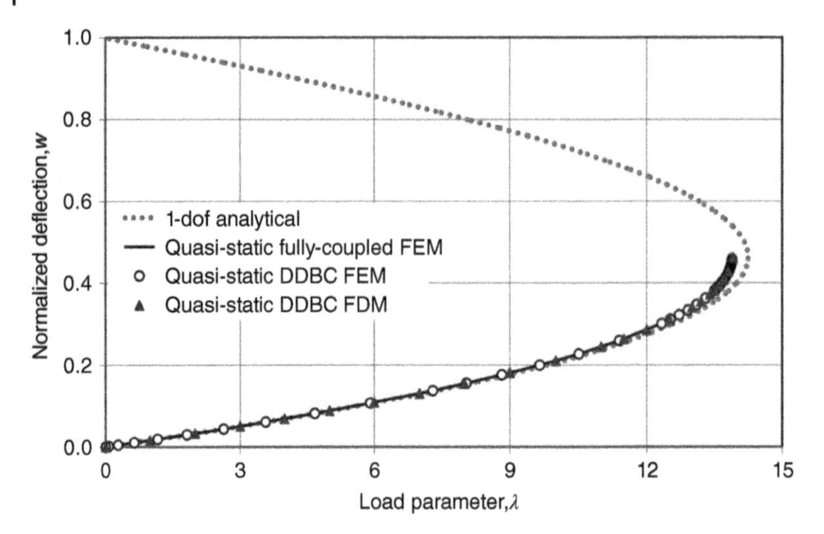

Figure 10.3 Nondimensional midpoint displacement versus load parameter: comparison between the solution of the one-degree-of-freedom model and more refined numerical procedures. DDBC, displacement-dependent boundary conditions; FDM, finite difference method; FEM, finite element method.

the effects of different geometric or mechanical parameters. For example, a circular pumping chamber with fixed radius $R = 2000$ μm is considered; the height of the pumping chamber is $h_f = 10$ μm, so that the initial volume is $v_0 = \pi R^2 h_f \approx 126$ nl. The diaphragm is realized in polysilicon (Young's modulus $E = 160$ GPa, Poisson's ratio $\nu = 0.22$), with thickness h_m ranging between 20 μm and 50 μm. The capacitor gap h_c, filled with air ($\epsilon_r = 1$), is variable between 4 μm and 10 μm. The actuation voltage is fixed at $\varphi = 60$ V. The ideal stroke volume, in the sense reported in Section 10.2, is plotted in Figure 10.4 for several values of the varying parameters. The maximum stroke volume is around 6 nl, which is quite low with respect e.g. to insulin micropumps (Dumont-Fillon *et al.*, 2014). Nevertheless, the most important parameter is represented by the flow rate, which can be increased by actuating at high frequency: as anticipated, the fluid–structure interaction plays a paramount role in obtaining some precise estimates.

10.3.2 Reliability Issues

10.3.2.1 Electrostatic Pull-In

The curves for $h_c = 4$ μm and $h_c = 5$ μm in Figure 10.4 do not cover the whole range of the parameter h_m. This is because, for some specific combinations of the geometric parameters, the applied voltage is beyond the pull-in threshold. Such a phenomenon, which has already been introduced in Section 2.5.2 with reference to parallel-plate capacitors, is now referred to the case of a deformable electrode over a rigid one. The jump-to-contact is clearly identified as the limit point in Figure 10.3: the load–displacement curve starts from the origin and follows a stable branch until the pull-in voltage is reached; after that, the displacement can be increased along an unstable path, which can only be computed by means of the analytical approach. In this quasi-static framework, the onset of the instability can be seen as the vanishing of the

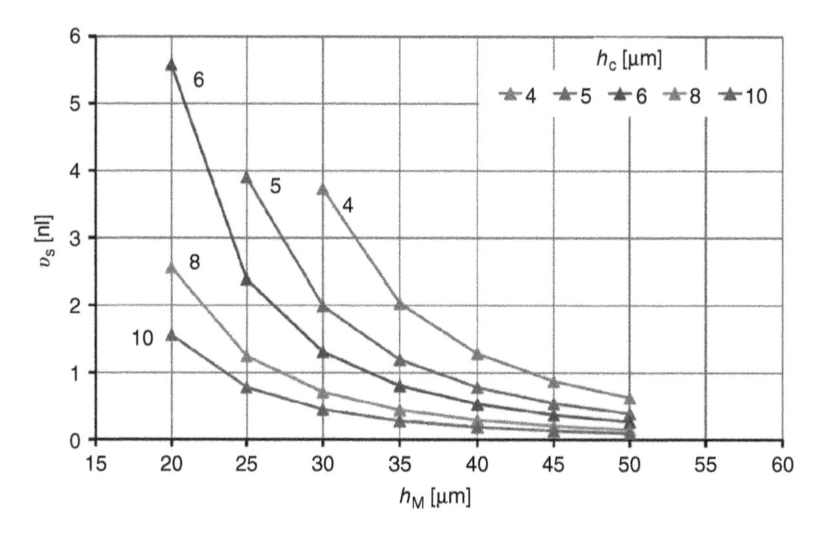

Figure 10.4 Parametric analysis of a micropump: variation of ideal stroke volume with respect to diaphragm thickness h_m and capacitor gap h_c.

systems stiffness, which is obtained through the first-order expansion of Equation 10.14:

$$k - \frac{\epsilon_r \epsilon_0 \varphi^2 R^2}{2\, h_c^2} \frac{\mathrm{d}f}{\mathrm{d}W} = 0. \tag{10.19}$$

Such an equation, combined with the equilibrium equation (10.14) can be solved analytically to obtain the dimensionless displacement and the load parameter at the pull-in limit:

$$\mathcal{W}_{\mathrm{pi}} = 0.4633, \quad \lambda_{\mathrm{pi}} = 14.24. \tag{10.20}$$

Pull-in happens (see Section 2.5.2) when the maximum plate displacement attains the value 0.4633 h_c, whereas in the case of the parallel-plate capacitor the critical displacement is equal to one-third of the capacitor gap: the difference can be well explained by considering that, in the present case, the electrode is deformable.

Equation 10.20 can be used to carry out a parametric investigation of the pull-in voltage; the results are shown in Figure 10.5. The critical load increases for high diaphragm thickness, since the structural stiffness has a positive impact on stability. Similarly, the larger the capacitor gap, the greater the pull-in voltage. It is worth noting that the present estimate of pull-in voltage is obtained in the quasi-static case; the actual behaviour of the diaphragm is represented by a dynamic model, as thoroughly discussed by Bertarelli *et al.* (2011a) and, for different types of microstructure, by Rochus, Rixen and Golinval (2005) and Elata and Bamberger (2006). The pull-in voltage for dynamic actuation is less than its static counterpart: for the considered microplate, Bertarelli *et al.* (2011a) report that $\lambda_{\mathrm{Dpi}} = 0.825\ \lambda_{\mathrm{pi}}$ in the ideal case of the absence of damping (the quality factor Q tends to infinity). The analyses can be repeated for different values of Q: the dynamic pull-in voltage is inversely proportional to Q and tends to the static value if $Q \to 0$ (however, for overdamped states, $Q < 1.2$, the assumption of constant damping coefficient is questionable and an improved damping model is required, see Rocha, Cretu and Wolffenbutte (2004)).

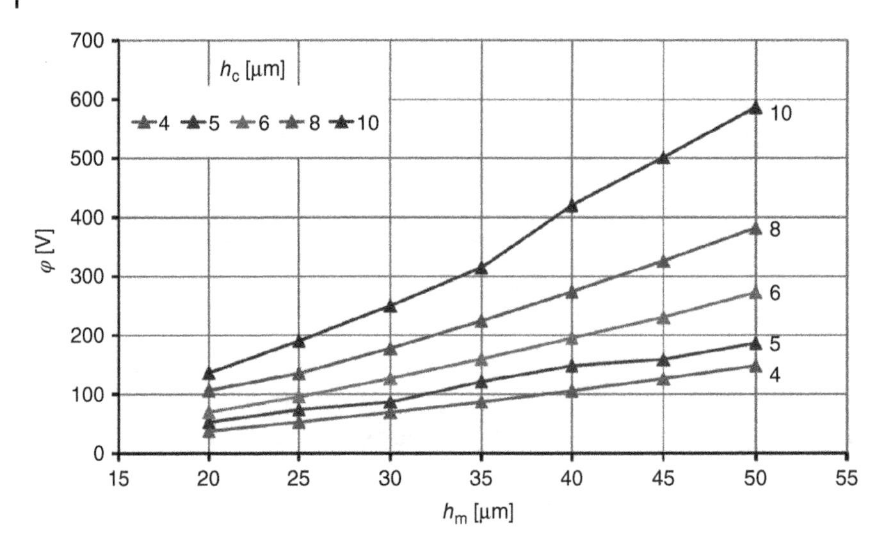

Figure 10.5 Parametric analysis of micropump: variation of pull-in voltage with respect to diaphragm thickness h_m and capacitor gap h_c.

10.3.2.2 Adhesion

The jump-to-contact of the diaphragm may lead to dangerous adhesion phenomena, which are described in detail in Chapter 16. Briefly, if the elastic restoring force is not sufficient to overcome the adhesion energy, the movable plate remains stuck on the rigid electrode after voltage deactivation. The elastic restoring force can be computed by solving analytically the governing equation (10.9) for an axisymmetric plate that is adhered to the substrate on a circular surface with radius a (see Figure 10.6). The final expression of the elastic energy depends on a dimensionless function $\alpha(R/a)$:

$$U_{\mathrm{E}} = \frac{1}{2} \frac{h_{\mathrm{c}}^2 E h_{\mathrm{m}}^3}{\alpha(R/a)\, R^2}. \tag{10.21}$$

The comparison with the specific adhesion energy can be obtained, in a straightforward manner, by dividing the elastic energy expressed in Equation 10.21 by the adhered area:

$$\Gamma > \frac{1}{2} \frac{h_{\mathrm{c}}^2\, E\, h_{\mathrm{m}}^3}{\pi R^4} \frac{(R/a)^2}{\alpha(R/a)}. \tag{10.22}$$

From analysis of Equation 10.22, one can easily find that the minimum adhesion energy is attained for $a/R \approx 0.357$. For example, if $h_{\mathrm{m}} = 30\ \mu\mathrm{m}$, $h_{\mathrm{c}} = 5\ \mu\mathrm{m}$,

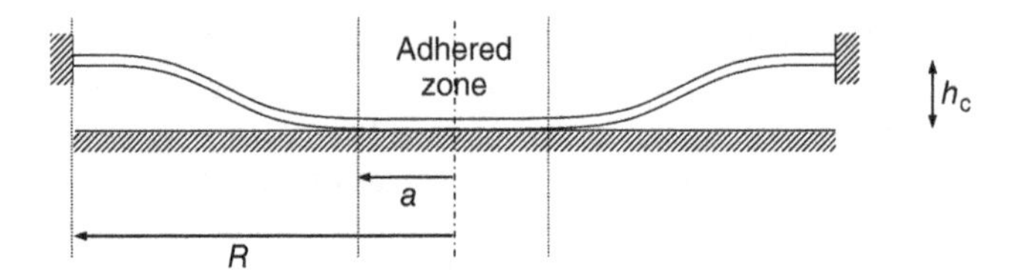

Figure 10.6 Scheme adopted for computation of elastic restoring energy after adhesion of microplate on substrate.

$R = 2000$ µm, the critical adhered state is reached when the adhesion energy is larger than about 156 mJ/m^2. Data reported by Ardito *et al.* (2016) and in Chapter 16 show that, for a typical polysilicon surface, the adhesion energy ranges between 1 mJ/m^2 and 70 mJ/m^2: since the values are smaller than the critical threshold, the considered structure is safe with respect to adhesion after pull-in.

10.3.2.3 Actuator Control

It has been rigorously proven that, under charge control, electrostatic actuators have, in general, a wider range of stability than under voltage control (Bochobza-Degani, Elata and Nemirovsky, 2003). The governing equation can be readily obtained if one considers that the overall charge is connected to the applied voltage by the definition of the global capacitance of the electrostatic system, i.e. $C = Q/\varphi$ (Bertarelli *et al.*, 2011b):

$$\nabla^4 w - \frac{\epsilon_r \epsilon_0}{2D} \frac{Q^2}{C^2} \frac{1}{(h_c - w)^2} = 0, \tag{10.23}$$

where

$$C = 2\pi \int_0^R \frac{\epsilon_r \epsilon_0}{h_c - w} r \, dr. \tag{10.24}$$

The approximate solution can be found, again using the principle of virtual power and the weighted residuals approach, with the assumption (10.13) for the displacement field. The one-dof equation now reads:

$$kW - \frac{\epsilon_r \epsilon_0 Q^2 R^2}{2C^2 h_c^2} f(W) = 0. \tag{10.25}$$

The global capacitance is computed by introducing the assumption (10.13) in the definition (10.24):

$$C = \frac{\pi \epsilon_r \epsilon_0 R^2}{h_c} \frac{2}{R^2} \int_0^R \frac{r}{1 - \frac{W}{h_c} \psi_w} dr = \frac{\pi \epsilon_r \epsilon_0 R^2}{h_c} \Theta(W), \tag{10.26}$$

where the dimensionless function $\Theta(W)$ has been introduced. The final form of the governing equation is

$$kW - \frac{Q^2}{2\pi^2 \epsilon_r \epsilon_0} \frac{f(W)}{R^2 \, \Theta(W)^2} = 0. \tag{10.27}$$

The load–displacement equation can be obtained in a closed form, as in the case of voltage control, but now the nondimensional load parameter is given by

$$\chi = \frac{Q^2}{2\pi^2 \epsilon_r \epsilon_0 D h_c}. \tag{10.28}$$

The solution is given by:

$$\chi = \frac{32}{3} \frac{\sqrt{\mathcal{W}^3} \ln^2 \left[\dfrac{1 - \sqrt{\mathcal{W}}}{1 + \sqrt{\mathcal{W}}} \right]}{\dfrac{\sqrt{\mathcal{W}}}{\mathcal{W} - 1} + \operatorname{arctanh}(\sqrt{\mathcal{W}})}. \tag{10.29}$$

By applying the same procedure as for the case of voltage control, one can find the dimensionless displacement and the load parameter at the pull-in limit:

$$\mathcal{W}_{\mathrm{pi}} = 0.641, \quad \chi_{\mathrm{pi}} = 23.46. \tag{10.30}$$

The charge control induces an interesting improvement, because the travel range of the diaphragm is increased by about 30% with respect to the case of voltage control: this means that the stroke volume can be increased as well.

As expected, the charge control exhibits a stabilizing effect with respect to the voltage drive, but this is not sufficient to achieve full-range stabilization within the considered device. Annular electrodes can be introduced to achieve complete stabilization, i.e. $\mathcal{W} = 1$ before the onset of instability. The case of electrodes that are extended only between R_{i} and R can be tackled with a similar procedure, with the only difference of extending the integrals in Equations 10.16 and 10.26 between R_{i} and R instead between of 0 and R (Bertarelli *et al.*, 2011c). Under voltage control, if a radial electrode with internal radius $R_{\mathrm{i}} = 0.63\ R$ is used to drive the system, full-range travel is theoretically achieved with a huge voltage penalty, estimated as more than three times the standard pull-in voltage. On the contrary, in charge control, when a ring electrode with internal radius $R_{\mathrm{i}} = 0.51\ R$ is adopted, the plate midpoint can travel the full gap if the injected charge is increased by a factor of only 1.77.

10.4 Multiphysics Model of an Electrostatic Micropump

The previous section focused on the quasi-static response of the electrostatic actuator, with the main purpose of achieving the ideal stroke volume and of pointing out some important reliability issues. This information is very useful in the initial design phase, to establish some basic features of the micropump. If the analyst requires a more refined evaluation of the pumping parameters, it is necessary to resort to multiphysics and fully coupled models, with the unavoidable use of numerical techniques, such as the finite element method. This section provides the reader with some hints on the set-up of the computational models and an example of a solution for an electrostatic micropump.

A circular pumping chamber with radius $R = 2000\ \mu\mathrm{m}$ and height $h_{\mathrm{f}} = 10\ \mu\mathrm{m}$ is considered. Inlet and outlet channels have the same depth as the chamber; they are 800 μm wide and 1000 μm long. The membrane thickness ranges between 20 μm and 50 μm and the capacitor gap ranges between 4 μm and 10 μm The actuation voltage is fixed at $\varphi = 60\ \mathrm{V}$. Inlet and outlet ports are placed along the perimeter of the pumping chamber, in diametrically opposed positions. The valves are not modelled in this case, to focus on the behaviour of the pumping chamber.

A 3D model should be adopted, using brick elements in both the electric domain (i.e. the space between the electrodes) and in the fluid domain (i.e. the space inside the pumping chamber and the inlet and outlet channels). For the structural domain, it would be desirable to adopt shell elements, in order to reduce the computational burden with respect to several layers of brick elements across the thickness. Such a modelling procedure can be pursued only if the finite element solver includes the fluid–structure interaction, not only for solid elements but also in the case of shell elements.

The mesh movement is described by the arbitrary Lagrangian–Eulerian method. A time-dependent solution is obtained by an implicit solver (backward Euler integration).

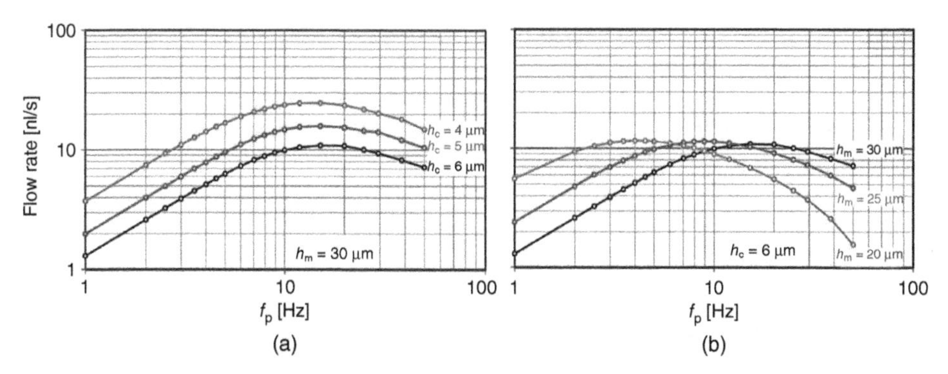

Figure 10.7 Flow rate of electrostatic micropump computed by means of the dynamic fully coupled electro-fluid-mechanical model: (a) effect of capacitor gap; (b) effect of membrane thickness.

The continuum hypothesis for fluid modelling, which is the base of the Navier–Stokes equations described in Section 2.4, is still valid at the size scale of the considered devices.

Suitable interface conditions should be defined between adjacent domains, in terms of electrostatic force and displacement for the electromechanical interaction and of boundary velocity and pressure for the fluid–structure interaction. Ideal valves are considered in this example, i.e. no pressure drop is introduced across the fully open channel and no leakage is considered if the valve is closed.

The actuation of the micropump is given by means of a square wave time history of the voltage on one electrode, while the other electrode is grounded. The voltage is applied for a half cycle, then it is switched off during the second half cycle. The analyses are referred to a wide range of frequencies of the actuation cycle, namely between 1 Hz and 50 Hz. The results are represented in Figure 10.7. The effect of the fluid behaviour is clearly evidenced by the nonmonotonic curve. Indeed, for low frequencies, roughly below 5 Hz, the full stroke is exploited and an almost linear increase in the flow rate is observed. A further increase in actuation frequency leads to a decrease in the stroke, because of the inertial effects of the fluid. This is initially overcompensated by faster actuation but, after a certain threshold, a further increase in actuation frequency leads to a decrease in the generated fluid flow. An optimal actuation frequency can be identified,

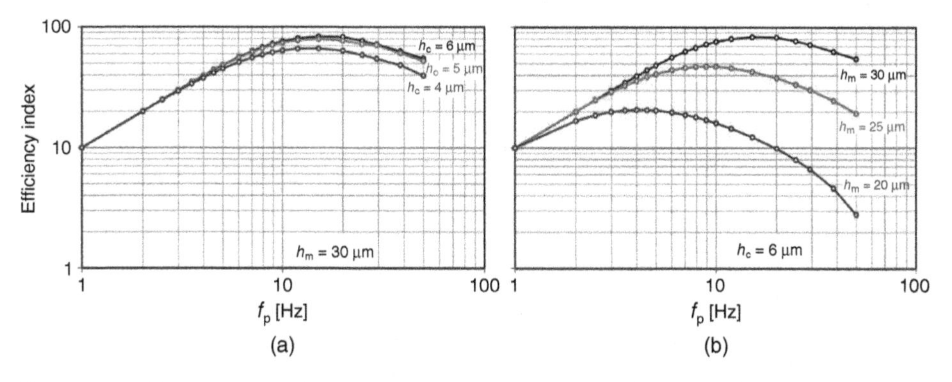

Figure 10.8 Efficiency index of electrostatic micropump computed using the dynamic fully coupled electro-fluid-mechanical model: (a) effect of capacitor gap; (b) effect of membrane thickness.

thus confirming the experimental results presented by Zengerle, Richter and Sandmaier (1992) and Machauf, Nemirovsky and Dinnar (2005). Evaluation of the actual flow can be used to compute the efficiency index, which is plotted in Figure 10.8. Parametric analysis leads to the conclusion that the efficiency is greatly influenced by the membrane thickness, whereas the capacitor spacing plays a minor role.

10.5 Piezoelectric Micropumps

Piezoelectric micropumps can be based on a layout that is similar to the one for electrostatic devices. As described in Figure 10.9, the pumping chamber can be obtained by means of a rigid substrate and a deformable diaphragm, which is now actuated by exploiting the so-called converse piezoelectric effect (i.e. transformation of electric energy into mechanical energy). This is exactly the opposite way of functioning with respect to energy harvesters, which are studied in Chapter 9. Thin-film piezoelectric materials can be used. The additional layer, depicted in Figure 10.9, is polarized in the vertical direction and works in the 31-mode: the application of an electric field in the vertical direction induces stretching in the horizontal plane and, because of the asymmetric position of the active layer, this entails the bending deformation of the circular plate.

10.5.1 Modelling of the Actuator

A complete model of the piezoelectric micropumps can be obtained only in a multiphysics framework, as better explained in the next section. Nevertheless, as for the electrostatic case, a simplified model can be obtained on the basis of some reasonable assumptions.

First of all, the presence of the additional layer induces the unavoidable coupling of membrane and bending problems in plate deformation. Consequently, it is not possible to limit the kinematic model to the bending case only and the additional kinematic field u should be introduced:

$$\begin{aligned} u_r &= u - z\,\frac{\mathrm{d}w}{\mathrm{d}r}, \\ u_\vartheta &= 0, \\ u_z &= w, \end{aligned} \tag{10.31}$$

The strain fields are now written as:

$$\begin{aligned} \varepsilon_r &= \frac{\mathrm{d}u}{\mathrm{d}r} - z\,\frac{\mathrm{d}^2w}{\mathrm{d}r^2} = \eta_r + z\chi_r, \\ \varepsilon_\vartheta &= \frac{u}{r} - z\,\frac{\mathrm{d}w}{r\,\mathrm{d}r} = \eta_\vartheta + z\chi_\vartheta. \end{aligned} \tag{10.32}$$

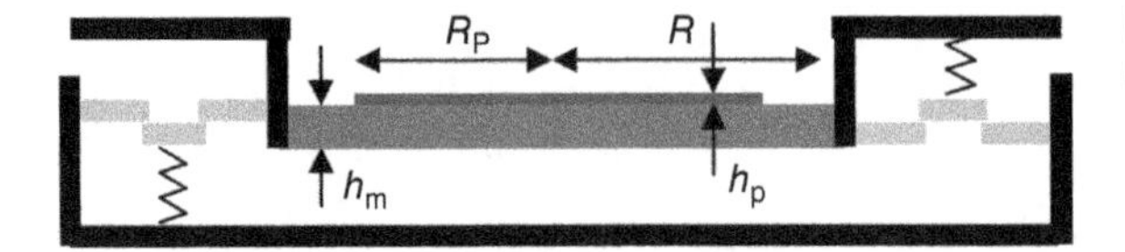

Figure 10.9 Piezoelectric diaphragm micropump.

It is necessary to introduce the radial and hoop axial forces:

$$N_r = \int_{-h_m/2}^{h_m/2} \sigma_r \, dz,$$

$$N_\vartheta = \int_{-h_m/2}^{h_m/2} \sigma_\vartheta \, dz. \tag{10.33}$$

No external load is considered in the piezoelectric case, so the principle of virtual power reads:

$$\int_0^R (N_r \delta \dot{\eta}_r + N_\vartheta \delta \dot{\eta}_\vartheta + M_r \delta \dot{\chi}_r + M_\vartheta \delta \dot{\chi}_\vartheta) \, r \, dr = 0. \tag{10.34}$$

The internal actions (bending moments and axial forces) should be defined on the basis of the piezoelectric constitutive load. In this case, where the converse effect is exploited, it is possible to decouple the piezoelectric problem artificially by assuming that the electric field is completely defined by the applied voltage across the active layer:

$$E_3 = -\frac{\varphi}{t_p}. \tag{10.35}$$

One obtains:

$$\begin{bmatrix} N_r \\ N_\vartheta \end{bmatrix} = \mathbf{B} \begin{bmatrix} \eta_r \\ \eta_\vartheta \end{bmatrix} + \mathbf{C} \begin{bmatrix} \chi_r \\ \chi_\vartheta \end{bmatrix} - N^p \begin{bmatrix} 1 \\ 1 \end{bmatrix}, \tag{10.36}$$

$$\begin{bmatrix} M_r \\ M_\vartheta \end{bmatrix} = \mathbf{C} \begin{bmatrix} \eta_r \\ \eta_\vartheta \end{bmatrix} + \mathbf{D} \begin{bmatrix} \chi_r \\ \chi_\vartheta \end{bmatrix} - M^p \begin{bmatrix} 1 \\ 1 \end{bmatrix}, \tag{10.37}$$

where the stiffness matrices are given by:

$$\mathbf{B} = \sum_k \int_{z_{k-1}}^{z_k} \frac{E_k}{1-\nu_k^2} \begin{bmatrix} 1 & \nu_k \\ \nu_k & 1 \end{bmatrix} dz, \tag{10.38}$$

$$\mathbf{C} = \sum_k \int_{z_{k-1}}^{z_k} \frac{E_k}{1-\nu_k^2} \begin{bmatrix} 1 & \nu_k \\ \nu_k & 1 \end{bmatrix} z \, dz, \tag{10.39}$$

$$\mathbf{D} = \sum_k \int_{z_{k-1}}^{z_k} \frac{E_k}{1-\nu_k^2} \begin{bmatrix} 1 & \nu_k \\ \nu_k & 1 \end{bmatrix} z^2 \, dz. \tag{10.40}$$

The contributions N^p and M^p are found by considering the integral on the piezoelectric layer only, taking account of the effective piezoelectric coefficient discussed in Section 10.1:

$$N^p = \int_{z_{p-1}}^{z_p} e_{\text{eff}} \, E_3 \, dz, \tag{10.41}$$

$$M^p = \int_{z_{p-1}}^{z_p} e_{\text{eff}} \, E_3 \, z \, dz. \tag{10.42}$$

An approximate solution can be found by considering, again, the weighted residuals approach with two fields:

$$u \cong U\psi_u, \quad w \cong W\psi_w, \tag{10.43}$$

where ψ_w is still the same as in Equation 10.13 and ψ_u should fulfil the kinematic boundary conditions for a fully clamped axisymmetric plate, i.e. null displacement in the centre and on the external boundary. For instance, one can choose

$$\psi_u = \frac{r^2}{R^2} - \frac{r}{R}. \tag{10.44}$$

By combining Equations 10.34 to 10.44, one finds the governing equation for the two-dof system:

$$\mathbf{k}\begin{bmatrix} U \\ W \end{bmatrix} = \mathbf{F}^p. \tag{10.45}$$

The analytical solution of an axisymmetric plate with a piezoelectric layer is very useful for different purposes. For example, it is possible to study the quasi-static response for different radii of the piezoelectric layer. Clearly, if $R_p \to 0$, no actuation is introduced and the displacement is zero, as well as the stroke volume. The same result is obtained if $R_p \to 1$: indeed, the piezoelectric effect is introduced as a uniform imposed strain that, for fully clamped boundary conditions, is completely absent from the governing equations. Consequently, there should be an optimal value of R_p, such that the transverse displacement is maximized (and, similarly, the stroke volume). Figure 10.10 shows the results for the specific case endowed with the following geometrical features: $R = 750$ μm, $h_\mathrm{m} = 14$ μm and $h_\mathrm{p} = 1.3$ μm. The elastic parameters of the silicon layer are $E = 160$ GPa and $\nu = 0.22$. For the piezoelectric material, $E = 70$ GPa, $\nu = 0.3$ and $e_\mathrm{eff} = -5.04$ N/Vm. It is worth noting that the optimal radius is different for the displacement and the stroke volume: to maximize the former, R_p should be equal to $0.64R$; for the latter, R_p should be equal to $0.73R$. Other parametric analyses can be carried out: one finds that the stroke volume is inversely proportional to both the silicon thickness and the piezoelectric thickness (Ardito *et al.*, 2013).

It is also interesting to compare the performance of a piezoelectric micropump with electrostatic micropumps. This can be done by considering the same geometric features, in terms of radius and silicon thickness; the capacitor gap in the electrostatic case is fixed at $h_\mathrm{c} = 4.6$ μm. The actuation voltages are computed with the objective of getting the

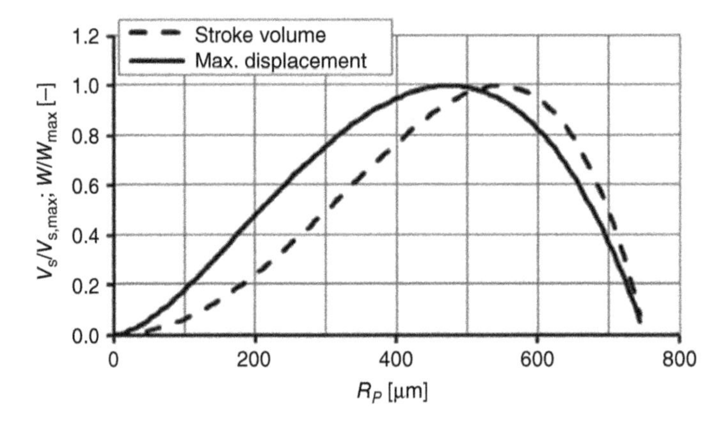

Figure 10.10 Parametric analysis of piezoelectric micropump: maximum displacement and stroke volume are plotted with respect to the radius of the piezoelectric layer.

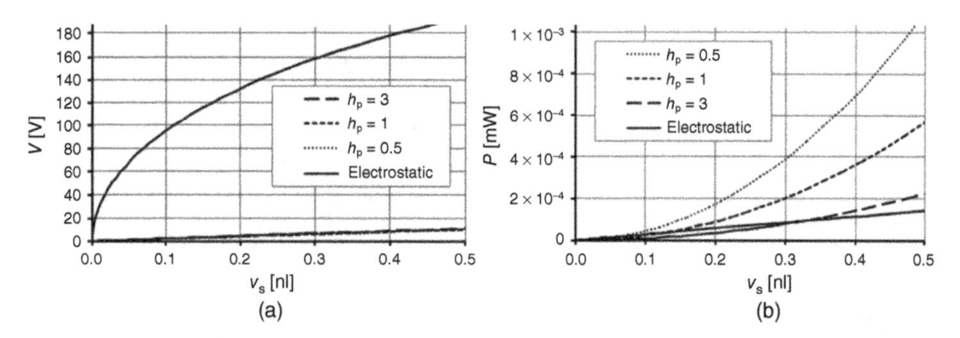

Figure 10.11 Comparison between electrostatic and piezoelectric actuation: (a) actuation voltage versus stroke volume; (b) electric power versus stroke volume.

same stroke volume for the two actuators. Figure 10.11 shows a comparison between the electrostatic case and different thicknesses of the piezoelectric layer: piezoelectric actuation is somewhat more expensive in terms of electric power, but electrostatic actuation requires very high voltages to get the same stroke volume.

10.5.2 Complete Multiphysics Model

In the case of piezoelectric actuation, the detailed response of the pump can also be found by considering a fully coupled analysis, which includes the piezoelectric behaviour and the fluid–structure interaction. Again, it is necessary to resort to numerical techniques, using complete 3D models of the solid and fluid domains. An example of a solution is considered herein, with reference to a pumping chamber with radius $R = 750$ μm and with optimal piezoelectric radius $R_p = 0.73R = 547.5$ μm. The pumping chamber is 20μm thick and the inlet and outlet channels are 100 μm long and 100 μm wide. The actuation voltage across the piezoelectric layer is 60 V and a square wave cycle is implemented.

The model that is considered in this section includes passive orthoplanar microvalves as well, i.e. silicon discs that are able to open and close the inlet and the outlet channels alternatively (Parise, Howell and Magleby, 2001; Nguyen *et al.*, 2004). Orthoplanar valves are fully developed in-plane, whereas the actuation mechanism is in the out-of-plane direction. Basically, the investigated microvalve consists of a central disc suspended by compliant elements between the upper chamber, which is connected to the pumping chamber, and a circular hole, which is connected to the inlet (or the outlet) channel. If the pressure in the upper chamber is higher than the pressure in the inlet channel, the disc is pushed on the hole and the inlet valve is closed; in the suction phase, the pressure in the upper chamber is reduced, so that the boss is lifted and the fluid is forced to flow in the pumping chamber. A similar concept can be applied to the outlet valve. Orthoplanar valves are advantageous for many reasons: compact size; easy manufacture; no need for active drive and timing control. Conversely, the valve should be carefully designed to avoid fluid leakage and the priming phase deserves special attention.

The most important result of multiphysics analyses is represented by the variation of the flow rate with respect to the actuation frequency, see Figure 10.12. In the present case, the analysis is extended to very low frequencies (less than 1 Hz). After the initial increase of the flow rate with frequency, an optimal situation is reached, as in the

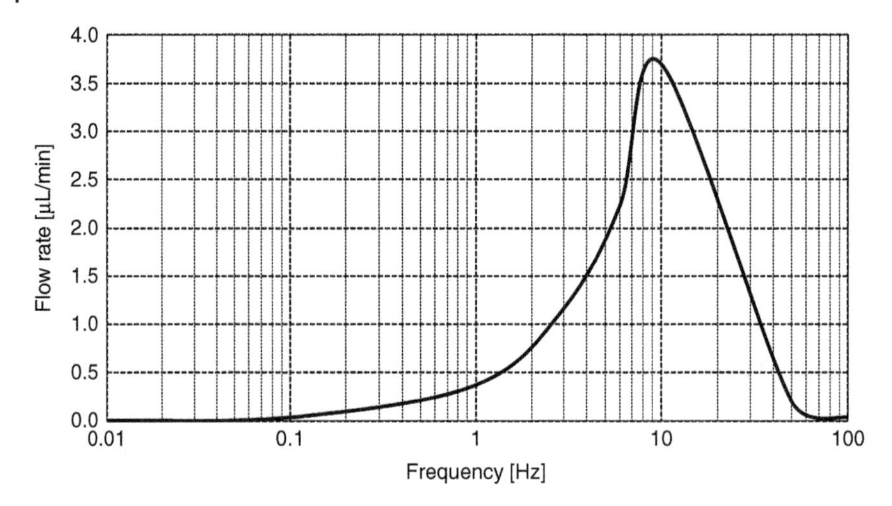

Figure 10.12 Flow rate for piezoelectric micropump, for actuation voltage equal to 60 V.

electrostatic case: the optimal flow rate is attained for the actuation frequency around 10 Hz.

The maximum flow rate is around 3.75 μl/min = 62.5 nl/s, which is by far larger than the values shown in Figure 10.7 for the electrostatic case, even though the pumping chamber is smaller than before. The reason is that, for equal voltage, the piezoelectric actuator is more effective than the electrostatic one. Moreover, it is worth remembering that no pull-in is encountered for piezoelectric actuation; hence, the displacement can be freely increased.

References

Ardito, R., Bertarelli, E., Corigliano, A. and Gafforelli, G. (2013) On the application of piezolaminated composites to diaphragm micropumps. *Composite Structures*, **99**, 231–240.

Ardito, R., Corigliano, A. and Frangi, A. (2013) Modelling of spontaneous adhesion phenomena in micro-electro-mechanical systems. *European Journal of Mechanics – A/Solids*, **39**, 144–152.

Ardito, R., Rizzini, F., Frangi, A. and Corigliano, A. (2016) Evaluation of adhesion in microsystems using equivalent rough surfaces modeled with spherical caps. *European Journal of Mechanics – A/Solids*, **57** (1), 121–131.

Batra, R., Porfiri, M. and Spinello, D. (2007) Review of modeling electrostatically actuated microelectromechanical systems. *Smart Materials and Structures*, **16** (6).

Bertarelli, E., Ardito, R., Bianchi, E. *et al.* (2010) A computational study for design optimization of an electrostatic micropump in stable and pull-in regime. *AES Technical Reviews International Journal Series*, **1** (1), 19–25.

Bertarelli, E., Ardito, R., Corigliano, A. and Contro, R. (2011a) A plate model for the evaluation of pull-in instability occurrence in electrostatic micropump diaphragms. *International Journal of Applied Mechanics*, **3** (1), 1–19.

Bertarelli, E., Ardito, R., Greiner, A. *et al.* (2011b) *Design Issues in Electrostatic Microplate Actuators: Device Stability and Post Pull-in Behaviour*, 12th International Conference on Thermal, Mechanical and Multi-Physics Simulation and Experiments in Microelectronics and Microsystems, EuroSimE, April 18–20, 2011, Linz, Austria, IEEE.

Bertarelli, E., Corigliano, A., Greiner, A. and Korvink, J.G. (2011c) Design of high stroke electrostatic micropumps: a charge control approach with ring electrodes. *Microsystem Technologies*, **17** (1), 165–173.

Bochobza-Degani, O., Elata, D. and Nemirovsky, Y. (2003) A general relation between the ranges of stability of electrostatic actuators under charge or voltage control. *Applied Physics Letters*, **82** (2), 302–304.

Cao, L., Mantell, S. and Polla, D. (2001) Design and simulation of an implantable medical drug delivery system using microelectromechanical systems technology. *Sensors and Actuators, A: Physical*, **94** (1–2), 117–125.

DelRio,F., De Boer, M., Knapp, J. *et al.* (2005) The role of van der Waals forces in adhesion of micromachined surfaces. *Nature Materials*, **4** (8), 629–634.

Dumont-Fillon, D., Tahriou, H., Conan, C. and Chappel, E. (2014) Insulin micropump with embedded pressure sensors for failure detection and delivery of accurate monitoring. *Micromachines*, **5** (4), 1161–1172.

Elata, D. and Bamberger, H. (2006) On the dynamic pull-in of electrostatic actuators with multiple degrees of freedom and multiple voltage sources. *Journal of Microelectromechanical Systems*, **15** (1), 131–140.

Garimella, S.V., Singhal, V. and Liu, D. (2006) On-chip thermal management with microchannel heat sinks and integrated micropumps. *Proceedings of the IEEE*, **94** (8), 1534–1548.

Ha, D.H., Phan, V.P., Goo, N.S. and Han, C.H. (2009) Three-dimensional electro-fluid-structural interaction simulation for pumping performance evaluation of a valveless micropump. *Smart Materials and Structures*, **18** (10).

Iverson, B.D. and Garimella, S.V. (2008) Recent advances in microscale pumping technologies: a review and evaluation. *Microfluidics and Nanofluidics*, **5** (2), 145–174.

Jiang, T.Y., Ng, T.Y. and Lam, K.Y. (2000) *Dynamic Analysis of an Electrostatic Micropump*, International Conference on Modeling and Simulation of Microsystems – MSM 2000, March 27–29, 2000, San Diego, CA, Nano Science and Technology Institute.

Kang, H.J. and Choi, B. (2011) Development of the MHD micropump with mixing function. *Sensors and Actuators, A: Physical*, **165** (2), 439–445.

Laser, D.J. and Santiago, J.G. (2004) A review of micropumps. *Journal of Micromechanics and Microengineering*, **14** (6).

Machauf, A., Nemirovsky, Y. and Dinnar, U. (2005) A membrane micropump electrostatically actuated across the working fluid. *Journal of Micromechanics and Microengineering*, **15** (12), 2309–2316.

Nguyen, N.T., Truong, T.Q., Wong, K.K. *et al.* (2004) Micro check valves for integration into polymeric microfluidic devices. *Journal of Micromechanics and Microengineering*, **14** (1), 69–75.

Nisar, A., Afzulpurkar, N., Mahaisavariya, B. and Tuantranont, A. (2008) MEMS-based micropumps in drug delivery and biomedical applications. *Sensors and Actuators, B: Chemical*, **130** (2), 917–942.

Parise, J.J., Howell, L.L. and Magleby, S.P. (2001) Ortho-planar linear-motion springs. *Mechanism and Machine Theory*, **36** (11–12), 1281–1299.

Rocha, L.A., Cretu, E. and Wolffenbutte, R.F. (2004) Behavioural analysis of the pull-in dynamic transition. *Journal of Micromechanics and Microengineering*, **14** (9),S37–S42.

Rochus, V., Rixen, D.J. and Golinval, J.C. (2005) Electrostatic coupling of MEMS structures: transient simulations and dynamic pull-in. *Nonlinear Analysis, Theory, Methods and Applications*, **63** (5–7), e1619–e1633.

Singh, S., Kumar, N., George, D. and Sen, A.K. (2015) Analytical modeling, simulations and experimental studies of a PZT actuated planar valveless PDMS micropump. *Sensors and Actuators, A: Physical*, **225**, 81–94.

Taylor, J. (2005) *Novel Insulin Delivery Systems. Is an Artificial Pancreas Possible?* UK Advanced Diabetes Course, Bosworth Hall, http://www.leicestershirediabetes.org.uk (accessed 17 July 2017).

Teymoori, M.M. and Abbaspour-Sani, E. (2005) Design and simulation of a novel electrostatic peristaltic micromachined pump for drug delivery applications. *Sensors and Actuators, A: Physical*, **117** (2), 222–229.

Timoshenko, S.P. and Woinosky-Krieger, S. (1959) *Theory of Plates and Shells*, McGraw-Hill.

Truong, T.Q. and Nguyen, N.T. (2004) A polymeric piezoelectric micropump based on lamination technology. *Journal of Micromechanics and Microengineering*, **14** (4), 632–638.

Tsui, Y.Y. and Lu, S.L. (2008) Evaluation of the performance of a valveless micropump by CFD and lumped-system analyses. *Sensors and Actuators, A: Physical*, **148** (1), 138–148.

van Lintel, H.T.G., van De Pol, F.C.M. and Bouwstra, S. (1988) A piezoelectric micropump based on micromachining of silicon. *Sensors and Actuators*, **15** (2), 153–167.

Voigt, P., Schrag, G. and Wachutka, G.K.M. (1998) Electrofluidic full-system modelling of a flap valve micropump based on Kirchhoffian network theory. *Sensors and Actuators, A: Physical*, **66** (1–3), 9–14.

Wijshoff, H. (2010) The dynamics of the piezo inkjet printhead operation. *Physics Reports*, **491** (4–5), 77–177.

Woias, P. (2005) Micropumps – past, progress and future prospects. *Sensors and Actuators, B: Chemical*, **105** (1), 28–38.

Wolff, A., Perch-Nielsen, I.R., Larsen, U.D. *et al.* (2003) Integrating advanced functionality in a microfabricated high-throughput fluorescent-activated cell sorter. *Lab on a Chip – Miniaturisation for Chemistry and Biology*, **3** (1), 22–27.

Zengerle, R., Geiger, W., Richter, M. *et al.* (1995a) Transient measurements on miniaturized diaphragm pumps in microfluid systems. *Sensors and Actuators: A. Physical*, **47** (1–3), 557–561.

Zengerle, R., Richter, A. and Sandmaier, H. (1992) *A Micro Membrane Pump with Electrostatic Actuation*, MEMS '92, Proceedings. An Investigation of Micro Structures, Sensors, Actuators, Machines and Robot, February 4–7, Travemünde, Germany, IEEE.

Zengerle, R., Ulrich, J., Kluge, S. *et al.* (1995b) A bidirectional silicon micropump. *Sensors and Actuators: A. Physical*, **50** (1–2), 81–86.

Zhang, C., Xing, D. and Li, Y. (2007) Micropumps, microvalves, and micromixers within PCR microfluidic chips: advances and trends. *Biotechnology Advances*, **25** (5), 483–514.

Zhang, Z. and Nagrath, S. (2013) Microfluidics and cancer: are we there yet? *Biomedical Microdevices*, **15** (4), 595–609.

Part III

Reliability and Dissipative Phenomena

11

Mechanical Characterization at the Microscale

11.1 Introduction

The first part of this book (Chapters 2 and 3) has been devoted to the fundamental knowledge necessary to understand how microsystems work and how they can be designed from a mechanical point of view. In the second part of this book (Chapters 4 to 10), a series of microsystems have been described, focusing on their mechanical behaviour and on the interaction of this mechanical behaviour with other physics.

From the knowledge of how MEMS work and of the high complexity of phenomena occurring at the microscale, it can be understood that a design process must take into consideration possible reliability problems from the beginning. This is why in MEMS a *design for reliability* approach must be used; this in turn implies being able to detect and avoid possible problems that can hinder the spread of microsystems.

In recent years, MEMS have become widely diffused industrial products, produced in billions of pieces. This large-scale industrial production obliges producers to focus more carefully on reliability issues related to various causes of failures, e.g. on mechanical failures, such as fatigue, and fracture induced by accidental drop. Reliability issues have become increasingly important when creating products that can have a real impact on the market.

This chapter, and the third part of the book, deals with issues related to MEMS reliability; particular issues, such as fracture, fatigue, rupture due to accidental impact, various sources of damping and stiction are discussed and accompanied by examples taken from the authors' direct experience.

The starting point for reliability studies is to be able to characterize, precisely, the materials from which microsystems are fabricated. It is of paramount importance to measure and control the mechanical properties of materials used in MEMS, primarily of polysilicon, which is by far the most widely used material in the production of microsystems. For this reason, in this and subsequent chapters, devoted to mechanical characterization, reference will be mainly made to polysilicon and to experimental techniques that can be adopted to capture its mechanical properties at the microscale.

The successful fabrication and the reliable use of structures with feature sizes in the range of 1 μm to 1 mm is strongly contingent on a sufficiently rigorous understanding of their length-scale-dependent and process-dependent mechanical properties. In turn,

Mechanics of Microsystems, First Edition. Alberto Corigliano, Raffaele Ardito, Claudia Comi, Attilio Frangi, Aldo Ghisi, and Stefano Mariani.

Companion website: www.wiley.com/go/corigliano/mechanics

such understanding requires the ability to measure the mechanical properties of microscale structures.

There exist today many different mechanical test techniques for polysilicon at the microscale; researchers have explored a variety of physical principles and experimental set-ups that enable the determination of mechanical properties of almost invisible structures (Ando, Shikida and Sato, 2001; Bagdahn and Sharpe, 2003; Ballarini *et al.*, 1997, 1998; Chasiotis and Knauss, 2002, 2003a,b; Chen, Ju and Fang, 2000; Chi and Wensyang, 1999; Cho and Chasiotis, 2007; Corigliano *et al.*, 2004, 2007; Greek *et al.*, 1997; Kahn *et al.*, 2000, 2002; Kahn, Ballarini and Heuer, 2004; Kramer and Paul, 2000; McCarty and Chasiotis, 2007; Muhlstein, Brown and Ritchie, 2001; Muhlstein, Howe and Ritchie, 2004; Oostemberg and Senturia, 1997; Sharpe, Yuan and Edwards, 1997; Sharpe, Turner and Edwards, 1999; Tsuchiya *et al.*, 1998; Yi, Li and Kim, 2000; Zhu, Corigliano and Espinosa, 2006).

A major distinction can be made between so-called *off-chip* (Ando, Shikida and Sato, 2001; Bagdahn, Sharpe and Jadaan, 2003; Chasiotis and Knauss, 2002, 2003a; Kramer and Paul, 2000; Sharpe, Yuan and Edwards, 1997; Sharpe, Turner and Edwards, 1999; Tsuchiya *et al.*, 1998) and *on-chip* (Ballarini *et al.*, 2003; Chi and Wensyang, 1999; Greek *et al.*, 1997; Kahn *et al.*, 2000, 2002; Muhlstein, Brown and Ritchie, 2001; Corigliano *et al.*, 2004) methodologies. In both cases, the microdevice is usually produced by deposition and etching procedures. An off-chip tensile test generally resorts to some sort of external gripping mechanism actuating the force together with an external sensor that measures the response of the specimen. On the contrary, on-chip test devices are real MEMS in which actuation and sensing is performed with the same working principles of MEMS.

The purpose of this chapter is to give a brief state-of-the-art review of the scientific literature relating to the experimental mechanical characterization of polysilicon at the microscale and to describe examples of devices conceived for the on-chip mechanical characterization of thin and thick polysilicon layers, which make use of electrostatic and thermomechanical actuation. A recent study specifically focused on the fracture strength of micro and nanoscale silicon components was conducted by DelRio, Cook and Boyce (2015). As in other chapters of this book, the examples here discussed are taken directly from the authors' experience.

Four different MEMS for on-chip testing are here discussed, which load thin (0.7 μm) or thick (15 μm) polysilicon specimens up to rupture under bending. The first device is based on a rotational electrostatic actuator, which contains a series of interdigitated comb fingers and loads a couple of thin polysilicon specimens in bending in the plane parallel to the substrate. The second device loads a couple of thin polysilicon specimens in the plane orthogonal to the substrate thanks to a parallel-plate actuator, which moves in the direction orthogonal to the substrate. The third device makes use of a thermomechanical actuator to load a thin polysilicon specimen in tensile loading until rupture. The fourth device was conceived to cause initiation and propagation of a crack in a thick polysilicon specimen and will also be discussed in Chapter 12, which more specifically deals with fracture and fatigue in microsystems; it is based on a large number of comb-finger electrostatic actuators, which load a notched specimen by means of a lever system. The device can be used for the mechanical characterization of the material in its elastic regime and also to obtain important parameters on the fracture behaviour, as described in Chapter 12.

In the four cases discussed, the data reduction procedure is based on measurement of the capacitance variation of a displacement sensor, from which it is possible to determine the effective Young's modulus of the various specimens and the maximum stress at rupture.

An outline of the chapter is as follows. Section 11.2 is dedicated to a brief state of the art concerning off-chip and on-chip methodologies recently proposed for MEMS mechanical characterization. Section 11.3 is devoted to a description of the Weibull approach for determining the failure probability of polysilicon microsystems. Section 11.4 contains a general introductory description of the data reduction procedure adopted in the on-chip experimental approach and of the four practical examples of the application of the on-chip approach for thin and thick polysilicon films, together with relevant experimental results.

11.2 Mechanical Characterization of Polysilicon as a Structural Material for Microsystems

11.2.1 Polysilicon as a Structural Material for Microsystems

Polysilicon is by far the most common structural material in MEMS. It is used for a huge variety of applications, mainly because of two factors: the existence of well-established deposition technologies, in which polycrystalline silicon has had a very important role since the beginning of the microelectronics era, and the excellent physical properties of this material. Its equivalent Young's modulus is higher than that of titanium and comparable with that of steel. The rupture strength at the microscale stands in the range of the one of the best construction steels, while its density is less than that of aluminium. Thermal properties, too, make it a very good material for high-temperature applications. Thermal conductivity is very high, while the thermal expansion coefficient is very small and the melting point is only one hundred degrees less than that of iron.

When referring to material properties of polysilicon, in many practical cases, one considers an equivalent, homogeneous, isotropic material with average material stiffness and strength; the main material stiffness parameter is still denoted the Young's modulus for simplicity. Of course, the truth of this assumption is highly dependent on the dimension and orientation of the grains.

Polysilicon is an aggregate of monocrystalline silicon grains; hence, its properties depend on the properties of the crystallites composing it, on their shapes, on their orientation and on the physical characteristics of the grain boundaries between different grains. This means that the overall properties of polysilicon are strongly influenced by the process used for the deposition. Processing conditions, such as deposition temperature and deposition pressure, influence the final properties of the material.

Besides the dispersion of the physical properties, owing to different processes, it must be stressed that at the microscale the measurement of these properties is a very difficult and challenging task. Earlier work conducted by several researchers revealed meaningful differences in the measured values of the elastic properties and of the nominal strength of polysilicon without providing in-depth explanations for such a variety of observations. A main question arose, therefore, as to whether the newly evolving test methodologies were adequately precise. In pursuing that question, a round-robin study

(Sharpe *et al.*, 1998) demonstrated the inconsistency of measured modulus and strength values, even when specimens from the same source were examined. The material in that work was fabricated in close physical proximity from the same wafer of the same run and in the same deposition reactor at the Microelectronics Center of North Carolina (now Cronos-JDS Uniphase). In that round-robin effort, values of the elastic modulus differed considerably, namely from 132 GPa to 174 GPa, and the strength also demonstrated a rather wide dispersion, ranging from 1.0 GPa, for specimens tested in tension, to 2.7 GPa for specimens tested in bending. A second round-robin examination, conducted on material fabricated at the Sandia National Laboratories (LaVan *et al.*, 2001) also showed signs of inconsistent rupture strengths, demonstrating a dependence on specimen size and measurement technique. A relatively new round-robin test was carried out by Tsuchiya *et al.* (2005). The specimens, produced with the same process, in the same wafer, were distributed to five different research groups. In this case, the obtained modulus varied from 134 GPa to 173 GPa, while fracture strength varied from 1.44 GPa to 2.51 GPa. As a result of these findings, it appears advisable for any microfabrication facility not to use the properties cited in the literature for final design and verifications but to identify the most feasible measurement technique and to conduct measurements for every fabrication run.

11.2.2 Testing Methodologies

Resorting to some definitions provided by the American Society for Testing and Materials standards for testing at the macroscale, among material properties of interest in the context of MEMS design are (see Chapter 2, Section 2.3.1): the Young's modulus; Poisson's ratio; the nominal rupture strength, i.e. the normal stress at the beginning of fracture; the response to cyclic loading in terms of a plot of the applied stress versus number of cycles at rupture (the S–N curve). To measure material properties, one should be able to construct a specimen according to a given design, apply an external input in terms of forces or displacements and measure the specimen response using direct procedures, in the sense that the variable of interest should be (almost) directly measured. All these steps are fully standardized at the macroscale and are currently applied for testing construction materials, such as steel and concrete. Unfortunately, these practices cannot be easily applied at the scale of MEMS. In particular, one has to resort to fully or partially indirect approaches. For example, to measure the Young's modulus, cantilever beams in bending are often utilized; deflection is measured and the property of interest is computed on the basis of an analytical or numerical model of the beam. Even during on-chip tension tests, some sort of inverse analysis must be performed since, in general, only capacitance variations are measured directly while deformations are obtained on the basis of a numerical model.

Many testing methodologies have been proposed in the scientific literature for the extraction of static mechanical properties of polysilicon (Sharpe, 2002). Restricting attention to silicon MEMS, as already mentioned, a first general classification of test procedures can be made between off-chip and on-chip devices. In both cases, the microdevice is generally produced by deposition and etching procedures.

On-chip test devices (Corigliano *et al.*, 2004) are real MEMS in which actuation and sensing is performed with the same working principles of MEMS. On-chip devices rely on the fact that all the mechanical parts needed to load the specimen and the majority of those (electrical or optical) needed to measure displacements and strains are built together with the specimen during the micromachining fabrication process. Two main parts (the actuator and the sensing devices) can usually be found in these structures. In many cases, these consist of a large number of capacitors that can be used to create an electric field; this in turn causes a force to act onto the specimen or enables measurement of a capacitance that is directly related to the displacement of the specimen itself.

The advantage of on-chip testing methods is linked to the ease of fabrication and of use (usually without costly equipment) and to the fact that complex handling and alignment of the specimen are avoided. The major drawback is that the force developed by on-chip actuators can be insufficient to break specimens in quasi-static conditions and that the maximum displacement is also limited, of the order of a micrometre. On-chip testing of MEMS devices is especially advocated since the thin-film microstructure and the state of residual stress are strong functions of the microfabrication process steps. Nevertheless this requires accurate modelling and numerical or analytical analyses of the whole device.

An off-chip test (Sharpe, 2002) generally resorts to some sort of external gripping mechanism actuating the force and an external sensor measuring the response of the specimen. All the experimental set-ups that use an external apparatus (load cells, microregulation screws, etc.) to create a stress state into the specimen are usually included in the category of off-chip testing. In this case, much attention has to be paid during the handling of tiny MEMS specimens, during the system–specimen alignment and to the specimen gripping systems. The challenge of picking a specimen only a few micrometres thick, placing it in a test machine and performing the test is formidable. The main advantage of off-chip methodologies is that the forces and the displacements can be relatively high, to break specimens even several micrometres thick in pure tension; moreover, many different configurations can be set up to create an a-priori desired multiaxial stress state in the specimen.

In principle, material parameters for MEMS and, primarily, the Young's modulus E can be determined by exploiting several test devices, including, among others, tension tests, bending of cantilever beams, resonant devices, bulge tests and buckling tests. Clearly, the most direct approach is the tension test; unfortunately, this is not always applicable, since it requires the deployment of considerable forces at the microscale to produce sensible deformation in the specimen. Hence, a wealth of alternative solutions have appeared in the literature.

In the following two sections, the experimental mechanical characterization of polysilicon as a structural material for MEMS is discussed. The most important experiments are grouped into two main families: quasi-static testing, used to characterize the Young's modulus, Poisson's ratio and fracture properties (Section 11.2.3) and high-frequency testing, used to characterize fatigue properties (Section 11.2.4).

11.2.3 Quasi-Static Testing

The very first tests conducted for the quasi-static mechanical characterization of silicon can be traced back to the 1980s (Chen and Leipold, 1980). The volumes of the specimens used were approximately 1 cm^3, huge by comparison with typical MEMS dimensions. From the first half of the 1990s, an increasing interest in the mechanical characterization of polysilicon arose and the consequence was that a large number of test configurations were designed.

In the rest of this section, a selection of test devices and set-ups considered as the most common and interesting is presented. The classification is based on the actuation mechanism and on the way in which the system response is read.

11.2.3.1 Off-Chip Tension Test

This is still a common technique for measuring mechanical properties for MEMS applications (Bagdahn, Sharpe and Jadaan, 2003; Chasiotis and Knauss, 2002, 2003a,b; Chasiotis, 2006; Oh *et al.*, 2005; Knauss, Chasiotis and Huang, 2003; Sharpe, Yuan and Edwards, 1997; Sharpe, Turner and Edwards, 1999; Sharpe, 2002; Tsuchiya, Shikida and Sato, 2002; Tsuchiya *et al.*, 2005). A MEMS specimen is produced and then placed on a testing system. Usually the specimen is gripped to the system with the aid of UV curing adhesives or via electrostatic gripping. This is the way to re-create common macroscale testing techniques at the microscale. The displacement is imposed on the specimen by the use of piezotransducers, with a resolution of the order of nanometres. The load cells read the applied load with an accuracy of some micronewtons. Specimen displacements can be measured using either optical systems (Oh *et al.*, 2005) or laser interferometry (Sharpe, Yuan and Edwards, 1997) or via digital image correlation (Chasiotis, 2006).

The results obtained using off-chip tension tests cover the most important quantities for mechanical design with polysilicon. Measurements of Young's modulus and in some cases of Poisson's ratio (Sharpe, Yuan and Edwards, 1997), together with rupture strength, are most common for all the research groups that worked with off-chip tension testing. Moreover, it is important to emphasize that with this kind of set-up it is possible to study scale effects, owing to specimen's dimension and the stress gradient acting in the specimen.

Chasiotis (2006) also found it possible to determine the fracture toughness K_{IC} of polysilicon. This result was achieved by means of a nanoindentation nearby the specimen that caused a crack to propagate through the substrate and partially involve the specimen. At the end of the process it was thus possible to have a pre-cracked specimen, necessary for complete fracture mechanics characterization (see Chapter 12).

Other possible configurations for off-chip tension tests (Sharpe, 2002) can be obtained, e.g. by first patterning a tensile specimen on the surface of a wafer and then exposing the gauge section by etching the wafer; the larger ends can then be gripped in a test machine. Alternatively, a specimen can be fixed to a die at one end and actuated by means of an electrostatic probe at the other end.

11.2.3.2 Off-Chip and On-Chip Bending Test

Out-of-plane bending of test specimens is generally performed via an off-chip apparatus, e.g. a diamond stylus that deflects a cantilever beam (Hollman *et al.*, 1995). The deflection of the free end is measured and the Young's modulus is obtained through inverse modelling of the cantilever beam. However, if the beam is long, forces are small and difficult to calibrate; if the beam is short, forces are higher but inverse analysis of the beam is more involved.

Doubly supported silicon micromachined beams can also be used to study the out-of-plane bending of materials. In this case a voltage is applied between the conductive polysilicon or micromachined beam and the substrate to pull the beam down. The voltage that causes the beam to make contact can be related to the beam stiffness. Residual stresses in the beams and support compliance cause significant vertical deflections, which affect the performance of these micromachined devices. Tests need to be supported by models of the devices that take into account the compliance of the supports and the geometrical nonlinear dependence of the vertical deflections on the stress in the beam.

In-plane bending is a classical test for MEMS since several structural parts of accelerometers are subjected to this kind of deformation. A typical on-chip layout is presented in Figure 11.1; a cantilever polysilicon beam attached to a moving mass (on the left) is subjected to bending induced by the fixed rectangular block (on the right). Actuation is performed by means of interdigitated comb-finger capacitors. This test is often conducted to establish the flexural strength of the cantilever beam as a structural component, but requires considerable care when employed for the evaluation of the Young's modulus, owing to uncertainties in the geometrical parameters and the model of the beam.

11.2.3.3 Test on Membranes (Bulge Test)

The bulge test is one of the earliest techniques used to measure the Young's modulus, Poisson's ratio or the residual stress of nonintegrated, free-standing thin structures. This testing method relies on the use of thin polysilicon membranes (circular, square, or rectangular in shape), which are relatively easy to design and realize, bonded along their periphery to a supporting frame (Jayaraman, Edwards and Hemker, 1998; Tabata *et al.*, 1989; Yang and Paul, 2002). Microfabrication techniques are particularly well suited for the creation of such test structures with reproducible and well-defined boundary conditions. During the test, the membrane is loaded with a pressure difference acting on the top and bottom surfaces. The membrane deforms and its profile is measured with a profilometer. Usually, the deflection at the centre is recorded as a function of the applied pressure. Several analytical or semi-empirical formulae exist, correlating deflection to elastic properties. From this test, it is possible to measure the biaxial elasticity modulus, Poisson's ratio, the nominal rupture strength and the internal stresses (measuring the buckled configuration of the membrane). A shortcoming of this methodology is that sometimes the membranes separate from the substrate before the end of the test. Moreover, since the mechanical response of the membrane is highly sensitive to the thickness and to the lateral dimensions, it is necessary to have a good fabrication technology and a very accurate measure of the thickness of the layer.

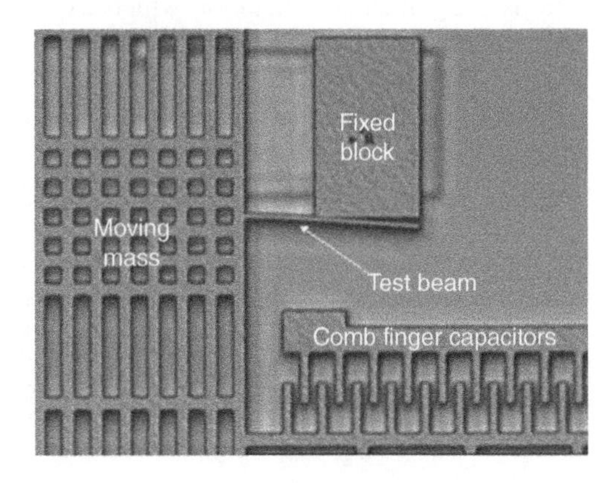

Figure 11.1 On-chip in-plane bending test. *Source*: Reproduced with permission of STMicroelectronics.

11.2.3.4 Nanoindenter-Driven Test

Nanoindenters are often used for the mechanical characterization of thin films. Basically, there are two ways of using nanoindenters: film nanoindentation and the use of a nanoindenter as an actuator to load MEMS structures. Hardness (indentation) tests are routinely used to characterize large-scale structures. In direct analogy, considerable effort has been made to develop nanoindentation techniques to characterize microscale structures, and commercial instruments have been developed. Nanoindentation experiments (Li and Bhushan, 1999; Ding, Meng and Wen, 2001; Kim *et al.*, 2003; Oh *et al.*, 2005) are performed, allowing the tip to penetrate the film under study. During the penetration in the layer, an elastic–plastic stress state is generated. This is the main reason why elastic characterization is done during the unloading phase, when the tip starts going backward to reach the rest position. The result of this experimental test is a force versus penetration depth plot and the projected area of contact under the indenter. The Young's modulus and fracture properties are computed using semi-empirical formulae. The main advantage of this method is that there is no need for an *ad-hoc* designed specimen; it is sufficient to have a portion of material large enough to apply the nanoindenter. However, the application of this technique to thin films is complicated by several factors, including substrate effects and pile-up of material around the indenter.

Another possibility is to use the nanoindenter tip as an actuator to perform a sort of off-chip test (Espinosa, Prorok and Fischer, 2003; Sundarajan and Bhushan, 2002). In these tests, the tip moves an extremity of the specimen, causing a stress state in it. The applied load is measured with a piezo scanner, while there are different ways to measure the displacement of the specimen, such as interferometry (Espinosa, Prorok and Fischer, 2003); alternatively, the displacement of the tip may be measured with the aid of a laser beam and photodiodes (Sundarajan and Bhushan, 2002). This methodology could be very accurate but it needs very expensive instrumentation.

11.2.4 High-Frequency Testing

As discussed in Chapters 6 to 8, polysilicon in MEMS technology is also used for the fabrication of resonators, gyroscopes and other devices that oscillate at high frequencies during their whole life. One of the most important failure mechanisms for such systems is fatigue. Fatigue (more deeply discussed together with fracture problems in Chapter 12), is usually interpreted as a phenomenon caused by the motion of dislocations present in the material, which can coalesce during the stress cycles to form microcracks. These microcracks then join together to form one or more macrocracks that cause the failure of the structure.

Polysilicon is a brittle material and there is no dislocation motion under temperatures of about 900 °C; therefore, it is not expected to be prone to fatigue in usual operating conditions. Nevertheless, in the second half of the 1990s, some groups started working on this subject and found that polysilicon can also undergo fatigue after a large number of cycles, typically more than 10^9, combined with high stress levels. To reproduce fatigue failures experimentally, it is very important to work with experimental set-ups that can allow relatively high-frequency testing (at least 1 kHz), which in turn allows a large number of cycles to be reached in a reasonable time. On-chip tests are usually the best choice for this kind of study because they can work at high frequencies and it is possible to perform several fatigue tests at the same time using some electrical control system.

On-chip test systems, in general, make use of electrostatic actuation between a fixed (stator) and a movable (rotor) part to load the specimen with a desired level of stress. The force developed by the actuator is proportional to the actuation area and inversely proportional to the gap between the rotor and the stator. It turns out that to have a force sufficiently large to induce fatigue in the specimens, one should have a highly scaled lithography, a thick polysilicon layer and a design area large enough for the thousands of capacitors needed for the actuation. These requirements are the main reason why on-chip testing is not common for quasi-static characterization. Loading the microsystem in a dynamic regime, it is possible to reach the stress level necessary for fatigue testing when loading the structure with a time-varying force at a frequency close to the resonant frequency of the system. This is what is usually done in fatigue tests for MEMS; in these cases, a reasonably low voltage is used to bring the specimen to fatigue rupture. Very interesting results have been obtained, e.g. by (Bagdahn and Sharpe, 2003).

11.2.4.1 Fatigue Mechanisms

As discussed in the previous section, it has been experimentally shown that fatigue in polysilicon is a possible failure mechanism. Nevertheless, the reasons why fatigue rupture occurs in polysilicon are not yet completely clear and understood. Among the most active groups in the study of fatigue in polysilicon, are those of Pennsylvania State University and Case Western Reserve University. These two groups proposed the most known and accepted interpretations for fatigue mechanisms in polysilicon.

The first group (Muhlstein, Brown and Ritchie, 2001; Muhlstein, Stach and Ritchie, 2002a,b; Muhlstein, Howe and Ritchie, 2004) proposed a mechanism called reaction-layer fatigue, which can be summarized as follows: when the polysilicon is first exposed to air, the native oxide is formed, as one of the final steps of the process; the oxide thickens in the highly stressed regions and becomes a site for environmentally assisted cracks, which grow in a stable way in the layer; when the critical size is reached, the silicon itself fracture catastrophically by transgranular cleavage.

The second group (Kahn *et al.*, 2000, 2002; Kahn, Ballarini and Heuer, 2004) demonstrated that even a high stress state cannot cause an appreciable growth of the native oxide, thus excluding pure environmentally assisted fatigue. Conversely, this group noticed that fatigue life decreases when the level of humidity is increased. They did not propose a specific fatigue mechanism for polysilicon.

Fatigue testing remains one open research area in the field of mechanical characterization of polysilicon. The mentioned studies are only examples of a much wider discussion now active in the scientific literature (see also, e.g., Ando, Shikida and Sato (2001)).

Section 12.4 is specifically dedicated to the study of fatigue of microsystems.

11.3 Weibull Approach

Before discussing, with some details, the on-chip approach to determining mechanical properties of materials at the microscale, it is useful to have an introduction to the Weibull approach, which is very popular for the study of failure in brittle materials. This introduction is also useful in providing a more complete and clear vision of fracture and fatigue in MEMS, as discussed in Chapter 12.

The Weibull approach (Weibull, 1951) is widely applied in the study of brittle materials, such as ceramics. This is why it has also recently been applied to the study of rupture phenomena in polysilicon MEMS (Greek *et al.*, 1997; Chasiotis and Knauss, 2003a,b; Corigliano *et al.*, 2004; Jadaan *et al.*, 2003). Weibull theory essentially provides a method to estimate the failure probability of a mechanical system, starting from the computation of the probability of failure of its weakest part; the theory is therefore also known as the *weakest-link approach*. The Weibull cumulative distribution function is found by applying the theorem of joint probability, by first computing the probability of survival of a system composed of a large number of elementary parts. This basic idea is extended to a general case, after introduction of limiting hypotheses. The choice of the function giving the survival probability of a single part was not originally based on a precise mechanical interpretation of the rupture process. An interesting discussion on the applicability of Weibull approach can be found in Bazant, Xi and Reid (1991).

By means of the Weibull approach, it is possible to take into account the experimental scatter of the strength values typical of brittle materials, the statistical size effect and the dependence of the probability of failure on the stress distribution.

The application of Weibull approach to a uniformly stressed uniaxial bar gives the following equation for the probability of failure P_{f}:

$$P_{\mathrm{f}} = 1 - \exp\left(-\frac{\Omega}{\Omega_{\mathrm{r}}}\left\langle\frac{\sigma - \sigma_u}{\sigma_0}\right\rangle_+^m\right), \tag{11.1}$$

where: Ω is the volume of the bar, Ω_{r} is a statistically uniform representative volume, σ_u, σ_0 and m are material parameters and $\langle f\rangle_+$ denotes the positive part of f. It is important to recall that Equation 11.1 is based on the assumption of statistical uniformity of every element Ω_{r}.

Let us insert $\sigma = \sigma_u + \sigma_0$, $\Omega = \Omega_{\mathrm{r}}$ into Equation 11.1:

$$P_{\mathrm{f}} = 1 - \exp\left(-\left\langle\frac{\sigma_0}{\sigma_0}\right\rangle_+^m\right) = 0.632. \tag{11.2}$$

From this result, the material parameter σ_0 can be interpreted as the increase of the stress level with respect to σ_u to which corresponds a probability of failure of 63.2% in a tensile specimen, uniformly stressed, having the volume Ω_{r}.

In a multiaxial, nonuniform stress state, it is usually assumed that cracks form in the planes normal to the principal stresses. Assuming a Cartesian reference frame with coordinates x_1, x_2, x_3, the principal stresses are denoted here as $\sigma_{\mathrm{p}1}, \sigma_{\mathrm{p}2}, \sigma_{\mathrm{p}3}$; the probability of failure can then be written as

$$P_{\mathrm{f}} = 1 - \exp\left(-\frac{1}{\Omega_{\mathrm{r}}}\int_{\Omega}\sum_{i=1}^{3}\left\langle\frac{\sigma_{\mathrm{p}i}(\mathbf{x}) - \sigma_u}{\sigma_0}\right\rangle_+^m \mathrm{d}\Omega\right). \tag{11.3}$$

Equation 11.3 is obtained from Equation 11.1 by assuming the hypothesis of statistical uniformity of every volume and computing the joint probability of survival for every infinitesimal volume.

The general expression (11.3) is here applied under the assumption that $\sigma_u = 0$, which means that all levels of stress have an influence on the probability of failure. Equation 11.3 is then rewritten in a more compact way, as

$$P_{\mathrm{f}} = 1 - \exp\left(-\frac{1}{\Omega_{\mathrm{r}}}\int_{\Omega}\left(\frac{\tilde{\sigma}(\mathbf{x})}{\sigma_0}\right)^m \mathrm{d}\Omega\right), \tag{11.4}$$

having defined the equivalent stress $\tilde{\sigma}(\mathbf{x})$ as

$$\tilde{\sigma}(\mathbf{x}) = \left(\left\langle \sum_{i=1}^{3} \sigma_{\mathrm{p}i}(\mathbf{x}) \right\rangle_{+}^{m} \right)^{\frac{1}{m}}. \tag{11.5}$$

These relations can be used to estimate the probability of failure P_{f} of a given structure or solid once the *Weibull parameters m* and σ_0 are known and the elastic distribution of stresses has been computed via analytical formulae or numerical solutions, e.g. by means of the finite element method.

The parameters m and σ_0 are usually determined experimentally, starting from a series of uniaxial tensile tests on cylindrical specimens of volume Ω. In this simple case, Equation 11.4 reduces to

$$P_{\mathrm{f}} = 1 - \exp\left(-\frac{\Omega}{\Omega_{\mathrm{r}}} \left(\frac{\sigma(\mathbf{x})}{\sigma_0} \right)^{m} \right). \tag{11.6}$$

The Weibull parameters can also be identified from a specimen or structure loaded in a multiaxial situation with a nonuniform stress distribution. Let us rewrite Equation 11.4 in a form similar to that of Equation 11.6:

$$\begin{aligned} P_{\mathrm{f}} &= 1 - \exp\left(-\frac{1}{\Omega_{\mathrm{r}}} \int_{\Omega} \left(\frac{\tilde{\sigma}(\mathbf{x})}{\sigma_0} \right)^{m} \mathrm{d}\Omega \right) \\ &= 1 - \exp\left(-\frac{\Omega}{\Omega_{\mathrm{r}}} \left(\frac{\sigma_{\mathrm{nom}}(\mathbf{x})}{\sigma_0} \right)^{m} \beta^{m} \right), \end{aligned} \tag{11.7}$$

where β has been defined as

$$\beta^{m} = \frac{1}{\sigma_{\mathrm{nom}}^{m} \Omega} \int_{\Omega} \sum_{i=1}^{3} \left\langle \sigma_{\mathrm{p}i}(\mathbf{x}) \right\rangle_{+}^{m} \mathrm{d}\Omega \equiv \frac{1}{\Omega} \int_{\Omega} (h(\mathbf{x}))^{m} \mathrm{d}\Omega \tag{11.8}$$

and σ_{nom} represents a nominal stress in the nonuniformly stressed specimen or structure, which acts as a scaling parameter for the elastic response.

Notice that function $h(\mathbf{x})$, defined by the second of relations (11.8), depends only on the normalized stress distribution in the linear elastic response and is therefore independent of the load level.

To compare the behaviour of different structures, it is possible to define a critical stress level as the nominal stress level σ_{nom} evaluated in the structure when the probability of failure is equal to 63.2%, equivalent to σ_0 for a uniaxially, uniformly loaded specimen. From the second of relations (11.7) thus follows:

$$\sigma_{\mathrm{nom0}} = \frac{\sigma_0}{\beta} \left(\frac{\Omega_{\mathrm{r}}}{\Omega} \right)^{\frac{1}{m}}. \tag{11.9}$$

Equation 11.9 states that σ_{nom0} depends on the volume Ω and on the stress nonuniformity through β. The variation of σ_{nom0} with volume determines the size effect related to the statistical uniform distribution of defects described by a Weibull approach. σ_{nom0} is inversely proportional to volume and this dependence increases with decreasing m; at the limit, by letting m be infinite, the statistical size effect disappears. Note also the dependence of σ_{nom0} on the parameter β, which in turn depends on the stress nonuniformity.

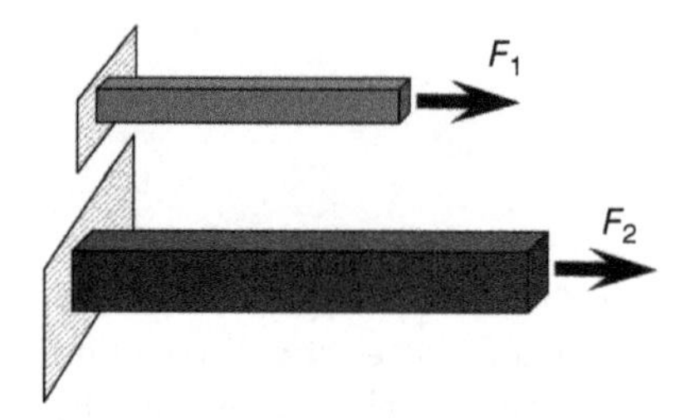

Figure 11.2 Two beams with different volumes subject to axial loading.

Given two structures, labelled A and B, built with the same material endowed with Weibull parameters σ_0 and m, it is therefore possible to write

$$\left(\sigma_{\text{nom0}}\right)_A = \frac{\sigma_0}{\beta_A}\left(\frac{\Omega_r}{\Omega_A}\right)^{\frac{1}{m}};$$

$$\left(\sigma_{\text{nom0}}\right)_B = \frac{\sigma_0}{\beta_B}\left(\frac{\Omega_r}{\Omega_B}\right)^{\frac{1}{m}};$$

$$\frac{\left(\sigma_{\text{nom0}}\right)_A}{\left(\sigma_{\text{nom0}}\right)_B} = \frac{\beta_B}{\beta_A}\left(\frac{\Omega_B}{\Omega_A}\right)^{\frac{1}{m}}. \tag{11.10}$$

Equation 11.10 allows for a direct comparison of the behaviour of structures with different volumes and stress distributions.

The effect of volume on the Weibull plot of two beams subject to an axial load is visualized in Figures 11.2 and 11.3. In the example of the figures, it is demonstrated that, plotting the two graphs using the same Weibull parameters, taking into account the different volumes in a situation with uniform stress distribution, such as the purely axial one, a reduction in strength is obtained for increasing volume.

The effect of nonuniform stress distribution on the Weibull plot of two beams having the same volume but subject to different loading conditions is visualized in Figures 11.4 and 11.5. In the example of the figures, it is demonstrated that plotting the two graphs with the same Weibull parameters, taking into account the different stress distribution at equal volume, the bending situation clearly gives a higher strength.

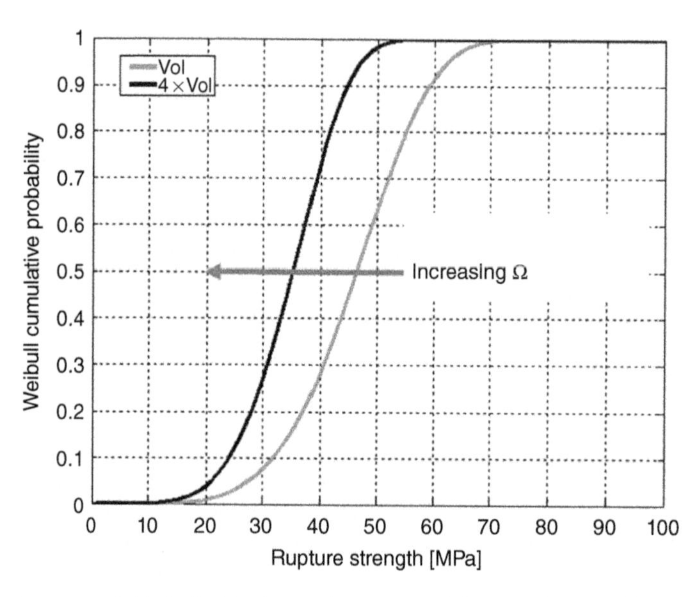

Figure 11.3 Weibull plots for two beams with different volumes subject to axial loading.

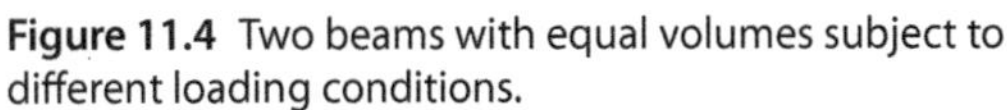

Figure 11.4 Two beams with equal volumes subject to different loading conditions.

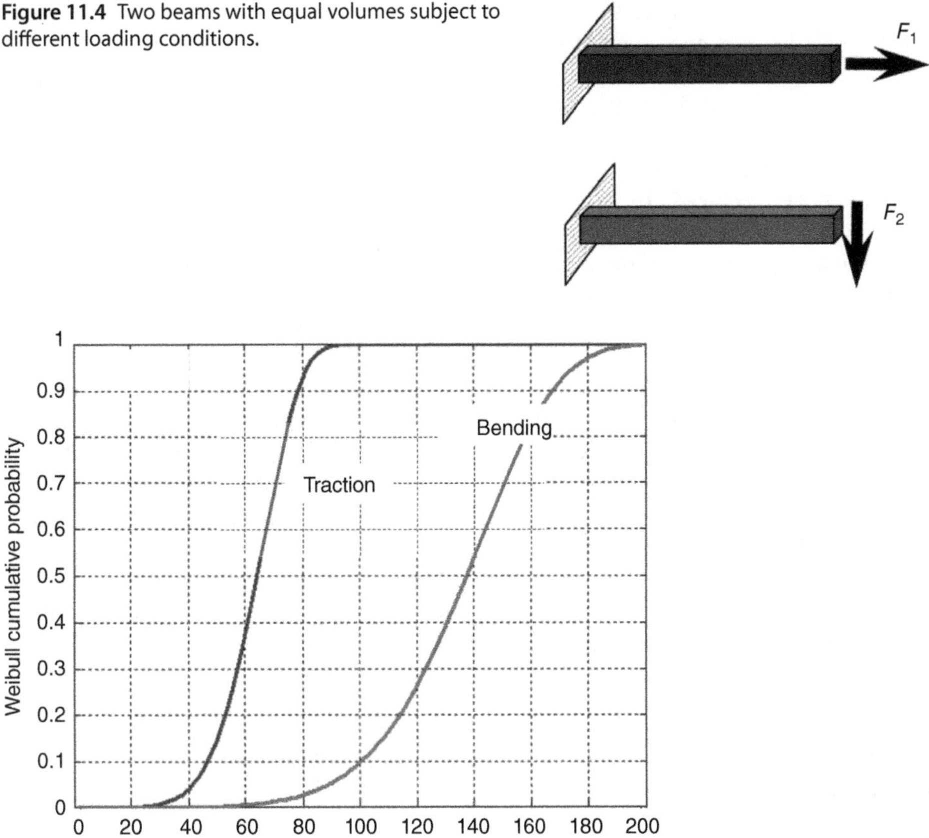

Figure 11.5 Weibull plots for two beams with equal volumes subject to different loading conditions.

11.4 On-Chip Testing Methodology for Experimental Determination of Elastic Stiffness and Nominal Strength

The purpose of this section is to describe, with some details, on-chip testing methodologies for the mechanical characterization of polysilicon at the microscale.

In the description of the devices, many problems and different fields of physics already discussed in the first part of the book appear again in a different perspective. The examples shown are by no means exhaustive of the proposals appearing in the literature; they are discussed here to show the potentialities of the on-chip testing approach, to underline the importance of a correct mechanical characterization of materials for MEMS and to give further examples of application of the theoretical basis described in Chapter 2.

In the test devices here described, the specimens are co-fabricated with the actuator in order to obtain precise alignment and gripping of the specimens and also to reduce the set-up size. The devices have an integrated system of actuation: electrostatic interdigitated comb-finger actuators in the case of the bending tests of Sections 11.4.1 and 11.4.4; electrostatic parallel-plate actuators in the case of the out-of-plane bending test of Section 11.4.2; electrothermomechanical actuation in the case of the tensile test of Section 11.4.3.

During the tests, an input voltage φ is applied to the actuator and a capacitance variation C is measured. The capacitance variation can be related to some significant displacement (or rotation) of the specimen through simplified analytical formulae or through electrostatic finite element simulations of the complete device. The force (or torque) exerted by the actuator (electrostatic or thermomechanical) can be determined as a function of the displacement (or rotation) from the correct interpretation of the device's working principle. In the case of the electrostatic actuator, the force (or torque) can be obtained as the derivative of the electrostatic energy with respect to the displacement (or rotation), which, in turn, is proportional to the derivative of the capacitance. In the case of the thermomechanical actuator, the force is deduced from the solution of the mechanical behaviour of the whole device, assuming knowledge of the elastic stiffness of the material used for the actuator. The scheme for data reduction is detailed in the rest of this section for each on-chip test structure.

In general, tests were carried out at room temperature and at atmospheric humidity, with a probe station mounted on an optical microscope. A slowly increasing voltage was applied, to create quasi-static loading conditions in the specimen. Details of the experimental set-ups can be found in the papers of the authors (see, e.g., Corigliano *et al.* (2004)).

11.4.1 On-Chip Bending Test through a Comb-Finger Rotational Electrostatic Actuator

11.4.1.1 General Description

The first on-chip device discussed in this chapter is shown in Figure 11.6 (see also Cacchione *et al.* (2005a,b, 2006)). It is made of a central ring connected to the substrate by means of two tapered 0.7 μm thick specimens (detail in the lower portion of Figure 11.6) which also act as the suspension springs of the whole device. Rigidly connected to the central ring are 12 arms with a total of 384 comb-finger capacitors, which move, owing to electrostatic attraction, towards the stators connected rigidly to the substrate.

When a voltage is applied to the device, the comb fingers develop a force distributed along the 12 arms, equivalent to a torque applied to the central ring. This, in turn, loads the two specimens in bending in the plane parallel to the substrate. The force developed by the set of comb fingers is sufficient to load the specimens up to rupture. The specimens are a pair of doubly clamped slender beams, with a length of 34 μm and a trapezoidal cross-section. Their width decreases linearly from 5.3 μm to 1.8 μm. This shape was designed *ad hoc* to localize the fracture of the specimen in a specified area through stress concentration.

11.4.1.2 Data Reduction Procedure

The rotational on-chip device was tested to determine, experimentally, the Young's modulus and the rupture strength of a 0.7 μm thick polysilicon film. The general data reduction procedure described at the beginning of this section was applied. In this case, simplified analytical formulae were used to transform the applied voltage into the

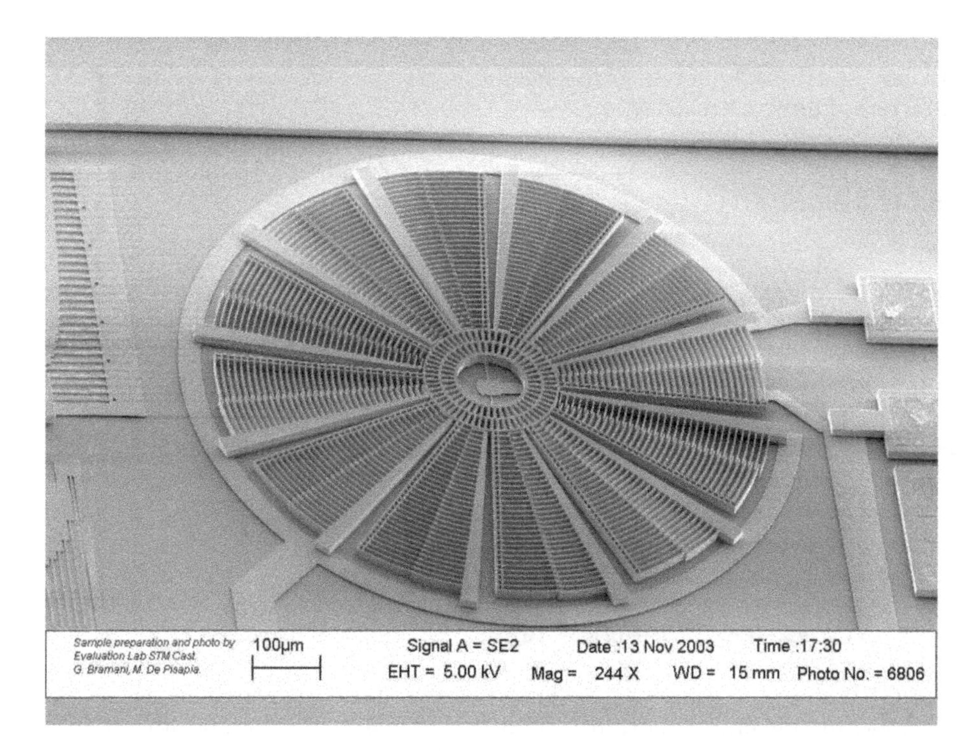

Figure 11.6 Rotational electrostatic actuator for on-chip bending tests: (top) general view; (bottom) detail of the bending specimens. *Source*: Corigliano *et al*. (2005), Figure 5. Reproduced with permission of Springer.

torque applied by the electrostatic actuator on the rotational device and the measured capacitance variation (with respect to a reference level) into the angle of rotation of the central ring of the rotational device.

The electrostatic attraction of the 32 comb fingers placed along each arm was assumed to be uniformly distributed along the arm, with a resultant F_{arm}; the total torque was therefore evaluated as

$$M = n_{\text{a}} \left(l_0 + \frac{l_{\text{arm}}}{2} \right) F_{\text{arm}} = n_{\text{a}} \left(l_0 + \frac{l_{\text{arm}}}{2} \right) n_{\text{cf}} \epsilon_0 \frac{t\varphi^2}{g}, \tag{11.11}$$

where n_{a} is the number of arms, l_0 is the distance between the external part of the central ring and the centre of the whole device, l_{arm} is the length of each arm, n_{cf} is the number of comb fingers for each arm, ϵ_0 is the permittivity of free space, t is the thickness of the arms in the direction orthogonal to the substrate, g is the gap between the rotor and the stator in each comb finger and φ is the applied voltage. Equation 11.11 allows the value of the global torque M to be computed for a given voltage φ.

Let us now consider the 32 comb fingers distributed along each arm defined by index $i = 0, 1, \ldots, 31$, with $i = 0$ the finger nearest to the centre. The contribution to the total capacitance of the ith comb finger is given by

$$C_i = C_{0i} + 2\frac{\epsilon_0 t}{g}\Delta x_i = 2\frac{\epsilon_0 t}{g}\theta(R_0 + i\Delta R), \tag{11.12}$$

where C_{0i} is the capacitance corresponding to the initial configuration, Δx_i is the displacement in the direction orthogonal to the arm due to electrostatic attraction, θ is the angle of rotation of the device, R_0 is the distance between the rotor finger $i = 0$ and the centre and ΔR is the distance between two stator fingers. The total capacitance of the rotational device can be computed by summing the contribution of each finger and of each arm:

$$\begin{aligned} C = n_{\text{a}} \sum_{i=0}^{31} C_i &= n_{\text{a}} \sum_{i=0}^{31} C_{0i} + 2n_{\text{a}} \frac{\epsilon_0 t}{g} \theta \sum_{i=0}^{31} (R_0 + i\Delta R) \\ &= C_0 + \left[2n_{\text{a}} \frac{\epsilon_0 t}{g} (32R_0 + 496\Delta R) \right] \theta. \end{aligned} \tag{11.13}$$

Using this equation, it is possible to compute the rotation of the device from the measured total capacitance C; since only the capacitance variations are relevant, the value of C_0 is assumed to be zero.

The total torque and total capacitance were computed starting from experimental values of voltage and capacitance variation by introducing the following data in Equations 11.11 and 11.13:

$$\begin{aligned} n &= 12; & n_{\text{cf}} &= 32; \\ l_0 &= 100\ \mu\text{m}; & l_{\text{arm}} &= 308\ \mu\text{m}; \\ t &= 15\ \mu\text{m}; & g &= 2.5\ \mu\text{m}; \\ R_0 &= 108\ \mu\text{m}; & \Delta R &= 9.5\ \mu\text{m}; \\ \epsilon_0 &= 8.854 \times 10^{-6}\ \text{pF}/\mu\text{m}. \end{aligned} \tag{11.14}$$

The resulting values are:

$$M = 5.18\ \varphi^2\ \mu\text{N}\,\mu\text{m}; \qquad C = C_0 + 10.41\ \theta\ \text{pF}. \tag{11.15}$$

The experimental capacitance versus applied voltage plots and the corresponding rotation versus capacitance and torque versus voltage plots are shown in Figure 11.7. Figure 11.8 collects the resulting torque versus rotation plots obtained after application of Equation 11.15.

Starting from the torque versus rotation plot, it was possible to calculate the Young's modulus and the rupture strength of the material with the aid of a linear elastic finite element analysis performed on a 3D finite element mesh. This result was achieved after using the following hypotheses: the deformation of the device is only due to the beam specimens at the centre, i.e. the external part built with a 15 μm thick polysilicon is assumed to be rigid; the specimen is linear elastic and homogeneous up to rupture; displacements and strains are small; the global behaviour is linear and the global stiffness is proportional to the Young's modulus of the specimen.

It is important to remark that, besides these mentioned hypotheses, the geometry of the specimen must be carefully reproduced in the finite element model. The final finite element mesh was therefore obtained from SEMs to reproduce the geometry exactly.

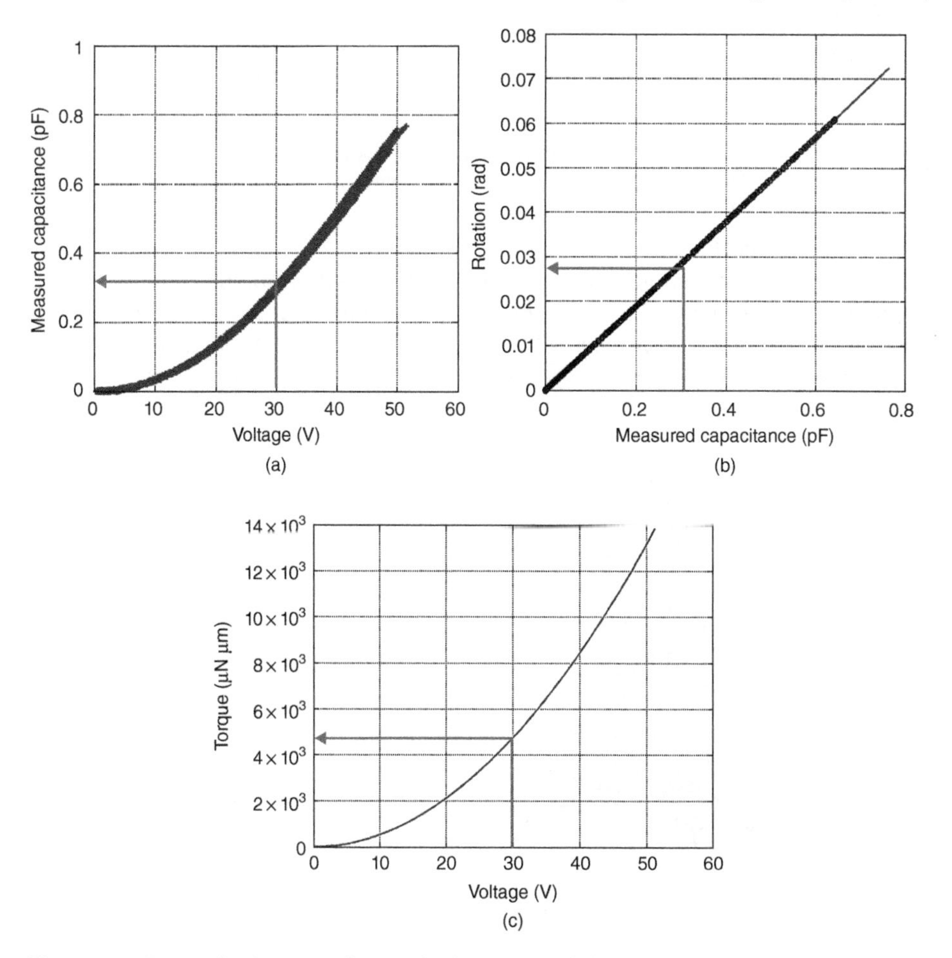

Figure 11.7 Data reduction procedure applied to rotational electrostatic actuator: (a) experimental capacitance versus voltage; (b) rotation versus measured capacitance; (c) torque versus applied voltage. *Source*: Corigliano *et al.* (2005), Figure 3. Reproduced with permission of Springer.

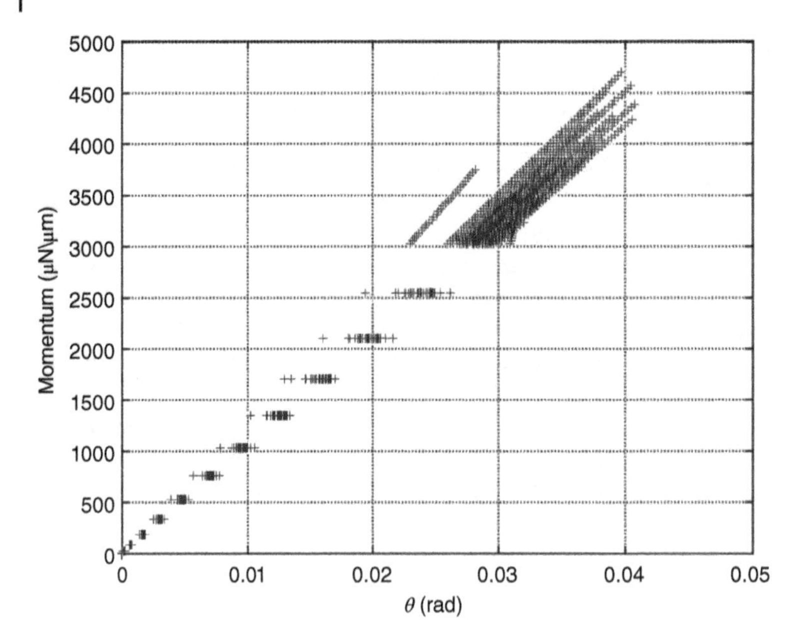

Figure 11.8 Rotational electrostatic actuator. Experimental torque versus rotation plots. *Source*: Corigliano *et al.* (2005), Figure 4. Reproduced with permission of Springer.

11.4.1.3 Experimental Results

Figure 11.9 shows the experimental distributions of the Young's modulus obtained using the rotational devices described in Section 11.4.1. The mean value is 178 GPa.

The Weibull approach briefly described in Section 11.3 was applied to 50 experimental results, computing the volume integral in Equation 11.8, starting from a linear

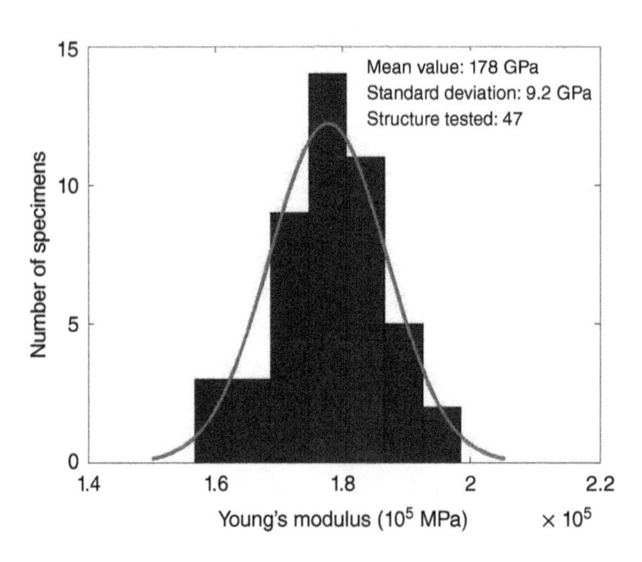

Figure 11.9 Rotational actuator for the in-plane bending test: distribution of Young's modulus. *Source*: Corigliano *et al.* (2005), Figure 10a. Reproduced with permission of Springer.

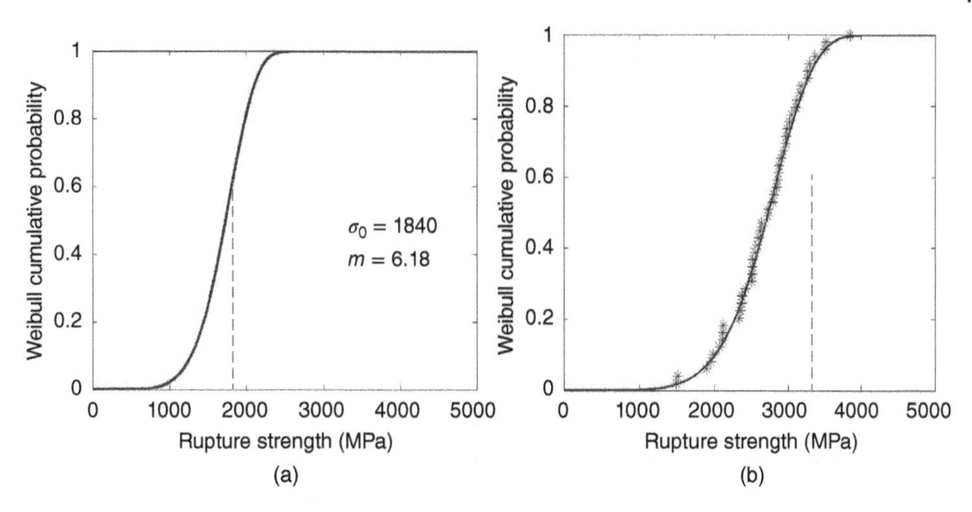

Figure 11.10 Rotational actuator for in-plane bending tests – Weibull cumulative probability densities: (a) equivalent Weibull plot for a uniaxial, unit volume specimen; (b) experimental data obtained from 50 tests on the rotational structure, $\sigma_{nom0} = 2.89$ GPa. *Source*: Corigliano *et al.* (2005), Figure 12. Reproduced with permission of Springer.

elastic finite element analysis. The volume integrals were computed by using a numerical Gaussian integration on each finite element.

The Weibull parameters, $\sigma_0 = 1.84$ GPa, $m = 6.2$, were obtained; the nominal stress value in Equation 11.9 was $\sigma_{\text{nom}0} = 2.89$ GPa. These results are shown in Figure 11.10.

11.4.2 On-Chip Bending Test through a Parallel-Plate Electrostatic Actuator

11.4.2.1 General Description

The second on-chip device discussed in this chapter is shown in Figure 11.11 (Cacchione *et al.*, 2005a,b, 2006). Figure 11.11a shows the whole device, and Figure 11.11b is an enlargement of the central part, in which the 0.7 μm thick beam specimens are placed. A perforated plate of 15 μm thick polysilicon is suspended over the substrate by means of four elastic springs placed at the four corners. The plate is also connected to the thin polysilicon film specimens placed at the centre, as shown in Figure 11.11. The two symmetric specimens are connected to the plate on one side and are rigidly connected to the substrate on the other side.

The two specimens are therefore equivalent to a pair of doubly clamped beams. The holes in the plate are formed in the etching process to eliminate the sacrificial layer, thus allowing for movement of the plate with respect to the substrate. The movement in the direction orthogonal to the substrate is obtained by electrostatic attraction of the plate towards the substrate. The whole plate and the substrate thus act as a parallel-plate electrostatic actuator. When the plate moves towards the substrate, the two specimens bend. It is important to remark that only the square part of the plate acts as an actuator (see Figure 11.11a), while the rectangular parts added on each side of the plate act as sensors. These in turn enable experimental determination of the capacitance variation and vertical movement, as discussed in Section 11.4.2. The length of each specimen is 7 μm. To force the rupture in a section, their cross-section changes linearly with the decrease in the width from 3 μm to 1 μm (see Figure 11.11b).

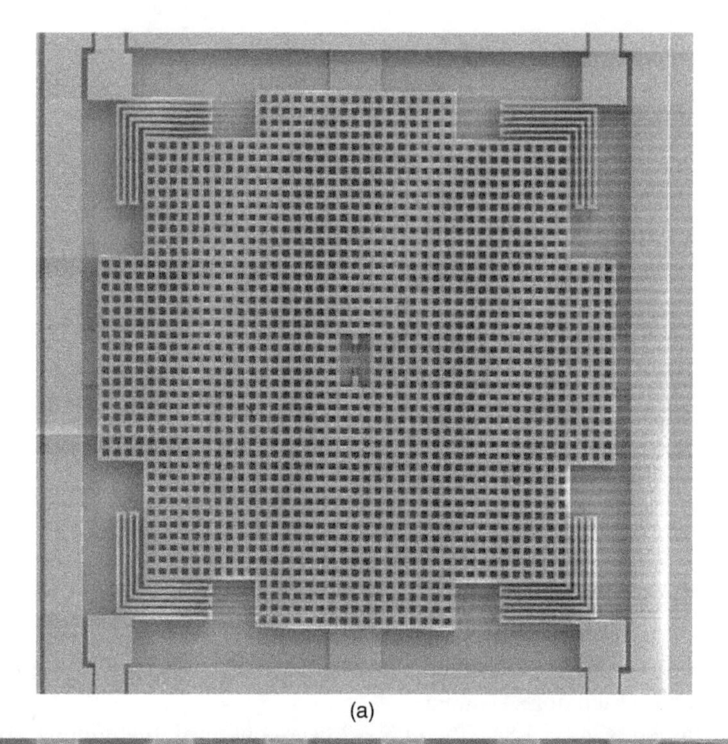

(a)

(b)

Figure 11.11 Parallel-plate actuator for out-of-plane bending tests: (a) general view; (b) detail of one of the two specimens. *Source*: Corigliano *et al*. (2005), Figure 7. Reproduced with permission of Springer.

11.4.2.2 Data Reduction Procedure

As for the device described in Section 11.4.1, the parallel-plate actuator was used to determine the Young's modulus and the rupture strength of the 0.7 μm thick polysilicon film.

The general data reduction procedure was again applied. The experimentally determined capacitance versus voltage plots were transformed into force versus displacement plots by use of the relationships between capacitance and displacement and between voltage and electrostatic force, respectively. These relations were determined in this case by means of electrostatic simulations obtained with the boundary element method in addition to the finite element method. In particular, a series of electrostatic boundary element method simulations on one of the lateral sensors enabled the determination of the capacitance variation versus vertical gap plot, which was used directly to convert experimental capacitance variation data to the vertical displacements.

Finite element method electrostatic simulations were used to obtain the vertical force of attraction on the square plate acting as a rotor. From the results of the finite element method electrostatic simulations it was deduced that the attractive vertical force can be computed by making use, with negligible error, of the analytical relation for a parallel-plate actuator with the same surface and with correction due to edge effects.

Starting from the force–displacement plot, the force acting on the specimens was obtained by subtracting the part equilibrated by the elastic suspension springs in the four corners of the plate (see Figure 11.11a). An elastic 3D finite element method solution of the specimen under bending in the vertical plane was then used to relate the global stiffness of the specimen to the Young's modulus and the force at rupture to the maximum tensile stress in the specimen. Experimental values of Young's modulus and rupture stress were therefore finally obtained.

A key point in the data reduction procedure is the sensitivity of the results to the value of the vertical gap between the plate and the substrate. The vertical gap cannot easily be measured on the real device and it can depend strongly on the quality of the etching process. In the results here presented, a mean value of 1.65 μm was used.

11.4.2.3 Experimental Results

Figure 11.12 shows the distribution of Young's modulus obtained using the device described in Section 11.4.2. The mean value is 174 GPa. It can be observed that the mean values of the Young's modulus (178 GPa and 174 GPa) obtained from the two devices, as described in Sections 11.4.1 and 11.4.2, differ less than the standard deviation of the two distributions. It can therefore be concluded that the two sets of experimental results give, in practice, the same value for the Young's modulus. The conclusion implies that the possible nonuniformity of the thin polysilicon film along its thickness does not influence, sensibly, the value of the obtained average Young's modulus.

In addition, the Weibull approach described in Section 11.3 was applied to the device. Figure 11.13 has the same meaning as Figure 11.10, obtained in the case of the rotational actuator. The obtained Weibull parameters are $\sigma_0 = 2.24$ GPa and $m = 5.1$ in this case, while the nominal stress value in Equation 11.9 is $\sigma_{nom0} = 3.03$ GPa.

The remarkable difference between σ_0 and σ_{nom0} obtained from both test structures demonstrates the importance of the effects of the stress distribution: in a highly nonuniformly stressed structure, such as those tested here, the apparent value of rupture is higher than in a uniformly stressed specimen.

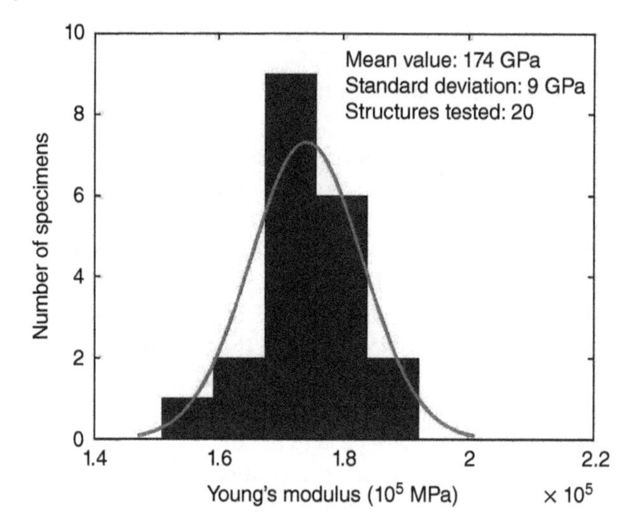

Figure 11.12 Parallel-plate actuator for out-of-plane bending tests: experimental distribution of Young's modulus. *Source*: Corigliano *et al.* (2005), Figure 10b. Reproduced with permission of Springer.

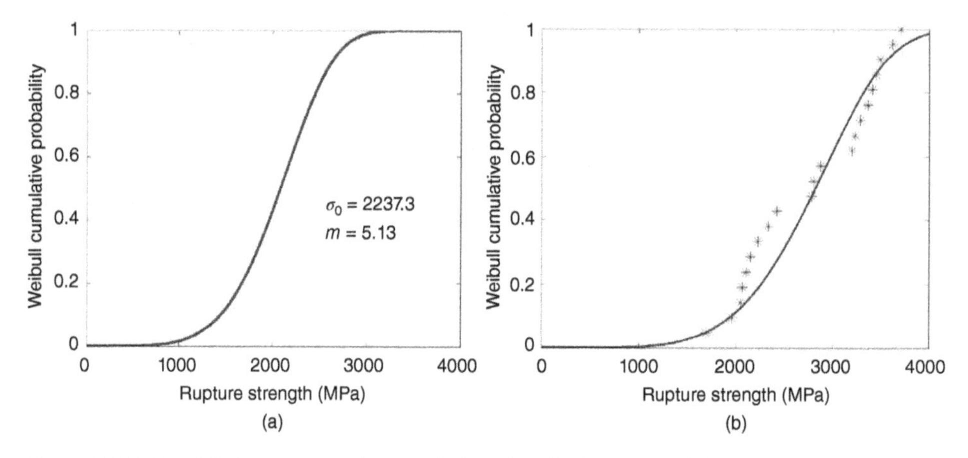

Figure 11.13 Parallel-plate actuator for out-of-plane bending tests – Weibull cumulative probability densities: (a) equivalent Weibull plot for a uniaxial, unit volume specimen; (b) experimental data obtained from 21 tests on the parallel-plate structure, $\sigma_{\text{nom0}} = 3.03$ GPa. *Source*: Corigliano *et al.* (2005), Figure 14. Reproduced with permission of Springer.

The difference in the Weibull parameters obtained from the two sets of experimental results may be explained in various ways. A first explanation could be the influence of different loading conditions (in-plane and out-of-plane), which could locally initiate different rupture mechanisms, related to the anisotropy and nonuniformity of distribution of the polysilicon film along its thickness. This conclusion is in partial contrast with the one derived with reference to the elastic behaviour.

A second explanation could be associated with the assumption of $\sigma_u = 0$ GPa, as introduced in Section 11.3, to simplify the Weibull approach. It is, in fact, known that a nonzero σ_u value has an influence on the final value of the Weibull parameters σ_0 and *m*. In addition, the possible discrepancies in the results obtained from the two on-chip tests could result from the influence of geometrical parameters, such as the gap between the stator and the plate in the device described in Section 11.4.2.

11.4.3 On-Chip Tensile Test through an Electrothermomechanical Actuator

11.4.3.1 General Description

The third device discussed in this chapter was built to perform tensile tests on the same thin polysilicon film tested using the devices of Sections 11.4.1 and 11.4.2.

Unlike the previous two cases, in the present device the actuation mechanism is based on a microelectrothermomechanical actuator, which is able to load a specimen until rupture in purely tensile conditions.

The elongation of the specimen is measured during the test through the capacitance variation of a set of parallel-plate capacitors, while the force in the specimen can be computed starting from the applied voltage. Through a unique loading cycle, in which the voltage is first increased until specimen rupture and then decreased, the device allows for the determination of the specimen's elastic stiffness and nominal tensile strength.

Figure 11.14 shows the microsystem designed for mechanical characterization of the on-chip material, presented and discussed by Corigliano, Domenella and Langfelder (2010). This is an evolution of devices presented previously (Zhu and Espinosa, 2005; Zhu, Corigliano and Espinosa, 2006; Corigliano *et al.*, 2007) and was produced by STMicroelectronics, exploiting the capabilities of the ThELMA process (Corigliano *et al.*, 2004).

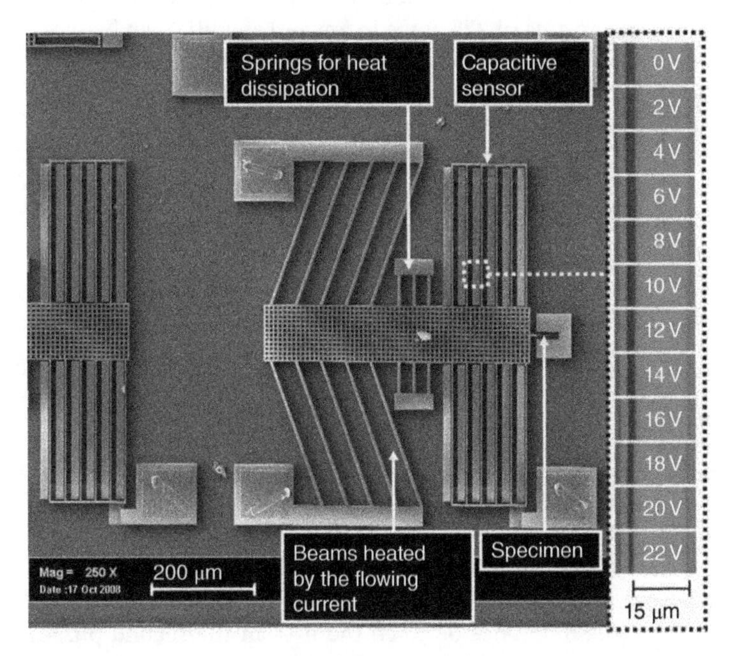

Figure 11.14 Electrothermomechanical actuator for on-chip tensile tests. On the right: the zoomed images of the capacitor's gap varying with the applied voltage. *Source*: Corigliano, Domenella and Langfelder (2010), Figure 1. Reproduced with permission of Springer.

The device has four main parts: the electrothermal actuator, the suspension springs, the displacement sensor and the specimen to be tested. The specimen is made of a 0.7 μm thick polysilicon film. The other parts of the device are all 15 μm thick polysilicon, to exploit the higher obtainable force for the actuator and the higher obtainable sensitivity of the electrostatic sensor.

The Electrothermal Actuator The electrothermal actuator is composed of five pairs of inclined beams, clamped to the substrate and to a central shuttle, which can be considered rigid. Each beam is 15 μm thick and 300 μm long, and is inclined at approximately 70° with respect to the shuttle side. In operation, a voltage difference is applied between the ends of the beams, mechanically connected to the substrate and electrically accessible by means of two external contacts. Owing to the Joule effect, heat develops along the beams, causing a temperature increase, which in turn elongates the beams. Owing to the mechanical constraints of the system, the shuttle moves in the horizontal direction, as represented in Figure 11.14.

The Suspension Springs An important part of the device is represented by the six suspension springs, 15 μm thick and 60 μm long, which connect the shuttle to the substrate. They represent both a mechanical constraint to the actuator movement, avoiding out-of-plane displacements of the shuttle, and thermal sinks, reducing the temperature variation of the electrostatic sensor and the specimen.

The Displacement Sensor The electrostatic displacement sensor consists of a set of 12 parallel-plate capacitors, rigidly connected at one side to the shuttle and at the other to the substrate. Each capacitor has an area of 275 μm × 15 μm and an initial gap between plates of 2.2 μm (including the process over-etches). As the gap distance changes with the shuttle movements, a measure of the capacitance variation enables the displacement of the shuttle from the initial position to be calculated.

The Specimen The part to be tested in the device shown in Figure 11.14 is represented by the specimen bar, designed to be loaded with a tensile force (an image of the whole specimen after rupture is shown in Figure 11.15). The bar is 0.7 μm thick, 45 μm long and has a varying width with a minimum value of 1.38 μm. One end of the specimen is rigidly connected to the shuttle, while the other end is rigidly connected to the substrate.

11.4.3.2 Data Reduction Procedure

The device here described allows the elastic stiffness and the nominal tensile strength of the specimen to be determined using a single on-chip test.

A typical experimental test is conducted in two phases: in the first, increasing voltage steps are applied between the electrothermal actuator contacts. As a result, the movable part is subjected to an increasing force and the displacement of the shuttle is measured by means of the capacitive displacement sensor. The maximum voltage level has been chosen to ensure rupture of the specimen. In the second phase, the voltage is gradually reduced. The sole difference between the first an the second phases is that in the second phase the measured displacement is of the shuttle without the constraint represented by the specimen, which was broken during the last part of the first phase.

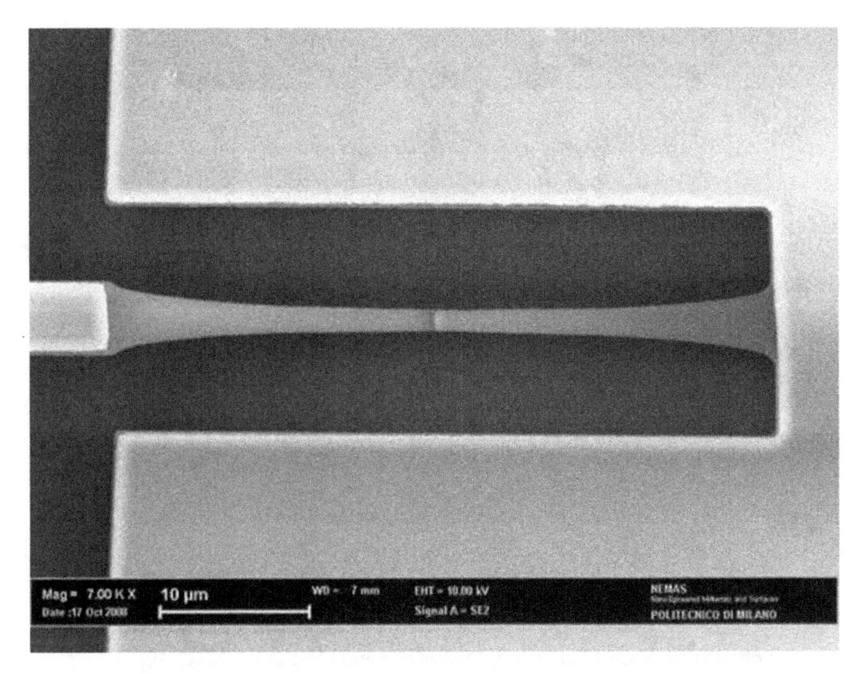

Figure 11.15 Electrothermomechanical actuator for on-chip tensile tests: broken specimen. *Source*: Corigliano, Domenella and Langfelder (2010), Figure 4a. Reproduced with permission of Springer.

Figure 11.16 reports a typical displacement versus applied voltage graph, from which the displacements in the first (continuous line) and second (dashed-dotted line) phases can be clearly distinguished. The rupture event of the specimen can also be clearly singled out at 15.5 V. As planned by the procedure, the voltage continues to increase until 17 V. It can be observed how, after the rupture, the mechanical response abruptly changes to follow a different displacement–voltage characteristic, defined by higher displacement values due to the reduced stiffness of the device, no longer constrained by the specimen.

Figure 11.15 shows an example of a broken specimen. As expected, the rupture occurs in the central, thinner part.

The nominal tensile strength of the specimen σ_s can be computed starting from the measured displacement at rupture d_{cr} as follows:

$$\sigma_s = \frac{F_{cr}}{A} = k\frac{d_{cr}}{A}, \tag{11.16}$$

where F_{cr} is the tensile force applied to the specimen at rupture, k is the specimen axial stiffness and A is the specimen cross-sectional area in the rupture section. The use of Equation 11.16 is based on knowledge of the specimen elastic stiffness, which can be determined as discussed by Corigliano, Domenella and Langfelder (2010). Moreover, Equation 11.16 is valid under the hypotheses that the displacement measured through the electrostatic sensor coincides with the specimen elongation and that the specimen behaviour is linear elastic until rupture.

As discussed by Corigliano, Domenella and Langfelder (2010), the specimen stiffness can be obtained provided that the displacement of the device before and after rupture

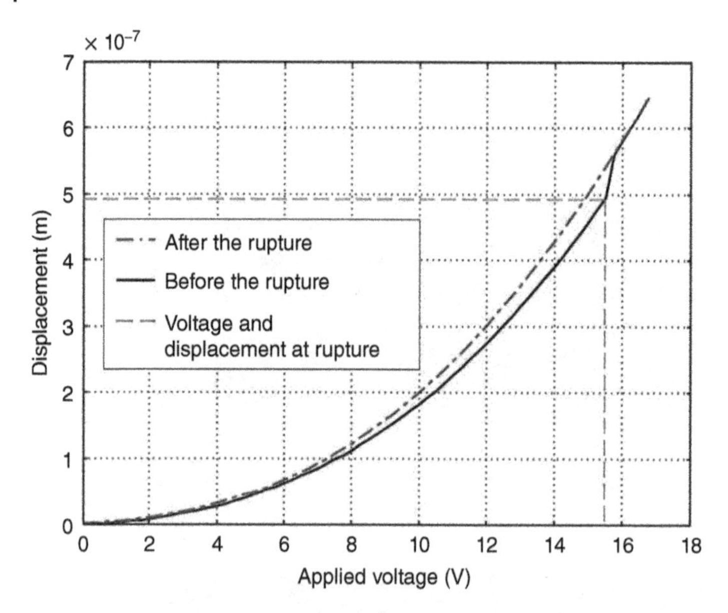

Figure 11.16 Electrothermomechanical actuator for on-chip tensile test: example of experimental displacement versus voltage plot. *Source*: Corigliano, Domenella and Langfelder (2010), Figure 3. Reproduced with permission of Springer.

is known for the same applied voltage, as shown in Figure 11.16 (i.e. for the same temperature increment along the beams) and that the elastic stiffnesses of the beams and springs are known.

The shuttle displacement during the on-chip tests is directly related to the capacitance variation measurements performed by means of the electrostatic sensing system embedded in the device.

A careful calibration procedure, based on the use of fixed capacitances, was implemented to find the precise conversion factor from the system read-out output voltage to the sensor capacitance value. From the formula for the capacitance of a parallel-plate capacitor (see Chapter 2, Section 2.5.1) it is then possible to obtain the gap:

$$C = \epsilon \frac{h_c l_c}{g} + C_{FE}. \tag{11.17}$$

where ϵ is the permittivity (taken as that of vacuum), h_c is the capacitor height (15 μm), l_c is the capacitor length (275 μm), g is the gap between the arms (2.2 μm at rest) and C_{FE} is the contribution due to the fringing effect.

Thanks to the large form factor of the designed capacitances, it is possible to consider C_{FE} constant during the displacement of the actuator. Moreover, since the displacement is related only to the capacitance variation from the rest value, the value of C_{FE} will be subtracted and is thus meaningless for our purpose.

11.4.3.3 Experimental Results

As discussed in the previous section, the nominal rupture stress of the specimen material can be computed after the evaluation of the elastic stiffness by applying Equation 11.16. This was done after careful measurement of the specimen rupture section, starting from SEMs, such as that shown in Figure 11.15. From Figure 11.16, a typical value of specimen displacement at rupture can be obtained, as $d_{cr} = 500\,\text{nm}$. Having computed the average value of the specimen stiffness, $k = 4977\,\text{N/m}$ (Corigliano, Domenella and Langfelder, 2010), it is then possible to apply Equation 11.16:

$$\sigma_s = \frac{F_{cr}}{A} = k\frac{d_{cr}}{A} = 2.58\ \text{GPa}. \tag{11.18}$$

Figure 11.17 summarizes the results obtained from 38 on-chip tests as a histogram of count of ruptures versus computed rupture stresses. The average value is 2.6 GPa, while the standard deviation is 248 MPa.

Starting from the results of Figure 11.17, a Weibull plot was obtained for the nominal rupture stress, by assuming that the tensile specimen is an ideal one with uniform stress distribution. The results are shown in Figure 11.18. The obtained Weibull parameters are $m = 11.8$ and $\sigma_0 = 2.68$ GPa.

These Weibull parameters can be compared with those obtained by means of the on-chip devices discussed in Sections 11.4.1 and 11.4.2.

11.4.4 On-Chip Test for Thick Polysilicon Films

11.4.4.1 General Description

The fourth on-chip device discussed in this chapter is shown in Figure 11.19. Compared with the devices of the previous sections, this structure presents many unique features. It is designed to test the mechanical properties of a so-called epitaxial polysilicon, *epipoly* layer deposited with the ThELMA process. This film is more than 20 times thicker than the thin film tested using the on-chip devices described in Sections 11.4.1, 11.4.2 and 11.4.3; therefore it requires a greater force to break the specimen. This explains the

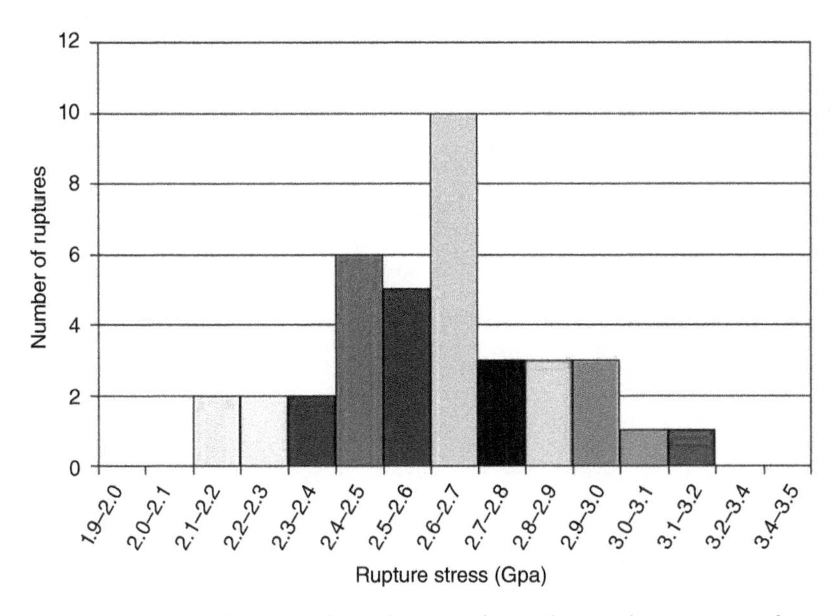

Figure 11.17 Electrothermomechanical actuator for on-chip tensile tests: count of specimen ruptures in 38 on-chip tests versus rupture stress. *Source*: Corigliano, Domenella and Langfelder (2010), Figure 13. Reproduced with permission of Springer.

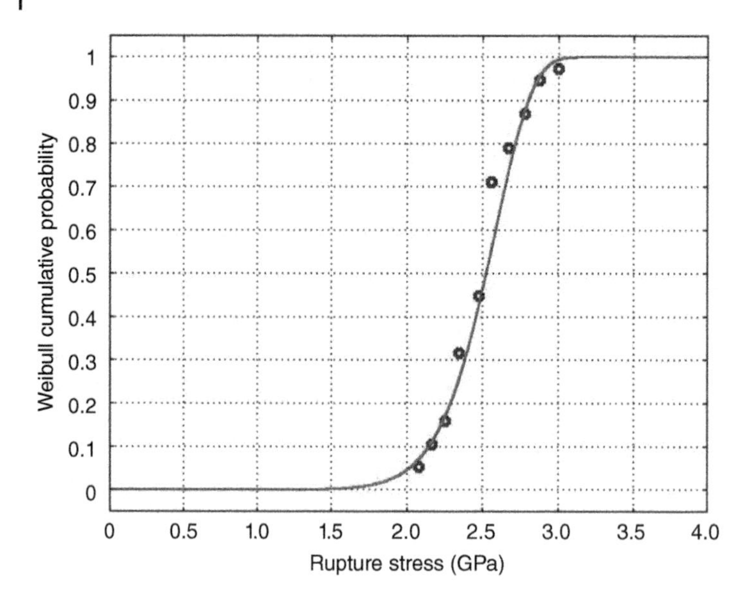

Figure 11.18 Electrothermomechanical actuator for on-chip tensile tests: Weibull plot of rupture stresses. *Source*: Corigliano, Domenella and Langfelder (2010), Figure 14. Reproduced with permission of Springer.

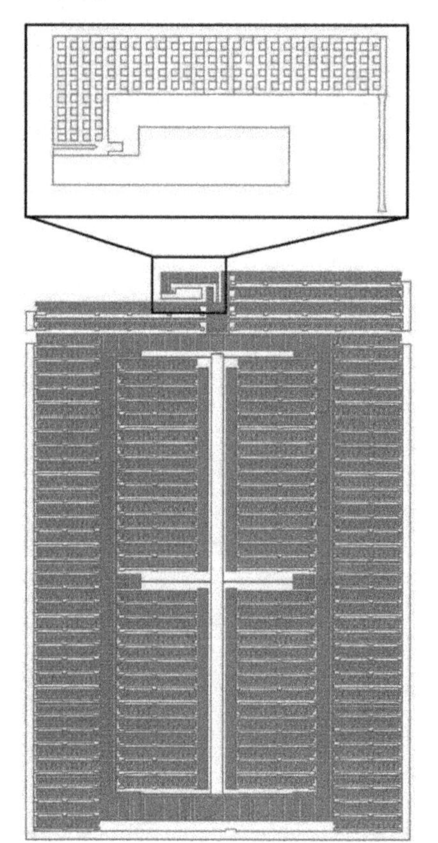

Figure 11.19 Test structure for thick polysilicon: top view of structure and enlargement to show specimen. *Source*: Corigliano, Cacchione and Zerbini (2008), Figure 15. Reproduced with permission of Springer.

reason for the relatively large dimensions of the structure, which takes up a rectangular area 1600 μm × 2250 μm^2.

The actuation mechanism causes a movement in the plane parallel to that of the wafer and is controlled by a number of comb-finger capacitors, a very common choice for MEMS structures.

The four parts of the device – actuator, frame, sensing and specimen – are discussed separately.

The Actuator Electrostatic actuation is realized by over four thousand comb-finger actuators, grouped on specific structures called arms.

Each arm contains 31 comb-finger actuators. Its capacitance is a function of the seismic mass displacement x, as

$$C_{\mathrm{arm}} = C_0 + 31C_{\mathrm{comb}} = C_0 + 31\frac{2\epsilon_0 t}{g}x, \tag{11.19}$$

where C_0 is the capacitance at rest of the system, C_{comb} is the capacitance of a single comb-finger actuator, t is equal to 15 μm, representing the thickness of the layer, and g is equal to 2.2 μm, representing the gap between the stator and rotor fingers. The force from each arm is

$$F_{\mathrm{arm}} = \frac{1}{2}\frac{\partial C_{\mathrm{arm}}}{\partial x}\varphi^2 = 31\frac{\epsilon_0 t}{g}\varphi^2, \tag{11.20}$$

where φ is the applied voltage. The total number of arms n_{a} is 130. Hence, the total force developed by the actuator can be expressed as

$$F_{\mathrm{tot}} = n_{\mathrm{a}}\frac{1}{2}\frac{\partial C_{\mathrm{arm}}}{\partial x}\varphi^2 = 4030\frac{\epsilon_0 t}{g}\varphi^2. \tag{11.21}$$

The Frame The frame is a suspended structure that supports the actuator arms (see Figure 11.20). Six suspension springs are placed in the central part of the frame. These springs avoid the collapse of the structure onto the substrate during the actuation. They are rectangular cross-sectioned 291 μm long slender beams. The in-plane width is 3.2 μm, and the out-of-plane thickness is 15 μm. Using a Young's modulus E =145 GPa, as determined by Corigliano *et al.* (2004), the linear stiffness of the six springs for a movement parallel to the substrate, as shown in Figure 11.20, can be computed as

$$K_{\mathrm{spring}} = 6\left(\frac{12EJ}{l^3}\right) = 17.35\ \mu\mathrm{N}/\ \mu\mathrm{m}. \tag{11.22}$$

With reference to Figure 11.19, the upper part of the frame is clamped to the specimen, which is thus loaded with the force created by the actuator that is not absorbed by the spring system. It is worth noting that the suspension system was designed to be as compliant as possible for in-plane measurements, to allow almost all the force (90% of the force developed) to load the specimen.

The Displacement Sensor In the upper part of the structure, shown in Figure 11.19, is the sensing system. It is made of six arms with 80 comb-finger capacitors on each. The total sensing capacitance, is a function of the displacement x of the specimen:

$$C_{\mathrm{sens}} = C_0^{\mathrm{sens}} + 480\frac{2\epsilon_0 t}{g}x. \tag{11.23}$$

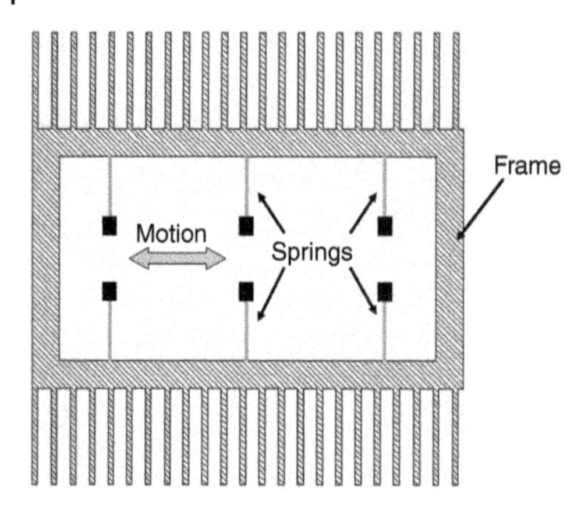

Figure 11.20 Test structure for thick polysilicon film: the frame is cross-hatched; the anchor points are shown in black; the six suspension springs are shown in grey. *Source*: Corigliano, Cacchione and Zerbini (2008), Figure 17. Reproduced with permission of Springer.

Using the finite element method for electrostatic analysis, the sensing system was simulated. The results of the simulation showed that the analytical formula holds for a displacement up to 10 μm, which was never attained during the experimental test.

The Specimen The specimen was designed to perform both quasi-static and fatigue tests (see Chapter 12). It consists of a lever system that causes a stress concentration in a localized region (Figure 11.21). The specimen can be divided into four parts: a beam, which is the physical link between the frame and the specimen; the lever, which transforms the axial action coming from the beam into a bending moment acting in the notched zone; a notch, which is the most stressed part, where the crack nucleates; and a part fixed to the substrate.

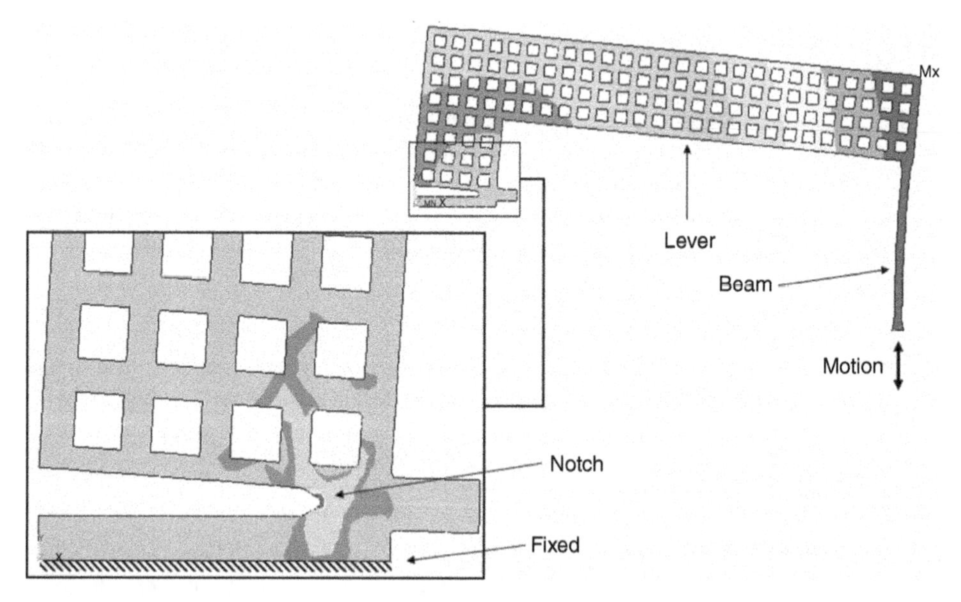

Figure 11.21 Test structure for thick polysilicon: deformed shape of the specimen and contour plot of principal tensile stress of the notched zone. *Source*: Corigliano, Cacchione and Zerbini (2008), Figure 19. Reproduced with permission of Springer. (*See color plate section for the color representation of this figure.*)

11.4.4.2 Data Reduction Procedure

The measured capacitance change (Figure 11.22) was used to compute the displacement imposed at the end of the load beam from Equation 11.23. This causes the rotation of the lever arm and the creation of the desired state of stress in the notch. The force produced by the actuator for each imposed voltage was computed using Equation 11.21. From this information, it was then possible to plot the force–displacement curves (Figure 11.23). Thus, it was possible to combine the experimental results and the finite element method simulations to determine the Young's modulus and the maximum tensile stress at which rupture occurred in every test.

11.4.4.3 Experimental Results

A total of 31 structures, deposited on the same wafer, were tested. As seen from Figure 11.22, the measurements were highly repeatable. The force–displacement plots shown in Figure 11.23 appear to be linear, implying that the electrostatic behaviour of the rotor and sensor parts could be described by the analytical formulae used in the data reduction procedure.

From the slope of the force–displacement plots, it was possible to compute the Young's modulus. The values deduced are in agreement with those obtained by Corigliano *et al.* (2004), confirming the overall quality of the data reduction procedure. The mean value measured was 143 GPa with a standard deviation of ±3 GPa.

Even in this case, the data concerning the rupture of the specimens were interpreted in the framework of the Weibull statistics, as discussed in Section 11.3. From Figure 11.24, the experimental results are clearly interpolated by the Weibull cumulative distribution

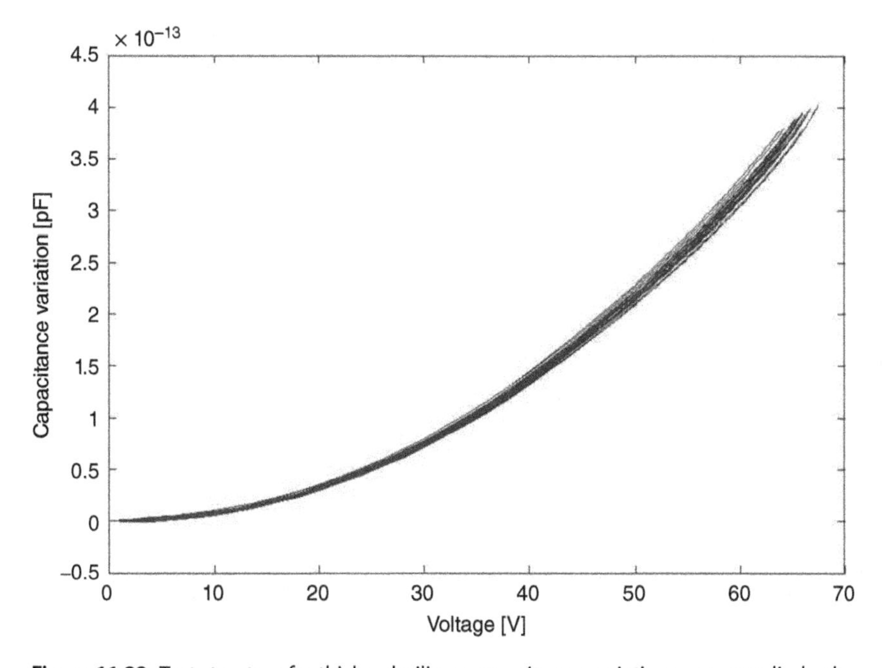

Figure 11.22 Test structure for thick polysilicon: capacitance variation versus applied voltage. *Source*: Corigliano, Cacchione and Zerbini (2008), Figure 20. Reproduced with permission of Springer.

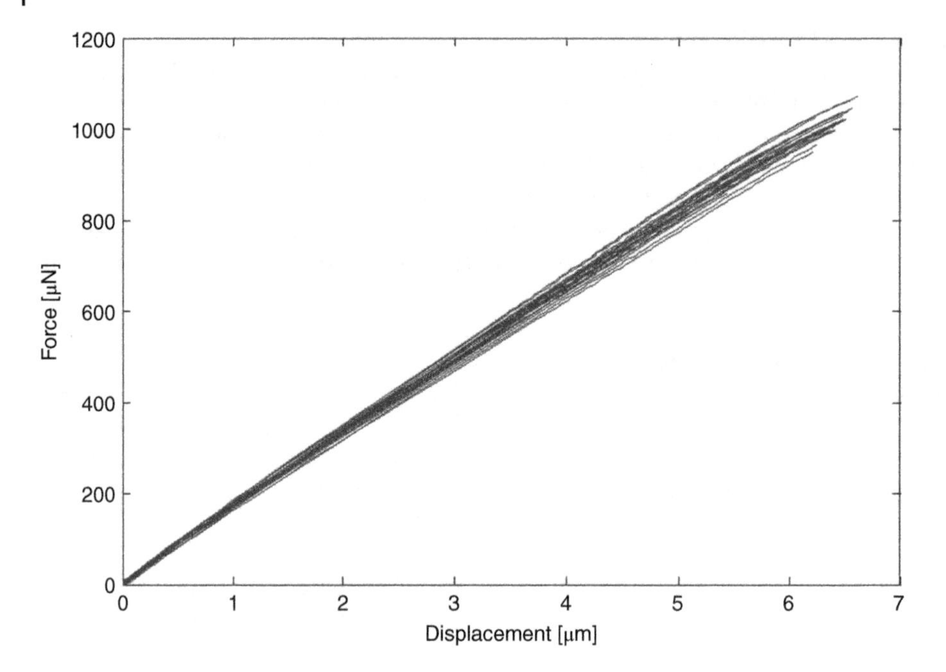

Figure 11.23 Test structure for thick polysilicon: force versus displacement. *Source*: Corigliano, Cacchione and Zerbini (2008), Figure 21. Reproduced with permission of Springer.

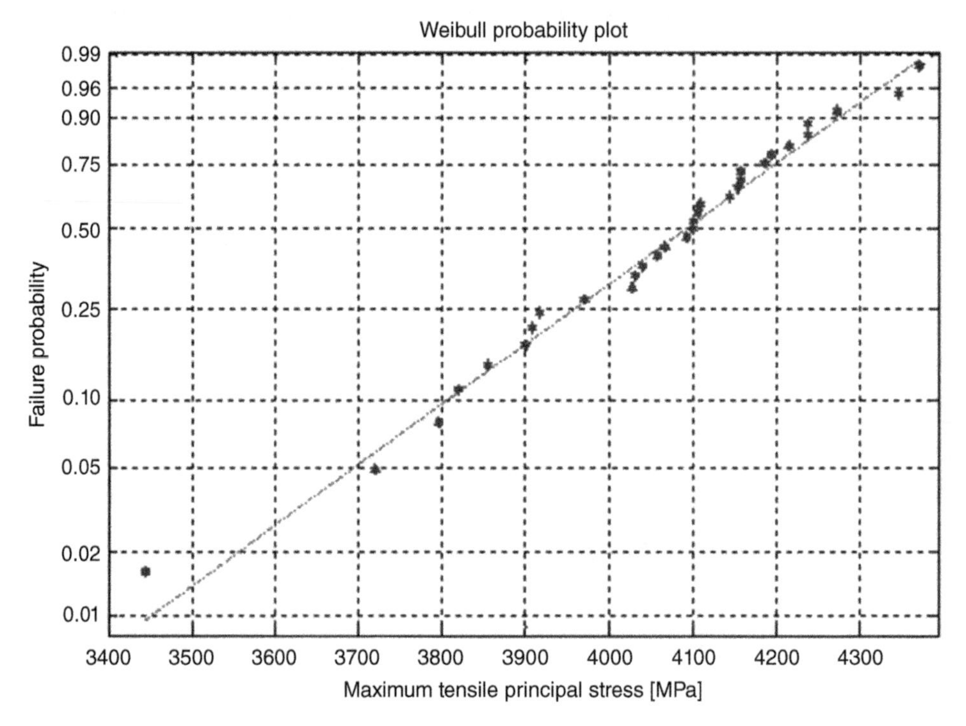

Figure 11.24 Test structure for thick polysilicon: Weibull plot of experimental data (asterisks) and interpolating Weibull cumulative distribution (dashed line). *Source*: Corigliano, Cacchione and Zerbini (2008), Figure 22. Reproduced with permission of Springer.

function. This allowed for the computation of the Weibull modulus, $m = 25.76$ and of the Weibull stress $\sigma_0 = 3.62$ GPa. It is worth recalling again that the Weibull stress represents the level of stress that gives a 63.2% probability of failure for a pure tension specimen with the same size as the reference volume.

After the testing, the specimens were investigated using an optical microscope. The images show that the fracture starts from the notch, as predicted from the finite element simulations carried out during the design phase. From the pictures shown in Figure 11.25, two main aspects emerge.

i) The crack path is often irregular and can be quite different from one structure to another. This difference can be due to the crystalline structure and grain orientation in the notched area. The grain morphology and orientation is different from one structure to another and has a very important impact on the crack propagation direction.
ii) The crack starting point is not always the same. This is because of the nonuniform flaw distribution on the notch surface, caused by the fabrication process. Flaw distribution is thought to be responsible of the scatter of the experimental fracture results.

Further comments on the use of this on-chip device for fracture and fatigue studies in microsystems are given in Chapter 12; some results were published by Langfelder *et al.* (2009); and Corigliano *et al.* (2011).

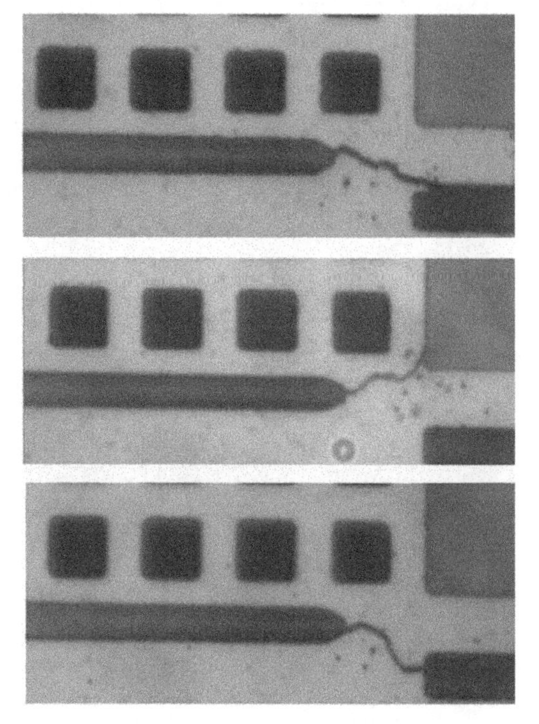

Figure 11.25 Test structure for thick polysilicon: Optical micrographs of broken specimens. *Source*: Corigliano, Cacchione and Zerbini (2008), Figure 23. Reproduced with permission of Springer.

References

Ando, T., Shikida, M. and Sato, K. (2001) Tensile-mode fatigue testing of silicon films as structural material for MEMS. *Sensors and Actuators A*, **93**, 70–75.

Bagdahn, J. and Sharpe, W.J. (2003) Fatigue of polycrystalline silicon under long-term cyclic loading. *Sensors and Actuators A*, **103**, 9–15.

Bagdahn, J., Sharpe, W.J. and Jadaan, O. (2003) Fracture strength of polysilicon at stress concentrations. *Journal of Microelectromechanical Systems*, **12**, 302–312.

Ballarini, R., Kahn, H., Heuer, A. *et al.* (2003) MEMS structures for on-chip testing of mechanical and surface properties of thin films, in *Comprehensive Structural Integrity* (eds I. Milne, R. Ritchie and B. Karihaloo), Elsevier, pp. 325–360.

Ballarini, R., Mullen, R., Kahn, H. and Heuer, A. (1998) The fracture toughness of polysilicon microdevices. *MRS Online Proceedings Library Archive*, **518**, 137.

Ballarini, R., Mullen, R., Yin, Y. *et al.* (1997) The fracture toughness of polysilicon microdevices: a first report. *Journal of Materials Research*, **12**, 915–922.

Bazant, Z., Xi, Y. and Reid, S. (1991) Statistical size effect in quasi brittle structures. I. Is Weibull theory applicable? *Journal of Engineering Mechanics*, **117**, 2609–2622.

Cacchione, F., Corigliano, A., Masi, B.D. and Riva, C. (2005a) Out of plane vs. in plane flexural behaviour of thin polysilicon films: mechanical characterization and application of the Weibull approach. *Microelectronics Reliability*, **45**, 1758–1763.

Cacchione, F., Masi, B.D., Corigliano, A. and Ferrera, M. (2006) Out of plane vs. in plane flexural behaviour of thin polysilicon films: mechanical characterization and application of the Weibull approach. *Sensor Letters*, **4**, 38–45.

Cacchione, F., Masi, B.D., Corigliano, A. *et al.* (2005b) *In-plane and Out-of-plane Mechanical Characterization of Thin Polysilicon*, NSTI Nanotechnology Conference and Trade Show, May 8–12, 2005, Anaheim, CA, Nano Science and Technology Institute.

Chasiotis, I. (2006) Fracture toughness and subcritical crack growth in polycrystalline silicon. *Journal of Applied Mechanics*, **73**, 714–722.

Chasiotis, I. and Knauss, W. (2002) A new microtensile tester for the study of MEMS materials with the aid of atomic force microscopy. *Experimental Mechanics*, **42**, 51–57.

Chasiotis, I. and Knauss, W. (2003a) The mechanical strength of polysilicon films. Part 1. The influence of fabrication governed surface conditions. *Journal of the Mechanics and Physics of Solids*, **51**, 1533–1550.

Chasiotis, I. and Knauss, W. (2003b) The mechanical strength of polysilicon films. Part 2. Size effects associated with elliptical and circular perforations. *Journal of the Mechanics and Physics of Solids*, **51**, 1551–1572.

Chen, G., Ju, M. and Fang, Y. (2000) Effects of monolithic silicon postulated as an isotropic material on design of microstructures. *Sensors and Actuators A*, **86**, 108–114.

Chen, M. and Leipold (1980) Fracture toughness of silicon. *Ceramic Bulletin*, **59**, 469–472.

Chi, S. and Wensyang, H. (1999) A microstructure for *in situ* determination of residual strain. *Journal of Microelectromechanical Systems*, **8**, 200–207.

Cho, S. and Chasiotis, I. (2007) Elastic properties and representative volume element of polycrystalline silicon for MEMS. *Experimental Mechanics*, **47**, 37–49.

Corigliano, A., Cacchione, F., Masi, B.D. and Riva, C. (2005) On-chip electrostatically actuated bending tests for the mechanical characterization of polysilicon at the micro scale. *Meccanica*, **40**, 485–503.

Corigliano, A., Cacchione, F. and Zerbini, S. (2008) Mechanical characterization of low-dimensional structures through on-chip tests, in *Micro and Nano Mechanical Testing of Materials and Devices* (eds F. Yang and J. Li), Springer, pp. 341–375.

Corigliano, A., Domenella, L., Espinosa, H. and Zhu, Y. (2007) Electro-thermal actuator for on-chip nanoscale tensile tests: analytical modelling and multi-physics simulations. *Sensor Letters*, **5**, 592–607.

Corigliano, A., Domenella, L. and Langfelder, G. (2010) On-chip mechanical characterization using an electro-thermo-mechanical actuator. *Experimental Mechanics*, **50**, 695–707.

Corigliano, A., Ghisi, A., Langfelder, G. *et al.* (2011) A microsystem for the fracture characterization of polysilicon at the micro-scale. *European Journal of Mechanics – A/Solids*, **30**, 127–136.

Corigliano, A., Masi, B.D., Frangi, A. *et al.* (2004) Mechanical characterization of polysilicon through on-chip tensile tests. *Journal of Microelectromechanical Systems*, **13**, 200–219.

DelRio, W., Cook, R. and Boyce, L.B. (2015) Fracture strength of micro- and nano-scale silicon components. *Applied Physics Reviews*, **2**, 021303.

Ding, J., Meng, Y. and Wen, S. (2001) Specimen size effect on mechanical properties of polysilicon microcantilever beams measured by deflection using a nanoindenter. *Materials Science and Engineering B*, **83**, 42–47.

Espinosa, H., Prorok, B. and Fischer, M. (2003) A methodology for determining mechanical properties of freestanding thin films and MEMS materials. *Journal of the Mechanics and Physics of Solids*, **51**, 47–67.

Greek, S., Ericson, F., Johansson, S. and Schweitz, J. (1997) *In situ* tensile strength measurement and Weibull analysis of thick film and thin film micromachined polysilicon structures. *Thin Solid Films*, **292**, 247–254.

Hollman, P., Alahelisten, A., Olsson, M. and Hogmark, S. (1995) Residual stress, Young's modulus and fracture stress of hot flamed deposited diamond. *Thin Solid Films*, **270**, 137–242.

Jadaan, O., Nemeth, N., Bagdahn, J. and Sharpe, W. (2003) Probabilistic Weibull behavior and mechanical properties of MEMS brittle materials. *Journal of Material Science*, **38**, 4075–4079.

Jayaraman, S., Edwards, R. and Hemker, K. (1998) Determination of the mechanical properties of polysilicon thin films using bulge testing. *MRS Online Proceedings Library Archive*, **505**, 623–628.

Kahn, H., Ballarini, R., Bellante, J. and Heuer, A. (2002) Fatigue failure in polysilicon not due to simple stress corrosion cracking. *Science*, **298**, 1215–1218.

Kahn, H., Ballarini, R. and Heuer, A. (2004) Dynamic fatigue of silicon. *Current Opinion in Solid State and Materials Science*, **8**, 71–76.

Kahn, H., Tayebi, N., Ballarini, R. *et al.* (2000) Fracture toughness of polysilicon MEMS devices. *Sensors and Actuators A*, **82**, 274–280.

Kim, J., Yeon, S., Jeon, Y. *et al.* (2003) Nano-indentation method for the measurement of the Poisson's ratio of MEMS thin films. *Sensors and Actuators A*, **108**, 20–27.

Knauss, W., Chasiotis, I. and Huang, Y. (2003) Mechanical measurements at the micron and nanometer scales. *Mechanics of Materials*, **35**, 217–231.

Kramer, T. and Paul, O. (2000) Surface micromachined ring test structures to determine mechanical properties of compressive thin films. *Sensors and Actuators A*, **92**, 292–298.

Langfelder, G., Longoni, A., Zaraga, F. *et al.* (2009) A new on-chip test structure for real time fatigue analysis in polysilicon MEMS. *Microelectronics Reliability*, **49**, 120–126.

LaVan, D., Tsuchiya, T., Coles, G. *et al.* (2001) Cross comparison of direct strength testing techniques on polysilicon films, in *Mechanical Properties of Structural Films, STP 1413* (eds C. Muhlstein and S. Brown), ASTM.

Li, X. and Bhushan, B. (1999) Micro/nanomechanical characterization of ceramic films for microdevices. *Thin Solid Films*, **144**, 210–217.

McCarty, A. and Chasiotis, I. (2007) Description of brittle failure of non-uniform MEMS geometries. *Thin Solid Films*, **515**, 3267–3276.

Muhlstein, C., Brown, S. and Ritchie, R. (2001) High-cycle fatigue and durability of polycrystalline silicon thin films in ambient air. *Sensors and Actuators A*, **94**, 177–188.

Muhlstein, C., Howe, R. and Ritchie, R. (2004) Fatigue of polycrystalline silicon for microelectromechanical system applications: crack growth and stability under resonant loading conditions. *Mechanics of Materials*, **36**, 13–33.

Muhlstein, C., Stach, E. and Ritchie, R. (2002a) A reaction-layer mechanism for the delayed failure of micron-scale polycrystalline silicon structural films subjected to high-cycle fatigue loading. *Acta Materialia*, **50**, 3579–3595.

Muhlstein, C., Stach, E. and Ritchie, R. (2002b) Mechanism of fatigue in micron-scale films of polycrystalline silicon for microelectromechanical systems. *Applied Physics Letters*, **80**, 1532–1534.

Oh, C., Lee, H., Ko, S. *et al.* (2005) Comparison of the Young's modulus of polysilicon film by tensile testing and nanoindentation. *Sensors and Actuators A*, **117**, 151–158.

Oostemberg, P. and Senturia, S. (1997) M-Test: a test chip for MEMS material property measurement using electrostatically actuated test structures. *Journal of Microelectromechanical Systems*, **6**, 107–118.

Sharpe, W. (2002) Mechanical properties of MEMS materials, in *The MEMS Handbook* (ed. M.G. el Hak), CRC Press, pp. 3.1–3.33.

Sharpe, W., Brown, J., Johnson, G. and Knauss, W. (1998) Round-robin tests of modulus and strength of polysilicon. *MRS Online Proceedings Library Archive*, **518**, 57.

Sharpe, W., Turner, K. and Edwards, R. (1999) Tensile testing of polysilicon. *Journal of Experimental Mechanics*, **39**, 162–170.

Sharpe, W., Yuan, B. and Edwards, R. (1997) A new technique for measuring the mechanical properties of thin films. *Journal of Microelectromechanical Systems*, **6**, 193–199.

Sundarajan, S. and Bhushan, B. (2002) Development of AFM-based techniques to measure mechanical properties of nanoscale structures. *Sensors and Actuators A*, **101**, 338–351.

Tabata, O., Kawahata, K., Sugiyama, S. and Igarashi, I. (1989) Mechanical property measurement of thin films using load-deflection of composite rectangular membranes. *Sensors and Actuators A*, **20**, 135–141.

Tsuchiya, T., Hirata, M., Chiba, N. *et al.* (2005) Cross comparison of thin-film tensile-testing methods examined using single-crystal silicon, polysilicon, nickel, and titanium films. *Journal of Microelectromechanical Systems*, **14**, 1178–1186.

Tsuchiya, T., Shikida, M. and Sato, K. (2002) Tensile testing system for sub-micrometer thick films. *Sensors and Actuators A*, **97–98**, 492–496.

Tsuchiya, T., Tabata, O., Sakata, J. and Taga, Y. (1998) Specimen size effect on tensile strength of surface micromachined polycrystalline silicon thin films. *Journal of Microelectromechanical Systems*, **7**, 106–113.

Weibull, W. (1951) A statistical distribution function of wide applicability. *Journal of Applied Mechanics*, **18**, 293–297.

Yang, J. and Paul, O. (2002) Fracture properties of LPCVD silicon nitride thin films from the load–deflection of long membranes. *Sensors and Actuators A*, **97–98**, 520–526.

Yi, T., Li, L. and Kim, C. (2000) Microscale material testing of single crystalline silicon: process effects on surface morphology and tensile strength. *Sensors and Actuators A*, **83**, 172–178.

Zhu, Y., Corigliano, A. and Espinosa, H. (2006) A thermal actuator for nanoscale *in-situ* microscopic testing: design and characterization. *Journal of Micromechanics and Microengineering*, **16**, 242–253.

Zhu, Y. and Espinosa, H. (2005) An electro-mechanical material testing system for *in situ* electron microscopy and applications. *Proceedings of the National Academy of Sciences*, **102**, 14503–14508.

12

Fracture and Fatigue in Microsystems

12.1 Introduction

Owing to their massive use for mobile applications, microsystems are often exposed to extreme loading conditions, possibly causing failure. Furthermore, the interaction through the package with the outer environment can become a source of catastrophic events at the micrometre or submicrometre scale.

Before assessing the links between failure occurrence and physical properties of all the materials having a mechanical role in packaged MEMS, in this chapter we provide some details of two specific issues to be taken into account in the design phase: fracture and fatigue. We mostly focus on the physics of failure related to such phenomena, showing also results of numerical approaches able to capture the relevant features; we do not discuss details of the currently available numerical procedures, and refer readers to the relevant literature.

The remainder of this chapter is organized as follows. As far as fracture mechanics is concerned, in Section 12.2 we give an account of some basic concepts, with a focus on brittle and quasi-brittle materials. We next provide in Section 12.3 an overview of possible failure modes (at all length scales) in packaged MEMS, discussing in more detail the cracking of single-crystal and polycrystalline silicon films. As far as fatigue is concerned, in Section 12.4.1 we first present common terminology and definitions adopted for fatigue in mechanics; then, in Section 12.4.2, we focus on the fatigue behaviour of polycrystalline silicon, due to its exposure to failure in resonant microsystems; furthermore, we briefly address in Section 12.4.3 the problem for metals at the microscale. Finally, some experimental methods used for fatigue testing are reviewed in Section 12.4.4.

12.2 Fracture Mechanics: An Overview

In this section, we provide a brief overview of fracture mechanics. We do not aim to discuss all the technical and theoretical details related to the cracking of materials across length scales; instead, we present a discussion of the main tools to be used later to interpret and understand the results relevant to some common crack-related failure

Mechanics of Microsystems, First Edition. Alberto Corigliano, Raffaele Ardito, Claudia Comi, Attilio Frangi, Aldo Ghisi, and Stefano Mariani.

Companion website: www.wiley.com/go/corigliano/mechanics

or malfunction of MEMS devices. A number of well-recognized books on fracture mechanics of either homogeneous or heterogeneous materials and structures can be suggested for further reading, e.g. Broek (1982), Freund (1990), Friedrich (1989), Broberg (1999) and Buehler (2008). In what follows, we basically refer to Bazant and Cedolin (1991), in which the energetic implications of fracture, and links with system stability at the continuum level are discussed in detail.

Although the inception and subsequent growth of a discontinuity within a continuum might look like nonsense, jumps in the fundamental fields of the formulation provided in Section 2.3 can be admitted if they occur in regions of zero measure, i.e. across (zero volume) surfaces in a 3D body. A discontinuity affecting the displacement field is a so-called strong discontinuity, and is commonly referred to as a crack; fracture mechanics provides the equations governing the stability of such a crack. If the discontinuity instead affects the strain field, e.g. owing to a mismatch in the orientation of crystallographic planes across a grain boundary of a polycrystalline material, it is considered a weak discontinuity; such a discontinuity is still governed by continuum mechanics in standard form, if the discontinuity locus does not evolve in time.

For linear elastic materials undergoing cracking, the problem is studied within the frame of linear elastic fracture mechanics. Asymptotically, in a sufficiently small region around the front of an already existing crack, the displacement $\boldsymbol{u}$ and stress $\boldsymbol{\sigma}$ fields are, respectively, given by or, better, dominated by the solution:

$$u_i = \sqrt{r}\left[K_{\mathrm{I}}\ \varphi_i^{\mathrm{I}}(\vartheta) + K_{\mathrm{II}}\ \varphi_i^{\mathrm{II}}(\vartheta) + K_{\mathrm{III}}\ \varphi_i^{\mathrm{III}}(\vartheta)\right], \tag{12.1}$$

$$\sigma_{ij} = \frac{1}{\sqrt{r}}\left[K_{\mathrm{I}}\ \psi_{ij}^{\mathrm{I}}(\vartheta) + K_{\mathrm{II}}\ \psi_{ij}^{\mathrm{II}}(\vartheta) + K_{\mathrm{III}}\ \psi_{ij}^{\mathrm{III}}(\vartheta)\right], \tag{12.2}$$

where: $i = 1, 2, 3$ or $i = x, y, z$, depending on the adopted reference frame; r, ϑ are, respectively, the radial and circumferential coordinates in a polar reference frame centred at the crack tip; $\varphi_i^{\mathrm{I}}, \varphi_i^{\mathrm{II}}, \varphi_i^{\mathrm{III}}$ and $\psi_{ij}^{\mathrm{I}}, \psi_{ij}^{\mathrm{II}}, \psi_{ij}^{\mathrm{III}}$ are functions of the ϑ coordinate only and $K_{\mathrm{I}}, K_{\mathrm{II}}, K_{\mathrm{III}}$ are the so-called stress intensity factors, which depend on the applied far-field load and on the crack length in the considered asymptotic frame. Any loading type at the tip of a crack can thus be given as a combination of the three fundamental and mutually orthogonal modes listed in Equation 12.1, and governed by $K_{\mathrm{I}}, K_{\mathrm{II}}, K_{\mathrm{III}}$; these three modes respectively induce opening, in-plane sliding and anti-plane sliding of the crack faces in the vicinity of the front. The presence of the term $1/\sqrt{r}$ in Equation 12.2 shows that, for $r \to 0$, the stress field increases to infinity, leading to a singularity that cannot be obviously withstood by any material with a finite strength. However, within this frame, a mode I crack is assumed to propagate only on the attainment at its tip of the critical threshold:

$$K_{\mathrm{I}} = K_{\mathrm{IC}}, \tag{12.3}$$

where K_{IC} is the material critical stress intensity factor or *fracture toughness*.

An energetic formulation can also be provided to study the evolution of a crack. Since crack formation locally requires a specific amount of energy per unit area, which is supposed to be a material-dependent parameter, propagation occurs if

$$\mathcal{G} = -\frac{\partial \Pi}{\partial A} = G_{\mathrm{c}}, \tag{12.4}$$

where $\mathcal{G}$ is the energy release rate, given by the derivative of the total potential energy Π of the load-structure system with respect to the crack area A, computed at constant applied displacements along the (far-field) boundary of the region considered; the critical energy release rate G_c is the aforementioned energy required to annihilate the bonding in the material and create a new surface of unit area. For linear elastic, isotropic materials, the energy release rate is one-to-one related to the stress intensity factors through

$$\mathcal{G} = \frac{K_I^2 + K_{II}^2}{E'} + \frac{(1+\nu)K_{III}^2}{E}, \tag{12.5}$$

where the accounted value E' of the Young's modulus is $E' = E$ if plane stress conditions take place along the crack front and $E' = E/(1-\nu^2)$ if plane strain conditions take place instead.

As mentioned, materials do not display infinite strength. Nonlinear fracture mechanics has therefore been developed to study the spreading of nonlinearities in the bulk of a cracking body, wherever the material strength is exceeded, and describe the interaction with the crack state. By contrast with the case of a brittle material behaving (in principle) according to linear elastic fracture mechanics, for which crack propagation, once incepted, almost surely leads to catastrophic failure due to brittleness, if the material has some additional toughness resources linked to bulk dissipation mechanisms, the crack can somehow be arrested during its growth. Depending on the material behaviour, the aforementioned bulk dissipation can be due to plasticity (typically in ductile, metallic materials) or damage (typically in quasi-brittle, ceramics-like materials). In the former case, the dissipation is spread in a region ahead of the crack front, whose shape indeed requires volumetric effects to be fully accounted for; in the latter case, the shape of the region wherein damage evolves is typically so elongated in a single direction, namely with a length largely exceeding the width, that the dissipation mechanism can be modelled as lumped along a line. Such a line is considered as a continuation of the real crack, and so termed the mathematical, or fictitious crack (Dugdale, 1960; Barenblatt, 1962). Its length ℓ_{pz} can be estimated as the locus where the stress intensification due to the presence of the crack would provide opening stress levels exceeding the material strength; accordingly (Irwin, 1964)

$$\ell_{pz} = \frac{EG_c}{\sigma_s^2}, \tag{12.6}$$

where σ_s is the material tensile strength. In practical terms, a material or, to be more precise, a structure can be classified as brittle or quasi-brittle depending on whether or not ℓ_{pz} can be disregarded in comparison with its dimensions.

Along the fictitious crack, an *ad-hoc* constitutive-like law needs to be provided to account for the local energy dissipation mechanism. Referring first to pure mode I loading, a function $\tau = \tau(\delta)$ is formulated to link the displacement jump δ across the crack and the relevant traction τ acting on the crack faces in the direction perpendicular to the crack. The traction τ accounts for the interaction between the two sides of the crack surface prior to the annihilation of the material bonding, and so provides the contribution to keep the crack closed; accordingly, $\tau(\delta)$ is termed the cohesive law. The overall, or integral effect of the cohesive law along the fictitious crack of length ℓ_{pz} has to provide $K_I = 0$, otherwise the finite strength property of the material would be exceeded somewhere in the bulk around the crack tip, with a solution inconsistent with the assumption of dissipation only within the process zone. The critical energy release rate or material toughness G_c turns out to be given by

$$G_c = \int_0^\infty \tau(\delta)\, d\delta. \tag{12.7}$$

In Equation 12.7, the upper limit of integration has generally been set to infinity, to account for possible long-range interactions between the two crack sides; it sometimes proves convenient to introduce a breakdown threshold distance, beyond which the mentioned interactions become so small that they can be disregarded, see e.g. Landis, Pardoen and Hutchinson (2000) and Corigliano, Mariani and Pandolfi (2003).

In a cohesive approach to quasi-brittle fracture (Corigliano, 1993), the following features of the law $\tau(\delta)$ need to be set: the toughness G_c; the (tensile) strength σ_s, already considered in Equation 12.6; the shape and, specifically, the initial slope

$$\kappa = \left.\frac{d\tau}{d\delta}\right|_{\delta=0^+}.$$

The material strength is represented by the maximum of the cohesive function; depending on whether the law is smooth or not, so depending on the law shape, such a maximum can be computed as a stationary point of the $\tau(\delta)$ function, or is provided a priori. While the shape of $\tau(\delta)$ has been reported to be almost irrelevant in some cases (Alfano, 2006), a major issue is instead represented by the initial slope κ. As for this last point, cohesive approaches can be classified as intrinsic or extrinsic (Kubair and Geubelle, 2003; Zhang, Paulino and Celes, 2007): in the first case, κ is positive as an initial hardening branch, characterized by increasing traction values for a growing crack opening, which precedes a softening branch leading to the final decohesion of the material; in the second case, the hardening branch is omitted and softening starts soon after the crack starts opening. Since cohesive models are supposed to provide only a description of the localized damaging process in quasi-brittle materials, the extrinsic approach appears to be more appropriate; indeed, as the initial elastic-like branch of the response is avoided, algorithmic problems arise and *ad-hoc* implementations are required (Ruiz, Pandolfi and Ortiz, 2001; Mariani *et al.*, 2011a). Moving from a purely theoretical perspective, the adoption of an intrinsic approach was instead proposed by referring to universal binding laws (Rose, Ferrante and Smith, 1981; Rose, Smith and Ferrante, 1983), and so smooth exponential functions were adopted (Xu and Needleman, 1994; Corigliano, Mariani and Pandolfi, 2003):

$$\tau = \kappa\delta \exp\left[-\frac{\delta}{\delta_s}\right], \tag{12.8}$$

where δ_s is the displacement jump corresponding to the peak traction $\sigma_s = \kappa\delta_s/\mathrm{e}$, where $\mathrm{e} = \exp[1]$. Since the fracture toughness reads, in this case, $G_c = \kappa\delta_s^2$, the three aforementioned main law parameters are linked through $\sqrt{(\kappa G_{(c)}}/\sigma_s = \mathrm{e}$; to avoid such spurious relationship for the cohesive model and enable fine tuning of all the parameters, a simple modification was proposed by Mariani (2009).

For mixed-mode cracking, more complex 3D cohesive crack models can be formulated (Corigliano, 1993); an effective displacement discontinuity is used to replace the pure opening δ and also to account for sliding motions; weighting factors are thus introduced to mix the three fundamental modes, and an energy-conjugate effective traction is accordingly introduced (Corigliano, 1993; Camacho and Ortiz, 1996; Ruiz,

Pandolfi and Ortiz, 2001). Further details relevant to the reversible or irreversible response of the fictitious crack are not discussed here; interested readers can find thorough discussions in Camacho and Ortiz (1996) and Corigliano, Mariani and Pandolfi (2006).

A final issue to be considered for the described cohesive approach to quasi-brittle fracture stems from the anisotropic behaviour of materials. So far, we have considered κ, σ_s and G_c to be independent of the crack orientation; for anisotropic materials, these properties must instead depend on the relative orientation of the crack plane and any privileged material direction, e.g. crystal lattice directions for single crystals or the fibre axis direction for unidirectional fibre-reinforced composites. Generally, if α represents such a relative orientation, functions $\kappa(\alpha)$, $\sigma_s(\alpha)$ and $G_c(\alpha)$ have to be set through *ad-hoc* experimental tests (Yu *et al.*, 2002); for complex materials, a single angle might not be enough, and so a properly defined set of Euler angles must be fully considered (Coffman *et al.*, 2008). Some results in this regard are discussed in Section 12.3.2, relevant to the fracture properties of single-crystal silicon (Tanaka *et al.*, 2006).

12.3 MEMS Failure Modes due to Cracking

In this section, we distinguish the possible failure modes of a packaged MEMS depending on the length scale at which failure is incepted. We first deal with fracture processes at the package level, and we also consider delamination between materials with different physical properties. Next, we go into the details of the micromechanics of failure of silicon films.

12.3.1 Cracking and Delamination at Package Level

Packaged devices are characterized by materials with different physical (not only mechanical) properties joined together. The interaction of the package as a whole with the outer environment can then result in failures. Most likely, failure events in case of semi-hermetic polymeric packaging can be induced by stresses linked to: (i) a mismatch of the coefficient of thermal expansion of the joined materials; (ii) coupled hygroscopic-thermal processes (Zhou *et al.*, 2012).

The former cause is well known to pose severe reliability issues in layered structures at all length scales (Zhang, van Driel and Fan, 2006). Whenever the temperature varies during the life-cycle of the device, stresses arise at the interface (or interphase) between adjacent layers, possibly causing delamination, i.e. decohesion between the layers themselves. This will later represent a privileged form of interaction between the MEMS and the environment, and so a possible source of malfunction.

The latter cause generally leads to the so-called popcorn effect (Kitano *et al.*, 1988; Zhou *et al.*, 2012; Chen and Li, 2011; Lim *et al.*, 1998; Dudek *et al.*, 2001). Moisture first diffuses into the moulding compound; when exposed to the temperature level required for reflow soldering, absorbed moisture is vaporized and leads to an internal pressure in holes and defects within which it is entrapped. Eventually, delamination between silicon and seal rings (Zhou *et al.*, 2012) or between the compound and the printed circuit board (Chen and Li, 2011) can occur, possibly accompanied by additional damage at the solder mask. Delamination is backed by cracking in the bulk materials, on either

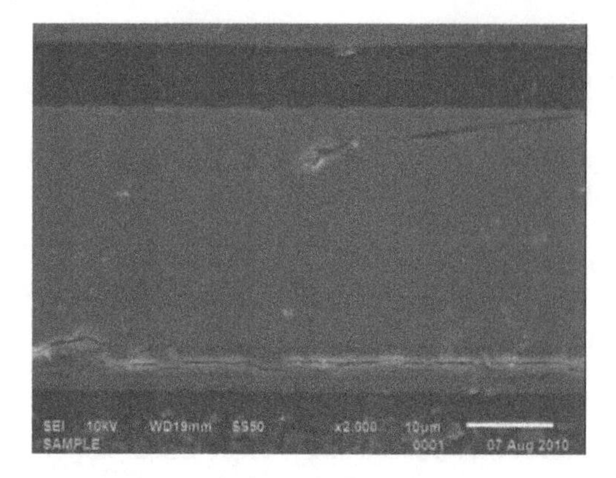

Figure 12.1 SEM showing popcorn cracking and delamination induced by humidity testing of a device. *Source*: Zhou *et al.* (2012), Figure 3(a). Reproduced with permission of IEEE.

the die or compound side, as shown in Figure 12.1. The entire process is enhanced by the miniaturization of the devices, leading to thinner and thinner layers of the compound.

Although delamination and cracking occur almost instantaneously during solder reflow and so cause the characteristic popcorning-like audible sound, they may only result in a partial damage of the device that can subsequently increase in time, owing to thermal cycles. To help detect popcorn failure modes just after reflow, Chen and Li (2011) proposed scanning acoustic microscopy. Figure 12.2 provides two images collected through scanning acoustic microscopy on two different packages.

12.3.2 Cracking at Silicon Film Level

Since movable parts of MEMS are commonly made of either single-crystal or polycrystalline silicon films, we now focus on the micromechanical issues linked to their cracking.

To enhance the performance indices of the devices, it has been shown that silicon films are often used for slender parts (beams or springs) connected to massive ones (plates or proof masses). Owing to external excitations, which are not necessarily shocks, as detailed in the next chapter, the slender parts may crack. To provide insights into the links between the types of crack-related failure and the geometry and morphology of the silicon film in standard devices, experimental results available in the literature are summarized next. Examples of on-chip testing procedures to assess the main properties

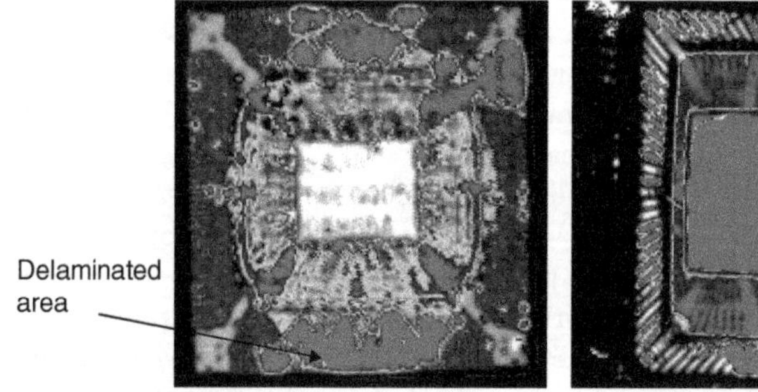

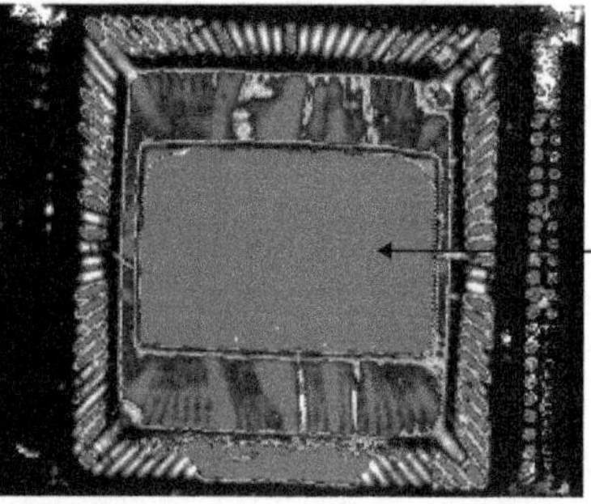

Figure 12.2 Scanning acoustic micrographs of two failed devices due to popcorn effect. *Source*: Chen and Li (2011), Figures 2 and 7. Reproduced with permission of IEEE.

of silicon, modelled as a quasi-brittle material, as discussed in Section 12.2, are also presented.

Silicon is an orthotropic material, with a face-centred-cubic crystal structure. Looking at its elastic behaviour, in a reference frame aligned with crystal lattice, it shows a slight deviation from isotropy (Mullen *et al.*, 1997; Mariani *et al.*, 2007, 2011b). Although weak, such deviation has been recognized as possibly leading to remarkable effects at the length scale typical of silicon films. As far as the post-elastic properties of silicon are concerned, a clear signature of the effects induced by strength or toughness anisotropy may be difficult to obtain through direct tests, owing to the mentioned brittleness of silicon, easily leading to catastrophic failure events.

For single-crystal silicon, Li *et al.* (2004) conducted on-chip tensile tests on films showing different relative orientations of crystal lattice and loading direction. Some results are collected in Figure 12.3 for single-edge notched specimens featuring an initial crack length greater than 0.5 μm (against a specimen width of 50 μm): depending on lattice orientation, cracking occurred along the different cleavage planes reported in the SEMs. The fracture toughness, in terms of critical value of the stress intensity factor K_{IC} (where mode I is considered for reference, as the tensile load mainly induces an opening of the notch), was shown to vary in the range $K_{IC} = 1.0–2.0$ MPa m$^{1/2}$ with maximum values for direction $\langle 1\ 0\ 0 \rangle$ aligned with the longitudinal axis of the specimen, and minimum values for direction $\langle 1\ 1\ 0 \rangle$.

Fujii *et al.* (2013) reported results obtained through tension–torsion–bending tests on beam specimens with a 100 μm × 100 μm cross-section. The test configuration was devised to better mimic the kind of loading conditions encountered in real structures, as

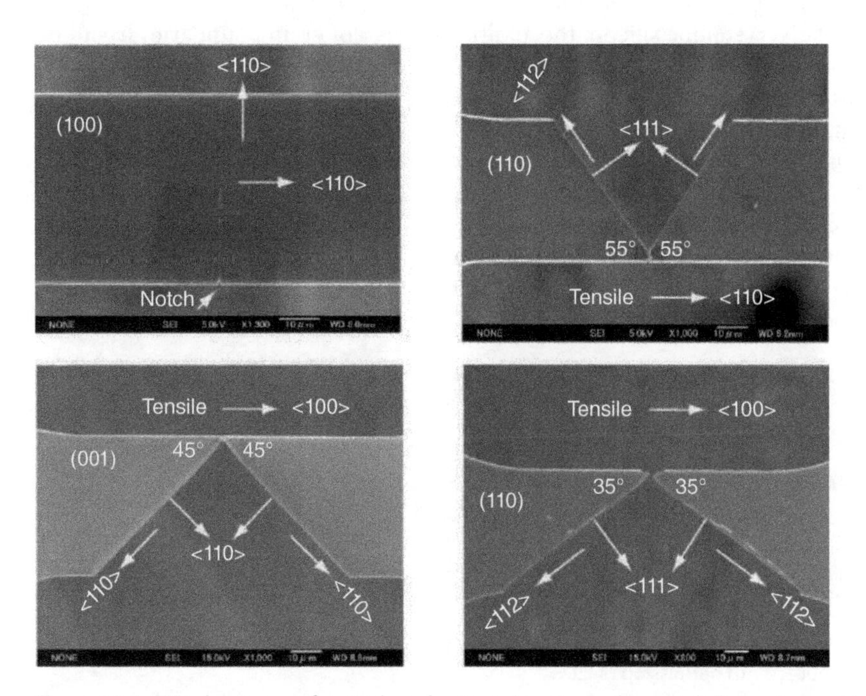

Figure 12.3 SEM showing preferential crack orientations in single-crystal silicon. *Source*: Li *et al.* (2004), Figure 3. Reproduced with permission of IEEE.

compared with the tensile conditions adopted in standard tests. For isotropic materials, this testing would not make much difference, but in the case of silicon it can enable assessment of lattice-induced anisotropy and better match the experimental data at device level. A loading-dependent strength of silicon in the range 800–1150 MPa was observed, with data obtained with different samples fitted quite well with a Weibull distribution (see Section 11.3). A variation of the fracture mode was also observed, and some relevant SEMs are collected in Figure 12.4. As a result of the etching process, fracture toughness may also depend on the thickness of the silicon film (Gaither *et al.*, 2013); Buchheit and Phinney (2015) obtained strength values in the range 1.0–3.5 GPa for 25 μm thick films, and in the range 0.6–2.2 GPa for 125 μm thick films.

The orientation-dependent fracture toughness of silicon was measured by Tanaka *et al.* (2006) with indentation tests through a conical indenter. The measured variation of K_{IC} with crack orientation is depicted in Figure 12.5 for three different crystallographic planes; values of K_{IC} almost all fall within the range 0.7–0.8 MPa $m^{1/2}$. Although tests, as well as atomistic models (Pérez and Gumbsch, 2000), did not clearly demonstrate that a smooth dependence of K_{IC} or G_c exists on the relative orientation of crack plane and crystal orientation, Mariani *et al.* (2011a) assumed that in the (0 0 1) plane the tensile strength of silicon is negligibly anisotropic, while the fracture toughness varies (in analogy with the elastic moduli) according to

$$G_c = G_{c[100]} - \Delta G_c(1 - \cos 4\alpha), \tag{12.9}$$

where the angle α defines the in-plane slope of the crack relative to the [1 0 0] crystal orientation, $G_{c[100]}$ is the value of G_c relevant to a crack aligned with the [1 0 0] crystal orientation and ΔG_c is half the maximum variation of G_c with α.

Moving to polycrystalline silicon, the additional effect of grain boundaries has to be taken into account. If the film is affected by stress intensification in a region large enough in comparison with the characteristic grain size, the overall effect of anisotropy at the level of a single grain is weakened or even lost (Cho and Chasiotis, 2007). Strength and fracture toughness can thus be considered smeared values at the film level, and may depend on silicon doping. Yagnamurthy, Boyce and Chasiotis (2015) showed that, while fracture toughness slightly varies around $K_{IC} = 1.0$ MPa $m^{1/2}$ with the doping level, the polysilicon tensile strength can be largely reduced (by even about 30%) in columnar films. Kahn *et al.* (2000) reported an average value of $K_{IC} = 1.1$ MPa $m^{1/2}$; several other strength values were also gathered and compared by Jadaan *et al.* (2003).

Figure 12.4 Effect of loading condition on cracking in single-crystal silicon. *Source*: Fujii *et al.* (2013), Figure 7. Reproduced with permission of IEEE.

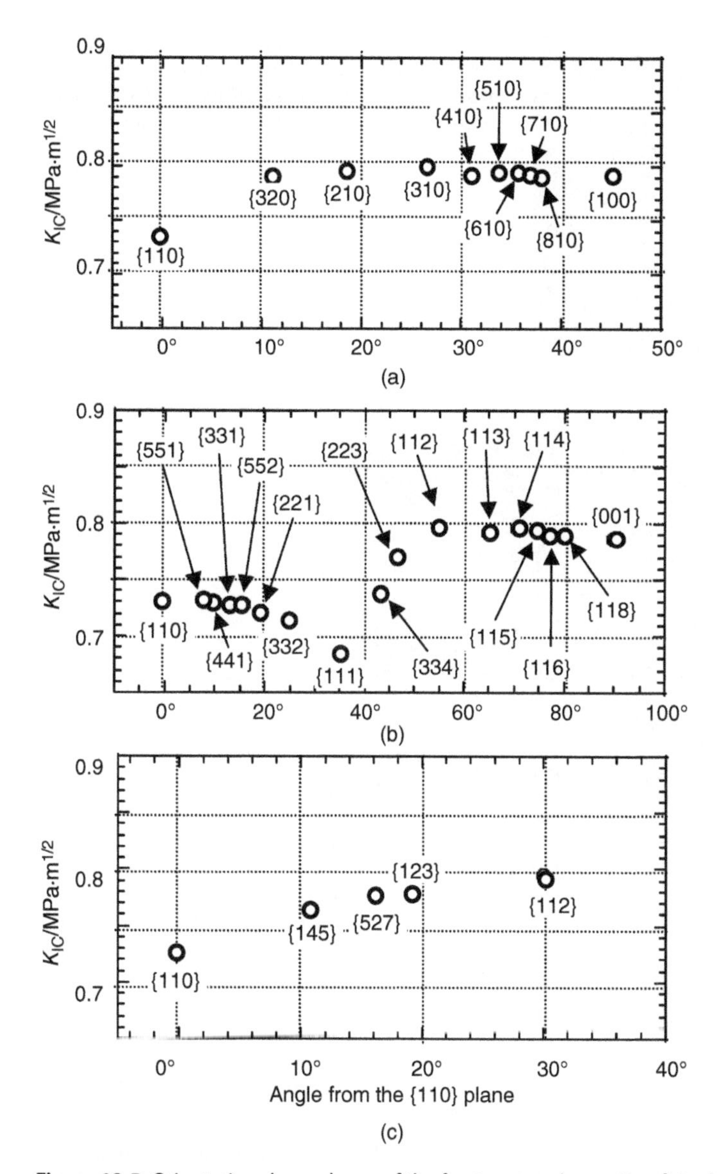

Figure 12.5 Orientation dependence of the fracture toughness K_{IC} of single-crystal silicon for (from top to bottom) crystallographic planes parallel to the ⟨0 0 1⟩, ⟨1 1 0⟩ and ⟨1 1 1⟩ directions. *Source*: Tanaka *et al.* (2006), Figure 6. Reproduced with permission of Springer.

Figure 12.6 shows a close-up of secondary cracking occurring when the main crack crosses a grain boundary between two grains with different lattice orientations; this effect is somehow linked to the very similar fracture toughness featured along different crystallographic planes. Figure 12.7 provides instead an overall view of the typical crack propagation path in polysilicon: owing to the different crystal lattice orientations of adjacent grains, the crack kinks nearly every time it crosses a grain boundary; owing

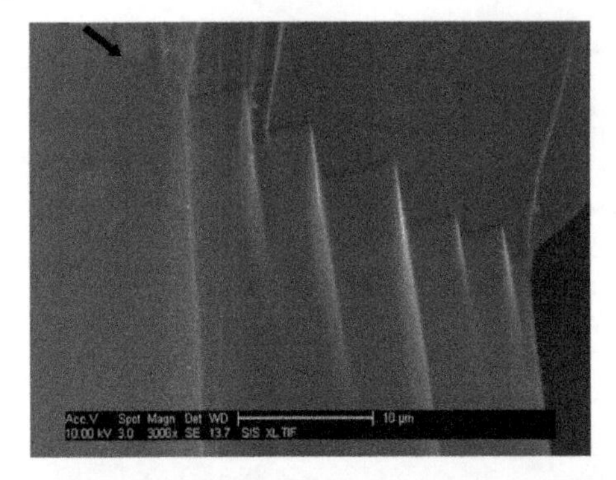

Figure 12.6 SEM showing secondary cracking at a polysilicon grain boundary (indicated by the black arrow in the picture). *Source*: Chen and Qiao (2007), Figure 2. Reproduced with permission of Elsevier.

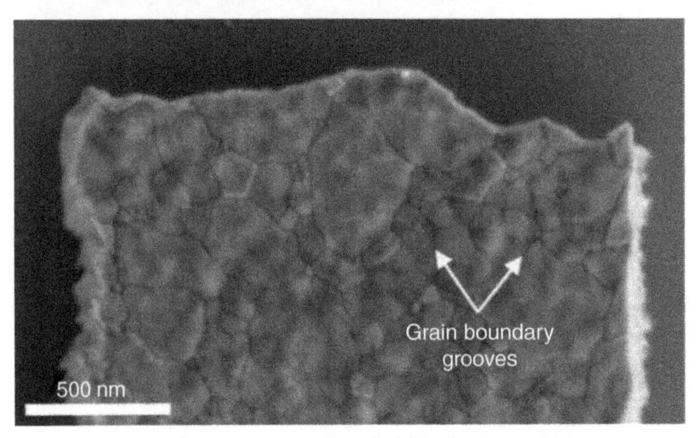

Figure 12.7 SEM of transgranular-dominated cracking in polysilicon. *Source*: Vayrette, Raskin and Pardoen (2015), Figure 12. Reproduced with permission of Elsevier.

to the higher toughness of the grain boundaries (Chasiotis and Knauss, 2003), it hardly grows along the grain boundaries themselves and so a highly transgranular-dominated cracking mode is typically observed.

The presented results were obtained thanks to on-chip testing structures. While for single-crystal silicon the loading condition affects the results, for polycrystalline films, if they are not excessively thin, the results are not much affected by the type of loading. Hence, simple tensile test set-ups were designed by Vayrette, Raskin and Pardoen (2015), Knauss, Chasiotis and Huang (2003) and Reedy *et al.* (2011), as reported in Figure 12.8; notches are usually added to induce local stress intensification and so localize the failure mode. Sometimes batches of specimens are actuated together, so as to give more insight into the stress–strain response of polysilicon, up to failure, with a single test. Outcomes of the tests obviously need to be post-processed, as through electrostatic actuation only $C-V$ plots are obtained. Independently of the film morphology, on-chip tests provided good repeatability and can be well interpreted through finite element simulations of the fracture process, explicitly accounting for the crystalline morphology and for the anisotropic properties of each grain, see Figure 12.9 and Corigliano *et al.* (2011), where the polysilicon critical stress intensity factor was found to be $K_{IC} = 1.43$ MPa m$^{1/2}$ by means of a combined experimental and numerical data reduction procedure.

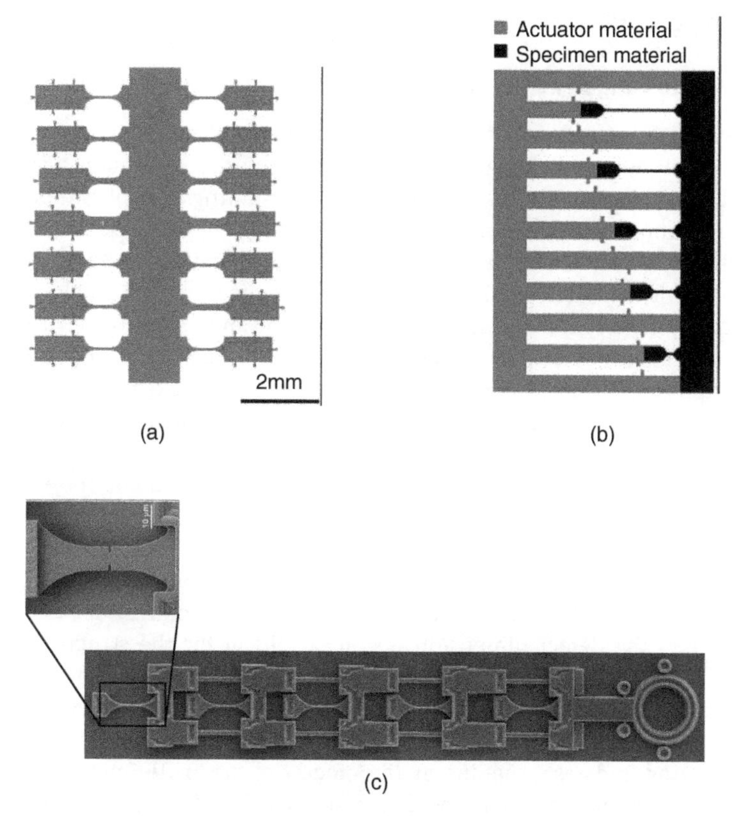

Figure 12.8 SEM of uniaxial, on-chip testing structures for polysilicon films. *Sources*: (a) Knauss, Chasiotis and Huang (2003), Figure 8. Reproduced with permission of Elsevier; (b) Vayrette, Raskin and Pardoen (2015), Figure 1. Reproduced with permission of Elsevier; (c) adapted from Reedy *et al.* (2011), Figures 2 and 13. Reproduced with permission of IEEE.

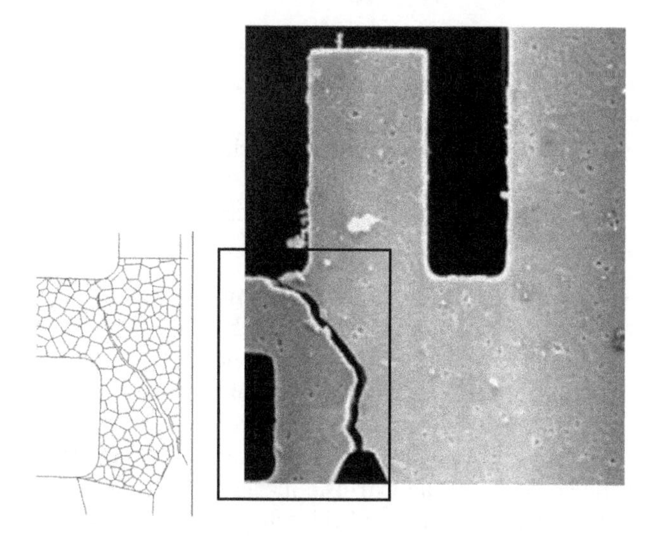

Figure 12.9 Failure of polysilicon obtained through on-chip testing, and relevant numerical prediction based on a cohesive approach. *Source*: adapted from Corigliano *et al.* (2011), Figures 12 and 14. Reproduced with permission of Elsevier.

12.4 Fatigue in Microsystems

12.4.1 An Introduction to Fatigue in Mechanics

The term *fatigue* in engineering refers to the damage and failure of materials under cyclic loads in many different forms. Fatigue can be induced by pure fluctuation in applied stresses or strains, this situation being called *mechanical fatigue*, or it can be coupled with changing temperatures (*thermomechanical fatigue*); the term *corrosion fatigue* refers, instead, to the case when loads are imposed in a chemically aggressive or embrittling environment; repeated loads combined with rolling contact produce *contact fatigue*, or *fretting fatigue*, when the stresses are oscillating and are coupled with a relative motion and sliding between two surfaces (Suresh, 1991). A further distinction is also related to the number of cycles: *low-cycle fatigue* is linked to stress levels able to create irreversible strain (usually plastic strain), while in *high-cycle fatigue* the nominal stresses remain far below the static elastic limit yet still failure occurs.

The decrease in elastic stiffness in the structure under cyclic loads is a commonly observed macroscopic display of fatigue damage; however, today it is widely known that fatigue is correlated to microscale damaging phenomena, namely to the nucleation of defects in an initially undamaged region. This phenomenon is then followed by the stable propagation of a dominant defect until a catastrophic fracture (unstable propagation) occurs, when the defect dimension becomes critical for the structure. It is understandable, therefore, that the term fatigue actually relates to different underlying micromechanisms: historically, it has first been recognized in ductile materials (more precisely in metals), where it is linked to the dislocation motion (slip) causing alternating blunting and resharpening at the crack tip; for brittle materials, instead, kinematically irreversible deformations at the microscopic level induce the fatigue (Suresh, 1991). Progression of fatigue can be therefore subdivided into the following stages:

1) At the microstructural level (or below), some changes induce permanent damage nucleation.
2) Microscopic cracks or flaws, varying in shape, arise.
3) Cracks or flaws grow and coalesce into a dominant feature, which we can call the *dominant crack*, for the sake of simplicity.
4) The dominant crack propagates stably for a very limited (brittle) or a significant (ductile) lifetime for the structure.
5) Structural instability or complete failure arises when a critical crack size has been reached.

To simplify, the first three items in the list refer to the crack initiation phase, the fourth to crack propagation and the last to fatigue failure. Ritchie (1999), mostly reasoning for fatigue observed at the macroscale, suggests that one can distinguish between intrinsic and extrinsic mechanisms. He argues that both of them compete to produce critical and subcritical crack extension: intrinsic damage mechanisms, such as cleavage fracture, microvoid coalescence or other inherent properties of the material, *promote* crack growth ahead of the crack tip, while extrinsic mechanisms *impede* the crack growth, mainly shielding the crack tip from far-field driving forces. Extrinsic mechanisms include crack deflection and meandering and zone and contact shielding.

Therefore, intrinsic mechanisms tend to toughen the material with respect to the crack growth, while extrinsic mechanism tend to decrease the toughness. In case of brittle materials, largely used in the MEMS industry, such as polysilicon, discussed in Section 12.4.2, the role of extrinsic mechanisms is more relevant.

While the phenomena leading to microscopic damage are manifold, from an engineering point of view, a global macroscopic vision is often preferred. Accordingly, it is common to refer to the number of cycles a material can be subjected to before failure, a quantity called *fatigue life*. In principle, fatigue life depends on several factors; by referring to the growth rate of a dominant defect whose size is a with respect to the number of cycles N, we can summarize the dependency as (Ritchie, 1999):

$$\frac{\mathrm{d}a}{\mathrm{d}N} = f(\sigma_{\max}, R, \omega, \text{wave form, environment}), \tag{12.10}$$

where $R = \sigma_{\min}/\sigma_{\max}$ is the stress ratio, i.e. the ratio between minimum and maximum principal stress at the material point in the cycle, and ω is the pulsation of the cycle load. Fatigue life is also related to the wave shape in environmentally assisted low-cycle fatigue (also known as stress corrosion cracking), and to environmental effects such as humidity or temperature. Figure 12.10a shows typical variables used in the description of fatigue, while Figure 12.10b presents the so-called *endurance curve* or *Wöhler curve* (Wöhler, 1860) or 'S–N' curve, i.e. a logarithmic graph of the failure stress versus the number of cycles. A considerable experimental effort is necessary to obtain the data for one S–N curve; typically, some materials (e.g. material A in Figure 12.10b, represented by solid squares and approximated by continuous straight lines) show an endurance limit, i.e. a stress level under which no failure is observed until a very high number of cycles is reached (e.g. 10^9 cycles), while other materials (e.g. material B in Figure 12.10b, represented by empty circles and approximated by dashed lines) do not present such characteristics. Each of the S–N curves can be influenced by the aforementioned variables in Equation 12.10.

Another way to look at fatigue takes account of the stable crack propagation phase, and is described by the so-called Paris–Erdogan law; which, in one of its simplest forms, is

$$\frac{\mathrm{d}a}{\mathrm{d}N} = C\ \Delta K^n, \tag{12.11}$$

where $\Delta K = K_{\max} - K_{\min}$ is the change in the stress intensity factor (see Section 12.1) induced during the cycling load (ranging between the stress levels $\sigma_{\max}$ and $\sigma_{\min}$) and C and n are the model parameters. Figure 12.10c, from Ritchie (1999), shows a typical representation of the Paris–Erdogan law (Paris and Erdogan, 1963), which has been subjected to many modifications over the years. ΔK_{th} is the threshold stress intensity factor necessary for the inception of crack propagation, below which no crack propagation is observed. ΔK refers to a single mode of far-field loading, typically opening mode I: e.g. for an edge-cracked fatigue test specimen, it would be $\Delta K = Y \Delta\sigma \sqrt{\pi\ a}$, with Y a geometrical factor depending on the ratio between the crack length a and the width of the specimen W, and $\Delta\sigma = \sigma_{\max} - \sigma_{\min}$. However, a stress intensity factor range ΔK_{II} or ΔK_{III} could be used in Equation 12.11, instead, for fracture mode II or III.

Paris' law is typically represented in a semi-logarithmic plane, see Figure 12.10c, where the straight line corresponds to the crack propagation phase and is typically obtained by a regression of the experimental data. Figure 12.10c actually depicts three

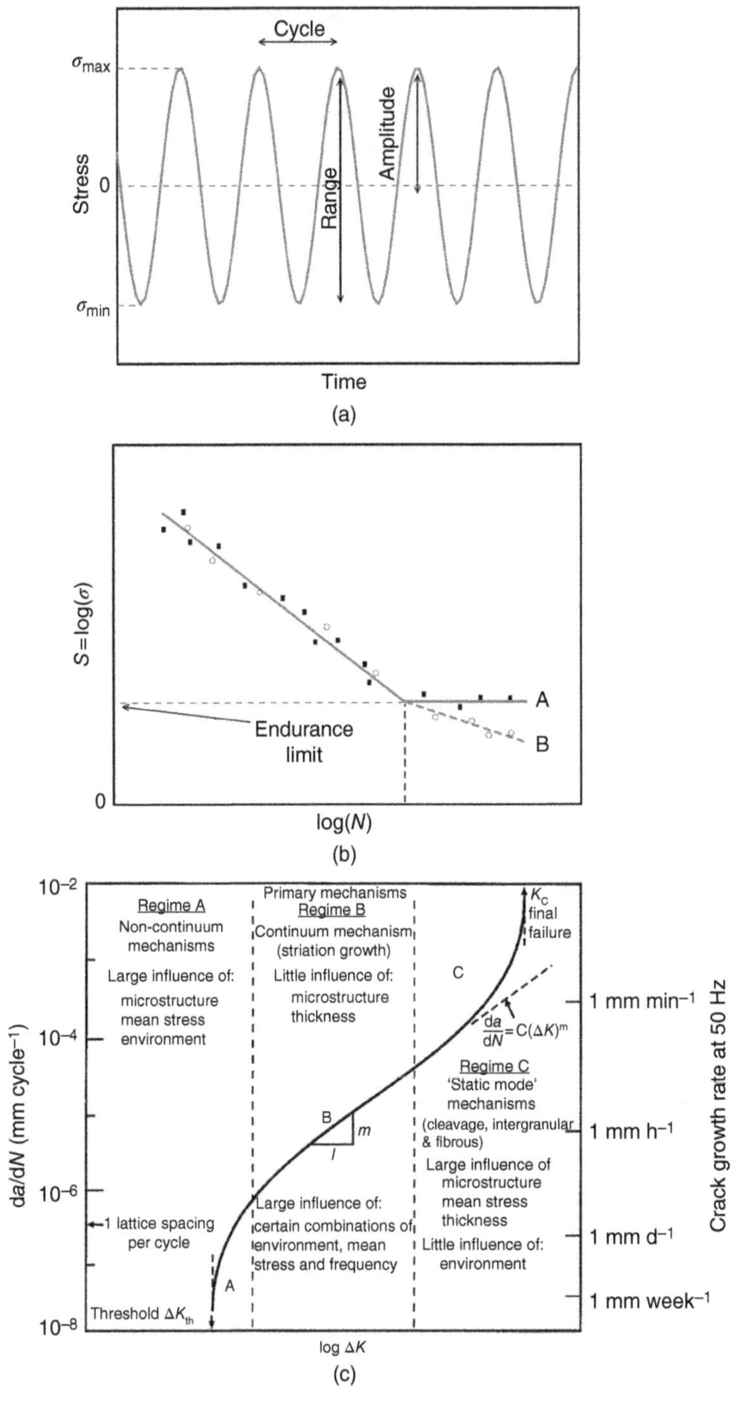

Figure 12.10 (a) Common terms used in mechanical fatigue; (b) *S–N* (Wöhler) curve; (c) Paris' law representation. *Source of image (c)*: Ritchie (1999), Figure 4. Reproduced with permission of Springer.

regimes: in region A, where negligible crack propagation is observed, the relevant features for determining fatigue behaviour are the mean stress inferable from non-continuum mechanics and the environment; in region B, corresponding to stable crack propagation during fatigue cycles, the microstructure effects are shadowed by mesoscale characteristics, captured by continuum mechanics, such as certain combinations of environment, mean stress and frequency; in region C, a catastrophic fracture finally happens, mainly depending on a combination of microstructure, mean stress and film thickness, mimicking a static mode fracture mechanism at the end of the fatigue life.

Different materials are characterized by different values oft ΔK, C and n. For ductile materials, the exponent n is typically between two and four, while for brittle materials n can be much higher (even up to 50) (Ritchie, 1999), so that in the latter case the grow rate sensitivity is more appropriately represented by K_{max} in Equation 12.11 instead of the difference ΔK.

12.4.2 Polysilicon Fatigue

While bulk (i.e. macroscopic) silicon is not affected by fatigue in practice, microscale films are (Connally and Brown, 1992). The underlying reason for the phenomenon is still controversial and, therefore, in this section only the current state of the research is summarized. As recalled, for brittle materials like polysilicon, the crack propagation mechanism due to fatigue is essentially unchanged with respect to the monotonic load condition; conversely, shielding extrinsic mechanisms, such as crack bridging, are degraded from the cycling load, so that the crack-driving force at the crack tip turns out to be larger than the monotonic case.

There are, at present, two different explanations for the fatigue crack propagation phenomenon in polysilicon, the main structural material for MEMS.

A first group of researchers interpret polysilicon fatigue as an environmentally assisted crack propagation within (superficial) silicon oxide layers (Alsem *et al.*, 2008). These layers are almost always present on the polysilicon surface; moreover, they can be induced or, perhaps, enhanced mechanically during the cyclic load. According to this view, the phenomenon has some similarities to (but it is not fully consistent with) stress corrosion cracking in silicon dioxide, and is also known as *reaction-layer fatigue* theory (Muhlstein, Stach and Ritchie, 2002b). Provided the silicon oxide layer is thick enough, a crack due to moisture or enhanced stress can arise inside the layer (environmental stress corrosion cracking is documented for silicon dioxide, see Boroch *et al.*, 2008); if the crack dimension becomes critical for the thin film, then the fatigue is initiated. This reasoning could explain why the fatigue is not observed at the macroscale, where the critical crack size, estimated on the order of tens of nanometres, is never comparable with the silicon oxide layer thickness. Other experimental evidences leading to this interpretation of the fatigue phenomenon in polysilicon are: i) a faster damage accumulation in high-humidity environments with respect to vacuum environments, where slower damage accumulation or no fatigue has instead been observed (Pierron and Muhlstein, 2006), at least for some data series; ii) an increased oxide thickness observed at high-resolution transmission electron microscope after fatigue loading (Muhlstein, Stach and Ritchie, 2002a). It is worth emphasizing that the mechanism of crack inception inside the *reaction layer* remains actually unexplained and the phenomenon does not seem to be related to local heating due to high-frequency loading, since high-resolution infrared imaging excludes any temperature increase larger than 1 K at the fatigue crack tip; hence, it would be a truly mechanical process (Muhlstein, Stach and Ritchie, 2002a).

A second view (known in the literature as the *mechanical* theory) assimilates polysilicon fatigue to a subcritical crack growth of microcracks (as in the usual materials fatigue) due, for example, to the asperities of the notch surfaces that enter into contact during compressive cycles and act as wedges to create areas of local tensions (Kahn *et al.*, 2008). This purely mechanical explanation for polysilicon fatigue is supported by: (i) the strong influence of the load ratio R on fatigue results, independent of air or vacuum environments, for some set of experimental data (Kahn *et al.*, 2002); (ii) the independence of frequency on cycling load (Bagdahn and Sharpe, 2003); (iii) the absence of propagation for sharp cracks under constant tension in high-humidity environments. This second group of researchers did not find evidence of environmental effects and also observed a strengthening effect after (high) cyclic loading (Kahn *et al.*, 2008; Boroch *et al.*, 2008), imputed to the intrinsic compressive stresses during cyclic loadings. This purely mechanical explanation, however, does not suggest a physical mechanism responsible for the crack inception: it could be grain boundary plasticity, as suggested by numerical simulation in nanocrystalline silicon (Demkowicz *et al.*, 2007), but this answer would leave unexplained the monocrystalline silicon case, where fatigue has also been observed (Pierron and Muhlstein, 2006).

Another research group (Allameh *et al.*, 2003; Shrotriya, Allameh and Soboyejo, 2004), focused on the appearance of the polysilicon surface after fatigue cracking. Atomic force microscope scanning in tapping mode of 2 μm × 2 μm and of 5 μm × 5 μm regions in the neighbourhood of the notch evidenced significant changes in surface topography during cycling load. The root mean square roughness in these regions increased as the zone near the bottom of the notch was approached, supposedly because of an interaction between water molecules in the air environment and the superficial silicon dioxide layer, assisted by stress, leading to silica dissolution. The details of the chemomechanical reaction remain unclear; however, these findings seem to corroborate the reaction-layer theory.

Kamiya *et al.* (2011) and Le Huy *et al.* (2012) claim that dislocations could be responsible for damage accumulation during fatigue in silicon, even at room temperature. In standard conditions, dislocations are typically ruled out in silicon except when higher temperatures are involved with respect to room temperature; however, it was found that a transition from brittle to ductile properties could actually be size dependent in silicon (Nakao *et al.*, 2008), and it can happen in the range of the scales adopted for (poly)silicon specimens: dislocations could therefore arise, according to this view, when the specimen size happens to be in the neighbourhood of the aforementioned transition. This could explain some of the ambiguity in the fatigue mechanism, since the dislocations change their nature during the brittle to ductile transition, as described by Shima, Izumi and Sakai (2010), passing from a glide set to a shuffle set dislocation type; the latter type seems to be linked to lower activation energies when both compressive and shear stresses are present. Ramping fatigue tests (see Section 12.4.4) executed by this research team showed that repeated loading induced some damage accumulation even in inert environments; they also claimed that humidity affects, by reducing it, the *static* strength of polysilicon thin films (Kamiya *et al.*, 2011); therefore, some other mechanism is responsible for the fatigue. These experiments can henceforth be ascribed as supportive of the mechanical theory.

Tanemura *et al.* (2012, 2013) observed that the reactive ion-etching procedure actually induces a certain roughness in the sidewall surfaces, determining fracture behaviour near notches. Therefore, the conclusions from experimental tests previously discussed are arguably representative of MEMS devices featuring membranes, such as microphones, pressure sensors and tunable filters. To avoid sidewall effects on experimental fatigue (and fracture) behaviour, Tanemura *et al.* (2012, 2013) proposed a membrane specimen (described in Section 12.4.4), in which the strength of the structure was related to the top membrane surface. By applying Weibull theory combined with Paris' law, (12.11), they correlated the cumulative distribution function F for fracture to the C and n parameters of Paris' law and to the Weibull parameters σ_0 and m (see Section 11.3), namely:

$$F = 1 - \exp\left\{-\left(\frac{\sigma}{\sigma_0}\right)^m \left[1 + \frac{n-2}{2} C \left(\frac{Y\sigma\sqrt{\pi}}{K_{IC}}\right)^2 N\right]^{\frac{m}{n-2}}\right\}. \tag{12.12}$$

With this expression it is possible to relate the fracture probability to the number of cycles to failure, as shown in Figure 12.11 for two membrane thicknesses ($t = 250$ nm and $t = 500$ nm) at different stress levels with respect to the static strength and at increasing relative environmental humidity. According to this reasoning, a lifetime prediction is also possible, leading to the following expression, from Tanemura *et al.* (2012), for the cycles to failure N_C:

$$N_C = 1 + \frac{a_0}{C}\frac{2}{2-n}\left(\frac{\sigma}{\sigma_0}\right)^{-2}\left[1 - \left(\frac{\sigma}{\sigma_0}\right)^{2-n}\right]. \tag{12.13}$$

The fatigue lifetime decreases with increased relative humidity in these tests. In Figure 12.11, it is shown that, for low relative humidity and low stress levels, the fatigue life increases significantly, up to practically infinity ($>10^{12}$ cycles); when the membrane thickness is $t = 250$ nm (left panels); the fatigue life is shorter for $t = 500$ nm (right panels). Figure 12.12 presents S–N curves for the two membrane thicknesses at different relative humidities at a given stress level. They also indicate the small influence of the choice of the reference stress in the Weibull expression, since the initial static strength σ_i turns out to be very close to the σ_0 value obtained at a low cycle number.

12.4.3 Fatigue in Metals at the Microscale

The behaviour of metals for MEMS applications at the microscale is still under discussion; we discuss here, briefly, as an example (taken from Pineau, Benzerga and Pardoen (2016)), only ultra-fine grained and nanocrystalline metals, i.e. when material grain size decreases to the order of hundreds (ultra-fine grained metals) or tens of nanometres (nanocrystalline metals), respectively, as obtained via a severe plastic deformation method. In this case, a low ductility, in the sense of reduced elongation to fracture and ultimate strain, is sometimes experienced with respect to the coarser grain macroscale material. In nanocrystalline metals obtained from severe plastic deformation, damage and fracture are typically intergranular mechanisms, because

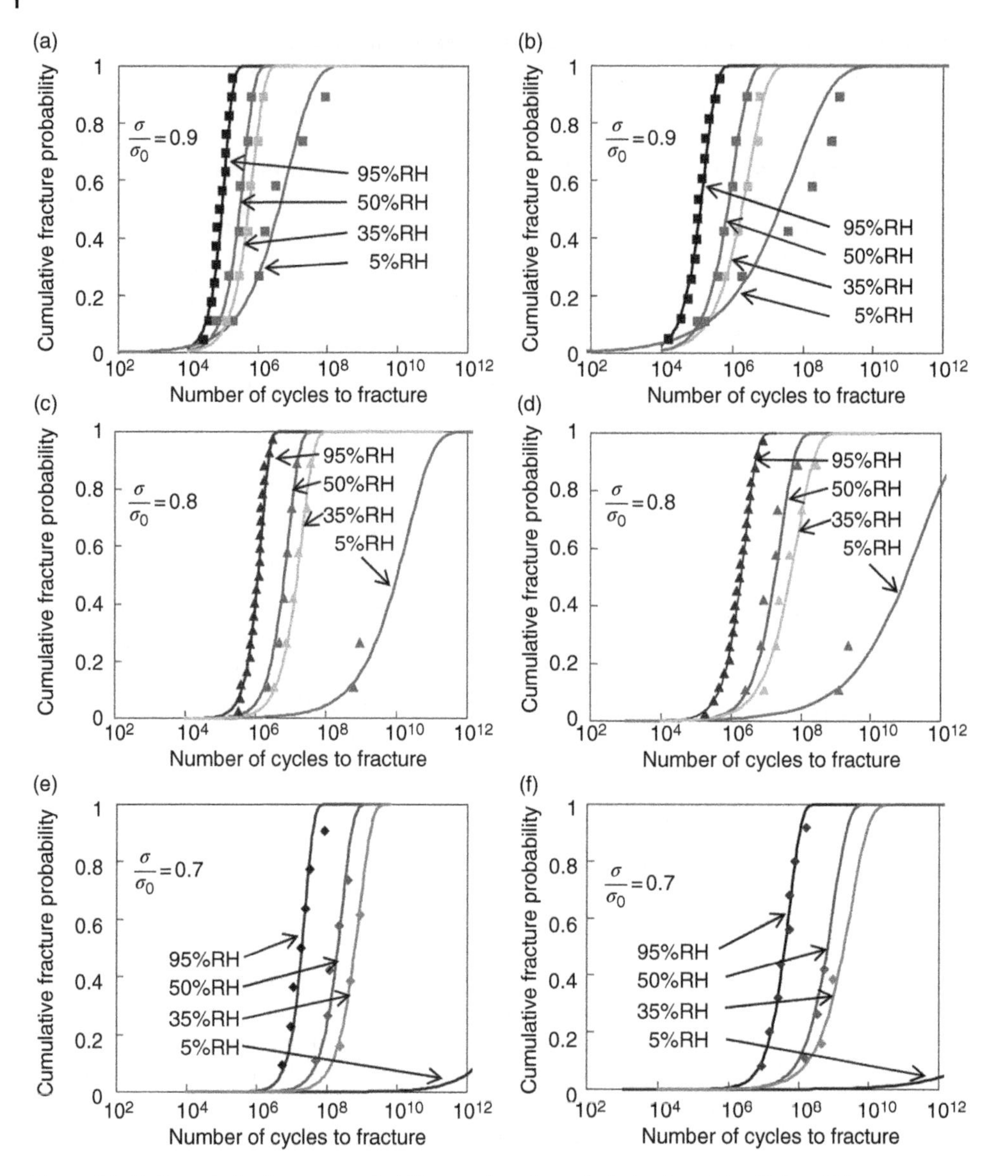

Figure 12.11 Cumulative fracture probability as a function of relative humidity (RH) at different stress level for polysilicon membranes with thickness equal to (left) $t = 250$ nm and (right) $t = 500$ nm. *Source*: Tanemura *et al.* (2013), Figure 11. Reproduced with permission of IOP publishing.

stress concentrations and defects are present at grain boundaries, which represent, in this case, a not negligible fraction of the whole domain. While there are some strategies to recover ductility (e.g. by introducing heterogeneities at multiple scales, by using bimodal or multimodal grain size distributions), this represents an obvious inconvenience and also influences the fatigue behaviour. The presence of defects is actually linked to the process necessary to obtain the fine grains: therefore, results from different sources in literature should be compared with care whenever the production details are not specified. Moreover, because of the severe plastic deformation process,

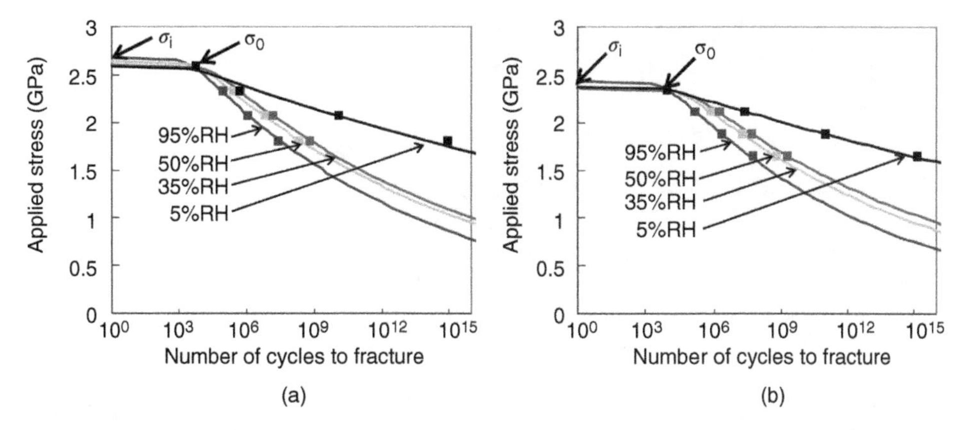

Figure 12.12 *S-N* plots including ramping-stress test results to evaluate initial fracture strength: (a) 250 nm thick polysilicon membrane; (b) 500 nm thick polysilicon membrane. *Source*: Tanemura *et al.* (2013), Figure 13. Reproduced with permission of IOP publishing.

macroscopic shear bands at intervals larger than the grain size can also be the source sites for fatigue cracks. Within this framework, it has been observed (Vinogradov *et al.*, 2001) that the Wöhler curve for ultra-fine grained metals shows a longer fatigue life (i.e. at a very large number of cycles) with respect to the coarse-grained counterpart, but the opposite trend is typically obtained at a small number of cycles. However, in observance of the previous warning, a general *increase* in fatigue life has instead been observed for nanocrystalline metals for all the levels of cycles (Hanlon, Tabachnikova and Suresh, 2005).

A good review of the behaviour of metals at the microscale under fatigue and fracture can be found in Pineau, Benzerga and Pardoen (2016).

12.4.4 Fatigue Testing at the Microscale

Some of the confusion in experimental fatigue results (e.g. in polysilicon) is due to the different adopted techniques and to the influence of the defects left on the external surfaces by the different production technologies. Therefore, some of the experimental approaches leading to the conclusions described in Section 12.4.2 are briefly summarized here. There are two main experimental philosophies for MEMS testing (see Chapter 11): in the on-chip approach, both the test specimen and the testing machine are built into the chip and an electrostatic (or, sometimes, thermal) actuation is adopted; alternatively, an external actuator at a higher scale is applied to a microscale specimen.

A good point to exploit in designing on-chip MEMS configurations for fatigue testing is the observation that resonating structures are very sensible to damage accumulation: after fatigue-induced damage, the resonating frequency of the structure tends to change. Actually, the first observation for polysilicon fatigue was found precisely by inspecting this feature (Connally and Brown, 1992).

Figure 12.13 shows the testing configuration adopted by the main exponents of the reaction-layer theory (Muhlstein, Stach and Ritchie, 2002a); the main components indicated in the figure are recurrent in on-chip fatigue testing: a perforated mass is actuated by comb-drive electrostatics and its rotation is sensed by comb fingers, with the objective of inducing a crack propagation at a notch. In this case, both actuation

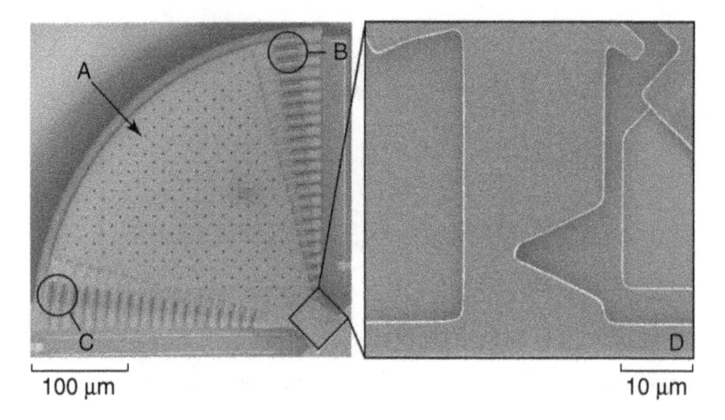

Figure 12.13 On-chip fatigue testing device. A, mass; B, comb drive; C, comb fingers; D, notch. *Source:* Muhlstein, Stach and Ritchie (2002a), Figure 1. Reproduced with permission of AIP Publishing.

and sensing are obtained through in-plane rotational resonance. The structure has a 2 μm thickness and a notch root radius of 1 μm and is fabricated via the MEMSCAP MUMPs process (boron-doped silicon). The idea for this structure derives from Arsdell and Brown (1999) and is also used by Shrotriya, Allameh and Soboyejo (2004) (but on n^+-type polysilicon).

The authors of this book and their research group (Langfelder *et al.*, 2009; Corigliano *et al.*, 2011) use an on-chip device, see Figure 12.14 and Section 11.4.4, based on the

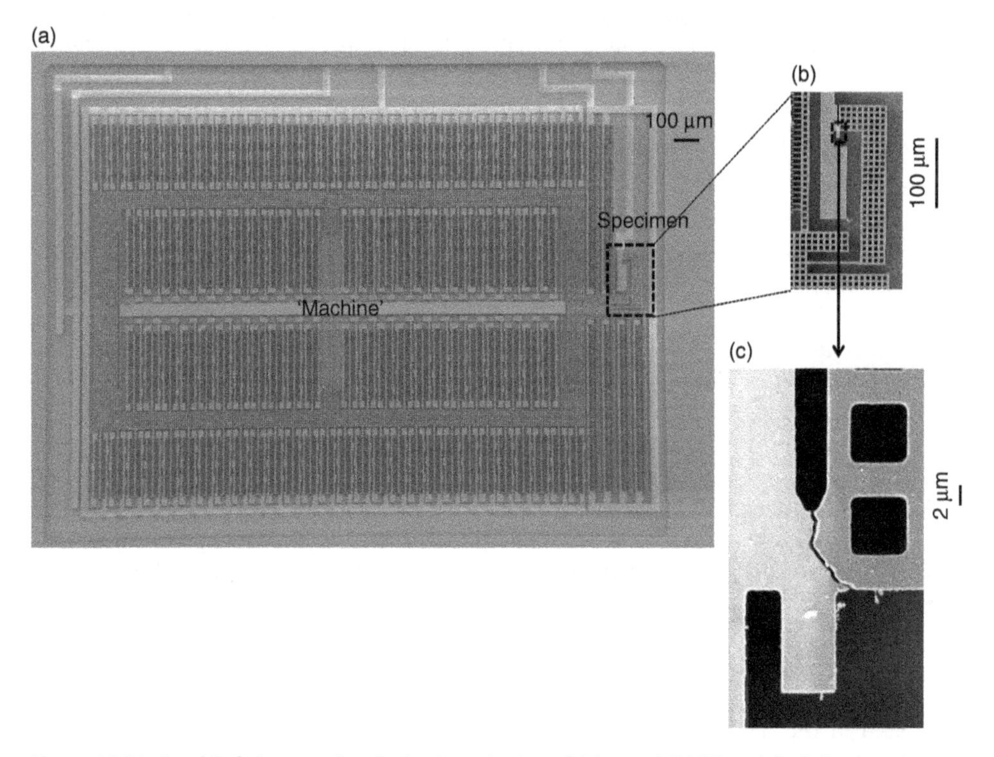

Figure 12.14 On-chip fatigue testing device (see also Langfelder *et al.* (2009) and Corigliano *et al.* (2011)).

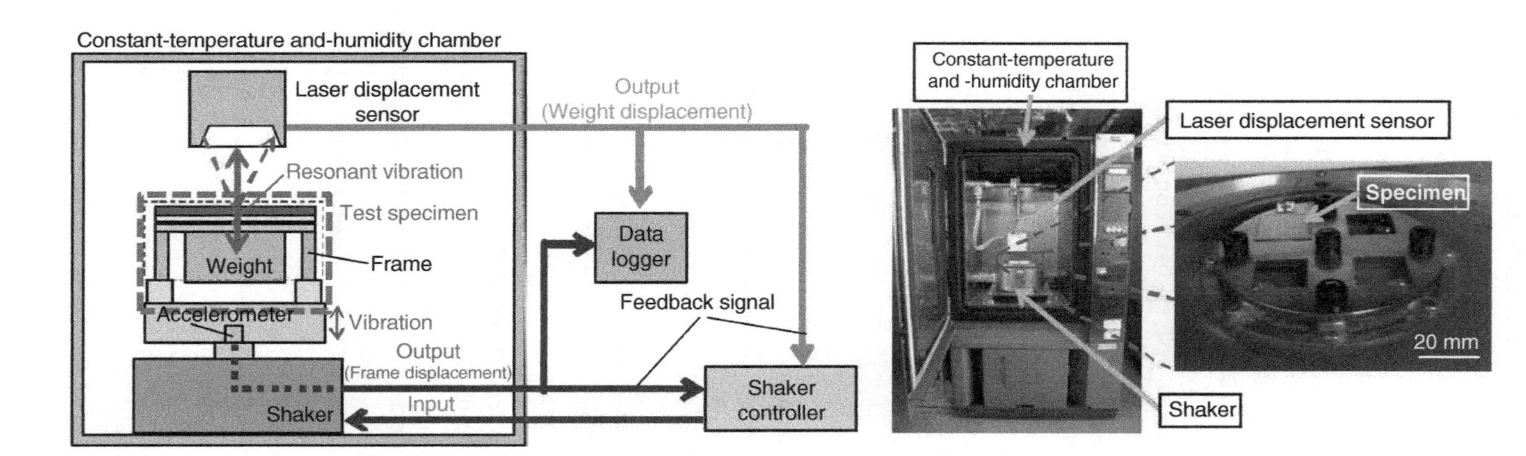

Figure 12.15 Off-chip fatigue testing device. *Source*: Tanemura *et al*. (2013), Figure 1. Reproduced with permission of IOP Publishing.

flexure induced on a notched region by a translation movement (along the horizontal direction in Figure 12.14). The actuator is represented, in this case, by the set of comb fingers acting on the moving mass, which tenses, through a slender beam, the specimen region: since the tension is eccentric, the notched region is subject to a cyclic bending moment, i.e. alternating stresses at the notch tip. The design is inspired by the device shown in Kahn *et al.* (2002).

An example of a thermal actuator for on-chip fracture testing can be seen in Kapels, Aigner and Binder (2000): it exploits an AC electrically powered thermal actuator, whose heat stretches two beams of different length connected to a plate to which a shorter beam is anchored. The displacements are optically measured; the applied frequency is 1 Hz with $R = 0$.

Many other groups use, instead, systems based on actuators at different scales with respect to the microscale specimen. We mention here, for space reasons, only the apparatus used in Tanemura *et al.* (2012, 2013). It consists of a circular membrane restrained at its supports with a weight at the centre, excited by an external shaker (see Figure 12.15). A laser displacement sensor measures the membrane deflection on the opposite side. Since the structure subject to fatigue is a thin membrane, this approach has the merit of avoiding the effects of sidewall imperfections created during the deep reactive ion-etching process.

References

Alfano, G. (2006) On the influence of the shape of the interface law on the application of cohesive-zone models. *Composites Science and Technology*, **66**, 723–730.

Allameh, S.M., Shrotriya, P., Butterwick, A. *et al.* (2003) Surface topography evolution and fatigue fracture in polysilicon MEMS structures. *Journal of Microelectromechanical Systems*, **12**, 31–324.

Alsem, D., Muhlstein, C., Stach, E. and Ritchie, R. (2008) Further considerations on the high-cycle fatigue of micron-scale polycrystalline silicon. *Scripta Materialia*, **59**, 931–935.

Arsdell, W.W.V. and Brown, S.B. (1999) Subcritical crack growth in silicon MEMS. *Journal of Microelectromechanical Systems*, **8**, 319–327.

Bagdahn, J. and Sharpe, W.N. (2003) Fatigue of polycrystalline silicon under long-term cyclic loading. *Sensors and Actuators A*, **103**, 9–15.

Barenblatt, G. (1962) The mathematical theory of equilibrium cracks in brittle fracture. *Advances in Applied Mechanics*, **7**, 55–129.

Bazant, Z.P. and Cedolin, L. (1991) *Stability of Structures*, Oxford University Press.

Boroch, R.E., Müller-Fiedler, R., Bagdahn, J. and Gumbsch, P. (2008) High-cycle fatigue and strengthening in polycrystalline silicon. *Scripta Materialia*, **59**, 936–940.

Broberg, K. (1999) *Cracks and Fracture*, Cambridge University Press.

Broek, D. (1982) *Elementary Engineering Fracture Mechanics*, Martnus Nijhoff Publishers.

Buchheit, T.E. and Phinney, L.M. (2015) Fracture strength characterization for 25 micron and 125 micron thick SOI-MEMS structures. *Journal of Micromechanics and Microengineering*, **25**, 075018.

Buehler, M. (2008) *Atomistic Modeling of Materials Failure*, Springer.

Camacho, G. and Ortiz, M. (1996) Computational modelling of impact damage in brittle materials. *International Journal of Solids and Structures*, **33**, 2899–2938.
Chasiotis, I. and Knauss, W.G. (2003) The mechanical strength of polysilicon films. Part 2: size effect associated with elliptical and circular perforations. *Journal of the Mechanics and Physics of Solids*, **51**, 1551–1572.
Chen, J. and Qiao, Y. (2007) Secondary cracking at grain boundaries in silicon thin films. *Scripta Materialia*, **57**, 1069–1072.
Chen, Y. and Li, P. (2011) *The "Popcorn Effect" of Plastic Encapsulated Microelectronic Devices and the Typical Cases Study*, International Conference on Quality, Reliability, Risk, Maintenance, and Safety Engineering (ICQR2MSE), June 17–19, 2011, Xi'an, China, IEEE.
Cho, S. and Chasiotis, I. (2007) Elastic properties and representative volume element of polycrystalline silicon for MEMS. *Experimental Mechanics*, **47**, 37–49.
Coffman, V.R., Sethna, J.P., Heber, G. *et al.* (2008) A comparison of finite element and atomistic modelling of fracture. *Modelling and Simulation in Materials Science and Engineering*, **16**, 065008.
Connally, J.A. and Brown, S.B. (1992) Slow crack growth in single-crystal silicon. *Science*, **256**, 1537–1539.
Corigliano, A. (1993) Formulation, identification and use of interface models in the numerical analysis of composite delamination. *International Journal of Solids and Structures*, **30**, 2779–2811.
Corigliano, A., Ghisi, A., Langfelder, G. *et al.* (2011) A microsystem for the fracture characterization of polysilicon at the micro-scale. *European Journal of Mechanics – A/Solids*, **30**, 127–136.
Corigliano, A., Mariani, S. and Pandolfi, A. (2003) Numerical modeling of rate-dependent debonding processes in composites. *Composite Structures*, **61**, 39–50.
Corigliano, A., Mariani, S. and Pandolfi, A. (2006) Numerical analysis of rate-dependent dynamic composite delamination. *Composites Science and Technology*, **66**, 766–775.
Demkowicz, M.J., Argon, A.S., Farkas, D. and Frary, M. (2007) Simulation of plasticity in nanocrystalline silicon. *Philosophical Magazine*, **87**, 4253–4271.
Dudek, R., Walter, H., Michel, B. *et al.* (2001) *Studies on Parameters for Popcorn Cracking*, First International IEEE Conference on Polymers and Adhesives in Microelectronics and Photonics, October 21–24, 2001, Potsdam, Germany, IEEE.
Dugdale, D. (1960) Yielding of steel sheets containing slits. *Journal of the Mechanics and Physics of Solids*, **8**, 100–104.
Freund, L. (1990) *Dynamic Fracture Mechanics*, Cambridge University Press.
Friedrich, K. (ed.) (1989) *Application of Fracture Mechanics to Composite Materials*, Elsevier.
Fujii, T., Yamagiwa, H., Inoue, S. and Namazu, T. (2013) *Tension–Torsion–Bending Combined Loading Test Technique for the Reliability of MEMS Structures*, Transducers Eurosensors XXVII: The 17th International Conference on Solid-State Sensors, Actuators and Microsystems, June 16–20, 2013, Barcelona, Spain, IEEE.
Gaither, M.S., Gates, R.S., Kirkpatrick, R. *et al.* (2013) Etching process effects on surface structure, fracture strength, and reliability of single-crystal silicon theta-like specimens. *Journal of Microelectromechanical Systems*, **22**, 589–602.

Hanlon, T., Tabachnikova, E. and Suresh, S. (2005) Fatigue behavior of nanocrystalline metals and alloys. *International Journal of Fatigue*, **27**, 1147–1158.
Irwin, G.R. (1964) Structural aspects of brittle fracture. *Applied Material Research*, **3**, 65–81.
Jadaan, O.M., Nemeth, N.N., Bagdahn, J. and Sharpe, W.N. (2003) Probabilistic Weibull behavior and mechanical properties of MEMS brittle materials. *Journal of Materials Science*, **38**, 4087–4113.
Kahn, H., Avishai, A., Ballarini, R. and Heuer, A. (2008) Surface oxide effects on failure of polysilicon MEMS after cyclic and monotonic loading. *Scripta Materialia*, **59**, 912–915.
Kahn, H., Ballarini, R., Bellante, J.J. and Heuer, A.H. (2002) Fatigue failure in polysilicon not due to simple stress corrosion cracking. *Science*, **298**, 1215–1218.
Kahn, H., Tayebi, N., Ballarini, R. *et al.* (2000) Fracture toughness of polysilicon MEMS devices. *Sensors and Actuators A: Physical*, **82**, 274–280.
Kamiya, S., Ikeda, Y., Gaspar, J. and Paul, O. (2011) Effect of humidity and temperature on the fatigue behavior of polysilicon thin films. *Sensors and Actuators A*, **170**, 187–195.
Kapels, H., Aigner, R. and Binder, J. (2000) Fracture strength and fatigue of polysilicon determined by a novel thermal actuator. *IEEE Transactions on Electron Devices*, **47**, 1522–1528.
Kitano, M., Nishimura, A., Kawai, S. and Nishi, K. (1988) *Analysis of Package Cracking During Reflow Soldering Process*, International Reliability Physics Symposium, April 12–14, 1988, Monterey, CA, IEEE, pp. 90–95.
Knauss, W.G., Chasiotis, I. and Huang, Y. (2003) Mechanical measurements at the micron and nanometer scales. *Mechanics of Materials*, **35**, 217–231.
Kubair, D.V. and Geubelle, P.H. (2003) Comparative analysis of extrinsic and intrinsic cohesive models of dynamic fracture. *International Journal of Solids and Structures*, **40**, 3853–3868.
Landis, C., Pardoen, T. and Hutchinson, J. (2000) Crack velocity dependent toughness in rate dependent materials. *Mechanics of Materials*, **32**, 663–678.
Langfelder, G., Longoni, A., Zaraga, F. *et al.* (2009) A new on-chip test structure for real time fatigue analysis in polysilicon MEMS. *Microelectronics Reliability*, **49**, 120–126.
Le Huy, V., Gaspar, J., Paul, O. and Kamiya, S. (2012) Statistical characterization of fatigue lifetime of polysilicon thin films. *Sensors and Actuators A*, **179**, 251–262.
Li, X., Kasai, T., Ando, T. *et al.* (2004) *Tensile Fracture Behavior of Single Crystal Silicon Film Having a Notch of Sub-micron-length*, 2004 International Symposium on Micro-Nanomechatronics and Human Science, and The Fourth Symposium Micro-Nanomechatronics for Information-Based Society, October 31–November 3, 2004, Nagoya, Japan, IEEE.
Lim, J.H., Lee, K.W., Park, S.S. and Earmme, Y.Y. (1998) *Vapor Pressure Analysis of Popcorn Cracking in Plastic IC Packages by Fracture Mechanics*, 2nd Electronics Packaging Technology Conference, December 10, 1998, Singapore, IEEE.
Mariani, S. (2009) Failure of layered composites subject to impacts: constitutive modeling and parameter identification issues, in *Strength of Materials* (eds G. Mendes and B. Lago),Nova Publishers, pp. 97–131.
Mariani, S., Ghisi, A., Corigliano, A. and Zerbini, S. (2007) Multi-scale analysis of MEMS sensors subject to drop impacts. *Sensors*, **7**, 1817–1833.

Mariani, S., Martini, R., Ghisi, A. *et al.* (2011a) Monte Carlo simulation of micro-cracking in polysilicon MEMS exposed to shocks. *International Journal of Fracture*, **167**, 83–101.

Mariani, S., Martini, R., Ghisi, A. *et al.* (2011b) Overall elastic properties of polysilicon films: a statistical investigation of the effects of polycrystal morphology. *International Journal for Multiscale Computational Engineering*, **9**, 327–346.

Muhlstein, C.L., Stach, E.A. and Ritchie, R.O. (2002a) Mechanism of fatigue in micron-scale films of polycrystalline silicon for microelectromechanical systems. *Applied Physics Letters*, **80**, 1532–1534.

Muhlstein, C.L., Stach, E.A. and Ritchie, R.O. (2002b) A reaction-layer mechanism for the delayed failure of micron-scale polycrystalline silicon structural films subjected to high-cycle fatigue loading. *Acta Materialia*, **50**, 3579–3595.

Mullen, R.L., Ballarini, R., Yin, Y. and Heuer, H. (1997) Monte Carlo simulation of effective elastic constants of polycrystalline thin films. *Acta Materialia*, **45**, 2247–2255.

Nakao, S., Ando, T., Shikida, M. and Sato, K. (2008) Effect of temperature on fracture toughness in a single-crystal-silicon film and transition in its fracture mode. *Journal of Micromechanics and Microengineering*, **18**, 015026.

Paris, P.C. and Erdogan, F. (1963) A critical analysis of crack propagation laws. *Journal of Basic Engineering*, **85**, 528–534.

Pérez, R. and Gumbsch, P. (2000) An *ab initio* study of the cleavage anisotropy in silicon. *Acta Materialia*, **48**, 4517–4530.

Pierron, O.N. and Muhlstein, C.L. (2006) The critical role of environment in fatigue damage accumulation in deep-reactive ion-etched single-crystal silicon structural films. *Journal of Microelectromechanical Systems*, **15**, 111–119.

Pineau, A., Benzerga, A.A. and Pardoen, T. (2016) Failure of metals III: fracture and fatigue of nanostructured metallic materials. *Acta Materialia*, **107**, 508–544.

Reedy, E.D., Boyce, B.L., Foulk, J.W. *et al.* (2011) Predicting fracture in micrometer-scale polycrystalline silicon MEMS structures. *Journal of Microelectromechanical Systems*, **20**, 922–932.

Ritchie, R.O. (1999) Mechanisms of fatigue-crack propagation in ductile and brittle solids. *International Journal of Fracture*, **100**, 55–83.

Rose, J., Ferrante, J. and Smith, J. (1981) Universal binding energy curves for metals and bimetallic interfaces. *Physical Review Letters*, **47**, 675–678.

Rose, J., Smith, J. and Ferrante, J. (1983) Universal features of bonding in metals. *Physical Review B*, **28**, 1835–1845.

Ruiz, G., Pandolfi, A. and Ortiz, M. (2001) Three-dimensional cohesive modeling of dynamic mixed-mode fracture. *International Journal for Numerical Methods in Engineering*, **52**, 97–120.

Shima, K., Izumi, S. and Sakai, S. (2010) Reaction pathway analysis for dislocation nucleation from a sharp corner in silicon: glide set versus shuffle set. *Journal of Applied Physics*, **108**, 063504.

Shrotriya, P., Allameh, S.M. and Soboyejo, W.O. (2004) On the evolution of surface morphology of polysilicon MEMS structures during fatigue. *Mechanics of Materials*, **36**, 35–44.

Suresh, S. (1991) *Fatigue of Materials*, Cambridge University Press.

Tanaka, M., Higashida, K., Nakashima, H. *et al.* (2006) Orientation dependence of fracture toughness measured by indentation methods and its relation to surface energy in single crystal silicon. *International Journal of Fracture*, **139**, 383–394.

Tanemura, T., Yamashita, S., Wado, H., Takeuchi, Y., Tsuchiya, T. and Tabata, O. (2013) Fatigue characteristics of polycrystalline silicon thin-film membrane and its dependence on humidity. *Journal of Micromechanics and Microengineering*, **23**, 035032.

Tanemura, T., Yamashita, S., Wado, H. *et al.* (2012) Fatigue testing of polycrystalline silicon thin-film membrane using out-of-plane bending vibration. *Japanese Journal of Applied Physics*, **51**, 11PA02.

Vayrette, R., Raskin, J.P. and Pardoen, T. (2015) On-chip fracture testing of freestanding nanoscale materials. *Engineering Fracture Mechanics*, **150**, 222–238.

Vinogradov, A., Stolyarov, V., Hashimoto, S. and Valiev, R. (2001) Cyclic behavior of ultrafine-grain titanium produced by severe plastic deformation. *Materials Science and Engineering A*, **318**, 163–173.

Wöhler (1860) Versuche über die Festigkeit der Eisenbahnwagen-Achsen. *Zeitschrift für Bauwesen*, **63**, 233.

Xu, X.P. and Needleman, A. (1994) Numerical simulation of fast crack growth in brittle solids. *Journal of the Mechanics and Physics of Solids*, **42**, 1397–1434.

Yagnamurthy, S., Boyce, B.L. and Chasiotis, I. (2015) Role of microstructure and doping on the mechanical strength and toughness of polysilicon thin films. *Journal of Microelectromechanical Systems*, **24**, 1436–1452.

Yu, C., Pandolfi, A., Ortiz, M. *et al.* (2002) Three-dimensional modeling of intersonic shear-crack growth in asymmetrically loaded unidirectional composite plates. *International Journal of Solids and Structures*, **39**, 6135–6157.

Zhang, G., van Driel, W. and Fan, X. (2006) *Mechanics of Microelectronics*, Springer.

Zhang, Z., Paulino, G. and Celes, W. (2007) Extrinsic cohesive modelling of dynamic fracture and microbranching instability in brittle materials. *International Journal for Numerical Methods in Engineering*, **72**, 893–923.

Zhou, J., Wan, L., Dai, F. *et al.* (2012) *Simulation analysis for interfacial failure of a poymer [sic] sealed MEMS device*, 13th International Conference on Electronic Packaging Technology and High Density Packaging (ICEPT-HDP), August 13–16, 2012, Guilin, China, IEEE.

13

Accidental Drop Impact

13.1 Introduction

Owing to the large use of portable devices, the mechanical failure of a microsystem caused by accidental drops is now recognized as of paramount importance in the design step. Experimental investigations (Varghese and Dasgupta, 2007; Luan *et al.*, 2007; Ghaffarian, 2006) were mainly focused on assessing the effects of impacts on boards or solder joints, as such effects can be dangerous for the throughput yield at packaged or naked die level (Ghisi *et al.*, 2009a). One of the difficulties in ascertaining the aforementioned effects of drops on a MEMS resides in the several length scales involved in the failure process, ranging from millimetres down to tens of nanometres (Srikar and Senturia, 2002; Cheng *et al.*, 2004; Wagner *et al.*, 2001; Li and Shemansky, 2000).

In accordance with standards, see e.g. JESD22-B111 (2003), it is customary to assess the MEMS response to shock loadings from vibration tests characterized by high acceleration peaks, or by drop tests where the device is let fall from a predefined height (Choa, 2005; Tanner *et al.*, 2000). Sensor failure or drifts in working characteristics are adopted as evidences of a problem induced by the drop, often on the basis of a pass–not pass classification. Through the information collected with such test protocols, the failure mechanism can be difficult to determine; hence, numerical simulations are necessary to assess the actual physics of failure.

In this chapter, we will refer to impacts, drops and shocks. These terms all describe phenomena leading to or related to the propagation of stress waves inside the system, but they are not strictly synonymous. In this context, they will be used with the following meanings:

- *Impact* refers to dynamic events characterized by one body coming into contact with (striking) another body.
- *Drop* is used for a free or guided fall of an object from a certain height, generally due to gravity; in real life, most drops are accidental, i.e. undesired.
- The term *shock* is adopted for both a pulse-induced travelling stress discontinuity and a loading condition obtained in tests where a given short-duration, high-g acceleration is imposed.

Moving from this introductory discussion, in Section 13.2 we discuss first the response of a single-dof mechanical system to dynamic loading. The two alternative perspectives

Mechanics of Microsystems, First Edition. Alberto Corigliano, Raffaele Ardito, Claudia Comi, Attilio Frangi, Aldo Ghisi, and Stefano Mariani.

Companion website: www.wiley.com/go/corigliano/mechanics

provided by structural dynamics, accounting for the vibration characteristics of the movable parts of the MEMS, and by wave propagation, accounting instead for the local deformation modes at the waves front, are then comparatively assessed. In Section 13.3, we briefly introduce an equation to provide the severity of a drop or impact on the basis of the induced acceleration peak, and we discuss its capability to classify impact severity according to criticisms available in literature. In Section 13.4, we instead discuss a multiscale procedure, proposed by Mariani *et al.* (2008, 2007), to go beyond the acceleration-based classification of drops, and hence to provide a one-to-one link between drop features and failure occurrence, if any. Results are finally gathered in Section 13.5 for some test cases referring to inertial devices, highlighting the effects of drop characteristics or sensor geometry on the possible failure event.

13.2 Single-Degree-of-Freedom Response to Drops

Li and Shemansky (2000) studied the relevance of impacts for packaged microsystems within the framework of structural mechanics. They considered the equation of motion of a single-dof mechanical system (see Equation 2.22 and Figure 13.1) as representative of the behaviour of an inertial microsystem after the impact.

They assumed that, during a drop:

1) The flight of the device from an initial height to the ground occurs without energy loss due to air drag;
2) Because values of inertia of the package and of the microsystem differ by orders of magnitude (typically four or five), the microsystem does not influence the motion of the package. Through a decoupling of the length scales, the dynamics of the MEMS can be studied in a second phase, by accounting for the results at the package level.

The former hypothesis neglects interaction between air and the falling object and is therefore reasonable only for drops from small heights, such as for mishandling. Accordingly, the computed velocity is an upper bound on the real velocity and can provide a kind of worst-case scenario. The main advantage of this approach is the simplicity in the velocity estimation, independent of the shape and, possibly, of the dynamics of the falling object. By equating the potential energy at the initial position for the device at rest at height h, $\Pi = mgh$ (where g is the acceleration due to gravity field), to the kinetic energy at the impact instant, $K = 1/2\ mv^2$, we obtain:

$$v = \sqrt{2\,g\,h} \tag{13.1}$$

where the same symbols adopted in Section 2.2.3 are used.

The latter hypothesis on length-scale separation appears reasonable, as long as no dissipative mechanisms take place inside the device.

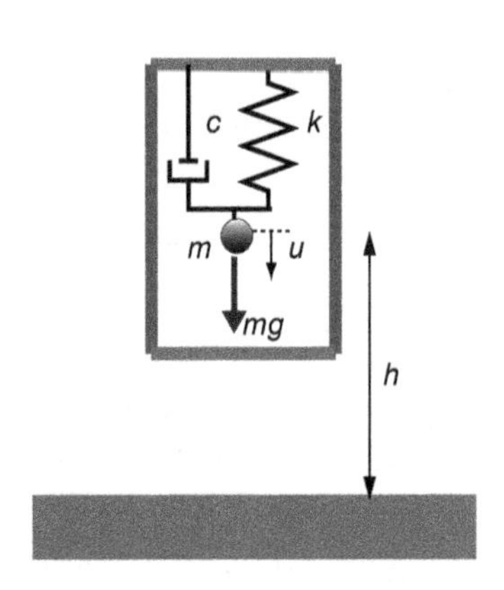

Figure 13.1 One-degree-of-freedom model of a falling microsystem.

The microsystem velocity just after the impact with the target surface is equal to $v^* = -rv$, where r is the restitution coefficient; $r = 0$ if the device sticks to the surface or $r = 1$ for a perfectly elastic rebound. The motion of the moving structure of the MEMS relative to the substrate is governed by Equation 2.31, including the weight term mg. The equation is here rewritten, together with the relevant initial conditions:

$$\ddot{u} + 2\xi\omega\dot{u} + \omega^2 u = g, \qquad u(0) = 0, \qquad \dot{u}(0) = v^*. \tag{13.2}$$

Symbols have the same meaning as described in Section 2.2.3. The response of the system obviously depends also on the damping factor ξ, as described in Equation 2.31. Li and Shemansky (2000) observed that, to obtain an analogous response in terms of maximum displacement, the system supposed to be initially at rest should be excited by a (very high) equivalent acceleration a. The problem would then be rewritten as:

$$\ddot{u} + 2\xi\omega\dot{u} + \omega^2 u = a, \qquad u(0) = 0, \qquad \dot{u}(0) = 0. \tag{13.3}$$

The value of a, obtained for a 1 m drop, depends on ξ and can be of the order of tens of thousands g, i.e. much larger than expected. While the one-dof system is a rough simplification, this observation points out that the problem is not trivial.

Srikar and Senturia (2002) observed that half-sine pulses can be a paradigm for the excitation of microsystems. The shock load should therefore be modelled by specifying the shape, amplitude, duration and spatial orientation of the input acceleration, see Figure 13.2. The rationale is that the nature of the shock load is determined by the interaction of the package with the environment during the impact, independently of the microstructure (as introduced by the second hypothesis given here). The shock load transmitted to the microsystem at the anchors can be described as a combination of acceleration half-sine pulses, each varying in time as:

$$a(t) = \bar{a} \sin\left(\frac{\pi t}{\tau}\right) \qquad \text{for} \quad 0 \leq t \leq \tau, \tag{13.4}$$

$$a(t) = 0 \qquad \text{for} \quad t \geq \tau, \tag{13.5}$$

where $\bar{a}$ is the acceleration amplitude and τ the period of the pulse.

With this approach, three characteristic times involved in the structural dynamic response are accounted for:

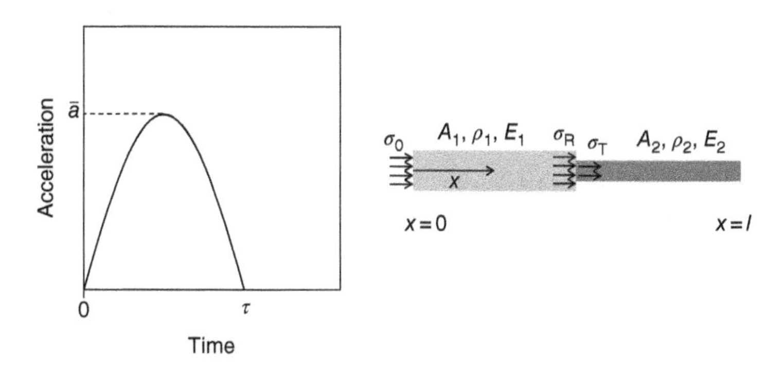

Figure 13.2 Half-sine acceleration (left) and uniaxial wave propagation problem for a two material-geometry domain (right).

- The acoustic traversal time, i.e. the time a dilatational wave needs to move across a characteristic length l of the device (e.g. the length of a beam in the microsystem), $\tau_{\mathrm{A}} = l/c_{\mathrm{d}}$, with $c_{\mathrm{d}} = \sqrt{E^{*}/\rho}$ the dilatational wave speed and

$$E^{*} = \frac{E(1-\nu)}{(1-2\nu)(1+\nu)},$$

where E is the material Young's modulus and ν the Poisson's ratio;
- The period of the sinusoidal excitation τ, see Equation 13.4;
- The vibration mode period T_n of the microsystem nearest to τ.

By comparing τ, τ_{A} and T_n, the dynamic problem can be classified. If τ is less than or of the same order of τ_{A}, a sequence of travelling elastic waves reasonably describes the problem. If τ becomes of the order of one of the vibration periods T_n, the mechanical response becomes oscillatory, as described in Section 2.2.3; in particular, the amplitude of the oscillations could be magnified for high quality factors, i.e. for small damping. If $\tau \gg T_n$, the response becomes quasi-static (the system is overdamped, according to the definition given in Equation 2.31). Table 13.1 provides a quick overview of this rationale. It is worth emphasizing that this discussion assumes an elastic linear behaviour for the system.

Let us now consider the case $\tau \leq \tau_{\mathrm{A}}$, so that a wave propagation problem is addressed. In principle, damage could be generated if the stress level induced by the wave exceeds the material strength. A travelling wave is an impulsive change in velocity, propagating from the boundary where the discontinuity is generated, e.g. the MEMS substrate subjected to $a(t)$: by assuming that the displacement direction coincides with the direction of the propagating wave (such as for dilatational or P-waves), then the system response can be written as the superposition of two counterpropagating, arbitrary displacement signals f_1 and f_2 along the x-axis:

$$u(x,t) = f_1(x - c_{\mathrm{d}}t) + f_2(x + c_{\mathrm{d}}t), \tag{13.6}$$

where c_{d} is the dilatational wave speed, as defined already (Davison, 2008). For uniaxial problems, the stress generated by the wave depends on the local velocity, according to $\sigma = \rho c_{\mathrm{d}} \dot{u}$. If the structure (package) is made of different materials joined together, whenever the wave hits an interface between two media featuring a contrast in their mechanical impedance, it is partially reflected. If a discontinuity occurs, either in terms of the transverse area A_i of the material, or of the relevant mechanical impedance,

Table 13.1 Definition of the problem depending on characteristic times.

Relationship between characteristic times	Problem type
$\tau \leq \tau_{\mathrm{A}}$	Wave propagation
$\tau \approx T_n$	Structural dynamics
$\tau \gg T_n$	Pseudo-static

$Z_i = \rho_i\, c_{di}$, the amplitudes of the original incident σ_I, transmitted σ_T and reflected σ_R stress components are linked through (see Figure 13.2):

$$\sigma_T = \sigma_I \frac{2A_1Z_2}{A_1Z_1 + A_2Z_2}, \qquad \sigma_R = \sigma_I \frac{A_2Z_2 - A_1Z_1}{A_1Z_1 + A_2Z_2}. \tag{13.7}$$

These relationships are obtained by imposing the momentum balance and the velocity continuity at the interface between the joined domains. When $A_2 = 0$, i.e. at a free boundary, the wave is reflected with the opposite sign; so a compressive stress wave becomes tensile. This condition is responsible for the so-called spallation fracture, when the tensile stress at a weak interface or near the boundary can produce a crack. If the variation in the cross-section is smooth instead of abrupt, the stress wave is amplified, as discussed by Srikar and Senturia (2002). Since the dilatational wave speed in silicon is roughly 8 km/s and the characteristic lengths in microsystems range between 100 μm and 1000 μm, the duration of a pulse necessary to induce spalling should be shorter than 10^{-7} s, a condition rather difficult to attain in practice.

In microsystems, the most common problem arising from a drop is characterized by $\tau \equiv T_n$, i.e. the structural response is oscillatory, as described by Equation 2.31. As discussed by Suhir (1997), it is always the stress and not the induced acceleration that is to be inspected in order to assess whether the microsystem can be damaged in case of drops. To justify this assertion, let us consider two systems, as in Suhir (1997). In system A, we have a simply supported beam modelling a slender connection, such as an electrical interconnect; in system B, we have instead a cantilever, such as a micro-sized energy harvester, with a concentrated mass at its free end. In both cases we assume that $\tau = T_0$, i.e. that the pulse duration is equal to the first, fundamental, vibration period of the structure. Moreover, we assume that the initial potential energy is completely transformed into the strain energy related to vibrations in accordance with such a fundamental mode. For a simply supported beam with a circular cross-section of diameter D, the maximum stress and the maximum acceleration turn out to be (Suhir, 1997)

$$\sigma_{\max} = \frac{4}{\sqrt{\pi}}\sqrt{E\rho g h}, \qquad \ddot{u}_{\max} = \frac{\pi^2}{2}\frac{D}{l^2}\sqrt{\frac{Egh}{\rho}}, \tag{13.8}$$

giving

$$\sigma_{\max} = 0.457\rho g \frac{l^2}{D}\frac{\ddot{u}_{\max}}{g}. \tag{13.9}$$

For the cantilever with a concentrated mass P/g at the free end, we have instead

$$\sigma_{\max} = \frac{4}{D}\sqrt{6PEh}, \qquad \ddot{u}_{\max} = \frac{gD^2}{8}\sqrt{\frac{6\pi Eh}{Pl^3}}, \tag{13.10}$$

giving

$$\sigma_{\max} = \frac{32}{\pi}\frac{Pl}{D^3}\frac{\ddot{u}_{\max}}{g}. \tag{13.11}$$

The two structures have maximum stresses actually dependent on the acceleration peak scaled by the gravity acceleration g, see Equations 13.9 and 13.11. However, for the two considered structures, the behaviour will be different, depending on the beam geometry (in terms of, e.g., length or diameter). In particular, to reduce the maximum stress, given the same maximum acceleration, the best solution would be to reduce the beam length for the simply supported beam (see Equation 13.9), or to increase the section diameter for the cantilever (see Equation 13.11).

This brief discussion leaves out many other cases, different for loading and boundary conditions; the reader can find the most common solutions in classical textbooks on structural vibration problems, e.g. Meirovitch (1975) and Weaver, Timoshenko and Young (1990). The important doubly clamped case has been studied in depth by Younis, Alsaleem and Jordy (2007). Nevertheless, in many industrial applications, the microsystem geometry is rather complex and can easily involve deformation modes not triggered by the working condition; moreover, nonlinear effects due to contact with the die or cap surfaces must be considered. The approach based on half-sine loading is an over-simplification, because stress waves propagating inside the package, sensor dynamics and micromechanics-driven failure modes give rise to complex interactions, see Mariani *et al.* (2008).

For this reason, the numerical approach seems to be the most accurate way to deal with the whole problem. In Sections 13.4 and 13.5 we go through the details of a multiscale approach to drop simulations, discussing its strengths and possible weaknesses.

13.3 Estimation of the Acceleration Peak Induced by an Accidental Drop

As already mentioned, the severity of a shock is often customarily defined in terms of the maximum acceleration felt by the sensor. Although such an approach would be accurate to define the input to the system coming from the impact against a deformable or rigid surface as the final outcomes of a drop, it does not prove correct if the said maximum acceleration is considered by itself as the indicator of a possible failure. In fact, such a criterion is not able to distinguish among all the possible falling orientations of the device; on the contrary, it was shown (Suhir, 1997; Ghisi *et al.*, 2009a; Mariani *et al.*, 2007, 2008, 2009) that the acceleration level depends very much on the way the package strikes the target surface. Since the response of the mechanical part of the MEMS depends not only on the input (i.e. on the evolution of the post-impact acceleration), but also on its own geometry, a one-to-one relationship between the acceleration peak and the stress field, which is actually responsible for MEMS failure, cannot be ascertained.

Before moving to the presentation of accurate procedures to assess the critical features of a drop, we briefly review here an estimate of the impact-induced acceleration provided by a simple analytical approach. The stress waves highlighted in Section 13.2 cannot be captured by this single-scale model: a rather low accuracy in terms of failure prediction is therefore expected (Li and Shemansky, 2000; Hauck *et al.*, 2006).

The overall effect of the strike against a target surface is defined in terms of the maximum amplitude of an acceleration pulse, or of the relevant period of variation in time. The former feature can be computed, e.g. according to Hertz theory (Falcon *et al.*, 1998), to obtain

$$\bar{a} \cong \sqrt[5]{\frac{\upsilon^6 R}{\left[m\left(\frac{1-\nu_t^2}{E_t}+\frac{1-\nu_d^2}{E_d}\right)\right]^2}}, \tag{13.12}$$

where υ is the velocity of the device when it strikes the target; m is the mass of the falling device; E_t and E_d are the Young's moduli of the target and device, respectively; ν_t and ν_d are the Poisson's ratios of the target and device, respectively, and R is a characteristic size of the device. With this approach, the package striking the target is considered as a sphere-like body. If handled as a loading condition at the MEMS scale, two major drawbacks can be singled out: (i) since the package length scale is not investigated in detail, the effects of drop orientation are totally missed; (ii) since the MEMS is wildly oscillating just after the impact (usually within a few milliseconds), the acceleration and stress peaks at the movable structure scale are not properly covered by such estimation, which provides too smooth a time evolution of the input.

The approach based on the formulation reported here leads to an underestimate of the maximum acceleration experienced by the sensor. In addition, since failure turns out to be triggered by the stress waves and not by the acceleration thresholds, an objective way to classify shock severity needs to be set accordingly. In Section 13.4, we then discuss a multiscale approach devised to investigate all the features of a drop or impact, down to the sensitivity of the failure process to the morphology of the polysilicon layer constituting the movable parts of the MEMS.

13.4 A Multiscale Approach to Drop Impact Events

Drop- or impact-induced failures of polysilicon MEMS involve processes taking place at several length scales. To simplify the analysis, three main scales can be singled out:

- A macroscopic one, at the whole package level;
- A mesoscopic one, at the sensor (or movable structure) level;
- A microscopic one, dealing with the actual failure process at the polysilicon film level.

The interaction of all the phenomena occurring at the different length scales leads to failure mechanisms characterized by the cracking of polysilicon in regions quite close to anchors, where an abrupt variation of the geometry (for example, in terms of the width of the film) exists, see Figure 13.3.

In what follows, some details are given for the analyses to be carried out. As a reference, in Figure 13.4, the geometry at the macro- and mesoscales of an inertial device (uniaxial accelerometer) is shown; this example will be further considered in Section 13.5 to provide some insights on the physics of the failure event.

13.4.1 Macroscale Level

At this scale, the analysis has to deal with the falling device striking the flat surface of a target body. The solution may be rather different, depending on the falling orientation,

Figure 13.3 Failure of a suspension spring of a MEMS accelerometer. *Source*: Hauck *et al.* (2006), Figure 2. Reproduced with permission of IEEE.

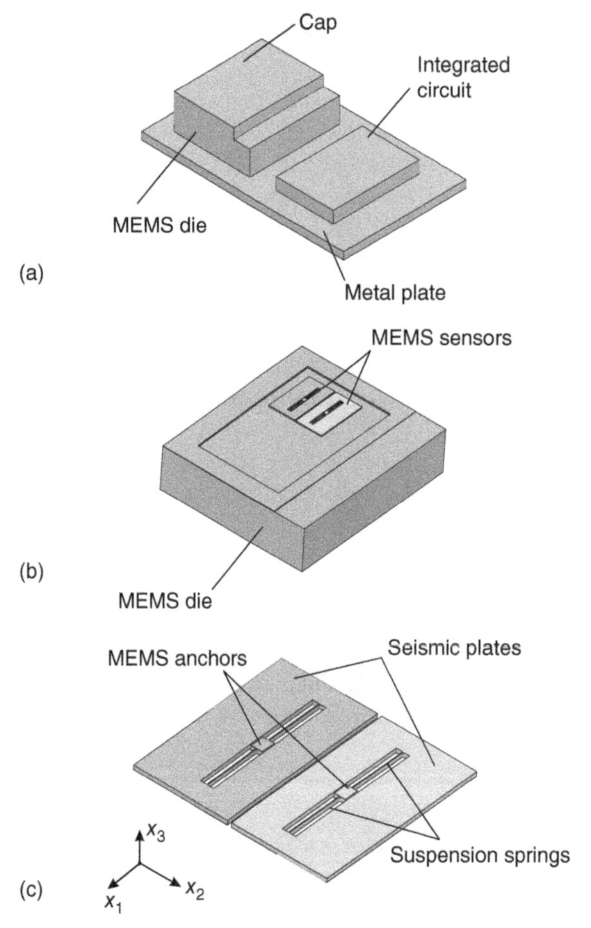

Figure 13.4 Reference device (uniaxial accelerometer):
(a) mould-free package; (b) uncapped die; (c) movable MEMS structure.
Source: Mariani *et al.* (2011a), Figure 1. Reproduced with permission of Nova Science Publishers.

e.g. depending on whether the device strikes the target with one of its outer surfaces or with a corner. Locally, the waves caused by the impact can induce different dynamics of the movable parts; also, the motion of the whole device after the impact while bouncing off the target is rather affected (Mariani *et al.*, 2007, 2011a).

The mentioned motion of the whole device is characterized by a long time scale (typically hundreds of microseconds or milliseconds); its effect on sensor failure can then be superseded by concurrent, much faster, phenomena linked to stress wave propagation inside the body.

13.4.2 Mesoscale Level

At this scale, vibrations of the movable parts of the MEMS induced by the interaction with the die or package at the anchors have to be analysed. As the suspended structure is shocked, its interaction with the surrounding fluid in the MEMS cavity is typically handled through a damping term, as described in Chapter 15. Moreover, possible contact with the inner die and cap surfaces needs to be accounted for, to provide an accurate estimate of the stress field in the possibly failing regions.

To model the dynamics of the polysilicon film, its effective mechanical properties must be considered. Mariani *et al.* (2007, 2009, 2008) and Ghisi *et al.* (2009a) proposed using values of elastic moduli and tensile (fracture) strength obtained through homogenization of representative volume elements of the film. This assumption leads to an additional issue, which is typically avoided at the macroscale level: statistical dispersion of the results (e.g. in terms of failure probability) in the most stressed film regions has a characteristic size comparable to that of the aforementioned representative volume elements (Mariani *et al.*, 2011d). It turns that the films can be considered as transversely isotropic materials at this scale, with the axis of transverse isotropy aligned with the preferential direction of crystal growth of the columnar polysilicon; hence, basically perpendicular to the substrate surface.

13.4.3 Microscale Level

At the microscale level, failure mechanisms need to be investigated in detail, fully accounting for the micromechanics of cracking in polysilicon and for the dispersion of results induced by it. Since the analysis must account for the morphology of the polycrystalline film through appropriate representations of it, the hypothesis of homogeneous bodies no longer holds true (Zavattieri and Espinosa, 2001; Espinosa and Zavattieri, 2003; Cho and Chasiotis, 2007; Boroch *et al.*, 2007).

A representative volume element of the polysilicon film is therefore tessellated, as depicted in Figure 13.5, where the two in-plane axes of elastic symmetry of each crystal grain are also reported (while the third axis is assumed to be perpendicular to the substrate). The grains are obtained via a (possibly regularized) Voronoi tessellation of the body, whereas lattice orientation is handled as a statistical variable sampled from a uniform distribution. Crack inception and growth is then simulated through a cohesive approach, as reported in Chapter 12.

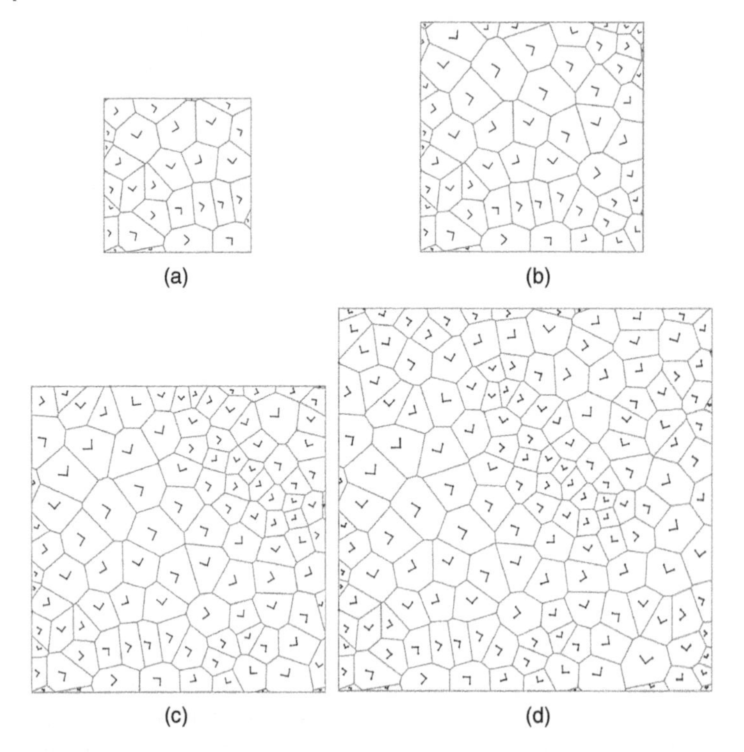

Figure 13.5 Grain morphologies for varying representative volume element size, L (average grain size 1 μm): (a) $L = 8$ μm; (b) $L = 12$ μm; (c) $L = 16$ μm; (d) $L = 20$ μm. *Source*: Mariani *et al.* (2011b), Figure 7. Reproduced with permission of Elsevier.

13.5 Results: Drop-Induced Failure of Inertial MEMS

In this section, we go through some results provided for the device shown in Figure 13.4. This uniaxial, z-axis (or out-of-plane) accelerometer is constituted of twin plates (see Figure 13.4c), each supported by a pair of slender polysilicon beams. Owing to its asymmetric geometry, the working mode of the device basically involves the torsional deformation of the beams, with moderate out-of-plane bending only in case of high-g excitations (Ghisi *et al.*, 2009b, 2012); the proof mass is instead designed to be displaced as a rigid body under loading. The two beams are connected at the anchor point to the die, as depicted in Figure 13.4(b); in turn, the die is placed on top of the metal plate, as shown in Figure 13.4(a).

In case of impact, waves travel inside the heterogeneous package until they hit the anchor point, and so a shaking of the movable structure is triggered. Although not reported in the pictures, stoppers can be placed all around the plate to prevent it from being displaced too far from its configuration at rest. In dynamics, the (contact) interaction of the polysilicon film with die or cap surfaces leads to the formation of further waves, which next interact with the dominant waves directly linked to the impact against the target. Keeping the elastic state as a reference, constraints on the motion of the plate–beam system relative to the substrate (or anchor) would reduce the deformation and stress fields in polysilicon and, ultimately, the failure probability under the given loading conditions.

The effect of tilting, i.e. of an imperfectly parallel strike of the falling device against the target surface, is reported in Figure 13.6. Here, tilting is measured by a rotation about axis x_2 only; as already noted in Section 13.4, the impact type can be further described by a rotation about axis x_1 if the strike happens at one corner only. The plot provides the time evolution of the envelope of the stress state in the beams; this envelope was built by keeping, at each time instant, the maximum value of the maximum principal stress all over the two springs. Through a comparison of this envelope with the polysilicon tensile strength, the time to failure and also the location of failure inception (not covered by this graph alone) can be assessed. It must be noted that, as detailed in Chapter 11, the strength of polysilicon cannot be set deterministically, as it depends on polycrystalline morphology (not considered in the analysis at this length scale), the type of loading and the size of the failing region; hence, the horizontal line depicted in the graph of Figure 13.6 for $\sigma = 4$ GPa must be considered to represent either a reference value (such as σ_{nom0} in Chapter 11) or, in a less conservative fashion, the value corresponding to an almost certain local failure of the material. In

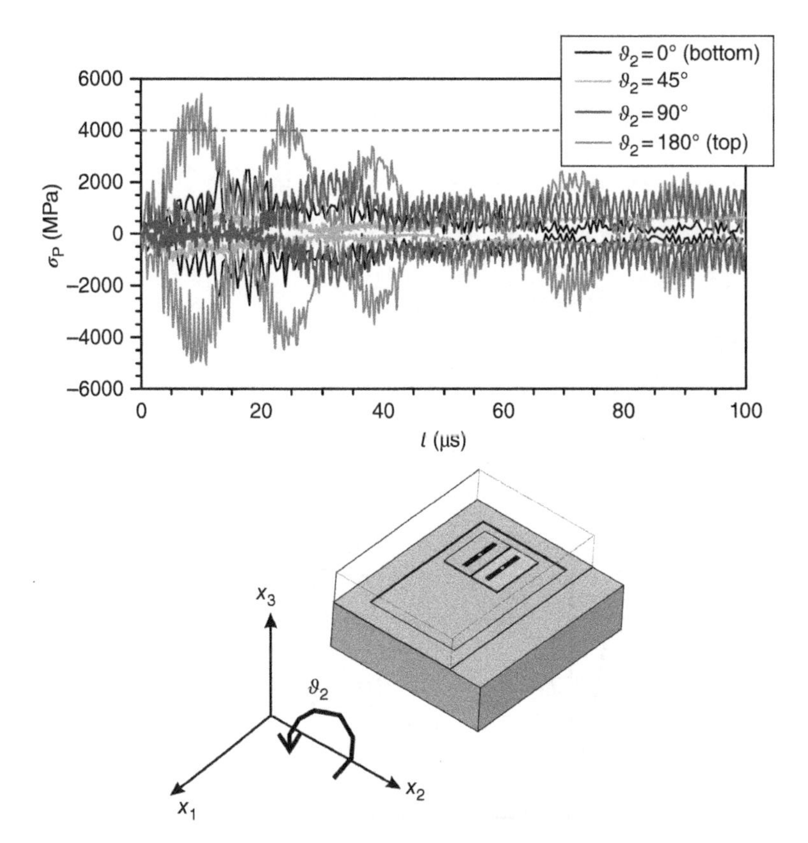

Figure 13.6 Effect of tilting on the stress envelope in the suspension springs. *Source*: adapted from Mariani *et al.* (2011a), Figure 14. Reproduced with permission of Nova Science Publishers. (*See color plate section for the color representation of this figure.*)

these analyses, owing to the brittleness of silicon (Chasiotis and Knauss, 2003; Cho, Jonnalagadda and Chasiotis, 2007), it can be assumed that, as soon as the tensile strength of the material is locally attained anywhere in the system, a cracking mechanism is incepted and unstably propagates to cause the failure of the movable structure as a whole.

The (Rankine-type) local failure criterion based on the maximum principal stress was adopted to avoid any micromechanical analysis at the polysilicon level. As stated in Section 13.4.2, the considered mechanical properties of the film then have to be tailored so as to represent the effective, or homogenized, properties. Such an approach can lead to quite accurate, yet fast to compute, solutions as far as average or representative features of the impact-carrying capacity of the system are concerned; if the spreading of the results due to polysilicon is instead of concern, microscale simulations need to be resorted to. An example is reported in Figure 13.7, where the region of the MEMS experiencing the higher stress intensification in case of drops is highlighted (at the suspension spring–anchor connection), and a digital polycrystalline morphology obtained in accordance with the discussion of Section 13.4.3 is also depicted. At the microscale, proper boundary conditions are to be adopted to reflect how the small portion of the device is deformed and stressed by the relative motion of the spring or plate and the substrate (Mariani *et al.*, 2011c).

Some representative results, in terms of the evolution of the cracking pattern in the highly stressed region of the movable structure, are shown in Figure 13.8. Independently of the polycrystalline morphology, the crack turned out to be always incepted at a

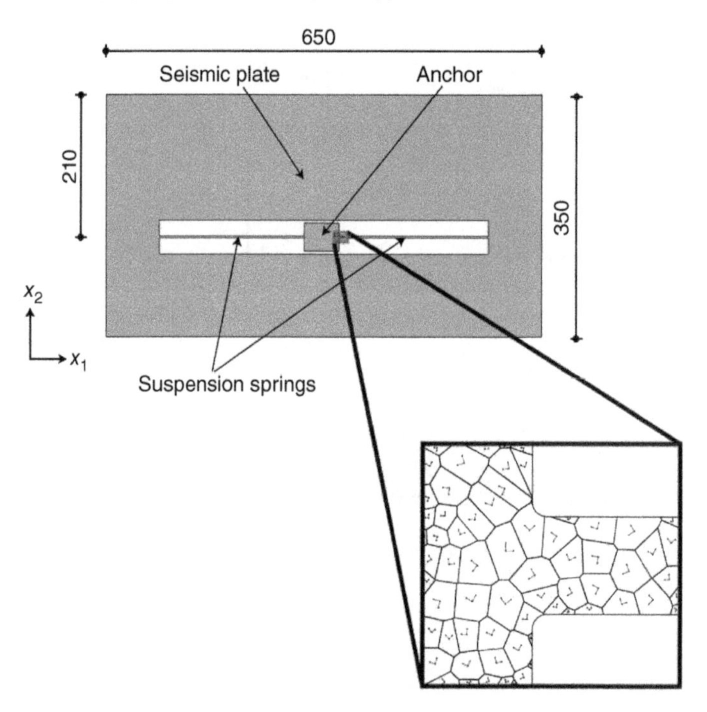

Figure 13.7 MEMS accelerometer; the enlarged panel shows the region prone to failure as studied in the micromechanical analysis. *Source*: Mariani *et al.* (2011c), Figure 7. Reproduced with permission of Springer.

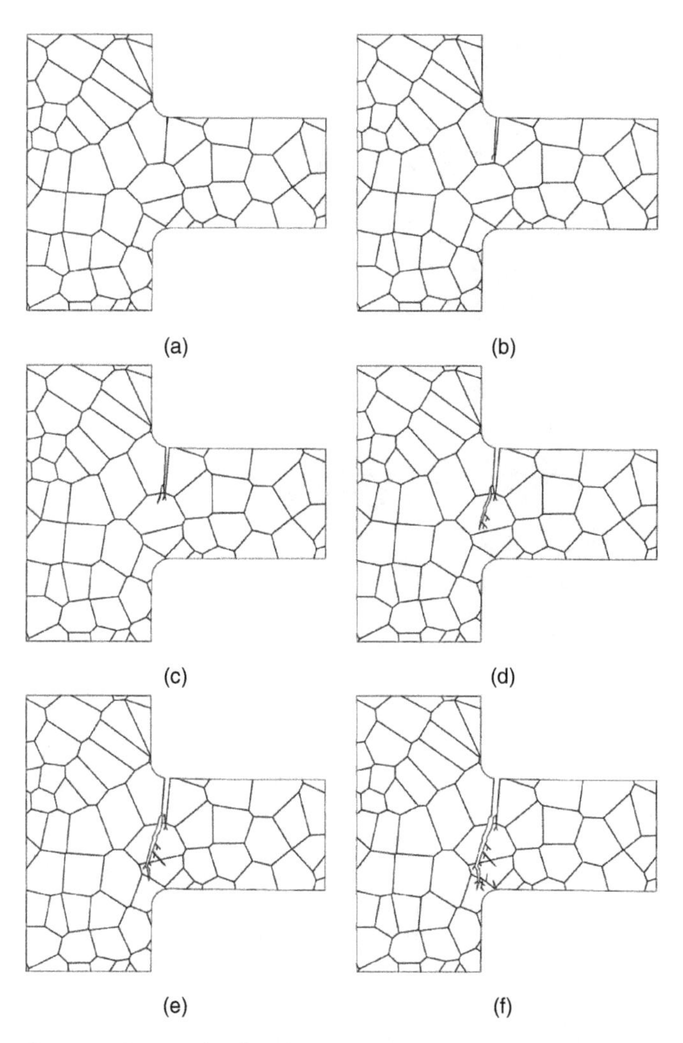

Figure 13.8 Example of crack propagation, up to percolation, in the most stressed region of the polysilicon film. *Source*: Mariani *et al.* (2011c), Figure 13. Reproduced with permission of Springer.

re-entrant corner; next, it was reported to grow almost straight ahead, kinking slightly at grain boundaries (owing to shielding or attraction phenomena, respectively, in the case of intergranular strength or toughness values greater or smaller than than the relevant transgranular values) but eventually leading to a percolation path towards the opposite re-entrant corner. This type of failure mechanism is reported whenever stress intensification is locally caused by the mentioned re-entrant corners in the in-plane structural geometry, often at the junction of thin supporting springs and massive seismic plates or anchors. To avoid biasing the results by a specific realization of the film morphology, Mariani *et al.* (2011c) proposed a Monte Carlo analysis at this microscale: a series of simulations, each featuring its own crystal morphology under the same loading conditions, was collected until convergence in the statistics of the failure mode (in terms of occurrence and topology) was achieved. The outcome is reported

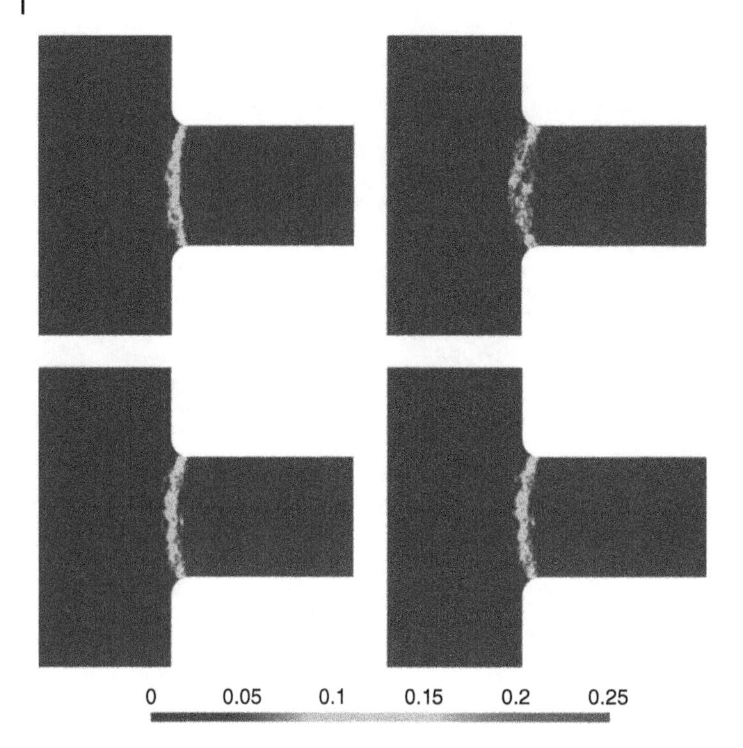

Figure 13.9 Statistical forecasts of microcracking pattern at failure. Top row: drop case A; bottom row: drop case B. Left column: perfect grain boundary case; right column: defective grain boundary case. *Source*: Mariani *et al.* (2011c), Figure 15. Reproduced with permission of Springer.

in Figure 13.9, where the level sets provide the local probability of cracking (namely, the probability that a crack passes through that point, if failure occurs) induced by the drops. Two different falling orientations were considered, see also Figure 13.7: drop case A is characterized by a fall aligned with axis x_1, while drop case B is characterized by a fall aligned with axis x_2. Furthermore, the case of defective polysilicon, featuring an intergranular strength less than the transgranular strength, was compared with the perfect polysilicon case. It was reported that fall orientation and grain boundary strength marginally affect the shape of the failure mode, which, in accordance with the discussion here, almost directly connects the two re-entrant corners at the opposite sides of the film.

Three-dimensional finite element analyses with the same rationale at the microscale, i.e. including an artificial Voronoi tessellation to represent the grain morphology, can currently also be carried out, see Figure 13.10(a). In this case, because of the aforementioned loading conditions and geometry, the conclusions are not different with respect to the 2D case. It is worth emphasizing that, however, the computational burden is significantly increased and more sophisticated numerical approaches must be invoked. In the example shown in Figure 13.10, a numerical technique called spatial domain decomposition was pivotal to obtaining results in reasonable times: in Figure 13.10(d), six blocks of elements have been defined and are sent to different processors during the calculation. More details are available in Confalonieri *et al.* (2014).

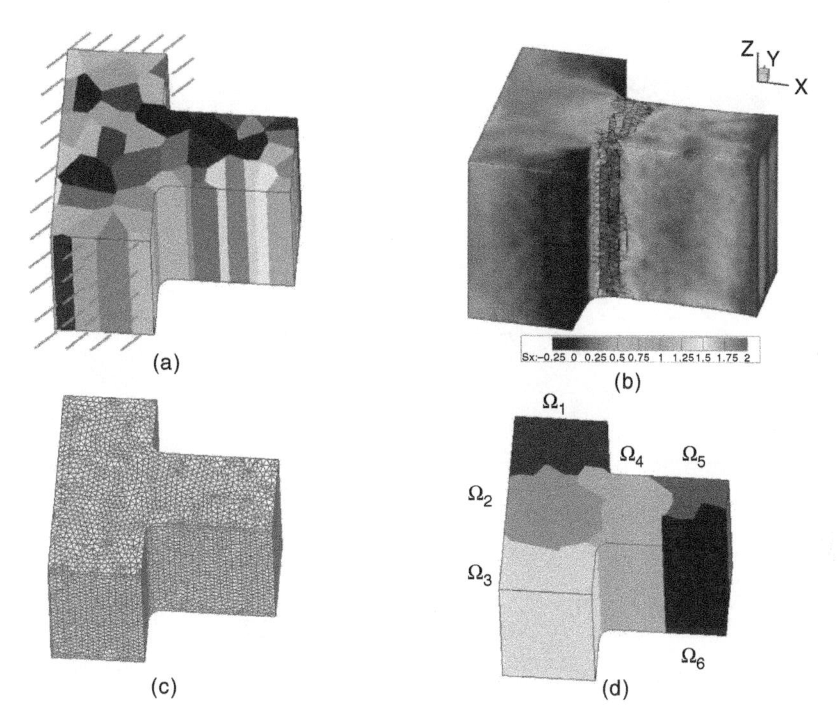

Figure 13.10 (a) 3D Voronoi tessellation of the most stressed polysilicon region; (b) longitudinal stress (in MegaPascals) and crack pattern under tensile loading directed along the *x*-direction; (c) example of discretization; (d) spatial decomposition in six domains. (*See color plate section for the color representation of this figure.*)

To gain more insight into the dynamics of the movable structure subjected to drops, and to understand the links between its motion and the stress field in the springs, Ghisi *et al.* (2009a) investigated the effect of packaging. Two different fall orientations were considered: the bottom drop is related to the case when the package or die strikes the target with its bottom surface; the top drop is instead related to a system falling up-side down, and so striking the target with its top surface (hence, with a rotation by 180° about axis x_2, as stated in Figure 13.6).

Figure 13.11 shows the time evolution of the relative displacement between the plate corners and the die or cap surfaces (represented by the bottom and top horizontal dashed lines). It was shown that the drops (not only the one considered in this graph) induce a major shaking linked to the working (torsional-dominated) mode of the suspension springs. Torsion is here evidenced by the different motions of points A and B (grey lines) and of points C and D (black lines). On top of this fundamental structural vibration, some additional higher frequency modes are excited by the spectrum of acceleration at anchor, as induced by the waves impinging upon that point. It is also shown that, owing to the motion of the whole package while bouncing off the target, the reported relative displacement components are not damped out within a few microseconds after the impact, but are instead further enhanced after around 75 μs.

As emphasized in the preceding sections, since polysilicon fails because of the attainment of a local stress state exceeding its strength capacity, the stress field must

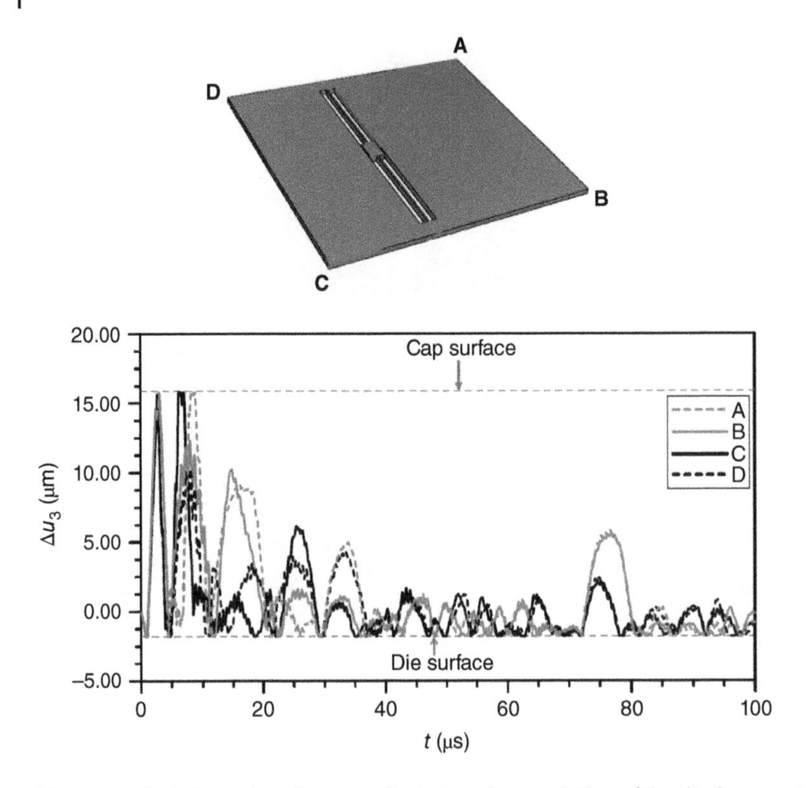

Figure 13.11 Package drop, bottom orientation: time evolution of the displacement at plate corners relative to die or cap surfaces. *Source*: adapted from Ghisi *et al.* (2009a), Figure 5. Reproduced with permission of Elsevier. (*See color plate section for the color representation of this figure.*)

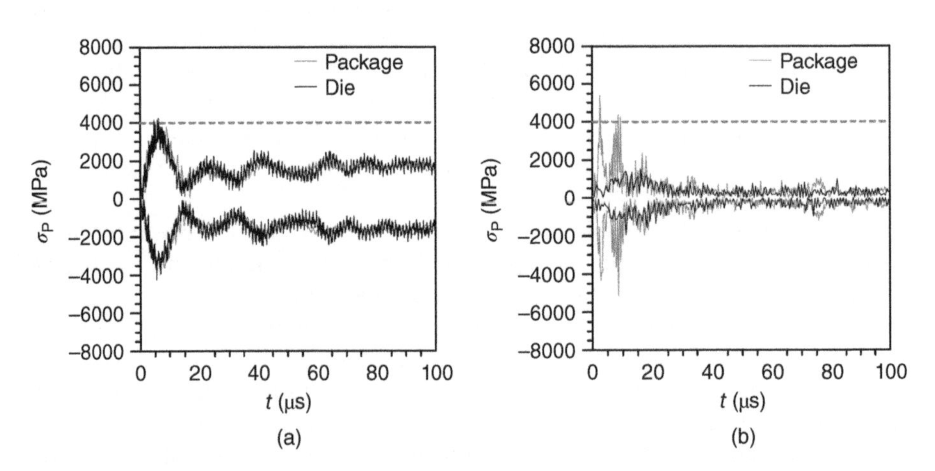

Figure 13.12 Comparison between stress envelopes in the suspension springs, as induced by die and package drops: (a) top drop; (b) bottom drop. *Source*: adapted from Ghisi *et al.* (2009a), Figure 7. Reproduced with permission of Elsevier.

be inspected in detail. As reported in Figure 13.12(a), a top drop from a height of 1.5 m will cause a failure of the device independently of the presence of a package, since the maximum tensile stress attained in the beams at around $t = 10$ μs is not affected much by it and reaches the reference threshold of 4 GPa (the curves of the case with package, displayed in grey, are almost superposed on those of the case without package, displayed in black). The other way around, from Figure 13.12(b) it emerges that, for a bottom drop from the same height, the peak stress values in the packaged device (grey lines) are larger than those experienced by a device supported by a naked die only. As explained by Ghisi *et al.* (2009a), such a counterintuitive result (as the package is supposed to protect the device, not only from the outer environment) is because, in the absence of appropriately designed stoppers, the different motions of the package and die while bouncing off the target provide a higher shaking of the plate and so critical conditions can be met in the beams.

Although the results collected in this section refer to a specific geometry, they can have a general validity for inertial sensors, since shocks and drops provide a frame within which inertial forces dominate the solution, and can give rise to unpredicted system failures.

References

Boroch, R., Wiaranowski, J., Müeller-Fiedler, R. *et al.* (2007) Characterization of strength properties of thin polycrystalline silicon films for MEMS applications. *Fatigue & Fracture of Engineering Materials & Structures*, **30**, 2–12.

Chasiotis, I. and Knauss, W.G. (2003) The mechanical strength of polysilicon films. Part 2: size effect associated with elliptical and circular perforations. *Journal of the Mechanics and Physics of Solids*, **51**, 1551–1572.

Cheng, Z., Huang, W., Cai, X., Xu, B., Luo, L. and X. Li (2004) *Packaging Effects on the Performances of MEMS for High-g Accelerometer: Frequency Domain and Time-Domain Analyses*, Proceeding of the Sixth IEEE CPMT Conference on High Density Microsystem Design and Packaging and Component Failure Analysis (HDP'04), July 3, 2004, Shanghai, China, IEEE, Shanghai, China.

Cho, S., Jonnalagadda, L. and Chasiotis, I. (2007) Mode I and mixed mode fracture of polysilicon for MEMS. *Fatigue & Fracture of Engineering Materials & Structures*, **30**, 21–31.

Cho, S.W. and Chasiotis, I. (2007) Elastic properties and representative volume element of polycrystalline silicon for MEMS. *Experimental Mechanics*, **47**, 37–49.

Choa, S.H. (2005) Reliability of vacuum packaged MEMS gyroscopes. *Microelectronics Reliability*, **45**, 361–369.

Confalonieri, F., Ghisi, A., Cocchetti, G. and Corigliano, A. (2014) A domain decomposition approach for the simulation of fracture phenomena in polycrystalline microsystems. *Computer Methods in Applied Mechanics and Engineering*, **277**, 180–218.

Davison, L. (2008) *Fundamentals of Shock Wave Propagation in Solids*, Springer-Verlag.

Espinosa, H. and Zavattieri, P. (2003) A grain level model for the study of failure initiation and evolution in polycrystalline brittle materials. Part I: theory and numerical implementation. *Mechanics of Materials*, **35**, 333–364.

Falcon, E., Laroche, C., Fauve, S. and Coste, C. (1998) Collision of a 1-D column of beads with a wall. *European Physical Journal B*, **5**, 111–131.

Ghaffarian, R. (2006) CCGA packages for space applications. *Microelectronics Reliability*, **46**, 2006–2024.

Ghisi, A., Fachin, F., Mariani, S. and Zerbini, S. (2009a) Multi-scale analysis of polysilicon MEMS sensors subject to accidental drops: effect of packaging. *Microelectronics Reliability*, **49**, 340–349.

Ghisi, A., Kalicinski, S., Mariani, S. *et al.* (2009b) Polysilicon MEMS accelerometers exposed to shocks: numerical–experimental investigation. *Journal of Micromechanics and Microengineering*, **19**, 035023.

Ghisi, A., Mariani, S., Corigliano, A. and Zerbini, S. (2012) Physically-based reduced order modelling of a uni-axial polysilicon MEMS accelerometer. *Sensors*, **12**, 13985–14003.

Hauck, T., Li, G., McNeill, A. *et al.* (2006) *Drop Simulation and Stress Analysis of MEMS Devices*, Proceedings of the International Conference on Thermal, Mechanical and Multiphysics Simulation and Experiments in Micro-Electronics and Micro-Systems (EuroSimE 2006), April 24–26, Como, Italy, IEEE.

JESD22-B111 (2003) *Board level drop test method of components for handheld electronic products*, JEDEC Solid State Technology Association.

Li, G. and Shemansky, F. (2000) Drop test and analysis on micro machined structures. *Sensors and Actuators A*, **85**, 280–286.

Luan, J., Tee, T.Y., Peck, E. *et al.* (2007) Dynamic responses and solder joint reliability under board level drop test. *Microelectronics Reliability*, **47**, 450–460.

Mariani, S., Ghisi, A., Corigliano, A. and Zerbini, S. (2007) Multi-scale analysis of MEMS sensors subject to drop impacts. *Sensors*, **7**, 1817–1833.

Mariani, S., Ghisi, A., Corigliano, A. and Zerbini, S. (2009) Modeling impact-induced failure of polysilicon MEMS: a multi-scale approach. *Sensors*, **9**, 556–567.

Mariani, S., Ghisi, A., Fachin, F. *et al.* (2008) A three-scale FE approach to reliability analysis of MEMS sensors subject to drop impacts. *Meccanica*, **43**, 469–483.

Mariani, S., Ghisi, A., Martini, R. *et al.* (2011a) Multi scale simulation of shock-induced failure of polysilicon MEMS, in *Advances in Electrical Engineering Research* (ed. T.E. Brouwer), Nova Science Publishers, pp. 267–291.

Mariani, S., Martini, R., Corigliano, A. and Beghi, M. (2011b) Overall elastic domain of thin polysilicon films. *Computational Materials Science*, **50**, 2993–3004.

Mariani, S., Martini, R., Ghisi, A. *et al.* (2011c) Monte Carlo simulation of micro-cracking in polysilicon MEMS exposed to shocks. *International Journal of Fracture*, **167**, 83–101.

Mariani, S., Martini, R., Ghisi, A. *et al.* (2011d) Overall elastic properties of polysilicon films: a statistical investigation of the effects of polycrystal morphology. *International Journal for Multiscale Computational Engineering*, **9**, 327–346.

Meirovitch, L. (1975) *Elements of Vibration Analysis*, McGraw-Hill.

Srikar, V. and Senturia, S. (2002) The reliability of microelectromechanical systems (MEMS) in shock environments. *Journal of Microelectromechanical Systems*, **11**, 206–214.

Suhir, E. (1997) Is the maximum acceleration an adequate criterion of the dynamic strength of a structural element in an electronic product? *IEEE Transactions on Components, Packaging and Manufacturing Technology*, **20**, 513–517.

Tanner, D.M., Walraven, A., Hegelsen, K. *et al.* (2000) *MEMS Reliability in Shock Environments*, 38th Annual International Reliability Physics Symposium, April 10–13, 2000, San Jose, CA, IEEE.

Varghese, J. and Dasgupta, A. (2007) Test methodology for durability estimation of surface mount interconnects under drop testing conditions. *Microelectronics Reliability*, **47**, 93–103.

Wagner, U., Franz, J., Schweiker, M. *et al.* (2001) Mechanical reliability of MEMS-structures under shock load. *Microelectronics Reliability*, **41**, 1657–1662.

Weaver, W., Timoshenko, S.P. and Young, D.H. (1990) *Vibration Problems in Engineering*, John Wiley & Sons, Inc.

Younis, M.I., Alsaleem, F. and Jordy, D. (2007) The response of clamped–clamped microbeams under mechanical shock. *International Journal of Non-Linear Mechanics*, **42**, 643–657.

Zavattieri, P. and Espinosa, H. (2001) Grain level analysis of crack initiation and propagation in brittle materials. *Acta Materialia*, **49**, 4291–4311.

14

Fabrication-Induced Residual Stresses and Relevant Failures

14.1 Main Sources of Residual Stresses in Microsystems

During microsystem fabrication, distortions or alterations in the atoms' positions are introduced when different materials are coupled, or even when the same material is used, because of different deposition techniques adopted for each layer. A nonzero initial stress configuration (i.e. before any external action is applied) is therefore obtained; in most cases, this condition entails undesired consequences, such as excessive strains, thin-film delamination or buckling and, in general, unforeseen changes in the microstructure.

It is common to adopt the expression *residual stresses* or *eigenstresses* when addressing the internal stress distribution arising in a microsystem in the absence of applied tractions at its boundary Γ or of thermal strains in the volume Ω.

At present, understanding of the link between the process variables and the residual stresses is, in most cases, qualitative, even if rather complex theories have been developed. Nonetheless, the demand for mechanical integrity remains a technology limiting factor for many microsystems, even when the structural behaviour is not specifically considered in the microsystem functionality, and therefore unsurprisingly the residual stresses play an important role in design decisions. In this chapter, the effects of the residual stress are addressed more from the point of view of the MEMS mechanical designer than from the scientist interpreting their physical or technological causes, such as how they arise in a manufacturing process.

A common but not definitive residual stresses classification (Freund and Suresh, 2003) distinguishes between *growth stresses* (also known as intrinsic stresses), due to the growth of films on substrates or adjacent layers, and other *induced stresses* (also known as extrinsic stresses), which are a consequence of physical changes occurring in the environment after the completion of grain growth. While the distinction helps to make order between numerous phenomena, the reader should recall that it is conventional.

Doerner and Nix (1988) listed some of the mechanisms responsible for the development of growth stress: they include surface or interface stress, cluster coalescence, grain growth or grain boundary area reduction, vacancy annihilation, grain boundary relaxation, shrinkage of grain boundary voids, incorporation of impurities, phase transformation and precipitation, moisture adsorption or desorption, epitaxy and damage to

Mechanics of Microsystems, First Edition. Alberto Corigliano, Raffaele Ardito, Claudia Comi, Attilio Frangi, Aldo Ghisi, and Stefano Mariani.

Companion website: www.wiley.com/go/corigliano/mechanics

the structure due to sputtering or other deposition processes at high energy. Just the length of this list shows the crowd of factors involved. However, the main mechanisms are related to the bonding of the deposit and the substrate, the mobility of atoms in the fluid phase during the deposition and the mobility of the grain boundaries during their growth.

A similarly long list can be provided for the induced stresses: they can arise because of different thermal expansion coefficients during temperature changes for bonded elements, because of piezoelectric or electrostrictive responses in an electric field, because of electrostatic or magnetostatic fields, gravitational or inertial forces, compositional segregation by bulk diffusion, electromigration, chemical reactions, stress induced phase transformations or plastic or creep deformation.

As an example of a typical intrinsic stress, the growth stress in epitaxial films due to the mismatch strain ε_m, also known as the elastic accommodation strain, can be calculated with formulae shown in Section 14.2 directly from the knowledge of lattice parameters, since it is

$$\varepsilon_m = \frac{a_s - a_f}{a_f}, \tag{14.1}$$

where a_f and a_s are the lattice spacing for the film and substrate material, respectively. (A film is growing epitaxially if atoms of the substrate material occupy natural lattice positions of the film material and vice versa.) Since the film is usually thin with respect to the substrate, all the elastic accommodation in Equation 14.1 is assumed to occur in the film. As an example taken from Freund and Suresh (2003), a thin $Si_{0.8}Ge_{0.2}$ film growing on a thick Si substrate at room temperature gives rise to the following mismatch strain:

$$\varepsilon_m = \frac{a_{Si} - a_{SiGe}}{a_{SiGe}} = \frac{0.5431 - 0.5476}{0.5476} = -0.0082, \tag{14.2}$$

where the lattice spacing for the $Si_{0.8}Ge_{0.2}$ alloy is calculated as $a_{SiGe} = 0.8\, a_{Si} + 0.2\, a_{Ge}$, with $a_{Ge} = 0.56574$ nm and $a_{Si} = 0.5431$ nm at room temperature.

An example of a common extrinsic stress is, instead, the thermal-induced stress arising from different thermal expansion coefficients for two adjacent layers. In this case the general expression for the mismatch strain is

$$\varepsilon_m = (\alpha_f - \alpha_s)(T - T_0) = (\alpha_f - \alpha_s)\,\Delta T, \tag{14.3}$$

where α_f and α_s are the isotropic thermal expansion coefficients for the film and substrate, respectively, and T and T_0 are the process and reference temperature, respectively.

For both these examples, the consequent residual stresses can be retrieved by using the formulae described in the next section.

14.2 The Stoney Formula and its Modifications

It was Stoney (1909) who found a simple relationship between the stress in a (metallic) thin film on a (metallic) substrate and its curvature, given the following assumptions.

1) Since the film and substrate thickness are much smaller than the lateral dimensions, the Kirchhoff thin plate theory is assumed to be valid for the film–substrate system;

in particular, straight material lines orthogonal to the substrate midplane remain straight and orthogonal to the deformed midplane. This structural theory can be therefore viewed as an extension of the Euler–Bernoulli beam theory, described in Section 2.3.2, to 2D structures; in particular, the normal stress in the z direction, orthogonal to the substrate (see Figure 14.1), is zero along all the thickness.
2) The film thickness h_f is smaller than the substrate thickness h_s.
3) The substrate material is homogeneous, isotropic and linear elastic; the film material is isotropic.
4) The edge effects near the boundary of the substrate are negligible, so that the solution is independent of a translation along the interface.
5) Strains, displacements and rotations are infinitesimally small.

Some of these hypotheses can be relaxed and the Stoney's conclusions can be extended to more general cases, as we will see in the following; however, holding the previous assumptions as true, now we consider an axisymmetric geometry, as shown in Figure 14.1 for generic materials, not only metals as in Stoney's original work. We assume a cylindrical coordinate system (r, θ, z) with a film subject to a membrane force f (positive or negative), whose dimension is a force over a length. Because of hypothesis 1, $h_s \ll R$; we assume that the substrate midplane is at $z = 0$, while the bonding between the substrate and the film is at $z = h_s/2$ and the substrate base is at $z = -h_s/2$.

We suppose that the change in magnitude of f due to the substrate deformation is small; f is assumed as given, for example because of a known assumed mismatch strain (such as Equation 14.3). The membrane force gives rise to an external equi-biaxial stress state ($\sigma_{\theta\theta} = \sigma_{rr}$) in the film (recall that $\sigma_{zz} = 0$ because of hypothesis 1); if we assume the film as ideally detached when the external condition creating the membrane force is acting, then, reassembling the strained film to the substrate after removing the external condition, the compatibility at the interface induces a curvature in the substrate, see Figure 14.2. We will consider that this curvature is uniform at the substrate midplane (mostly because of hypotheses 1 and 5), where a uniform isotropic strain ε_0 can also be imagined. The axisymmetric geometry assures a spherical (i.e. constant curvature) deformed shape for the midplane, therefore simplifying the following analytical expressions. The more fertile idea for the comprehension of the phenomena is to derive the Stoney formula through an energy approach: starting from the definition of the elastic specific (i.e. per unit of volume) strain energy U, we have

$$U(r,z) = \frac{G_s}{1-\nu_s}\left(\varepsilon_{rr}^2 + \varepsilon_{\theta\theta}^2 + 2\,\nu_s\varepsilon_{rr}\varepsilon_{\theta\theta}\right), \tag{14.4}$$

where the subscript 's' refers to substrate isotropic elastic variables and all the other strain components are zero under the previous assumptions. Under hypothesis 5, the strains ε_{rr} and $\varepsilon_{\theta\theta}$ can be expressed as

$$\varepsilon_{rr}(r,z) = u'(r) - z\,w''(r),$$
$$\varepsilon_{\theta\theta}(r,z) = \frac{1}{r}u(r) - \frac{z}{r}\,w'(r), \tag{14.5}$$

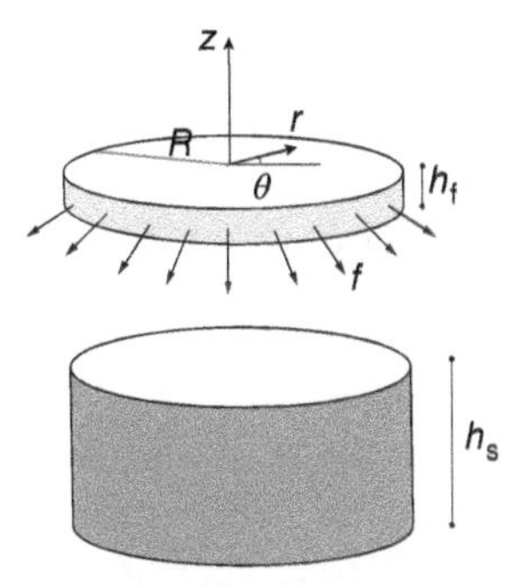

Figure 14.1 Axisymmetric reference system frame adopted for a thin film on a thick substrate.

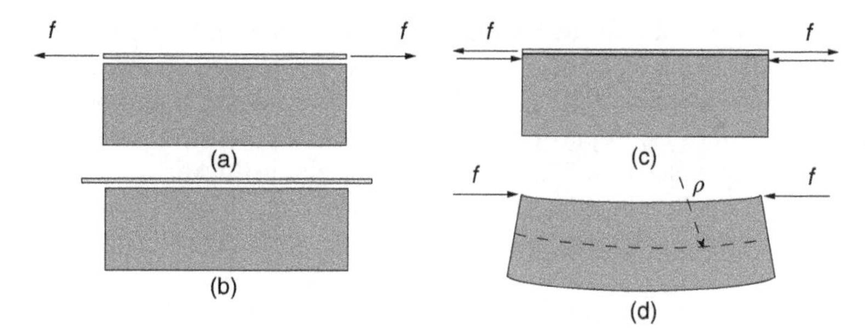

Figure 14.2 Axisymmetric reference system frame adopted for a thin film on a thick substrate.

with $u(r)$ and $w(r)$ the radial and the out-of-plane displacement of a point at the midplane, respectively. The approach starts with a reasoned choice for the displacements u and w, depending on some parameters, and the subsequent imposition of the principle of stationary potential energy, which fixes the aforementioned parameters. By assuming decoupling between radial and out-of-plane deformation, the displacements at the midplane are chosen so that the radial displacement u is growing with the distance from the reference centre point, while the out-of-plane displacement w depends only on the constant curvature χ, namely

$$u(r) = \varepsilon_0 r, \qquad w(r) = \frac{1}{2}\chi r^2. \tag{14.6}$$

Because of the axisymmetry and of the isotropic linear elasticity, $\varepsilon_{\theta\theta} = \varepsilon_{rr}$, and it follows that

$$\varepsilon_{rr} = \frac{\sigma_{rr}}{E_s} - \nu_s \frac{\sigma_{rr}}{E_s} = \frac{\sigma_{rr}}{M_s}, \tag{14.7}$$

where the quantity introduced, $M_s = E_s/(1 - \nu_s)$, is the (substrate) biaxial modulus. At the generic distance z from the midplane, the strains turn out to be $\varepsilon_{rr} = \varepsilon_{\theta\theta} = \varepsilon_0 - \chi z$, since the constant component at the midplane ε_0 combines with the linear contribution owing to bending identified by the constant curvature χ. These considerations lead to a simple form for the strain energy density:

$$U(r, z) = M_s(\varepsilon_0 - \chi z)^2, \tag{14.8}$$

which must be inserted into the total potential energy expression for the substrate,

$$V(\varepsilon_0, \chi) = 2\pi \int_0^R \int_{-h_s/2}^{h_s/2} U(r, z) r \mathrm{d}r \mathrm{d}z + 2\pi f u_r(R, h_s/2)R \tag{14.9}$$

$$= \pi R^2 M_s h_s \left(\varepsilon_0^2 + \frac{1}{2}\chi^2 h_s^2\right) + 2\pi R^2 f \left(\varepsilon_0 + \frac{1}{2}\chi h_s\right), \tag{14.10}$$

where the second term in the addition is the external work due to the membrane force f, see Figure 14.1. By imposing the stationary total potential energy, namely $\partial V/\partial \varepsilon_0 = 0$ and $\partial V/\partial \chi = 0$, from Equation 14.9, not only it follows that $\varepsilon_0 = -f/(M_s h_s)$ but, most importantly, also the Stoney formula is derived:

$$\chi = \frac{6f}{M_s h_s^2}. \tag{14.11}$$

The importance of this equation relies, besides its simplicity, on the fact that it is independent of the *film* material properties, mostly because of hypothesis 2.

It follows naturally that in the substrate the mean stress and strain are

$$\sigma_{\mathrm{m}} = \frac{f}{h_{\mathrm{s}}}, \qquad \varepsilon_{\mathrm{m}} = \frac{f}{M_{\mathrm{s}} h_{\mathrm{s}}}, \tag{14.12}$$

while the neutral plane in the substrate, i.e. where $\varepsilon_{rr}(r,z) = 0$, is unequivocally (i.e. independently of the sign of f) found to be at the position $z_{\mathrm{n}} = \varepsilon_0/\chi = -1/6h_{\mathrm{s}}$. The implication is that the curvature for the midplane, for the neutral plane and for the top and bottom substrate surfaces is almost the same, a conclusion largely exploited in the experimental determination of substrate curvature, as shown in Section 14.3.

As an example, we calculate the curvature of a 100 mm radius and 750 μm thick (100) silicon substrate over which a thin gold film, 1 μm in thickness, is deposited at a temperature of 100 °C. A reference room temperature of 20 °C is assumed. We need the following data for the gold material: Young's modulus $E_{\mathrm{f}} = E_{\mathrm{gold}} = 79$ GPa, Poisson's ratio $\nu_{\mathrm{f}} = \nu_{\mathrm{gold}} = 0.4$, coefficient of thermal expansion $\alpha_{\mathrm{f}} = \alpha_{\mathrm{gold}} = 14 \times 10^{-6}$ °C^{-1}. For the silicon substrate instead, we have: $M_{\mathrm{s}} = M_{\mathrm{Si}} = 181$ GPa, $\alpha_{\mathrm{s}} = \alpha_{\mathrm{Si}} = 3 \times 10^{-6}$ °C^{-1}. We assume the system as stress-free at the deposition temperature and we look for the curvature at room temperature. First of all, we calculate the mismatch strain, by assuming that it depends totally on the thermal strain:

$$\varepsilon_{\mathrm{m}} = (\alpha_{\mathrm{f}} - \alpha_{\mathrm{s}})\Delta T = 8.8 \times 10^{-4}. \tag{14.13}$$

The stress in the thin film is equi-biaxial and is equal to

$$\sigma_{\mathrm{m}} = M_{\mathrm{f}} \varepsilon_{\mathrm{m}} = 115.87 \text{ MPa}, \tag{14.14}$$

a relevant value. Finally, the curvature is retrieved with the Stoney formula (14.11):

$$\chi = \frac{6f}{M_{\mathrm{s}} h_{\mathrm{s}}^2} = \frac{6\sigma_{\mathrm{m}} h_{\mathrm{f}}}{M_{\mathrm{s}} h_{\mathrm{s}}^2} = 6.828 \times 10^{-3} \text{ m}^{-1}, \tag{14.15}$$

corresponding to a curvature radius of about 146 m.

Freund, Floro and Chason (1999) extended the Stoney formula to the case when the film thickness becomes comparable with that of the substrate, loosening hypothesis 2. If h_{f} and h_{s} are of the same order of magnitude, then the contribution of the film strain is added to that of the substrate in the strain energy density (14.4). The strain in the film has a similar expression to Equation 14.7, with the addition of a given mismatch strain ε_{m} (whose physical origin is not relevant for this analysis); then, the stationary total potential energy is sought with respect to the parameters (ε_0, χ), giving the following expression for the curvature:

$$\chi = \chi_{\mathrm{St}} \cdot \left[\frac{1+h}{1+hm(4+6h+4h^2)+h^4m^2} \right], \qquad \chi_{\mathrm{St}} = \frac{6\varepsilon_{\mathrm{m}}}{h_{\mathrm{s}}} hm, \tag{14.16}$$

where the nondimensional quantities $h = h_{\mathrm{f}}/h_{\mathrm{s}}$ and $m = M_{\mathrm{f}}/M_{\mathrm{s}}$ have been introduced, and χ_{St} is the curvature given by the Stoney formula (14.11). To appreciate the entity of the original Stoney approximation with respect to this generalization, the range of parameters for which the difference between χ and χ_{St} is less than 10% is shown in Figure 14.3. Both the stiffness ratio m and the thicknesses ratio h concur with the quality of the Stoney approximation.

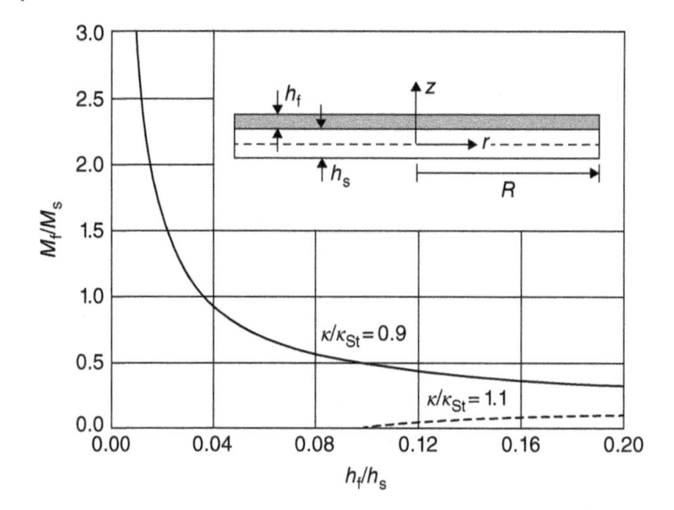

Figure 14.3 The continuous (dashed) line represents the locus of points where the Stoney formula (14.11) overestimates (underestimates) the curvature by 10%, taking into account the actual thickness ratio h_f/h_s. *Source*: Freund, Floro and Chason (1999), Figure 1. Reproduced with permission of AIP Publishing.

A second extension to the Stoney formula is also given in Freund, Floro and Chason (1999) and in Freund (2000): it considers the contribution of finite rotations w' into the radial strain (violating assumption 5), namely:

$$\varepsilon_{rr}(r,z) = u'(r) - z\, w''(r) + \varepsilon_m + \frac{1}{2} w'(r)^2, \tag{14.17}$$

therefore the choice for the symmetric midplane radial displacement becomes

$$u(r) = \varepsilon_0 r + \varepsilon_m r + \varepsilon_1 r^3. \tag{14.18}$$

Taking the derivatives of the stationary total potential energy with respect to $(\varepsilon_0, \chi, \varepsilon_m)$ leads to a complex, nonlinear system; however, when assumption 2 holds true, i.e. $h_f \ll h_s$, then a relative simple relationship can be obtained for the nondimensional expressions of the mismatch strain S versus the curvature X:

$$S = X\left(1 + (1 - \nu_s)X^2\right), \tag{14.19}$$

where the normalized quantities are defined as $S = 3/2\varepsilon_m R^2 h_f M_f / h_s^3 M_s$ and $X = 1/4R^2 \chi / h_s$.

Figure 14.4 shows in continuous lines the linear relationship between S and X (ignoring the nonlinear kinematics) together with the nonlinear law defined in Equation 14.19. It is important to know that the effect of finite rotations becomes appreciable when $S > 0.3$, which corresponds to mismatch strains that can be observed in practice. The dashed lines in Figure 14.4 refer to a set of nonlinear finite element analyses (with $R/h_s = 50$, $h_f/h_s = 0.01$, $M_f/M_s = 1$ and no assumptions on the deformation behaviour) to evidence the effect of nonconstant curvature. It is arguable, in fact, that the substrate midplane would give rise to a spherical deformed shape when nonlinear

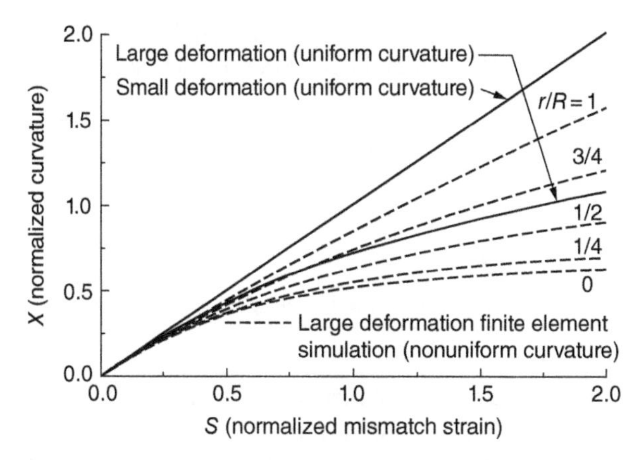

Figure 14.4 Continuous lines represent, under the constant-curvature hypothesis, the linear relationship between normalized mismatch strain or normalized curvature and the nonlinear law by Equation 14.19. Dashed lines are the results of nonlinear finite analyses accounting for nonconstant curvature along the radial distance *r* on the substrate. *Source*: Freund, Floro and Chason (1999), Figure 3. Reproduced with permission of AIP Publishing.

effects are accounted for; a changing curvature is instead expected. Therefore, the values in Figure 14.4 should refer to local mean curvature values calculated at different positions along the radial distance from the centre of the substrate (spaced $r/R = 0, 1/4, 1/2, 3/4, 1$): with increasing mismatch S, the midplane surface tends to be flat towards the centre of the substrate and curves more near the edges.

Huang and Zhang (2006) extended the Stoney formula to the case where the residual stress in the film is no longer uniform, but also depends on z, in other words it presents a gradient. In this case, a homogenization theory developed in the composite material field can be applied to multilayered microsystems. We give here only a sketch of the approach, which assumes infinitesimally small, pure bending strains. It starts from the following assumption on the expression of the residual stress:

$$\varepsilon_{\rm res}(z) = \varepsilon_{\rm con}(z) + \varepsilon_{\rm ben}(z) = \varepsilon_{\rm con}(z) + (\varepsilon_0 - \chi_{\rm ben} z), \tag{14.20}$$

where $\varepsilon_{\rm con}$ is the (intrinsic or extrinsic) mismatch strain applied when curvature is made impossible by some constraint, while $\varepsilon_{\rm ben}$ is the multilayered system deformation consequent to the removal of the constraint, assumed as a pure bending deformation, linearly varying with z. The actual residual stress is a superposition of these causes; while $\varepsilon_{\rm con,0}$ can be interpreted as a lattice mismatch or thermal strain, the gradients along z are instead caused by more localized effects, such as atomic diffusion, grain size variation through the film thickness, interstitial or substitutional defects or atomic peening (Thornton and Hoffman, 1989).

The general expression of the constrained strain in the film can be written as a kth order polynomial function:

$$\varepsilon_{\rm con}(z) = \sum_{k=0}^{\alpha} \varepsilon_{\rm con,\,k} \left(\frac{z - h_{\rm s}}{h_{\rm f}} \right)^k, \qquad h_{\rm s} \le z \le h_{\rm s} + h_{\rm f} \tag{14.21}$$

and the curvature and the strain at the midplane can therefore be expressed as a function of the k coefficients $\varepsilon_{\text{con, k}}$ and of the stiffness ratio m and the thickness ratio h, as follows:

$$\chi_{\text{ben}} = \frac{1}{h_{\text{s}}} \frac{6mn}{1 + 4mn + 6mn^2 + 4mn^3 + m^2n^4} \cdot \sum_{k=0}^{\alpha} \varepsilon_{\text{con, k}} \frac{mn^2k + 2n(k+1) + k + 2}{(k+1)(k+2)}, \tag{14.22}$$

$$\varepsilon_0 = \frac{2mn}{1 + 4mn + 6mn^2 + 4mn^3 + m^2n^4} \cdot \sum_{k=0}^{\alpha} \varepsilon_{\text{con, k}} \frac{(k-1)mn^3 + 3kmn^2 + 3(k+1)n + k + 2}{(k+1)(k+2)}. \tag{14.23}$$

All the previous generalizations assume an isotropic material behaviour, but the case of an anisotropic film or substrate is not uncommon in practice. If the substrate is isotropic and only the film shows anisotropy, a simple generalization of the Stoney formula is (Freund and Suresh, 2003):

$$\chi_x = \frac{6\left(f_x - \nu_s f_y\right)}{E_s h_s^2}, \quad \chi_y = \frac{6\left(f_y - \nu_s f_x\right)}{E_s h_s^2}, \quad \chi_{xy} = (1 - \nu_s)\frac{6 f_{xy}}{E_s h_s^2}, \tag{14.24}$$

where x, y, z have been assumed as both the coordinates and elasticity axis coordinates, and where the resultant forces

$$f_i = \int_{h_s/2}^{h_s/2+h_f} \sigma_{m,i}(x_3)\mathrm{d}x_3, \qquad i = x, y, xy \tag{14.25}$$

have been introduced. The latter forces derive from the residual stresses $\sigma_{m,i}$ obtained from the corresponding mismatch strains $\varepsilon_{m,i}$ through the relationships shown in Section 2.3.1. Different curvatures along the x- and y-directions are now present and a twist χ_{xy} of the substrate midplane occurs.

A very common case covers an elastically anisotropic substrate, such as a single-crystal silicon wafer. If a (001) silicon wafer is considered as a substrate, the Stoney equation (14.11) reads (Nix, 1989; Janssen *et al.*, 2009):

$$\chi = \frac{6\sigma_f t_f}{M_{001}^{\text{Si}} h_s^2}, \tag{14.26}$$

where the biaxial modulus is expressed as a function of the components of the silicon compliance tensor c_{ij} (see Section 2.3 and Hopcroft, Nix and Kenny (2010)) as

$$M_{001}^{\text{Si}} = \frac{1}{c_{11} + c_{12}} = 1.803 \times 10^{11}\ \text{Pa}. \tag{14.27}$$

Similarly, for (111) silicon wafers, it turns out to be

$$\chi = \frac{6\sigma_f t_f}{M_{111}^{\text{Si}} h_s^2} \tag{14.28}$$

with

$$M_{111}^{\text{Si}} = \frac{6}{4c_{11} + 8c_{12} + c_{44}} = 2.291 \times 10^{11}\ \text{Pa}. \tag{14.29}$$

When both substrate and film materials are not isotropic, the problem obviously becomes more involved, and the curvatures and in-plane extensional strains along x, y

can be retrieved only by solving a system of linear algebraic equations. The interested reader can find them in Freund and Suresh (2003), Section 3.7.2.

14.3 Experimental Methods for the Evaluation of Residual Stresses

Many experimental approaches have been envisaged to determine the residual stresses in materials; limiting the discussion to microsystems, we can distinguish between methods for thin films on substrates, for free-standing films and for the characterization of residual stresses induced by the packaging.

Starting from the first category, it is important to understand that, more appropriately, the residual stresses for thin films on a substrate are obtained from a direct measure of strain or displacement, typically based on the evaluation of changes in the substrate curvature. Between the approaches apt to measure curvature, a further distinction is made according to the physical principle followed: mechanical methods, capacitance methods, X-ray diffraction methods and optical methods. Excluding X-ray diffraction, these methods aim to measure the out-of-plane displacement. In the mechanical method, a stylus contacting the microsystem surface records the out-of-plane displacement w as a function of the distance r from a reference point and converts it into a curvature radius ρ or a curvature χ through

$$\chi = \frac{1}{\rho} = \frac{\mathrm{d}^2 w(r)}{\mathrm{d}r^2}. \tag{14.30}$$

To account for an originally uneven plane surface, the curvature must be measured before and after film deposition; typically the Stoney formula (14.11) for elastic isotropic materials is used, so the average residual stress is obtained as

$$\sigma_\mathrm{m} = \frac{M_\mathrm{s} h_\mathrm{s}^2}{6 h_\mathrm{f}} \left(\frac{1}{\rho_\mathrm{ad}} - \frac{1}{\rho_\mathrm{orig}} \right), \tag{14.31}$$

where ρ_orig and ρ_ad are the radii of curvature before and after the deposition, respectively.

Capacitance methods are instead based on the change in capacitance observed by a non-contacting probe testing the surface; the change is due to the varying gap between the probe and the surface; it is first related to the out-of-plane displacement and then converted to the curvature and stresses via Equations 14.30 and 14.31.

While both mechanical and capacitance methods appear well-founded, they lack accuracy in the presence of temperature changes or at high temperatures or when space is restricted, precisely all the conditions found in deposition chambers or, in general, during the foundry processes for many microsystems. This motivated the recourse to X-ray diffraction methods (Cullity, 1956), which are based on the determination of the Bragg angle (i.e. the angle between the incident beam and the crystalline plane) corresponding to the maximum diffraction intensity at different points on a test surface. The rationale is that the strains induced by the residual stresses modify the spacing of the atomic planes in the crystal structure; this spacing is the objective of the analysis of the diffraction phenomenon. Starting from a reference point and moving radially, a peak in the diffraction intensity signals the Bragg angle, linked to the particular crystallographic texture of the substrate: from the radial distance of two locations

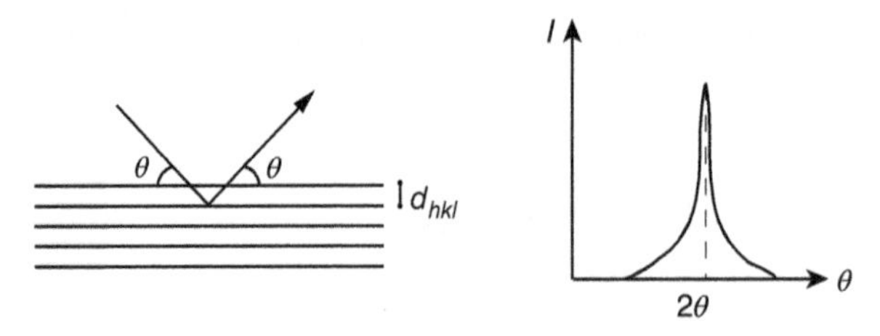

Figure 14.5 Experimental determination of crystallographic spacing d_{hkl} (eventually modified by a residual stress) through interferometric measurement of the Bragg angle, θ.

where the diffraction measurements are taken, divided by the difference of the angles corresponding to the position of the intensity maxima, the curvature of the substrate is retrieved. The spacing d_{hkl} between crystallographic planes is then determined through Bragg's law:

$$d_{hkl} = \frac{N\,\lambda}{2\,\sin\theta}, \tag{14.32}$$

where λ is the X-ray wavelength, θ the incident and reflecting angle, and N is the order of the reflected beam (typically $N = 1$). As shown in Figure 14.5, the diffraction intensity $I(\theta)$ is extracted as a function of 2θ. From the information on the spacing, the stress state could be inferred as well, for example by calculating the mismatch strain as $\varepsilon_{\rm m} = (d_{hkl} - d_0)/d_0$ with d_0 the (assumed known) unstressed original spacing. The drawbacks of the X-ray diffraction method are the cost and the required safety precautions; moreover, similar results can be obtained using optical methods.

The most common optical method (see Figure 14.6) considers laser beam scanning (see, but the list is far from being complete, Retajczyk and Sinha (1980); Pan and Blech (1984); Flinn, Gardner and Nix (1987); Volkert (1991); Floro *et al.* (1997); Floro and Chason (1996); Shull and Spaepen (1996)): by monitoring, via a mirror, the (laser beam) reflecting angle θ of the substrate surface as a function of the radial distance r along a straight line from a reference point, the curvature is retrieved as

$$\chi = \frac{1}{\rho} = \frac{{\rm d}^2 w(r)}{{\rm d}r^2} = \frac{{\rm d}\theta(r)}{{\rm d}r}. \tag{14.33}$$

As said for the mechanical methods, the effect of the initial curvature is considered by repeating the measurements before and after the deposition process. This method

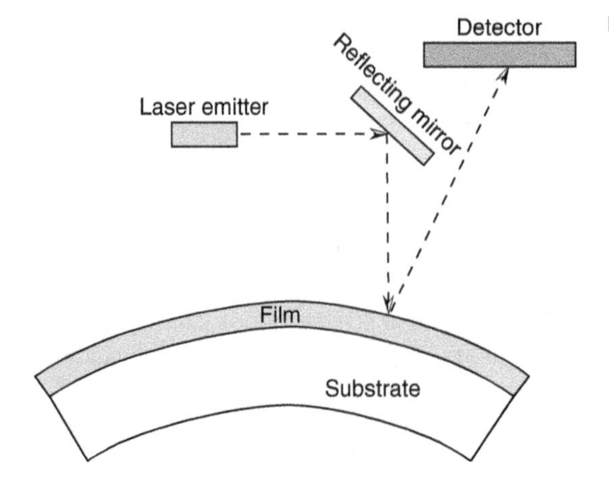

Figure 14.6 Laser scanning method.

is actually so widespread that it is sometimes referred to simply as the wafer curvature measurement method; a commercial laser scanning apparatus can measure radii of curvature as large as kilometres. Its main drawbacks are represented by the time involved in the measurement of the whole wafer and the sensibility to vibrations, such as in high-vacuum chambers. Some of these problems can be addressed by adding more instrumentation, such as a beam splitter (Schell-Sorokin and Tromp, 1990), and a charge coupled device (CCD) camera (Martinez, Augustyniak and Golovcheko, 1990). With a multibeam optical device, in fact, there is no necessity for multiple position detectors to avoid vibration problems, since the original laser beam is split into several beams thanks to a spatial filter with a focusing objective lens and an etalon (see Figure 14.7). The reflected beams are then captured by a CCD camera; this information, typically digitalized, is retrieved as a grid of reflective spots, representative of the mapping of the curvature substrate. Even if vibrations are present, because of the large area involved by the CCD, the consequent grid translation (the lasers move in unison during vibration) does not modify the measurement. The drawback of this method resides in the monochromatic nature of the beams, which is problematic for transparent films. Moreover, a reflective surface is obviously mandatory.

Among the other optical methods are the so-called grid reflection method (Finot *et al.*, 1997) and the coherent gradient sensor method (Rosakis *et al.*, 1998). The former is based on the reflection of a grid from the surface of a curved substrate; the latter is instead based on the interference fringe patterns generated by laterally shearing an incident wavefront. Both methods are able to measure full-field nonuniform curvatures (i.e. considering curvature as a tensor varying on the substrate surface); although they require *ad-hoc* software and a dedicated computer for the calculations (as with the coherent gradient sensor method), they have the advantage of measuring *in-situ* curvatures even during thermal cycles and they can overcome the limitation of a monochromatic beam. However, their full description would be too extensive for this short survey and we address the interested reader to the cited references and to Chapter 2 in Freund and Suresh (2003).

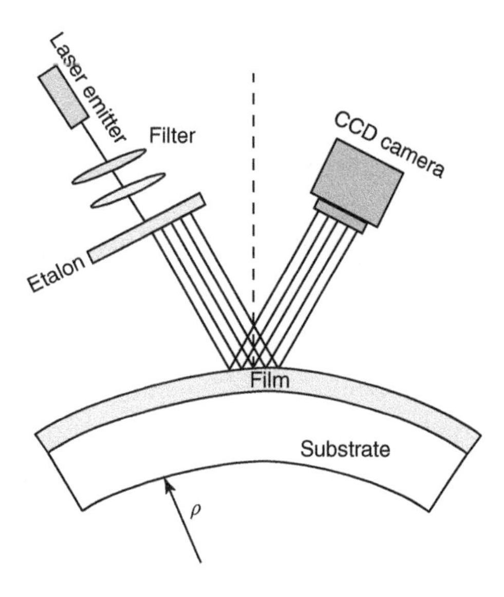

Figure 14.7 Multibeam inspector for residual stress determination.

The previously mentioned methods are suitable for thin films on a substrate. We present in the following, instead, two methods suitable to identify residual stresses in free-standing films (i.e. without a substrate underneath). The first method is the bulge test, based on a pressurization of the thin film and the monitoring of its out-of-plane displacement. The main advantages of the method are the almost homogeneous stress state, its relativity simplicity and the isothermal behaviour of the system during the test. While the first application of the method dates backs to Beams (1959), it has been recognized (Xiang, 2005) that, for rectangular samples with an aspect ratio higher than four, the plane strain condition can be held as true at the centre of the membrane. Therefore, the membrane deflection can be investigated in a 2D section case, as depicted in Figure 14.8, where the film deformation is assumed cylindrical. Given the applied pressure p, the radius R of the circular cross-section, and the membrane thickness h, then the tangential stress, homogeneous in the membrane, is simply

$$\sigma_t = \frac{pR}{h}. \tag{14.34}$$

The tangential strain is also easily defined as

$$\varepsilon_t = \frac{l - l_0}{l_0} = \frac{R\theta - a}{a}, \tag{14.35}$$

with l, l_0 the final and initial lengths, respectively, θ the central angle and a the half span. Referring to Figure 14.8, the geometrical variables can be expressed in terms of the measurable quantities as

$$R^2 = a^2 + (R - \delta)^2 \Rightarrow R^2 = \frac{a^2 + \delta^2}{2\delta}, \qquad \theta = \arcsin\frac{a}{R}, \tag{14.36}$$

where δ is the maximum out-of-plane deflection observed for the the membrane. It is then possible to rewrite Equations 14.34 and 14.35 as a function of these measurable quantities:

$$\sigma_t = \frac{p(a^2 + \delta^2)}{2\delta h}, \qquad \varepsilon_t = \frac{p(a^2 + \delta^2)}{2\delta a} \cdot \arcsin\left(\frac{2a\delta}{a^2 + \delta^2}\right) - 1. \tag{14.37}$$

The residual strains can be accounted for by adding them to Equation 14.37. The quality of the bulge equations can be improved by including moderately thick films (Van Eijden, 2008); the bulge test can be also used for plastic and fracture characterization (Merle, 2013).

Another widely employed method for free-standing characterization is nanoindentation, (Zhu *et al.*, 2015). By coupling instrumented indentation and numerical simulation, typically carried out using the finite element method, the residual stresses can also be extracted. These stresses, however, significantly affect the load–depth curve: in its loading part, for a fixed penetration depth, the loading curve for tensile residual stresses is lower than the stress-free case, while for a compressive residual stress case the load curve is higher than the

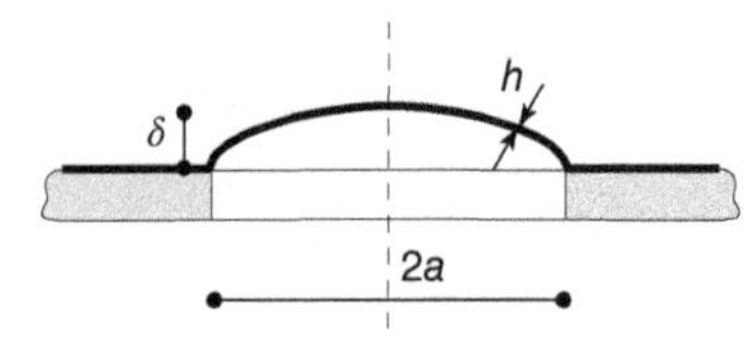

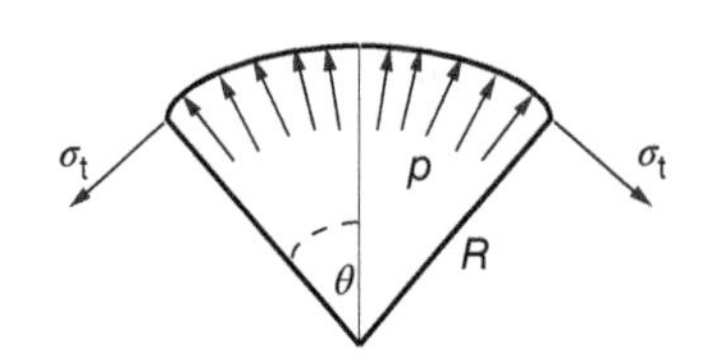

Figure 14.8 2D interpretation of the bulge test.

stress-free case; under unloading, the relationship is reversed, since the compressive state, which induced a higher peak load, generates more elastic recovery than the tensile state. Moreover, the plasticity is increased for compressive residual stresses and decreased for the tensile case, including the pile-ups around the indentation footprint, i.e. the plastic deformation due to the plastic incompressibility condition. The true contact area for the indenter is also influenced by the residual stresses, increasing with compressive stresses and decreasing for tensile stresses. There is still no complete agreement on the models able to interpret residual stresses for all materials. Some models can be found in Suresh and Giannakopoulos (1998), Lee and Kwon (2004) and Swadener, Taljat and Pharr (2001).

Micro-Raman spectroscopy is a type of molecular spectroscopy based on inelastically scattered light: 'inelastic' means that the frequency of monochromatic light, typically generated by a laser, changes on interaction with the sample. It allows for the interrogation and identification of the vibrational states of molecules. Therefore, it can be used for monitoring changes in the molecular bond structure, e.g. state changes, strains and, after some calculation, stresses. A good introduction to this technique is found in De Wolf (1996), where it is recalled that, for single-crystal silicon, there are three optical Raman modes: for example, associated with an incident optical direction aligned with the crystallographic axis [001], there is a mode in the same direction and two more transverse modes in the other ([100] and [010]) directions. All the three modes have a very recognizable peak for silicon at a frequency of ω_R=520 R cm^{-1} when there is no stress. (This unit R cm^{-1} indicates relative cm^{-1}, because the frequency is measured with respect to (i.e. relative) to the frequency of the laser light.) When there is backscattering from the (100) plane, the Raman signal is due to the longitudinal mode only; the frequencies of vibration of the three modes shift from ω_R when there is strain in the sample. The technique is now widely used; as examples, some recent applications of the micro-Raman spectroscopy can be seen in Krishna *et al.* (2015), Kociniewski, Moussodji and Khatir (2014) and Li *et al.* (2010).

As a last approach to the characterization (or the exploitation!) of residual stresses, we consider the realization of microstructures developing clearly recognizable phenomena, such as buckling or other instabilities, from which the entity of the residual stresses can be estimated with great accuracy. We show the idea here just with an example, taken from Fachin *et al.* (2001). Let us consider a micro-bridge whose beam, possibly layered, is supported at its ends; the boundary flexibility can be also taken into account by applying flexural springs at the ends. The equations expressing the vertical displacement w in accordance with the Euler–Bernoulli theory and considering the axial contribution to bending is

$$\overline{E\,I}w^{\mathrm{IV}}(x) + \left[\overline{A}\sigma_{\mathrm{mean}} - \frac{\overline{E\,A}}{2L}\int_0^L w^{\mathrm{I}}(x)^2\mathrm{d}x\right] w^{\mathrm{II}}(x) = 0, \tag{14.38}$$

where the quantities with a bar over are effective (i.e. homogenized), L is the beam length and the residual stress inside the beam is assumed to be

$$\sigma = \sigma_{\mathrm{mean}} + \sigma_{\mathrm{grad}}\frac{z}{\overline{h}}, \tag{14.39}$$

with σ_{mean} the mean value for the residual stress in a section and σ_{grad} the gradient stress, varying linearly with z along the effective beam height $\overline{h}$. The solution of the buckling problem can be obtained both analytically and numerically, and in general it can be found that the post-buckling displacement is described by the relationship

$$w(x) = f(\sigma_{\text{mean}}, \sigma_{\text{grad}}, K), \tag{14.40}$$

where K is the ratio between the considered, flexible, boundary and a clamped boundary (it is fully clamped if $K = 1$). Three independent experimental measurements of the displacement are therefore necessary in the post-buckling regime to characterize the parameters, for example they can be taken at different locations, such as at $x = \{L/6, L/3, L/2\}$.

14.4 Delamination, Buckling and Cracks in Thin Films due to Residual Stresses

While residual stresses are not always negative in microsystem design (they could be exploited e.g. to identify other material properties, provided the residual stresses have been accurately measured experimentally), in some cases they undeniably cause undesired consequences, such as film delamination, buckling or cracks. In this section, we briefly describe these three phenomena, pointing out to the reader how they are related to the theory developed previously; we warn, however, that for further reference it will be necessary to refer to the cited bibliography.

Let us consider first the delamination problem due to tensile residual stresses in the film. When the edge effects for a thin film deposited on a substrate are neglected, as for hypothesis 4 in Section 14.2, we can assume that the system is invariant-by-translation, and that there are no normal tractions across the film–substrate interface. If, instead, we focus our attention to the edge of the film–substrate interface, the stress concentration could give rise to the tractions responsible for interface opening, i.e. delamination. We use in the following the Griffith approach to the fracture problem (see Chapter 12) and we concentrate on the example shown in Figure 14.9, see Freund and Suresh (2003), where the only source of the phenomenon is the residual biaxial stress σ_m arising from a mismatch strain whose physical origin is left undefined. In the delaminated region, the stress in principle reduces to σ_a, whose value is zero if the problem scale considers a straight delamination front, or different from zero if the delamination front would be

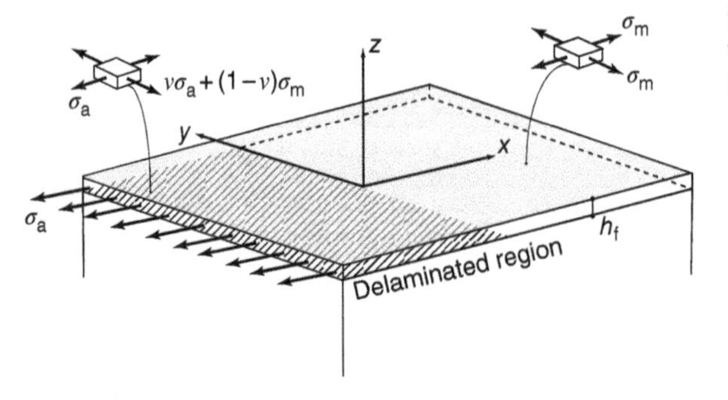

Figure 14.9 Delamination front in thin film with equi-biaxial residual stress field.

curved (in this case, Figure 14.9 would show an enlargement of a region local to the front). In general, the energy release rate becomes

$$\mathcal{G} = \frac{1-\nu_f^2}{2E_f}(\sigma_m - \sigma_a)^2 h_f \tag{14.41}$$

and after the imposition of Griffith's criterion $\mathcal{G} = \Gamma$, with Γ the interface fracture energy, it follows that

$$\frac{1-\nu_f^2}{2E_f}\sigma_m^2 h_f = \Gamma, \tag{14.42}$$

or, equivalently,

$$\frac{(1+\nu_f)E_f}{2(1-\nu_f)}\varepsilon_m^2 h_f = \Gamma. \tag{14.43}$$

Since the fracture energy Γ is a material property ('material' being here the interface itself), the relationship (14.43) can be solved for the thickness, giving

$$h_{f,cr} = \frac{2(1-\nu_f)\Gamma}{(1+\nu_f)E_f\varepsilon_m^2} = \frac{2E_f\Gamma}{(1+\nu_f)\sigma_m^2}. \tag{14.44}$$

Therefore, a thickness larger than the critical value $h_{f,cr}$ would induce a spontaneous (i.e. without the imposition of any external load other than the mentioned mismatch strain) delamination of the thin film on the substrate. With this rather simplified model, the substrate contribute limits to the value of Γ, and the delamination would include the whole film.

To exemplify the spontaneous delamination phenomenon, let us consider (Freund and Suresh, 2003) a thin gold layer placed on a substrate of Al_2O_3 at 320 °C in a clean environment, subsequently the temperature is decreased to 20 °C. The relevant material properties are as follows: for the interface, the fracture energy is equal to $\Gamma = 10$ J/m^2, while for the film, the isotropic elastic parameters are $E_f = 79$ GPa and $\nu_f = 0.4$. The critical thickness for the gold layer can be calculated from Equation 14.44 and is equal to $h_{f,cr} = 6.15$ μm. A gold film whose thickness is equal to or larger than this value will (theoretically) spontaneously delaminate from the Al_2O_3 substrate. Let us assume now that the gold layer has been set, prudently, to 2 μm, but that a contaminant or moisture has found its way into the deposition chamber, decreasing the quality of the interface properties. If a spontaneous delamination were observed in the system, then an upper bound for the interface fracture energy could be estimated from Equation 14.43 with $h_f = 2$ μm: in this case, the estimate would be $\Gamma \leq 3.25$ J/m^2.

In practice, however, it is unlikely that the film completely separates from the substrate, because a curved delamination front actually has a lower energy release rate (when $\sigma_a \neq 0$ in Equation 14.41, it follows that $\mathcal{G}_f$ will decrease with respect to the case for $\sigma_a = 0$), and because the energy release rate near an edge, e.g. at a distance of h_f from the edge itself, becomes lower than the energy release rate $\mathcal{G}_{ss}$ necessary for a steady-state crack propagation. The transition from the steady-state crack propagation to the edge condition is a nontrivial problem: it has been addressed effectively by Yu, He and Hutchinson (2001). In that work, a very different behaviour is recognized for a film edge aligned with the substrate edge (labelled 'corner edge') and, conversely, a film edge interior to the substrate (labelled 'interior edge'). In the first case, the advancing crack finds a barrier to propagation, since the ratio between the local and the steady-state energy release rate $\mathcal{G}/\mathcal{G}_{ss}$ becomes significantly lower than when the crack length to thickness ratio a/h tends to zero; when, instead an interior edge is considered, the same ratio decreases only marginally; the barrier to propagation becomes less relevant, and it manifests itself at a smaller distance from the edge. These calculations have been carried out in statics, so the locution 'advancing crack' or 'the crack is converging to the edge' should be considered within the static condition. As shown in Yu, He and Hutchinson (2001), the actual behaviour is dependent on the substrate and film elastic properties through the Dundurs parameters, defined as

$$D_1 = \frac{\overline{E}_f - \overline{E}_s}{\overline{E}_f + \overline{E}_s}, \qquad D_2 = \frac{G_f(1-2\nu_s) - G_s(1-2\nu_f)}{2G_f(1-\nu_s) + G_s(1-\nu_f)}, \tag{14.45}$$

with $\overline{E}_{(\cdot)} = E_{(\cdot)}/(1-\nu^2_{(\cdot)})$ and $G_{(\cdot)} = E_{(\cdot)}/(1-\nu^2_{(\cdot)})$. Note that for a film and substrate of the same material $D_1 = D_2 = 0$, while in the present considerations it is assumed that $D_2 = 0$, as is typical in fracture mechanics.

A similar study, see again Yu, He and Hutchinson (2001), has been carried out to investigate the proclivity to delamination for a system intact except for a *crack* at a corner or at an interior edge: there is a clear barrier to propagation for the corner edge, while for the interior edge, the barrier is almost negligible (i.e. the fracture energy $\mathcal{G}_f$ is almost equal to the steady-state crack propagation value $\mathcal{G}_{ss}$ when the crack distance from the interior edge tends to zero).

The tractions obtained for an interior edge in a semi-infinite film are shown in Figure 14.10 as a function of the distance a/h from the edge and of the Dundurs parameters. Note that the normal stress changes sign: it is tensile near the edge and becomes compressive at a distance less than the thickness; moreover, a stiffer substrate ($D_1 = -0.5$) induces a higher restraint and therefore a higher traction near the edge. Figure 14.11 shows, instead, the shear and normal stresses for a corner edge without a crack. These solutions do not pretend to solve the singularity very near to the crack tip (the case assuming no elastic mismatch, i.e. $D_1 = D_2 = 0$, is a classical problem in elasticity): to capture very local effects the asymptotic solution must be explored more

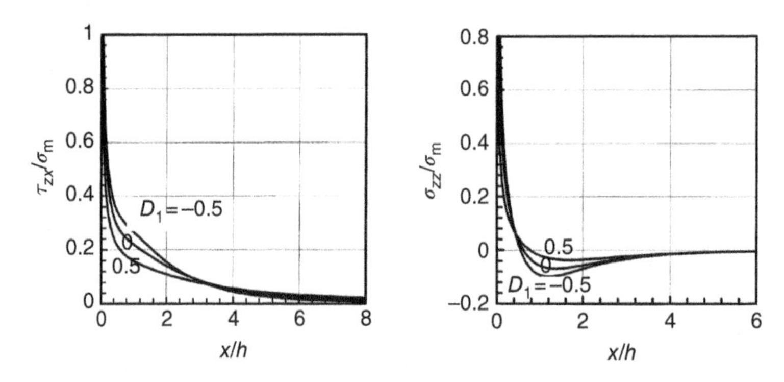

Figure 14.10 Tangential (τ_{zx}) and normal (σ_{zz}) interface stresses for an interior crack. *Source*: Yu, He and Hutchinson (2001), Figure 10. Reproduced with permission of Elsevier.

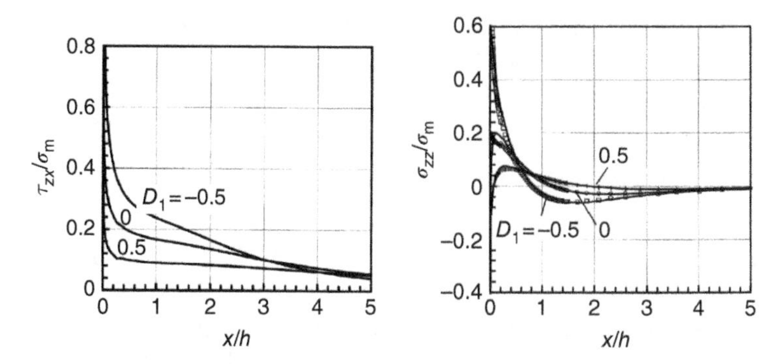

Figure 14.11 Tangential (τ_{zx}) and normal (σ_{zz}) interface stresses for an edge crack. *Source*: Yu, He and Hutchinson (2001), Figure 12. Reproduced with permission of Elsevier.

deeply, e.g. as explained by Liu, Suo and Ma (1999) or, alternatively, an accurate finite element analysis must be carried out.

Typically, however, the deposition process creates film strips of finite width, and one edge could interact with another, see Figure 14.12. This long-range interaction occurs not only through the film but also through the substrate; according to Yu, He and Hutchinson (2001), the ratio between the average in-plane stress and the mismatch stress modifies with the distance x from each edge, as depicted in Figure 14.12, and as a function of the ratio $2L/h$ and for the two values of the Dundurs parameter $D_1 = 0.5$ and $D_1 = -0.5$. As a guide, an estimate of the middle cross-sectional stress in the film has also been suggested by Yu, He and Hutchinson (2001) as follows:

$$1 - \frac{\sigma_{\text{avg}}}{\sigma_{\text{m}}} = \frac{4h\overline{E}_{\text{f}}}{\pi L\overline{E}_{\text{s}}}. \tag{14.46}$$

For the buckling problem induced by the residual stresses, an obvious phenomenon is the buckling of slender MEMS parts, such as beams or membranes. More interesting is the wrinkling of thin films on soft elastic substrates (Bowden *et al.*, 1998): it occurs

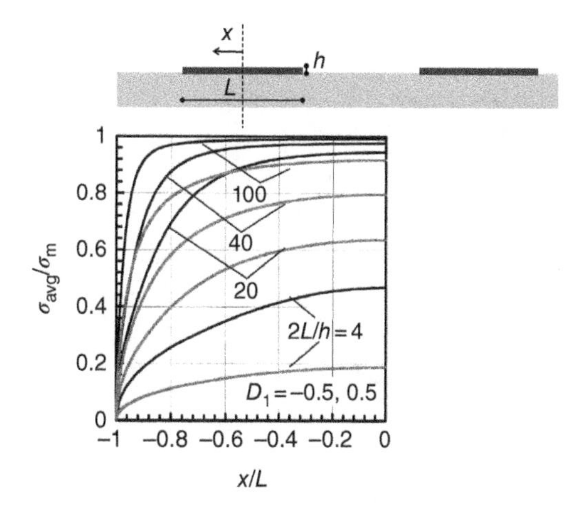

Figure 14.12 Average of the residual in-plane stress (see Equation 14.46) between edges of film strips. *Source*: Yu, He and Hutchinson (2001), Figure 13. Reproduced with permission of Elsevier.

without delamination between film and substrate if, theoretically, the compressive stress in the film is larger than (Freund and Suresh, 2003; Mei *et al.*, 2007)

$$\sigma_{\mathrm{m}} = \left(\frac{3}{8}\right)^{2/3} \overline{E}_{\mathrm{f}}^{1/3} \overline{E}_{\mathrm{s}}^{2/3} \tag{14.47}$$

and the wavelength of the wrinkled surface is

$$\frac{\lambda}{h_{\mathrm{f}}} = \pi \left(\frac{8\overline{E}_{\mathrm{f}}}{3\overline{E}_{\mathrm{s}}} \right)^{1/3}. \tag{14.48}$$

Another configuration where buckling manifests itself is the out-of-plane deflection of a thin film that is initially not bonded or is delaminated from the thick substrate, see Figure 14.13. Prior to buckling, an equi-biaxial mismatch stress $\sigma_{\mathrm{m}} < 0$ is present in the film, including the debonded region $-a < x < +a$, $-\infty < y < +\infty$ (see Figure 14.13). After buckling in the delaminated region, the film stress resultant $t_{\mathrm{m}} = \sigma_{\mathrm{m}} h_{\mathrm{f}} (< 0)$ decreases and becomes $t_a - t_{\mathrm{m}}$, with $t_a = t_x(a) > 0$ (positive). At the still-attached edge both the tensile force t_a and a bending moment $m_x(x)$ are acting on the cross-section. To calculate the stress leading to the buckling of the debonded region, the thin plate theory is invoked, with its extension to finite rotations. After recalling that

$$m_x(x) = \frac{\overline{E}_{\mathrm{f}} h_{\mathrm{f}}^3}{12} w''(x), \tag{14.49}$$

where the apex indicates the derivative with respect to the spatial coordinate x, introducing $u(x)$ and $w(x)$ as the displacements along the x and z coordinates, respectively, the system to be solved is

$$\left[u'(x) + 1/2 w'(x)^2\right]' = 0 \tag{14.50}$$

$$w^{\mathrm{IV}}(x) - \frac{12 t_x(x)}{\overline{E}_{\mathrm{f}} h_{\mathrm{f}}^3} w''(x) = 0, \qquad \text{for } -a < x < +a. \tag{14.51}$$

From the first equation, it follows that t_x must be constant, i.e. $t_x = t_x(a) = t_a$, and therefore the problem simplifies to

$$w^{\mathrm{IV}} - \frac{12 t_a}{\overline{E}_{\mathrm{f}} h_{\mathrm{f}}^3} w'' = 0, \qquad \text{for } -a < x < +a \tag{14.52}$$

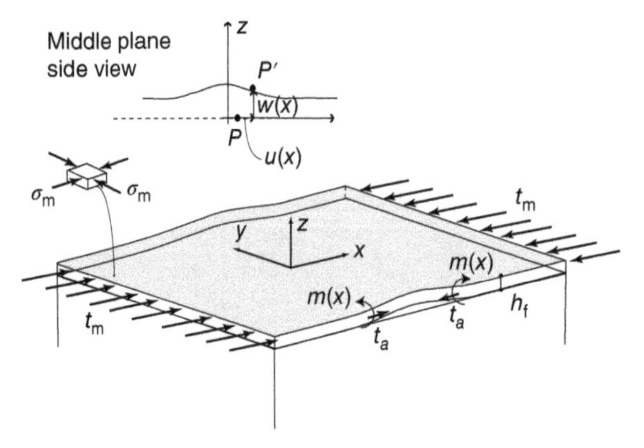

Figure 14.13 Buckling-induced delamination due to a residual stress in a thin film.

under the boundary conditions for clamped edges $w(\pm a) = 0$ and $w'(\pm a) = 0$. The eigenvalue problem gives the critical stress (recall that $t_{\mathrm{m}} = \sigma_{\mathrm{m}} h_{\mathrm{f}}$)

$$\sigma_{\mathrm{m}} = -\frac{\pi^2 \overline{E}_{\mathrm{f}}}{12}\left(\frac{h_{\mathrm{f}}}{a}\right)^2. \tag{14.53}$$

Alternatively, Equation 14.53 can be solved for a_{m}, i.e. the critical size of the debonded region for a given mismatch stress, giving

$$a_{\mathrm{m}} = \frac{\pi h_{\mathrm{f}}}{2}\sqrt{\frac{\overline{E}_{\mathrm{f}}}{3|\sigma_{\mathrm{m}}|}}. \tag{14.54}$$

A thorough discussion of the approach and its implications can be found in Freund and Suresh (2003), Chapter 5.

As a final undesired effect due to residual stresses presented in this section, cracking inside a film and (possibly) its substrate is considered (Freund and Suresh, 2003). We start by assuming that there is no elastic mismatch between a film and its substrate. The work W_{c} necessary to overcome the fracture resistance for a crack advancing in a perfectly vertical plane inside the film thickness h_{f}, under linear elastic fracture mechanics (see Figure 14.14), is

$$W_{\mathrm{c}} = h_{\mathrm{f}}\Gamma_{\mathrm{f}}\begin{cases} a_{\mathrm{c}}/h_{\mathrm{f}}, & a_{\mathrm{c}} \leq h_{\mathrm{f}}, \\ 1 + (a_{\mathrm{c}}/h_{\mathrm{f}} - 1)\Gamma_{\mathrm{s}}/\Gamma_{\mathrm{f}}, & a_{\mathrm{c}} > h_{\mathrm{f}}. \end{cases} \tag{14.55}$$

where a_{c} is the critical crack length.

The driving external work W_{m}, given by the background elastic field and responsible for the crack advancement, must be compared with the work resistance W_{c} defined previously; W_{m} can be calculated by considering a crack advancing orthogonally to the free surface in an elastic half-space, according to the expression given by Tada, Paris and Irwin (1985):

$$W_{\mathrm{m}} = \int_0^{a_{\mathrm{c}}} \mathcal{G}(\eta)\mathrm{d}\eta, \tag{14.56}$$

with the energy release rate provided by the rather convoluted expression

$$\mathcal{G}(\eta) = \frac{h_{\mathrm{f}}\sigma_{\mathrm{m}}^2}{\overline{E}_{\mathrm{f}}}\begin{cases} \pi c_{\mathrm{e}}^2 \dfrac{\eta}{h_{\mathrm{f}}}, & 0 < \eta < h_{\mathrm{f}}, \\ \dfrac{4}{\pi}\dfrac{\eta}{h_{\mathrm{f}}}\left(1.69 - 0.47\dfrac{h_{\mathrm{f}}}{\eta} + 0.032\dfrac{h_{\mathrm{f}}^2}{\eta^2}\right)\arcsin^2\dfrac{h_{\mathrm{f}}}{\eta}, & h_{\mathrm{f}} < \eta, \end{cases} \tag{14.57}$$

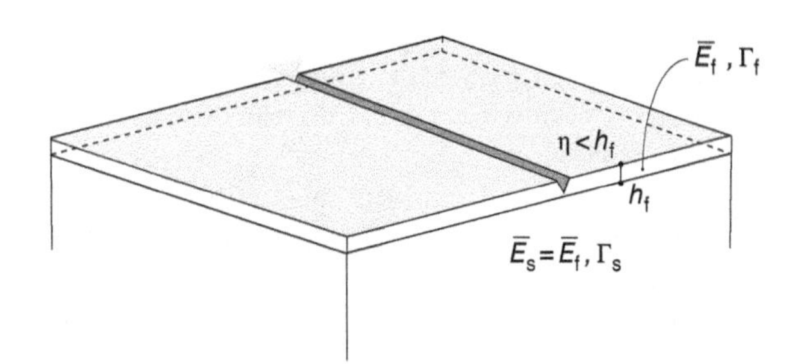

Figure 14.14 Advancing crack into a thin film due to residual stresses.

with c_e=1.1215 a nondimensional coefficient necessary to describe the stress singularity at the edge crack tip. The critical thickness leading to the film cracking implied by $W_c = W_m$ is

$$h_{fcr} = \frac{2}{\pi c_e^2} \frac{\Gamma_f \overline{E}_f}{\sigma_m^2}. \tag{14.58}$$

Now let us consider the possibility of the crack penetrating the substrate. We first assume that the fracture energy of the substrate is Γ_s (with respect to the film fracture energy Γ_f); the elastic properties are instead assumed to be the same for the film and substrate. It can be shown that the crack will not attempt to advance into the substrate when $2\Gamma_f \leq \Gamma_s$ (Freund and Suresh, 2003). However, when more realistic systems are investigated, such as those with different elastic properties for film and substrate, then the response is quite different from the aforementioned no-mismatch case. From a qualitative point of view, a substrate stiffer than the film would constrain the advancing crack more than the no-mismatch case, because the stored elastic energy would be released with more difficulty; conversely, a soft substrate, permitting a relaxation of the residual stress on a larger area, will involve more stored energy to drive the crack ahead towards the substrate. Quantitative conclusions have also been drawn in the context of linear elastic fracture mechanics (Beuth, 1992).

References

Beams, J. (1959) Mechanical properties of thin films of gold and silver, in *Structure and Properties of Thin Films* (eds C. Neugebauer, J. Newkirk and D. Vermilyea), John Wiley & Sons, Inc., p. 183.

Beuth, J. (1992) Cracking of thin bonded films in residual tension. *International Journal of Solids and Structures*, **29**, 1657–1675.

Bowden, N., Brittain, S., Evans, A.G. *et al.* (1998) Spontaneous formation of ordered structures in thin films of metals supported on elastomeric polymer. *Nature*, **393**, 146–149.

Cullity, B.D. (1956) *Elements of X-ray Diffraction*, Addison Wesley.

De Wolf, I. (1996) Micro-Raman spectroscopy to study local mechanical stress in silicon integrated circuits. *Semiconductor Science and Technology*, **11**, 139–154.

Doerner, M.F. and Nix, W.D. (1988) Stresses and deformation processes in thin films on substrates. *CRC Critical Reviews in Solid State and Materials Science*, **14**, 225–268.

Fachin, F., Nikles, S.A., Dugundji, J. and Wardle, B.L. (2001) Analytical extraction of residual stresses and gradients in MEMS structures with application to CMOS-layered materials. *Journal of Micromechanics and Microengineering*, **21**, 095017.

Finot, M., Blech, I.A., Suresh, S. and Fujimoto, H. (1997) Large deformation and geometric instability of substrates with thin-film deposits. *Journal of Applied Physics*, **81**, 3457–3464.

Flinn, P.A., Gardner, D. and Nix, W.D. (1987) Measurement and interpretation of of stress in aluminum-based metallization as a function of thermal history. *IEEE Transactions on Electron Devices*, **ED34**, 689–699.

Floro, J.A. and Chason, E. (1996) Measuring Ge segregation by real-time stress monitoring during $Si_{1-x}Ge_x$ molecular beam epitaxy. *Applied Physics Letters*, **69**, 3830–3832.

Floro, J.A., Chason, E., Lee, S.R. *et al.* (1997) Real-time stress evolution during $Si_{1-x}Ge_x$ heteroepitaxy: dislocations, islanding and segregation. *Journal of Electronics Materials*, **26**, 969–979.

Freund, L.B. (2000) Substrate curvature due to thin film mismatch strain in the nonlinear deformation range. *Journal of Mechanics and Physics of Solids*, **48**, 1159–1174.

Freund, L.B., Floro, J.A. and Chason, E. (1999) Extensions of the Stoney formula for substrate curvature to configurations with thin substrates or large deformations. *Applied Physics Letters*, **74**, 1987–1989.

Freund, L.B. and Suresh, S. (2003) *Thin Film Materials. Stress, Defect Formation and Surface Evolution*, Cambridge University Press.

Hopcroft, M.A., Nix, W.D. and Kenny, T.W. (2010) What is the Young's modulus of silicon? *Journal of Microelectromechanical Systems*, **19**, 229–238.

Huang, S. and Zhang, X. (2006) Extension of the Stoney formula for film–substrate systems with gradient stress for MEMS applications. *Journal of Micromechanics and Microengineering*, **16**, 382–389.

Janssen, G., Abdalla, M., van Keulen, F. *et al.* (2009) Celebrating the 100th anniversary of the Stoney equation for film stress: developments from polycrystalline steel strips to single crystal silicon wafers. *Thin Solid Films*, **517**, 1858–1867.

Kociniewski, T., Moussodji, J. and Khatir, Z. (2014) μ-Raman spectroscopy for stress analysis in high power silicon devices. *Microelectronics Reliability*, **54**, 1770–1773.

Krishna, R., Jones, A.N., Edge, R. and Marsden, B.J. (2015) Residual stress measurements in polycrystalline graphite with micro-Raman spectroscopy. *Radiation Physics and Chemistry*, **111**, 14–23.

Lee, Y.H. and Kwon, D. (2004) Estimation of biaxial surface stress by instrumented indentation with sharp indenters. *Acta Materialia*, **52**, 1555–1563.

Li, Q., Qiu, W., Tan, H. *et al.* (2010) Micro-Raman spectroscopy stress measurement method for porous silicon film. *Optics and Lasers in Engineering*, **48**, 1119–1125.

Liu, X.H., Suo, Z. and Ma, Q. (1999) Split singularities: stress field near the edge of a silicon die on a polymer substrate. *Acta Materialia*, **47**, 67–76.

Martinez, R.E., Augustyniak, W.M. and Golovcheko, J.A. (1990) Direct measurement of crystal surface stress. *Physical Review Letters*, **64**, 1035–1038.

Mei, H., Huang, R., Chung, J.Y. *et al.* (2007) Buckling modes of elastic thin films on elastic substrates. *Applied Physics Letters*, **90**, 151902.

Merle, B. (2013) *Mechanical properties of thin films studied by bulge testing*, Ph.D. thesis, Universität Erlangen.

Nix, W.D. (1989) Mechanical properties of thin films. *Metallurgical Transactions A*, **20A**, 2217—2245.

Pan, J.T. and Blech, I.A. (1984) *In situ* measurement of refractory silicides during sintering. *Journal of Applied Physics*, **55**, 2874–2880.

Retajczyk, T.F. and Sinha, A.K. (1980) Elastic stiffness and thermal expansion coefficient of BN films. *Applied Physics Letters*, **36**, 161–163.

Rosakis, A.J., Singh, R.B., Tsuji, Y. *et al.* (1998) Full field measurements of curvature using coherent gradient sensing – application to to thin-film characterization. *Thin Solid Films*, **325**, 42–54.

Schell-Sorokin, A.J. and Tromp, R.M. (1990) Mechanical stress in (sub) monolayer epitaxial films. *Physical Review Letters*, **64**, 1039–1042.

Shull, A.L. and Spaepen, F. (1996) Measurements of stress during vapor deposition of copper and silver thin films and multilayers. *Journal of Applied Physics*, **80**, 6423–6256.

Stoney, G.G. (1909) The tension of metallic films deposited by electrolysis. *Proceedings of the Royal Society of London*, **A82**, 172–175.

Suresh, S. and Giannakopoulos, A. (1998) A new method for estimating residual stresses by instrumented sharp indentation. *Acta Materialia*, **46**, 5755–5767.

Swadener, J.G., Taljat, B. and Pharr, G. (2001) Measurement of residual stress by load and depth sensing indentation with spherical indenters. *Journal of Materials Research*, **16**, 2091–2102.

Tada, H., Paris, P. and Irwin, G. (1985) *The Stress Analysis of Cracks Handbook*, Dell Research Corp.

Thornton, J.A. and Hoffman, D.W. (1989) Stress-related effects in thin film. *Thin Solid Films*, **171**, 5–31.

Van Eijden, R. (2008) *Numerical–experimental analysis of an improved bulge test for thin films*, Ph.D. thesis, Eindhoven University of Technology.

Volkert, C.A. (1991) Stress and plastic flow in silicon during amorphization by ion-bombardment. *Journal of Applied Physics*, **70**, 3521–3527.

Xiang, Y. (2005) *Plasticity in Cu thin films: an experimental investigation of the effect of microstructure*, Ph.D. thesis, Harvard University.

Yu, H.H., He, M.Y. and Hutchinson, J.W. (2001) Edge effects in thin film delamination. *Acta Materialia*, **49**, 93–107.

Zhu, L.N., Xu, B.S., Wang, H.D. and Wang, C.B. (2015) Measurement of residual stresses using nanoindentation method. *Critical Reviews in Solid State and Materials Sciences*, **40**, 77–89.

15

Damping in Microsystems

15.1 Introduction

The technology of MEMS is being rapidly developed for the production of miniaturized sensors and actuators that can be successfully applied in a broad variety of situations. In many cases, MEMS are composed of solid parts in relative motion and a proper design should therefore include a prediction of the dynamic response and, in particular, of the dissipation induced by various factors. Traditionally, the dissipation level in a micromachined sensor is measured by means of the so-called quality factor Q. The quality factor of a vibrating structure can be defined as

$$Q = \frac{2\pi W}{\Delta W}, \tag{15.1}$$

where ΔW and W are the energy lost per cycle and the maximum value of energy stored in the device, respectively. Dissipation is determined by many causes, each connected to an amount of energy loss and the total ΔW is computed as the sum of the different contributions, assuming perfect decoupling. The sources of dissipation can be roughly divided into two categories: the former ('fluid damping') embraces phenomena related to the interaction of the solid parts with the surrounding gases; the latter is connected to loss mechanisms in the solid material, and for this reason is often called 'intrinsic damping' or 'solid damping'. At least two factors increase the complexity of the task: (i) MEMS are fully 3D microstructures, which cannot always be reduced to simple 1D or 2D models; (ii) working pressures are spread over a large range (1 bar to 10^{-6} bar). If the quality factor of a MEMS is measured in a chamber with controlled pressure (assuming constant temperature) a plot like the one drawn in Figure 15.1 is expected, in which different regimes can be identified.

A large class of inertial sensors, like accelerometers, work at pressures in the range of 0.5–1.0 bar. Here, the fluid can be treated as a continuum and standard (Navier–)Stokes models can be applied with corrected 'slip' boundary conditions. These will be briefly reviewed in Section 15.2.

Other MEMS are packaged at lower pressures and rarefaction effects must be accounted for with techniques of rarefied gas dynamics (Chapman and Cowling, 1960; Cercignani, 1988; Gad-el-Hak, 2002; Karniadakis and Beskok, 2002). A measure of rarefaction is provided by the Knudsen number $\text{Kn} = \lambda/L$, where λ is the molecule

Mechanics of Microsystems, First Edition. Alberto Corigliano, Raffaele Ardito, Claudia Comi, Attilio Frangi, Aldo Ghisi, and Stefano Mariani.

Companion website: www.wiley.com/go/corigliano/mechanics

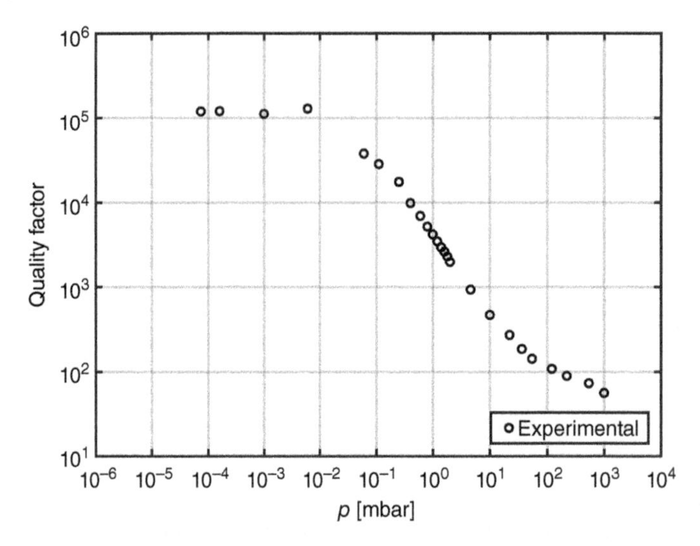

Figure 15.1 Comb-finger device: quality factor versus pressure down to near vacuum. *Source:* Frangi 2006. Reproduced with permission of Wiley.

mean free path (proportional to $1/p$) and L is a typical dimension of the flux, e.g. the gap between plates in Poiseuille (squeeze) flow. As a rule of thumb, for $0.1 < \mathrm{Kn} < 10$, the flow develops in the so-called transition regime, which can be addressed only by means of kinetic theories (e.g. the Boltzmann equation). Since perturbations are small, simplified models can be utilized, such as the one discussed in Section 15.3. These provide accurate predictions but are very expensive in terms of computing time.

However, many MEMS, such as gyroscopes, resonators or Lorentz force magnetometers are packaged with a getter, which guarantees pressures of the order of 1 mbar, i.e. exactly in the linear part of the plot shown in Figure 15.1. This regime is known as free-molecule flow and occurs typically at $\mathrm{Kn} > 10$, where collisions between molecules can be neglected. In this case simplified tools are preferred, as discussed in Section 15.4. The dissipation is proportional to pressure $\Delta W_{\mathrm{fm}}(p) = p\Delta \tilde{W}_{\mathrm{fm}}$ and becomes progressively negligible.

Eventually the plateau of Figure 15.1 is reached. Here, 'solid' losses dominate, which generate a $\Delta W_{\mathrm{solid}}$ that is independent of pressure. These include thermoelastic damping, anchor losses and other still unidentified sources of dissipation, as discussed in Sections 15.5 to 15.7.

15.2 Gas Damping in the Continuum Regime with Slip Boundary Conditions

Let us now focus on gas dissipation at ambient pressure. The squeeze flow between two parallel plates has been thoroughly analysed in the literature by means of the Reynolds equations and interesting results are summarized in the review of Bao *et al.* (2002). However real MEMS are fully 3D and we prefer to focus here on approaches that allow this particular feature to be addressed.

Obtaining a quantitative estimate of damping for a general 3D MEMS remains a formidable task unless suitable working hypotheses are introduced, as discussed in the following. A first remark is that the typical velocities of movable parts are always a fraction of the speed of sound, hence the Mach number M is small with respect to unity: $M \ll 1$. This implies that the incompressibility condition $\mathrm{div}\mathbf{u} = 0$ will be assumed in the following, where $\mathbf{u}$ is the fluid velocity. It is worth stressing that this does not imply that density is constant everywhere, but only that density is constant along a streamline.

A second crucial parameter is the Reynolds number $\mathrm{Re} = UL/\nu$, where U is the flow velocity, and ν is the kinematic viscosity. In a periodic movement with maximum amplitude D and frequency f, an estimate of the Reynolds number can be provided as $\mathrm{Re} = f\,DL/\nu$. If $\mathrm{Re} \ll 1$, the convective nonlinear terms in the Navier–Stokes equations can be neglected. For instance, in the case of the biaxial accelerometer of Figure 15.2, $L = 2.6$ μm is the gap between the plates, $f \simeq 4400$ Hz is the resonating frequency, the maximum displacement is approximately one-tenth of the gap and, hence, the Reynolds number is of the order of 10^{-4}. It is here worth stressing that, typically, the condition $\mathrm{Re} \ll 1$ is met by a very large category of vibrating MEMS.

Moreover, the ratio of inertia forces to viscous forces is generally measured by means of the nondimensional Stokes number $\mathrm{St} = fL^2/\nu$. If $\mathrm{St} \ll 1$, a quasi-static approach applies and inertia terms can be neglected. This condition is respected by many inertial MEMS, and in particular by the accelerometer shown in Figure 15.2, but is not essential in what follows.

Finally, since the amplitude of vibration is generally small with respect to the gap, the viscous force can be assumed to be independent of the position of the shuttle and

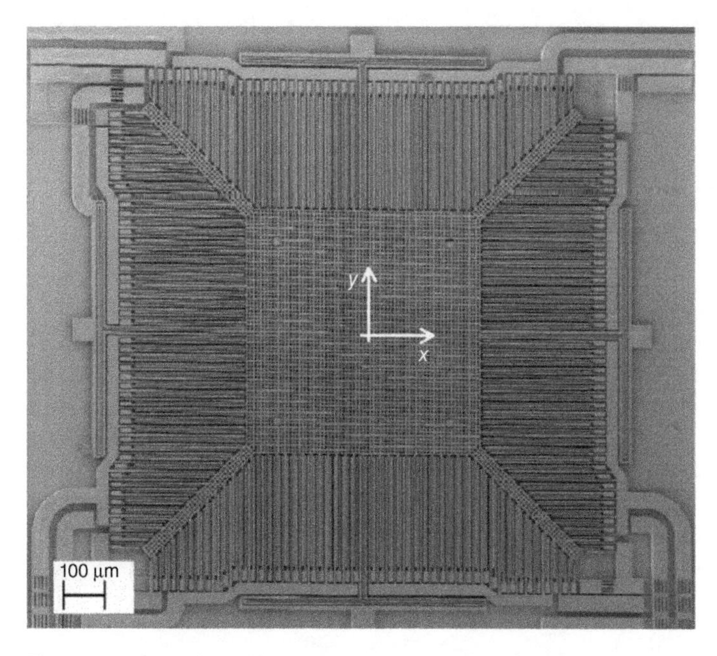

Figure 15.2 Biaxial accelerometer. *Source:* Frangi 2006. Reproduced with permission of Wiley.

the analyses can be run in the configuration of static equilibrium. Couette (shear) flow clearly represents an exception to the gap versus amplitude condition, but in this case the viscous force can be assumed approximately independent of the position of the shuttle finger.

As a consequence of all these remarks, unless addressing specific conditions violating these assumptions, an incompressible Stokes formulation guarantees very accurate results for Kn < 0.1 and is by far more stable and agile than a full Navier–Stokes compressible model.

Given a field velocity $\mathbf{u}$ in the fluid, global stresses $\boldsymbol{\sigma}$ and surface tractions $\mathbf{t}$ are defined by

$$\sigma = -p\mathbf{1} + \eta \left(\text{grad } \mathbf{u} + \text{grad }^{\mathrm{T}}\mathbf{u}\right), \qquad \mathbf{t} = \sigma \cdot \mathbf{n},$$

where $\mathbf{u}$ is the fluid velocity and $\mathbf{n}$ is the unit vector orthogonal to the surface, which points outside the fluid domain. Tractions $\mathbf{t}$ hence represent the opposite of the forces per unit area exerted by the fluid on the solid surface area S.

While for applications at the macroscopic scale and ambient pressure no-slip boundary conditions generally apply, i.e. $\mathbf{u}(\mathbf{x}) = \mathbf{g}(\mathbf{x})$ ($\mathbf{g}$ denotes the velocity of solid surfaces), $\mathbf{x} \in S$, this is not the case at the scale of MEMS. With typical gaps between structures of a few micrometres, the flow mainly develops in the slip regime even at 1 bar and hence slip boundary conditions must be utilized.

Let $\mathbf{t}^S$ denote the surface components of tractions:

$$\mathbf{t}^S(\mathbf{x}) = [\mathbf{1} - \mathbf{n}(\mathbf{x}) \otimes \mathbf{n}(\mathbf{x})] \cdot \mathbf{t}(\mathbf{x}),$$

where $\mathbf{1} - \mathbf{n}(\mathbf{x}) \otimes \mathbf{n}(\mathbf{x})$ is the surface projector tensor. Boundary slip conditions should be expressed in terms of $\mathbf{t}^S$ (and not in terms of the normal derivative of $\mathbf{u}$):

$$\mathbf{u}(\mathbf{x}) = \mathbf{g}(\mathbf{x}) - c_{\mathrm{t}}\mathbf{t}^S(\mathbf{x}), \qquad c_{\mathrm{t}} := \frac{2-\sigma}{\sigma}\frac{\lambda}{\eta}, \tag{15.2}$$

where σ is the tangential momentum accommodation coefficient (Karniadakis and Beskok, 2002).

With these definitions, the complete formulation of the exterior Stokes problem with slip boundary conditions reads:

$$\rho\frac{\partial \mathbf{u}}{\partial t} - \text{grad } p(\mathbf{x}) + \eta\nabla^2\mathbf{u}(\mathbf{x}) = \mathbf{0}, \qquad \text{div}\mathbf{u}(\mathbf{x}) = 0 \quad \text{in} \quad \Omega_\infty - \Omega, \tag{15.3}$$

$$\mathbf{u}(\mathbf{x}) = \mathbf{g}(\mathbf{x}) - c_{\mathrm{t}}\mathbf{t}^S \quad \text{on} \quad S. \tag{15.4}$$

Equations 15.3 and 15.4 can be implemented in any finite element method code (including suitable provisions for the incompressibility constraint), but when a quasi-static assumption holds and fluid inertia can be neglected, integral equation approaches are highly competitive with respect to classical finite element methods, since they reduce the dimensionality of the domain to be meshed. Moreover, the output of interest from the analysis is the damping force on the movable shuttle and the boundary element method is more accurate, since the traction field exerted by the fluid on the MEMS surface is a direct unknown of the numerical formulation (Frangi, 2005; Frangi, Spinola and Vigna, 2006).

15.2.1 Experimental Validation at Ambient Pressure

We now present two validations with experimental results. The first test has been performed on the silicon biaxial accelerometer of Figure 15.2, produced by STMicroelectronics and discussed in Frangi, Spinola and Vigna (2006).

The accelerometer consists of a central suspended shuttle ('rotor') and four series of external 'stators' attached to the substrate. Fixed and movable parts are conventionally termed stators and rotors, respectively, even in those cases where a purely translatory motion occurs. Both parts are endowed with a series of long and thin plates interdigitated into capacitors serving both as actuators and sensors. The length of the longest plates is 277 μm, the in-plane width is 3.9 μm, the height is 15 μm and the air gaps between plates are 2.6 μm. The gap between the rotor and the substrate is 4.2 μm. The rotor is attached to the substrate by means of silicon springs, which are stiff in the out-of-plane (z) direction and very compliant in the xy plane. Hence, the rotor is essentially free to move parallel to the xy plane and is otherwise constrained.

During the experimental tests, the accelerometer is set in oscillation along the y-direction by means of electrostatic actuation in a wide range of frequencies centred at the undamped resonating frequency $f_0 = 4400$ Hz. The actuating forces are such that the maximum amplitude of oscillation of the plates is much smaller than the air gaps. A set of plates, i.e. those parallel to the x-direction, mainly induce a Poiseuille-like flow, while those parallel to the y-direction induce a Couette-like flux.

In these working conditions, the accelerometer can be effectively represented by a linearized 1D model:

$$m\ddot{y} + b\dot{y} + ky = F\cos\omega t \qquad \text{or} \qquad \ddot{y} + 2\xi\omega_0\dot{y} + \omega_0^2 y = \frac{F}{M}\cos\omega t, \tag{15.5}$$

where m denotes the mass of the structure, k the equivalent stiffness due to springs and b the damping coefficient. Since the amplitude of the oscillation is small with respect to the plate gap, the coefficients can be reasonably taken as independent of y. All the comparisons will be presented in terms of $b = 2\xi\omega_0 M$, which represents the viscous force on the shuttle (in the 1D model) at unit velocity.

The highly repetitive layout and the simple movement along the y-direction allow to restrict the analysis to simple 'units' and extrapolate the results to the overall structure. Hence, the focus is set, initially, on the geometry of Figure 15.3, where a rotor plate (darker) and two stator plates (lighter) are considered. Two different situations are addressed, corresponding to Poiseuille and Couette flow, by simply imposing a unit velocity orthogonal and parallel to the plates, respectively.

Table 15.1 collects the results concerning the force exerted on the shuttle. As anticipated, the global force is obtained by simply scaling the force obtained for the simple unit by the global length of the rotor plates in squeeze and Couette flow, respectively. A third analysis has been performed on a sample of the central bulky mass with holes to estimate the contribution to the global force at ambient pressure. As expected, Poiseuille flow provides the most important contribution. The values of these forces are computed at ambient pressure both both no-slip and slip boundary conditions, assuming $\sigma = 0.9$ and a mean free path of $\lambda = 0.064$ μm. While stick boundary conditions overestimate the force on the rotor, an excellent agreement is obtained with slip boundary conditions.

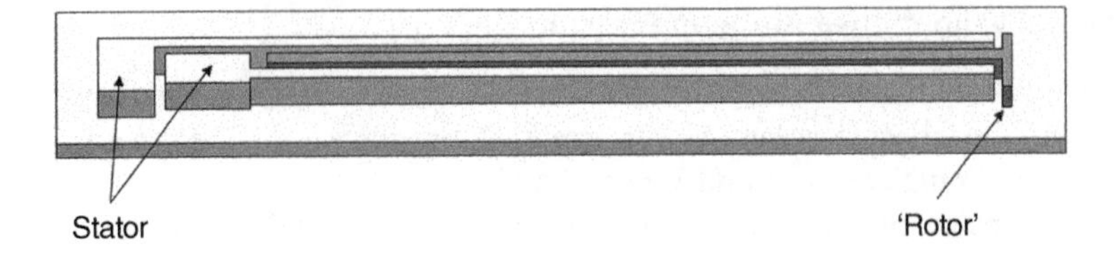

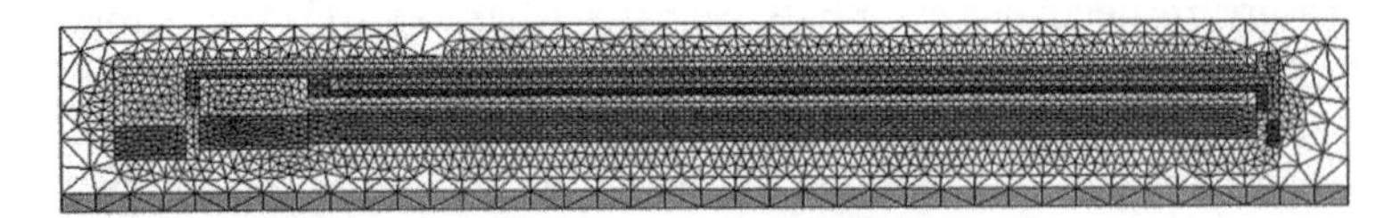

Figure 15.3 Single unit used for simplified squeeze and Couette analysis: geometry and finest mesh. *Source:* Frangi 2006. Reproduced with permission of Wiley.

Table 15.1 Comparison of experimental results with numerically computed damping forces (for unit velocity) at ambient pressure: contribution from different parts of the rotor and global results.

	Numerical 'no-slip'	Numerical 'slip'	Experimental
Poiseuille flow	2.32×10^{-4} N	2.10×10^{-4} N	–
Couette flow	7.37×10^{-6} N	7.03×10^{-6} N	–
Mass with holes	2.10×10^{-6} N	1.94×10^{-6} N	–
Total force	2.41×10^{-4} N	2.19×10^{-4} N	2.21×10^{-4} N

15.2.2 Effects of Decreasing Working Pressure

The approach presented here is not limited to those MEMS that can be considered as a collection of rigid structures. For instance, in the case of the Tang resonator shown in Figure 15.4, the deformability and the contribution to damping of the flexible springs cannot be neglected (Cercignani *et al.*, 2007).

Both the shuttle and the stators are endowed with a series of thin plates interdigitated into capacitors (Figure 15.5). The length of each plate is 20 μm, the in-plane width is 3.2 μm and the air gaps are 2.6 μm wide; the length of each spring is 405 μm, while its width is 2.2 μm. The out-of-plane dimension of the whole structure is 15 μm. The gap between the shuttle and the substrate is 1.8 μm. This structure also has a first resonating frequency of approximately 4400 Hz.

The coefficient identified can thus be interpreted as the 'experimentally measured' force exerted by the gas on the shuttle when the shuttle has a unit velocity along $\mathbf{e}_2$ at the specific pressure considered. The same procedure can be repeated at different pressures, yielding the evolution of the measured force.

Different pressures can be simulated by setting the correct value of the c_t slip coefficient in Equation 15.2. Indeed, λ is the mean free molecular path, which is inversely proportional to pressure, with $\lambda \simeq 0.064$ μm at 1 bar. Considering the gaps between

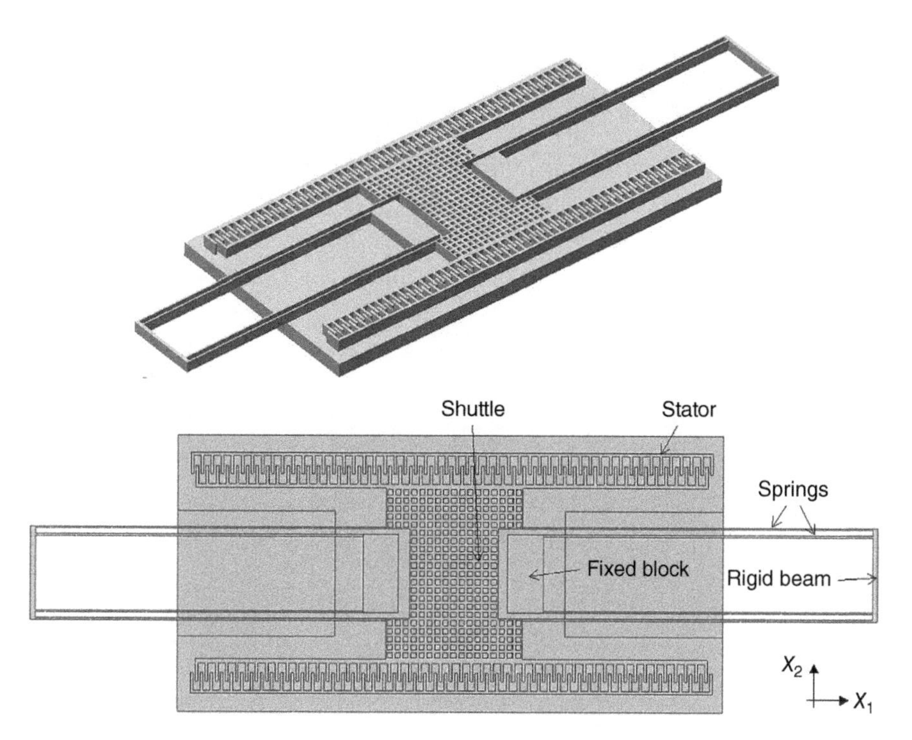

Figure 15.4 Comb-finger resonator: 3D view and 2D layout. *Source:* Cercignani *et al.* (2007), Figure 1. Reproduced with permission of Elsevier.

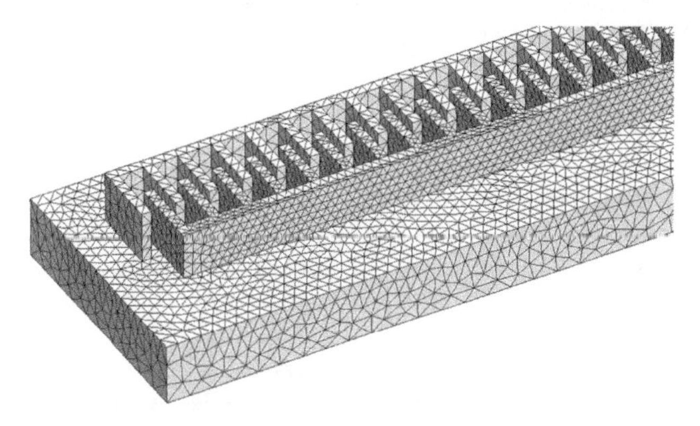

Figure 15.5 Comb-finger resonator: detail of the finest mesh adopted for the boundary element method approach. *Source:* Cercignani *et al.* (2007), Figure 3. Reproduced with permission of Elsevier.

the structure and the substrate and between the fingers, the transition regime starts at approximately $p = 0.1$ bar. At higher pressures, the Stokes model with slip boundary conditions is thus expected to be accurate. This is confirmed by the results plotted in Figure 15.6, which demonstrate the discrepancy at low pressures, dictated by the increasing importance of rarefaction effects in the flow.

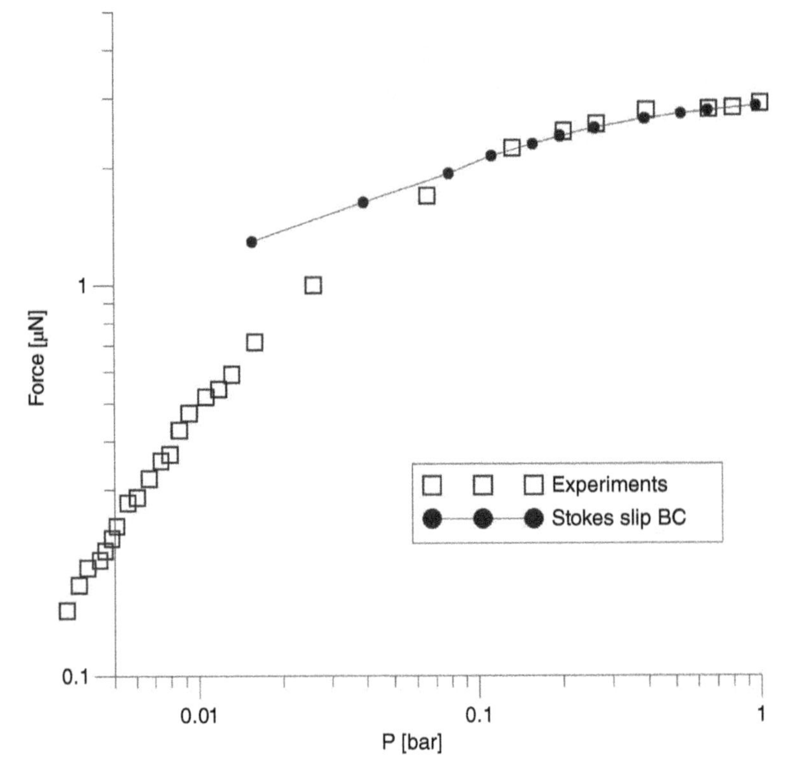

Figure 15.6 Boundary element method results with slip boundary conditions (BC): comparison with experiments. Viscous force exerted on the shuttle for a unit velocity. *Source:* Cercignani *et al.* (2007), Figure 4. Reproduced with permission of Elsevier.

15.3 Gas Damping in the Rarefied Regime

When the gas flow enters the transition regime ($\text{Kn} \gtrsim 0.1$), corrections due to rarefaction issues are mandatory. For instance, in a MEMS with parallel plates having a gap of 2 μm, this may occur for pressures $p < 0.1$ bar. Viscous forces decrease but, unless almost-vacuum conditions are reached, they still represent the main source of mechanical dissipation.

Most contributions in the literature have focused on the analysis of the squeeze flow between two parallel plates, since this relatively simple geometry enables the inherent complexity of the problem to be reduced and since in some MEMS the squeeze flow dissipation dominates over other sources. In this context, models for the continuum regime, and in particular the Reynolds equation of lubrication theory, are applied with a modified viscosity to account for rarefaction effects (Veijola *et al.*, 1995; Veijola, 2004).

It is known that the effective viscosity approach applies rigorously to the two simple idealized Poiseuille and Couette flows between two infinite parallel plates (Cercignani and Daneri, 1963; Cercignani, 1988). However, even in these cases, the two modified viscosity laws are different; as a consequence, the definition of a suitable modified viscosity becomes critical in the 3D context of a real MEMS (Cercignani *et al.*, 2007). The comparison with 'exact' approaches in the case of extreme rarefaction shows, however, that the accuracy of these techniques is rather limited, even though they provide the correct order of magnitude of dissipative terms and can be very useful during the design phase.

15.3.1 Evaluation of Damping at Low Pressure using Kinetic Models

In principle, a correct theoretical description of gas flow in the transition regime can be obtained by solving the Boltzmann equation (Chapman and Cowling, 1960; Cercignani, 1988), a complex nonlinear integro-differential equation providing the distribution function of molecular velocities at any flow field location. Mathematical difficulties prevent closed-form solutions of the Boltzmann equation from being obtained in cases of practical interest, but efficient Monte Carlo methods have been developed for its numerical treatment (Bird, 1994). Unfortunately, in most MEMS flows, reference Mach numbers and deviations from local equilibrium are small and difficult to capture using traditional statistical Monte Carlo methods. A number of possible modifications to statistical particle schemes have been proposed (Cai *et al.*, 2000), but research in this direction is still very active.

A deterministic approach to the numerical solution of kinetic equations becomes viable if the complicated collision integral in the Boltzmann equation is replaced by a simpler expression. As described next, in the BGK model kinetic equation (Bhatnagar, Gross and Krook, 1954), the term giving the collisional rate of change of the distribution function is simply proportional to the departure from local equilibrium. In its simpler and more useful form, the model contains a single disposable function (the collision frequency), which depends on local density and temperature and assigns the same decay rate to all kinetic modes. Hence, the hydrodynamic limit of the model is only partially correct, since the collision frequency can be tuned to obtain either the correct fluid viscosity or the correct thermal conductivity, but not both. In spite of its shortcomings, the BGK model is often more accurate than expected, particularly in problems where momentum and heat transport are not equally important and the collision frequency can be adjusted to match the most important transport coefficient. Its applications to rarefied gases date back to the first semi-analytical solutions (Cercignani and Daneri, 1963; Sone, 1972) for Poiseuille and Couette flow. Subsequently, many different numerical applications have been presented in the literature (Aoki, Kanba and Takata, 1997), focusing especially on high speed applications. More recently, a number of papers have appeared, reporting studies of low-speed flows of various complexity using deterministic numerical solutions of the BGK model kinetic equation (Sharipov, 1996).

Let $f(\mathbf{x}, \boldsymbol{\xi})$ denote the velocity distribution function of molecules, where $\mathbf{x}$ are space coordinates and $\boldsymbol{\xi}$ is molecular velocity. If grad denotes the gradient with respect to $\mathbf{x}$, the BGK model of the Boltzmann equation reads:

$$\frac{\partial f}{\partial t} + \boldsymbol{\xi} \cdot \operatorname{grad} f = \mathrm{c}(\rho, T)(f_{\mathrm{M}} - f), \tag{15.6}$$

where the right-hand side relaxation term replaces the collision operator of the Boltzmann equation, which accounts for binary collisions between particles; $\mathrm{c}(\rho, T)$ is the collision frequency, which is assumed independent of $\boldsymbol{\xi}$ and f_{M} is the local equilibrium Maxwellian:

$$f_{\mathrm{M}} = \frac{\rho}{(2\pi \mathcal{R} T)^{3/2}} \exp\left(-\frac{|\boldsymbol{\xi} - \mathbf{u}|^2}{2\mathcal{R}T}\right). \tag{15.7}$$

where the macroscopic velocity $\mathbf{u}$, density ρ and temperature T are moments of f in the velocity space:

$$\rho = \int_{R^3} f \mathrm{d}\boldsymbol{\xi}, \qquad \rho\mathbf{u} = \int_{R^3} f\boldsymbol{\xi}\mathrm{d}\boldsymbol{\xi}, \qquad T = \frac{1}{3\mathcal{R}\rho}\int_{R^3} f|\boldsymbol{\xi} - \mathbf{u}|^2\mathrm{d}\boldsymbol{\xi}. \tag{15.8}$$

Finally, $\mathcal{R}$ is the specific gas constant (the universal constant divided by the molar mass). It can be shown (Cercignani, 1988) that the correct fluid viscosity $\eta(T)$ in the hydrodynamic limit can be obtained from Equation 15.6 by setting

$$\mathrm{c}(\rho, T) = \frac{\rho\mathcal{R}T}{\eta(T)}. \tag{15.9}$$

The BGK model (15.6) can thus be interpreted as a relaxation towards a local Maxwellian equilibrium state. The gas interacts with both fixed and movable surfaces immersed in a virtually unconfined domain, which is often truncated at a sufficient distance from the structures. To simplify the gas–wall interaction, it is assumed that the scattering from the wall is either diffused or specular, or a combination of the two. As a matter of fact, silicon surfaces originating from etching procedures are very rough, so that, as a first approximation, diffuse reflection will be assumed in the following. If $\mathbf{n}$ denotes the unit normal vector to the solid surface S_{R} pointing inside the fluid domain, diffuse reflection (Cercignani, 1988) means that molecules are re-emitted from the surface according to the wall distribution function,

$$f = \frac{\rho_{\mathrm{w}}}{(2\pi\mathcal{R}T_{\mathrm{w}})^{3/2}}\exp\left(-\frac{|\boldsymbol{\xi} - \mathbf{g}|^2}{2\mathcal{R}T_{\mathrm{w}}}\right), \qquad \text{for} \quad (\boldsymbol{\xi} - \mathbf{g})\cdot\mathbf{n} > 0 \quad \text{on } S_{\mathrm{R}}, \tag{15.10}$$

where T_{w} is the wall temperature, $\mathbf{g}$ is the wall velocity and ρ_{w} is fixed so as to impose zero net flux across the surface:

$$\rho_{\mathrm{w}} = \left(\frac{2\pi}{\mathcal{R}T_{\mathrm{w}}}\right)^{1/2}\int_{R^3,(\boldsymbol{\xi}-\mathbf{g})\cdot\mathbf{n}<0} (|\boldsymbol{\xi} - \mathbf{g}|\cdot\mathbf{n})f\mathrm{d}\boldsymbol{\xi}. \tag{15.11}$$

On the outer fictitious boundary S_{F}, on the contrary, the inflow is forced to have the same features as the far region assumed at equilibrium with $f = f_0$, where f_0 is the Maxwellian at rest defined by temperature T_0, density ρ_0 and zero macroscopic velocity:

$$f = f_0 = \frac{\rho_0}{(2\pi\mathcal{R}T_0)^{3/2}}\exp\left(-\frac{|\boldsymbol{\xi}|^2}{2\mathcal{R}T_0}\right), \qquad \text{for} \quad \boldsymbol{\xi}\cdot\mathbf{n} > 0 \quad \text{on } S_{\mathrm{F}}. \tag{15.12}$$

Equations 15.6, 15.10 and 15.12 represent a nonlinear large-scale problem, since the velocity distribution f generally depends on the two 3D vector variables $\mathbf{x}$, $\boldsymbol{\xi}$. However, if we limit ourselves to 2D situations in which the geometry is invariant with respect to translations along x_3 and $v_3 = 0$, following Aoki, Kanba and Takata (1997), we introduce two new distribution functions,

$$\chi = \int_R f\mathrm{d}\xi_3, \qquad \psi = \int_R \xi_3^2 f\mathrm{d}\xi_3, \tag{15.13}$$

associated with the equations:

$$\frac{\partial\chi}{\partial t} + \boldsymbol{\xi}\cdot\mathrm{grad}\,\chi = \frac{\rho}{\sigma}(\chi_{\mathrm{M}} - \chi), \qquad \frac{\partial\psi}{\partial t} + \boldsymbol{\xi}\cdot\mathrm{grad}\,\psi = \frac{\rho}{\sigma}(\psi_{\mathrm{M}} - \psi), \tag{15.14}$$

which can be obtained from Equation 15.6 by integrating over R with respect to ξ_3 (after multiplying by ξ_2^3 in the latter case). In Equation 15.14:

$$\chi_\mathrm{M} = \int_R f_\mathrm{M} \mathrm{d}\xi_3 = \frac{\rho}{2\pi \mathcal{R} T} \exp\left(-\frac{|\xi - \mathbf{u}|^2}{2\mathcal{R}T}\right),$$

$$\psi_\mathrm{M} = \int_R \xi_3^2 f_\mathrm{M} \mathrm{d}\xi_3 = \frac{\rho}{2\pi} \exp\left(-\frac{|\xi - \mathbf{u}|^2}{2\mathcal{R}T}\right).$$

In Equation 15.14, and in the following, bold letters denote 2D vectors, e.g. $\xi = \xi_1 \mathbf{e}_1 + \xi_2 \mathbf{e}_2$, and integrations are limited to the 2D space. Density, mean velocity and temperature can be expressed in terms of χ and ψ as:

$$\rho = \int_{R^2} \chi \mathrm{d}\xi, \qquad \rho \mathbf{u} = \int_{R^2} \xi \chi \mathrm{d}\xi,$$

$$T = \frac{1}{3\mathcal{R}\rho} \int_{R^2} |\xi - \mathbf{u}|^2 \chi \mathrm{d}\xi + \frac{1}{3\mathcal{R}\rho} \int_{R^2} \psi \mathrm{d}\xi.$$

Far-field conditions (normal $\mathbf{n}$ pointing towards fluid domain) become $\chi = \chi_0, \psi = \psi_0$ for $\xi \cdot \mathbf{n} > 0$ on S_F, where

$$\chi_0 = \frac{\rho_0}{2\pi \mathcal{R} T_0} \exp\left(-\frac{|\xi|^2}{2\mathcal{R}T_0}\right), \qquad \psi_0 = \frac{\rho_0}{2\pi} \exp\left(-\frac{|\xi|^2}{2\mathcal{R}T_0}\right),$$

while diffused reflection at rigid walls S_R reads:

$$\chi = \frac{\rho_\mathrm{w}}{2\pi \mathcal{R} T_\mathrm{w}} \exp\left(-\frac{|\xi - \mathbf{g}|^2}{2\mathcal{R}T_\mathrm{w}}\right), \quad \text{for} \quad (\xi - \mathbf{g}) \cdot \mathbf{n} > 0, \tag{15.15}$$

$$\psi = \frac{\rho_\mathrm{w}}{2\pi} \exp\left(-\frac{|\xi - \mathbf{g}|^2}{2\mathcal{R}T_\mathrm{w}}\right), \quad \text{for} \quad (\xi - \mathbf{g}) \cdot \mathbf{n} > 0, \tag{15.16}$$

with

$$\rho_\mathrm{w} = \frac{(2\pi)^{1/2}}{(\mathcal{R}T_\mathrm{w})^{1/2}} \int_{R^2, (\xi - \mathbf{g}) \cdot \mathbf{n} < 0} |(\xi - \mathbf{g}) \cdot \mathbf{n}| \chi \mathrm{d}\xi.$$

15.3.2 Linearization of the BGK Model

Despite simplifications, the numerical solution of these equations is still a very demanding task, which can be greatly simplified by the following key hypotheses. Perturbations in inertial MEMS are generally small enough so as to apply a linear expansion to the distribution functions,

$$\chi \simeq \chi_0(1 + \chi_1), \qquad \psi \simeq \psi_0(1 + \psi_1). \tag{15.17}$$

Moreover, if the nondimensional Stokes number $\mathrm{St} = fL^2/\nu \ll 1$, a quasi-static approach applies and the velocity of rigid walls is enforced as a boundary condition while the time derivatives in Equation 15.14 can be dropped. It is worth stressing that these hypotheses permit a straightforward numerical solution of the model and fully describe many real structures. However, a general solution without these restrictive assumptions is always possible at a greater computational cost.

Moreover, since all the walls are assumed to remain at $T_\mathrm{w} = T_0$, the linear term in the expansion for β can be dropped, with the consequence that the equations for χ and ψ

decouple and can be solved independently. If the aim of our analyses is the evaluation of forces exerted on the structures by the fluid, only χ is required and the focus is set, henceforth, on the single 2D equation:

$$\tilde{\xi} \cdot \text{grad}\ \chi_1 + \tilde{\mathbf{n}}_0 \chi_1 = \tilde{\mathbf{n}}_0 \left[M^{\chi_0} + 2\tilde{\xi} \cdot \mathbf{M}^{\chi_1} \right], \tag{15.18}$$

where $\tilde{\mathbf{n}}_0 = \beta_0^{1/2} \rho_0 \mathcal{R} T_0 / \eta_0$, $\beta_0 = 1/(2\mathcal{R}T_0)$, $\tilde{\xi} = \beta_0^{1/2} \xi$ and:

$$M^{\chi_0} = \frac{1}{\pi} \int_{R^2} \exp\left(-|\tilde{\xi}|^2\right) \chi_1 \mathrm{d}\tilde{\xi}, \tag{15.19}$$

$$\mathbf{M}^{\chi_1} = \frac{1}{\pi} \int_{R^2} \tilde{\xi} \exp\left(-|\tilde{\xi}|^2\right) \chi_1 \mathrm{d}\tilde{\xi}. \tag{15.20}$$

Accordingly, boundary conditions are also linearized. On the far-field boundary S_F, we have:

$$\chi_1 = 0 \quad \text{for} \quad \xi \cdot \mathbf{n} > 0. \tag{15.21}$$

If we assume that $T_w = T_0$ and that

$$\tilde{\mathbf{g}} = \frac{\mathbf{g}}{\sqrt{2\mathcal{R}T}} \ll 1,$$

the linearized equation of diffuse reflection on S_R reads:

$$\chi_1 = \rho_{w1} + 2\tilde{\xi} \cdot \tilde{\mathbf{g}}, \quad \text{for} \quad \xi \cdot \mathbf{n} > 0, \tag{15.22}$$

with

$$\rho_{w1} = \sqrt{\pi} \tilde{\mathbf{g}} \cdot \mathbf{n} - \frac{2}{\sqrt{\pi}} \int_{R^2, \tilde{\xi} \cdot \mathbf{n} < 0} (\tilde{\xi} \cdot \mathbf{n}) \exp\left(-|\tilde{\xi}|^2\right) \chi_1 \mathrm{d}\tilde{\xi}.$$

Once the distribution function has been solved, the global force acting on the solid walls can be computed by integrating the stress tensor $\boldsymbol{\sigma}$ on the wall (Chapman and Cowling, 1960) (only in-plane components have been retained):

$$\sigma = \int_{R^2} \chi(\xi - \mathbf{u}) \otimes (\xi - \mathbf{u}) \mathrm{d}\xi \simeq \int_{R^2} \chi_0 \xi \otimes \xi \mathrm{d}\xi + \int_{R^2} \chi_0 \chi_1 \xi \otimes \xi \mathrm{d}\xi.$$

After some algebraic manipulations:

$$\begin{aligned} \sigma &= \frac{\rho_0}{2\beta_0} \mathbf{1} + \frac{\rho_0}{\pi \beta_0} \int_{R^2} \exp\left(-|\tilde{\xi}|^2\right) \tilde{\xi} \otimes \tilde{\xi} \chi_1 \mathrm{d}\tilde{\xi} \\ &= \rho_0 \mathcal{R} T_0 \left[\mathbf{1} + \frac{2}{\pi} \int_{R^2} \exp\left(-|\tilde{\xi}|^2\right) \tilde{\xi} \otimes \tilde{\xi} \chi_1 \mathrm{d}\tilde{\xi} \right]. \end{aligned} \tag{15.23}$$

15.3.3 Numerical Implementation

The usual approach for addressing steady-state problems with the nonlinear BGK model is to solve an unsteady problem with steady boundary conditions and let time evolve to infinity. This guarantees, in general, robust convergence properties. However, the linearized quasi-steady BGK model (15.18) can be solved directly by means of a semi-implicit iterative approach with no recourse to time evolution. Moreover, since MEMS layouts are intrinsically very regular, owing to process requirements, the finite volumes (or finite differences) technique can be applied without restriction and is here preferred to other domain methods, such as finite elements. For simplicity, the region of interest is meshed with a structured Cartesian grid of N_C rectangular cells and χ_1 has been assumed constant over each cell. The 2D velocity space is 'truncated' and only the square surface $|\xi_1| \leq \tilde{U}, |\xi_2| \leq \tilde{U}$ is considered; $\tilde{U}$ is set such that the contributions from larger velocities are negligible. The truncated velocity space is partitioned into N_U^2 equal square velocity cells of size $\Delta\tilde{U}$; χ_1 is assumed to be constant over each velocity cell and equal to the value in the cell centre. The global number of unknowns is thus $N_C N_U^2$.

Let $x^{(i)}, y^{(j)}$ be the coordinates of the centre of a generic space cell (in row i and column j of the Cartesian grid) and $\xi^{(k)}, \xi^{(\ell)}$ be the coordinates of the centre of a generic velocity cell. Also let $\Phi_{i,j,k,\ell}$ denote $\chi_1(x^{(i)}, y^{(j)}, \xi^{(k)}, \xi^{(\ell)})$. Adopting a classical first-order upwind scheme, the discretized version of Equation 15.18 reads:

$$2s_k\xi^{(k)}\frac{\Phi_{i,j,k,\ell} - \Phi_{i-s_k,j,k,\ell}}{\Delta H_i + \Delta H_{i-s_k}} + 2s_\ell\xi^{(\ell)}\frac{\Phi_{i,j,k,\ell} - \Phi_{i,j-s_\ell,k,\ell}}{\Delta V_j + \Delta V_{j-s_\ell}} + \tilde{\mathbf{n}}_0\Phi_{i,j,k,\ell}$$
$$= \tilde{\mathbf{n}}_0\left[M^{\chi_0} + 2\xi^{(k)}M_1^{\chi_1} + 2\xi^{(\ell)}M_2^{\chi_1}\right], \qquad (15.24)$$

where $s_m = \text{sign}[\xi^{(m)}]$ and ΔH_i, ΔV_j are the dimension of the rectangular cell of centre $x_1^{(i)}, x_2^{(j)}$. It should be remarked that, in order to introduce numerical diffusion only in flow direction, the upwind scheme depends on the sign of the molecular velocity (Aoki, Kanba and Takata, 1997). These equations, supplemented by the boundary conditions (15.21, 15.22) yield a linear system of large dimensions, which is solved with a semi-implicit iterative scheme (Frangi, Frezzotti and Lorenzani, 2007). Thanks to the upwind discretization utilized, the numerical solution of the system can be recast into a triangular sparse form by a simple renumbering of the unknowns. This can be solved at a cost that is proportional to the number of unknowns $N_C N_U^2$. The procedure converges rapidly and this steady-state semi-implicit approach turns out to be much faster than an explicit time-dependent technique.

Unfortunately, the space cell size depends on the free molecular path λ. Good results can be obtained if, in the regions of high gradients, the mesh size is some fraction of λ. The proposed numerical tool is thus ideal for analyses at low pressures, as evidenced in the following, but in view of the microdimensions of MEMS, it still maintains a practical applicability up to the lower limits of the slip-flow region, above which alternative 'continuum-like' approaches should be applied.

15.3.4 Application to MEMS

The biaxial accelerometer of Figure 15.2, already utilized to test the performance of 3D Stoke solvers, has been selected to validate the proposed approach. Quantitatively correct estimates can be obtained by analysing and scaling results obtained on a single 'unit' like the one depicted in Figure 15.7, in which the motion of the shuttle occurs in the direction orthogonal to the long plates and induces a Poiseuille flow. Indeed, the contribution to damping from the central mass and from the other sets of interdigitated plates can be reasonably neglected, as a first approximation. The advantage of this simplification is that the plates in Figure 15.7 are sufficiently long to ignore 3D effects and permit a 2D analysis by focusing the attention on section A, where the structure is depicted in dark grey and the light grey region is, by contrast, the fluid domain analysed.

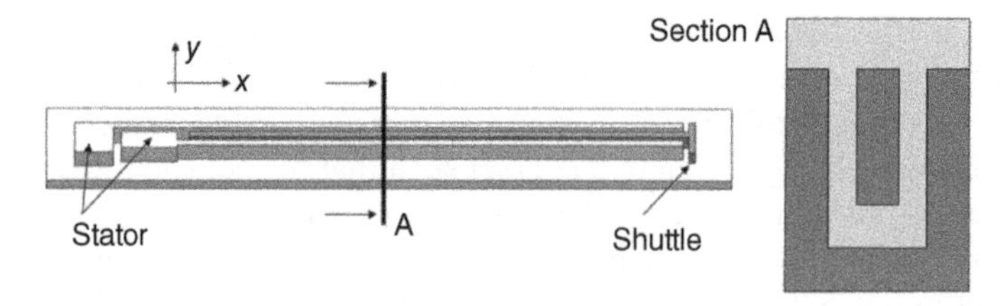

Figure 15.7 Single unit used to analyse the structure. *Source:* Frangi, Frezzotti and Lorenzani (2007), Figure 2. Reproduced with permission of Elsevier.

Section A has been discretized with several structured meshes; the air region above the section is rather limited in extent, but different analyses using 'larger' domains did not demonstrate significant variations of the forces on the shuttle. The force acting on the shuttle cross-section (of depth 1 μm), when the shuttle moves with unit velocity towards the stator, is considered the primary output.

Table 15.2 presents numerical results for different meshes and pressures. Far-field and wall temperature has been set to $T_0 = 293$ K and the dynamic viscosity to $\eta_0 = 1.8 \times 10^{-5}$ in SI units. The meshes utilized are M1, M2, M3, M4 and M5, with typical cell sizes of 0.4 μm, 0.2 μm, 0.1 μm, 0.05 μm and 0.025 μm, respectively. The analysis stops when the relative norm of the increment over one iteration: $\|\mathbf{\Phi}^{(n+1)} - \mathbf{\Phi}^{(n)}\| / \|\mathbf{\Phi}^{(n)}\|$ becomes smaller than the fixed tolerance, 10^{-5}. The speed of convergence depends, as largely expected, on the value of the far-field pressure, p_0, imposed. For mesh M3, the numbers of iterations and the global computing times at different pressures are collected in Table 15.3. All the analyses were run on a Dell Precision 490 with a Xeon 3.2 GHz dual core processor implementing a serial Fortran90 code.

As recalled in the previous section, the size of the cells to be utilized in kinetic approaches depends on the mean free molecular path, in the sense that at least four or five cells per molecular path should be utilized. The cost of the approach thus rapidly increases with p_0. The width of the channel between the stator and the shuttle is 2.6 μm,

Table 15.2 Forces [μN] acting on the a cross-section of the shuttle of unit depth.

	1 bar	10^{-1} bar	10^{-2} bar	10^{-3} bar	10^{-4} bar	10^{-5} bar
M1	0.448^{-1}	0.688^{-2}	0.106^{-2}	0.112^{-3}	0.112^{-4}	0.112^{-5}
M2	0.280^{-1}	0.635^{-2}	0.105^{-2}	0.112^{-3}	0.113^{-4}	0.113^{-5}
M3	0.190^{-1}	0.615^{-2}	0.104^{-2}	0.112^{-3}	0.113^{-4}	0.113^{-5}
M4	0.144^{-1}	0.607^{-2}	0.103^{-2}	0.113^{-3}	–	–
M5	0.123^{-1}	0.600^{-2}	0.103^{-2}	0.113^{-3}	–	–

Table 15.3 Mesh M3: number of iterations and CPU time.

Pressure	1 bar	10^{-1} bar	10^{-2} bar
Iterations for mesh M3	2269	211	106
CPU time	521 s	49 s	24 s

which explains the rapid convergence for pressures below 0.1 bar and the coarse prediction at ambient pressure, where $\lambda = 0.064$ μm. However, it is worth stressing once more that the approach proposed is intended for medium- to low-pressure applications, i.e. for $p \leq 0.1$ bar, since other simpler approaches can be applied in the slip-flow regime, such as the Stokes model of previous sections.

Hence, the analyses have been here performed up to 1 bar only for the purpose of comparing results with experimental data, as presented in Figure 15.8 for M5. This motivated the use of extremely fine meshes, which would otherwise be unnecessary. The finest mesh contains 321,152 cells of almost uniform size and $\tilde{U} = 4$ with 20^2 cells in the velocity space.

While at high pressures a coarser discretization of the velocity space should be sufficient, at high Knudsen numbers, the use of a larger number of velocities is generally recommended. However, according to the numerical tests performed in this work, the damping forces (the only forces of interest for the analysis at hand) all differ by a few percent when increasing the number of velocities. This is possibly because the output of interest, the overall viscous force, is a global measure. It should also be recalled that the maximum speeds of MEMS at these frequencies are still very low, and always below 1 m/s.

However, although applications of this approach to 3D problems have been reported (Frangi, Ghisi and Frezzotti, 2008), they still remain very challenging from a numerical point of view.

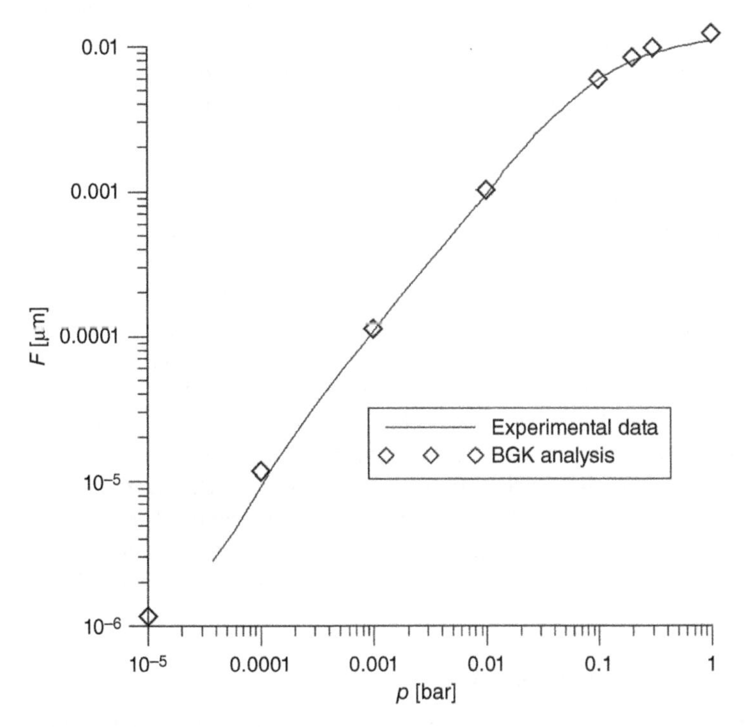

Figure 15.8 Biaxial accelerometer: comparison between experimental and numerical forces [μN] for the finest mesh M5. *Source:* Frangi, Frezzotti and Lorenzani (2007), Figure 4. Reproduced with permission of Elsevier.

15.4 Gas Damping in the Free-Molecule Regime

When the working pressure of MEMS further decreases, important simplifications can be adopted. A large number of MEMS available on the market, such as gyroscopes or Lorentz force magnetometers, operate at pressures in the range of 1 mbar and at frequencies just above the audio band (i.e. around 20 kHz). Gas damping is often still dominant with respect to classical sources of solid dissipation. In air, the mean free path of molecules is $\lambda = 64$ μm at 1 mbar and, considering a typical flow length d of 2 μm, one has $\mathrm{Kn} \simeq 30$. According to standard classifications of rarefied gas dynamics (Cercignani, 1988; Chapman and Cowling, 1960; Bird, 1994) collisions between molecules can be neglected and the assumption of free-molecule flow is fully justified.

Excluding the modified viscosity approach already discussed in the introduction to Section 15.3, the available approaches can be classified in two general categories: (i) the test particle Monte Carlo method; (ii) integral equation techniques based on the collisionless Boltzmann equation.

The test particle Monte Carlo method is a statistical approach belonging to the family of direct simulation Monte Carlo techniques (Bird, 1994). A large number of typical molecular trajectories is computed and these collectively predict the behaviour of the real system. Since intermolecular collisions can be neglected, these trajectories are independent of one another and may be generated serially. However, the stochastic nature of this approach is a major drawback for the relatively low-frequency applications at hand, where the thermal speed of molecules dominates over perturbation terms, generating an annoying statistical noise (Cai *et al.*, 2000; Frangi, Ghisi and Coronato, 2009). As a consequence, 3D simulations are very challenging. Apart from the choice of the molecule–wall interaction law, the test particle Monte Carlo method is exact and generates reference numerical data.

The second family, related to the angular coefficient method of the Russian school, enforces, via an integral equation formulation, the kinetic approaches of Section 15.3.1, setting $c(\rho, T) = \mathbf{0}$ in Equation 15.6 according to the assumption of no collisions between molecules. This technique fully generalizes the approach of Christian (1966), to which it reduces when the interactions of the vibrating shuttle with nearby walls can be neglected. Preliminary work (Kádár *et al.*, 1996; Bao *et al.*, 2002) in this direction has simulated the experiments reported by Zook *et al.* (1992) on a doubly clamped thin plate. Some limiting assumptions, pointed out by Hutcherson and Ye (2004), have been removed (Frangi, Ghisi and Coronato, 2009; Frangi, 2009), leading to a fully 3D deterministic approach based on first principles and on the simple diffuse model for wall–molecule interaction, which is realistic for polysilicon surfaces. The approach has been validated against analytical data, the test particle Monte Carlo method and known experimental results. Recent improvements in the implementation based on multicore parallelism and concepts of computer graphics allow almost full-scale realistic MEMS to be addressed.

15.4.1 Boundary Integral Equation Approach

The deterministic model implemented for the simulation, proposed in Frangi, Ghisi and Coronato (2009) and Frangi (2009), is an application of the collisionless Boltzmann equation, which rests on the following assumptions: (i) the mean free-molecule path is much larger than the typical dimension d of the flow (i.e. Kn $\gg$ 1); (ii) molecules and solid surfaces interact according to the diffuse-reflection model, which basically states that molecules are re-emitted by the walls according to the wall equilibrium Maxwellian; (iii) perturbations are small so that quadratic terms in the expansion of variables can be neglected. Moreover, if the frequency f of perturbations is slow with respect to thermal velocity, i.e. $d \times f / \sqrt{2\mathcal{R}T_0} \ll 1$ (where $\mathcal{R}$ is the universal gas constant divided by the molar mass and T_0 is the temperature), the formulation can be further simplified leading to the model discussed herein (see, however, Frangi (2009) for applications to high frequencies). Let $J(\mathbf{x}, t)$ denote the linear term in the expansion of the flux of incoming molecules at point $\mathbf{x}$ of the MEMS surface. This scalar unknown is governed by the following 'quasi-static' integral equation:

$$
\begin{aligned}
J(\mathbf{x}, t) = \sqrt{\pi}\tilde{g}_n(\mathbf{x}, t) - \frac{1}{\pi}\int_{S^+} J(\mathbf{y}, t)\,(\mathbf{r}\cdot\mathbf{n}(\mathbf{x}))\,(\mathbf{r}\cdot\mathbf{n}(\mathbf{y}))\,\frac{1}{r^4}\mathrm{d}S \\
+ \frac{3}{2}\frac{1}{\sqrt{\pi}}\int_{S^+} (\mathbf{r}\cdot\tilde{\mathbf{g}}(\mathbf{y}, t))\,(\mathbf{r}\cdot\mathbf{n}(\mathbf{x}))\,(\mathbf{r}\cdot\mathbf{n}(\mathbf{y}))\,\frac{1}{r^5}\mathrm{d}S, \qquad (15.25)
\end{aligned}
$$

where $r = \| \mathbf{y} - \mathbf{x} \|$; $\mathbf{n}$ is the outward normal to the surface; S^+ denotes the portion of surface visible from $\mathbf{x}$ and $\tilde{\mathbf{g}}(\mathbf{y}, t) = \mathbf{g}(\mathbf{y}, t)/\sqrt{2\mathcal{R}T_0}$ is the normalized velocity of surfaces, with $\tilde{g}_n = \tilde{\mathbf{g}} \cdot \mathbf{n}$ its projection along $\mathbf{n}$. A posteriori, a second integral equation provides the distribution of perturbation forces $\mathbf{t}$ on the structure:

$$
\begin{aligned}
-\frac{\mathbf{t}(\mathbf{x}, t)}{\rho_0 2\mathcal{R}T_0} = \frac{1}{4}J(\mathbf{x}, t)\mathbf{n}(\mathbf{x}) + \frac{1}{\sqrt{\pi}}\tilde{g}_n(\mathbf{x}, t)\mathbf{n}(\mathbf{x}) + \frac{1}{2\sqrt{\pi}}\tilde{\mathbf{g}}_t(\mathbf{x}, t) \\
- \frac{3}{8\pi}\int_{S^+} \mathbf{r}(\mathbf{r}\cdot\mathbf{n}(\mathbf{x}))(\mathbf{r}\cdot\mathbf{n}(\mathbf{y}))\frac{1}{r^5}J(\mathbf{y}, t)\mathrm{d}S \\
+ \frac{2}{\pi^{3/2}}\int_{S^+} \mathbf{r}(\mathbf{r}\cdot\tilde{\mathbf{g}}(\mathbf{y}, t))(\mathbf{r}\cdot\mathbf{n}(\mathbf{x}))(\mathbf{r}\cdot\mathbf{n}(\mathbf{y}))\frac{1}{r^6}\mathrm{d}S, \qquad (15.26)
\end{aligned}
$$

where $\tilde{\mathbf{g}}_t = \tilde{\mathbf{g}} - \tilde{g}_n\mathbf{n}$. For instance, if a thin plate of dimensions $H \times V \times t$ ($t \ll H, V$) is moving with velocity $\mathbf{g}$ far from any reflecting surface so that $J = 0$, the total force acting on the plate is

$$
\mathbf{F} = \rho_0\sqrt{2\mathcal{R}T_0}\left[1/\left(2\sqrt{\pi}\right)\mathbf{g}_t + \left(\sqrt{\pi}/2 + 2/\sqrt{\pi}\right)g_n\mathbf{n}\right]HV. \qquad (15.27)
$$

In general, however, the contribution of reflecting surfaces cannot be neglected. Apart from exceptions, such as the case of two opposing plates in sliding motion (Park, Bahukudumbi and Beskok, 2004), where an analytical expression for $\mathbf{t}$ is available, the numerical solution of Equations 15.25 to 15.26 is mandatory. A large-scale parallel implementation of the two integral equations allows simulation of realistic MEMS structures. Geometrically, surfaces are represented as a collection of nonoverlapping 'large' planar quadrangles. Next, each quadrangle is meshed with triangular elements with the required degree of refinement and J is modelled as piecewise constant over each triangle with visibility tested only between centres of mass of the elements. The output of the analysis is the force $\mathbf{t}$ exerted on the surfaces along any direction, which can later be post-processed as follows. The displacement $\mathbf{s}$ of the deformable MEMS is assumed to be proportional to a given vector shape function: $\boldsymbol{\psi}(\mathbf{x})q(t)$. Hence, one has that $\mathbf{t}(\mathbf{x},t) = \mathbf{f}(\mathbf{x})\dot{q}(t)$, where $\mathbf{f}(\mathbf{x})$ is given by Equation 15.26, replacing $\mathbf{g}$ with $\boldsymbol{\psi}$. The equivalent damping term in the 1D reduced-order model finally becomes

$$\int_S \mathbf{t}(\mathbf{x},t)\boldsymbol{\psi}(\mathbf{x})\mathrm{d}S = \left(\int_S \mathbf{f}(\mathbf{x})\boldsymbol{\psi}(\mathbf{x})\mathrm{d}S\right)\dot{q} = b\dot{q},$$

where the integral is extended over the whole surface of the MEMS. The constant b can be conveniently expressed as

$$b = \tilde{b}\rho_0\sqrt{2\mathcal{R}T_0} = \tilde{b}p_0\sqrt{2/(\mathcal{R}T_0)}, \tag{15.28}$$

where $\tilde{b}$ is a coefficient with the dimensions of a surface area, which depends only on the problem geometry.

As a trivial example, in the case of the isolated plate of Equation 15.27 undergoing a rigid-body translation U in the direction $\mathbf{e}$, the vector shape function is $\boldsymbol{\psi}(\mathbf{y}) = \mathbf{e}$ and

$$\tilde{b} = HV\left[1/\left(2\sqrt{\pi}\right)\mathbf{g}_t\cdot\mathbf{e} + \left(\sqrt{\pi}/2 + 2/\sqrt{\pi}\right)g_n\mathbf{n}\cdot\mathbf{e}\right]. \tag{15.29}$$

From Equation 15.28, it can be appreciated that, at constant temperature, the viscous coefficient predicted is proportional to the working pressure, as expected in a free-molecule flow. Indeed, it also depends on the gas type: e.g. $\mathcal{R} = 287$ J/(kg K) for air and $\mathcal{R} = 208$ J/(kg K) for argon.

For a free plate vibrating orthogonally to its surface, Bao *et al.* (2002) predicts $\tilde{b} = 4/\sqrt{\pi}$, to be compared with the exact value $\tilde{b} = 2/\sqrt{\pi} + \sqrt{\pi}/2$ from Equation 15.29 (also validated against the test particle Monte Carlo method by Frangi, Ghisi and Coronato (2009)). The origin of this 12.5% discrepancy has been individuated by Hutcherson and Ye (2004) under the assumption of constant molecular velocity made by Bao *et al.* (2002). The overall formulation with multiple interacting surfaces has been validated by Frangi, Ghisi and Coronato (2009) against the analytical solution of Park, Bahukudumbi and Beskok (2004) for two facing plates and sliding velocity and against experimental data for a perforated MEMS. For the sake of completeness, in Figure 15.9 we present a comparison between the present model and the predictions of Bao *et al.* (2002) and the original model of Veijola *et al.* (1995). Two parallel rectangular identical plates move orthogonally to their surface separated by a gap of 4 μm. The longer side is fixed at $H = 100$ μm, while the shorter side V is varied over the range 10–100 μm. As often remarked in the literature, Bao's prediction largely underestimate the damping force, while the model of Veijola *et al.* (1995) performs better but is still far from the 'exact' results of the integral equation approach or of the test particle Monte Carlo method.

15.4.2 Experimental Validations

One of the first validations of the methodology has been discussed by Frangi, Ghisi and Coronato (2009) and addresses the out-of-plane rotational resonator of Figure 15.10 produced by STMicroelectronics. It is basically a plane structure (plane $\mathbf{e}_1$–$\mathbf{e}_2$) with uniform out-of-plane (along $\mathbf{e}_3$) thickness $t = 15$ μm. The gap between the resonator and the substrate is $g = 1.8$ μm. The highly perforated central mass is attached to the substrate via a set of four flexible beams connect to the solid circle in the middle and (almost) rigidly vibrates around the in-plane axis $\mathbf{e}_2$ so that the whole structure can be

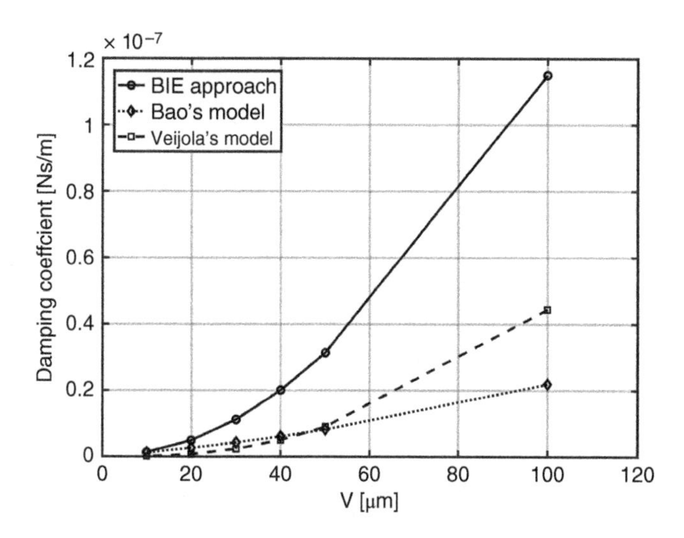

Figure 15.9 Damping force between two parallel identical rectangular plates at $T = 273$ K; $p = 1$ mbar; gap $g = 4$ μm; longer side of the plates $H = 100$ μm. Comparison between the b coefficients predicted by the integral equation approach, the model of Veijola *et al.* (1995) and the model of Bao *et al.* (2002). *Source:* Frangi *et al.* (2016), Figure 7. Reproduced with permission of IEEE.

approximated by a one-degree-of-freedom model in terms of the rotation angle ϑ, the displacement being $x\theta(t)\mathbf{e}_3$. The first resonating frequency is $f \simeq 4000$ Hz.

As can be appreciated from Figures 15.10 and 15.11, all the holes are squares of constant size and are aligned along circles so as to have almost constant spacing all over the MEMS. Dissipation occurring in the comb-finger capacitors can be reasonably neglected. According to the assumptions discussed already, gas dissipation is proportional to $\dot{\vartheta}$ and the governing equation is

$$\rho t\ddot{\vartheta}(t)\int_S x^2 \mathrm{d}S + \dot{\vartheta}(t)\int_{\partial\Omega} x t_3(\mathbf{x})\mathrm{d}S + k\vartheta(t) = C(t), \tag{15.30}$$

where S is the in-plane distribution of masses in the MEMS, $C(t)$ denotes the electrostatic forcing couple and k accounts for the the elastic stiffness of the four central springs.

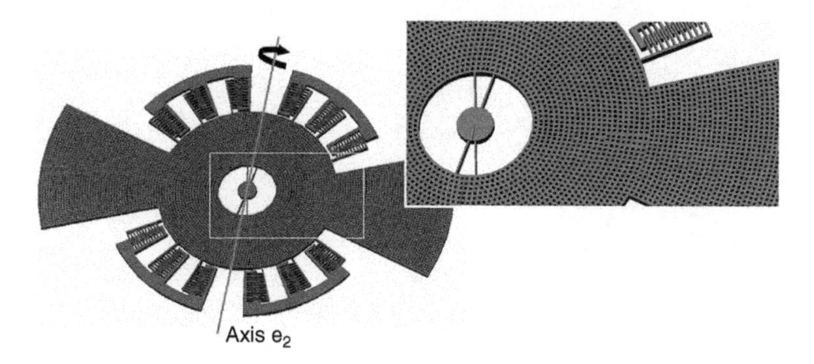

Figure 15.10 Out-of-plane rotational resonator: detail of the perforated mass. *Source:* Frangi, Ghisi and Coronato (2009), Figure 5. Reproduced with permission of Elsevier.

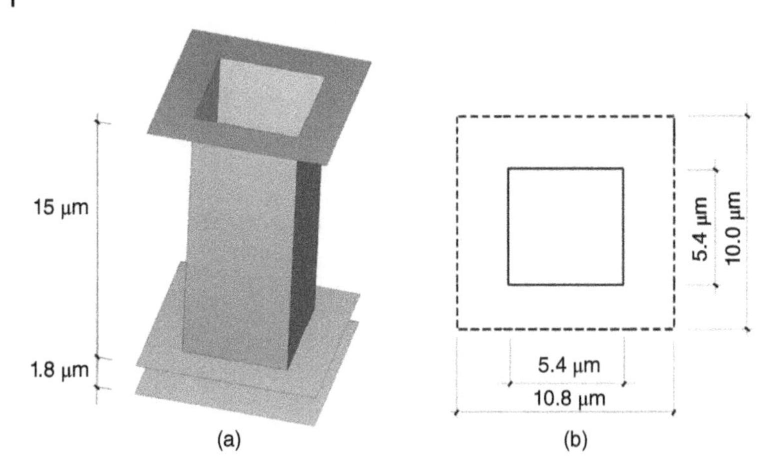

Figure 15.11 Physical bounding surfaces of a fluid elementary cell: (a) 3D view; (b) in-plane dimensions. *Source:* Frangi, Ghisi and Coronato (2009), Figure 6. Reproduced with permission of Elsevier.

At this stage, we introduce several realistic hypotheses to further simplify the problem at hand. The MEMS is assumed to be made of an assemblage of N identical units containing one hole each and having infinitesimal dimensions. Moreover, the cells near the axis $\mathbf{e}_2$ contribute negligibly to the damping forces. Finally, if we denote by $\mathbf{x}_I$ the centre of the Ith cell and by X_I its distance from the axis of rotation, the distribution of tractions is expressed as the product of X_I and of the same function $T_3(\mathbf{x}-\mathbf{x}_I)$ over any cell $t_3(\mathbf{x}) = X_I T_3(\mathbf{x}-\mathbf{x}_I)$. It should be remarked that T_3 is the out-of-plane damping force exerted on the surface of the unit due to the enforced velocity $\mathbf{e}_3$. The integrals in Equation 15.30 can now be simplified with good accuracy, since:

$$\rho t \int_S x^2 \mathrm{d}S \simeq m\left(\sum_{I=1} X_I^2\right) = mJ,$$

$$\int_{\partial\Omega} t_3(\mathbf{x}) x \mathrm{d}S \simeq \int_{\partial U} T_3(\mathbf{y}) \mathrm{d}S \left(\sum_{I=1} X_I^2\right) = bJ,$$

where m is the mass of the typical unit, ∂U the surface of the unit, and $J = \sum_{I=1} X_I^2$. Finally,

$$mJ\ddot{\vartheta}(t) + bJ\dot{\vartheta}(t) + k\vartheta(t) = C(t) \quad \text{or} \quad \ddot{\vartheta}(t) + 2\xi\omega_0\dot{\vartheta}(t) + \omega_0^2\vartheta(t) = \frac{C(t)}{mJ}$$

with

$$\omega_0 = \sqrt{\frac{k}{mJ}}, \qquad \frac{b}{m} = 2\xi\omega_0.$$

The results obtained with a mesh of $\simeq$6000 elements are presented in Figure 15.12 and compared with experimental tests conducted by STMicroelectronics on several different resonators, showing surprisingly good agreement.

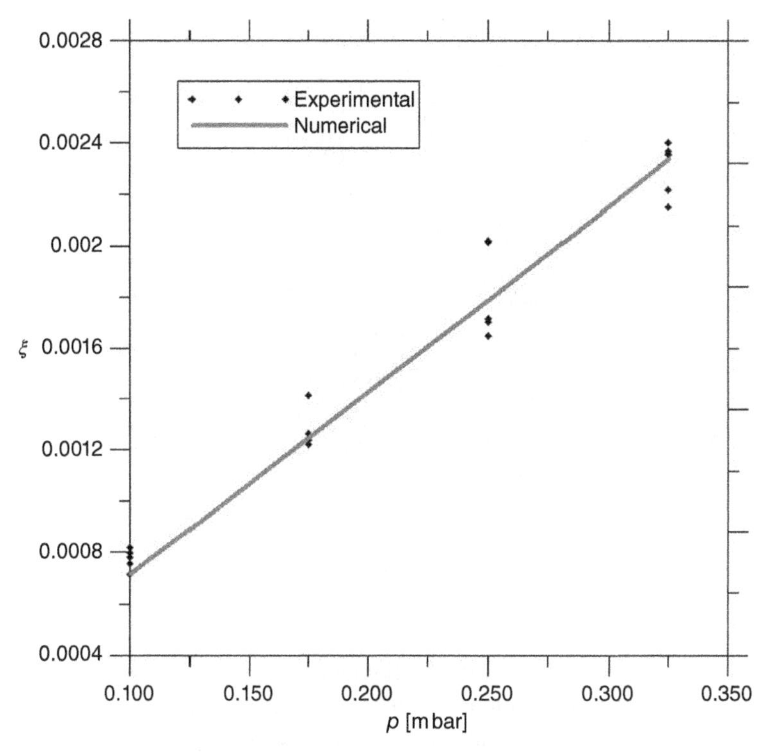

Figure 15.12 Damping coefficients: comparison of experiments and simulations in the free-molecule regime. *Source:* Frangi, Ghisi and Coronato (2009), Figure 8. Reproduced with permission of Elsevier.

Recently, an extensive experimental campaign has been reported (Frangi *et al.*, 2016), where three different families of MEMS have been designed, fabricated, tested and simulated. A total of 292 devices of 36 different geometries have been addressed, to test the ability of the numerical model to capture, for a fully generic 3D flow, the dependence on geometrical parameters that are typical of fabricated MEMS. The groups represent standard building blocks of MEMS working in these conditions. For instance, limiting our attention to gyroscopes, parallel-plate structures are utilized for the differential read-out of a z-axis gyro; comb-finger devices are applied for the drive modes of all axes and perforated-mass units are often exploited for the x- and y-axis sense modes. Refined numerical simulations have been performed to obtain the best possible estimate of the quality factors.

The three families of device addressed in this contribution have been fabricated with the ThELMA process by STMicroelectronics.

For instance, in the comb-finger devices (see Figures 15.13 and 15.14), a perforated shuttle frame with folded springs supports a set of N shuttle fingers interdigitated with equivalent stator sets. The oscillation of the shuttle frame induces sliding between the fingers. The 11 devices are obtained from variations, as reported in Table 15.4, of the

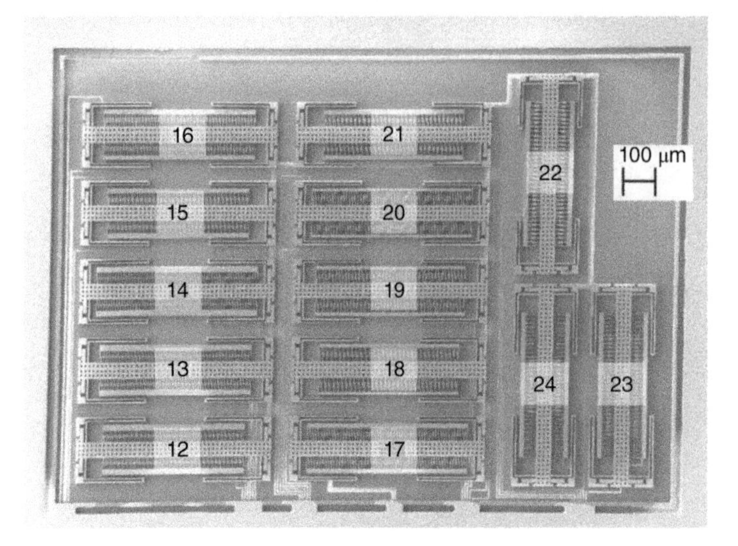

Figure 15.13 SEM of comb-finger structures. *Source:* Frangi *et al.* (2016), Figure 2. Reproduced with permission of IEEE.

nominal gap g, the overlap between stator and shuttle fingers V, the number of shuttle fingers N and the nominal resonant frequency f_0. It is worth stressing that, as can be argued from Figure 15.13, varying the gap among the fingers also modifies the overlap between the springs and the stators.

Some of the results are collected in Figures 15.1, 15.15 and 15.16. Figure 15.1 plots the quality factor over the whole pressure range for a selected comb-finger device. It should be remarked that, as expected, at very low pressures, Q reaches a plateau associated with solid dissipation. Since Q in this region is always of the order of 10^5 for all the tested devices, this contribution to dissipation can be safely neglected in the pressure range of interest, i.e. around 1 mbar. A zoom of this region with the comparison between simulations and experiments is plotted in Figure 15.15.

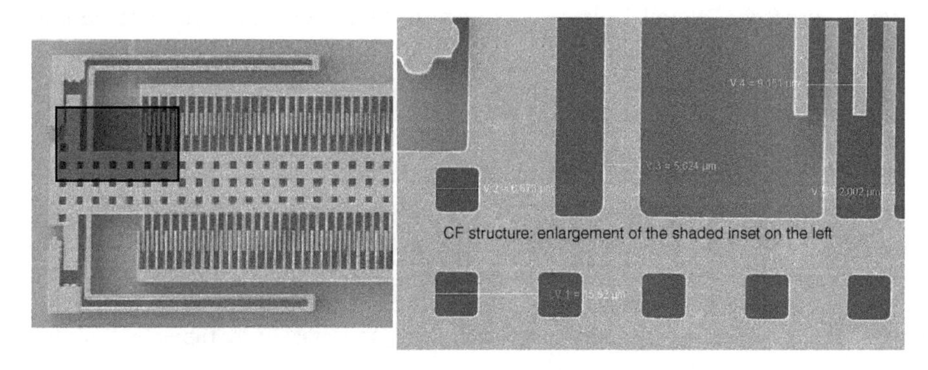

Figure 15.14 Typical comb-finger structures. The shaded portion is enlarged on the right with indications of typical dimensions after production. *Source:* Frangi *et al.* (2016), Figure 4b. Reproduced with permission of IEEE.

Table 15.4 Comb-finger structures: nominal gap g; overlap between stator and shuttle fingers V; nominal resonating frequency f_0; number of shuttle fingers N.

ID	g [μm]	V [μm]	f_0 [kHz]	N
12	1.9	10	20	130
13	2.2	10	20	130
14	2.5	10	20	130
15	1.9	15	20	130
16	2.2	15	20	130
17	2.5	15	20	130
20	2.5	20	20	130
21	1.9	15	25	130
22	1.9	15	30	130
23	1.9	15	20	90
24	1.9	15	20	50

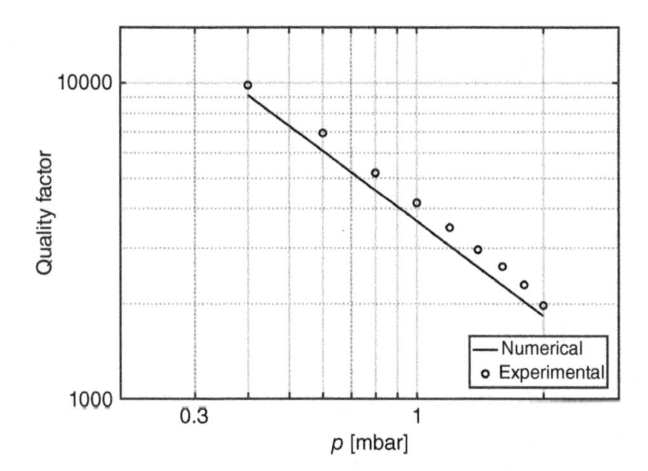

Figure 15.15 Comb-finger device 13. Quality factor versus pressure: zoom in the linear region. *Source:* Frangi *et al.* (2016), Figure 11. Reproduced with permission of IEEE.

Finally, Figure 15.16 shows a comparison between numerical and experimental data (three samples per device ID) for all the tested comb-finger devices at a fixed pressure of $p = 1.94$ mbar.

15.5 Solid Damping: Thermoelasticity

Thermoelasticity is possibly the most classical solid dissipation mechanism for small-scale mechanical devices (Lifshitz and Roukes, 2000). MEMS with important bending deformations, such as cantilevers or tuning forks, are typically dominated by

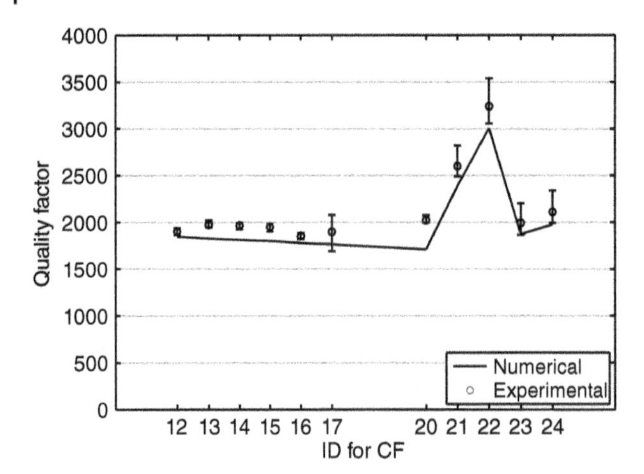

Figure 15.16 Numerical and experimental data for the comb-finger devices at $p = 1.94$ mbar. *Source:* Frangi *et al.* (2016), Figure 12. Reproduced with permission of IEEE.

thermoelastic losses. The dissipation is caused by the complex interaction of acoustic modes with the thermally excited modes in the crystalline lattice. It is common practice to evaluate the thermoelastic quality factor by means of simplified formulae, such as the well-known Zener's expression (Zener, 1937). It is worth recalling that such an approach can be used only for a certain range of geometrical conditions, while poor results are obtained when simplified methods are adopted for nonstandard cases.

The equations of fully coupled thermoelasticity, introduced in Section 2.7.2, are rewritten here for convenience, assuming zero body forces:

$$\rho\frac{\partial^2 \mathbf{u}}{\partial t^2} = \mathrm{div}\sigma, \qquad \sigma = \mathbf{d}(\varepsilon - \alpha(T - T_0)\mathbf{1}), \tag{15.31}$$

$$\rho c_h\frac{\partial T}{\partial t} = \mathrm{div}(k\ \mathrm{grad}\ T) - \alpha T_0\frac{E}{1-2\nu}\frac{\partial \mathrm{tr}\varepsilon}{\partial t}. \tag{15.32}$$

Let us consider an elongated cantilever beam of length L, with the axis aligned with the x coordinate; let y be the coordinate along the direction of vibration (thickness h) and v denote the u_y displacement. In the special case of a vibrating beam, some simplifications can be introduced in the problem formulation. In particular, the usual Euler–Bernoulli kinematic hypothesis is assumed and heat conduction is limited to the transverse direction.

According to the Euler–Bernoulli hypothesis, the axial strain ε and the bending moment M are expressed in terms of the beam curvature by

$$\varepsilon = -\frac{\partial^2 v}{\partial x^2}y, \qquad M = EI\frac{\partial^2 v}{\partial x^2} + bE\alpha\int_{-h/2}^{h/2}(T - T_0)y\mathrm{d}y,$$

where b is the out-of-plane thickness of the beam. If one neglects heat conduction along the longitudinal direction of the beam, the governing equations for thermoelastic vibrations are

$$\rho hb\frac{\partial^2 v}{\partial t^2} + \frac{\partial^2}{\partial x^2}\left[EI\frac{\partial^2 v}{\partial x^2} + bE\alpha\int_{-h/2}^{h/2}(T - T_0)y\mathrm{d}y\right] = 0 \tag{15.33}$$

$$\rho c_h \frac{\partial T}{\partial t} = \alpha T_0 E \frac{\partial^3 v}{\partial x^2 \partial t} y + k \frac{\partial^2 T}{\partial y^2}. \tag{15.34}$$

We solve the coupled thermoelastic problem for the case of harmonic vibrations of the form

$$v(x,t) = U(x)\mathrm{e}^{\mathrm{i}\omega t}.$$

Assuming adiabatic boundary conditions at $y = \pm h/2$, the solution in terms of temperature is sought in the form:

$$T - T_0 = T_n \mathrm{e}^{\mathrm{i}\omega t} \sin \frac{\pi y}{h}.$$

Substituting into the heat equation, multiplying it by $\sin(\pi y/h)$ and integrating from $-h/2$ to $h/2$, after some algebra, one obtains the temperature profile across the beam, with

$$T_n = \frac{\mathrm{i}\omega}{1 + \mathrm{i}\omega\tau_z} \frac{\alpha T_0}{2kb} EIU'', \qquad \tau_z = \frac{h^2}{\pi^2 k/(\rho c_\mathrm{h})}.$$

Inserting this expression into Equation 15.33, the eigenfrequency ω^2 is a solution of

$$-\rho h b \omega^2 U + EIU^{\mathrm{IV}} \left(1 - \frac{\mathrm{i}\omega\tau_z}{1 + \mathrm{i}\omega\tau_z} \frac{E\alpha^2 T_0}{\rho c_\mathrm{h}}\right) = 0.$$

Whatever the approximation for U, the quality factor Q is given by the ratio

$$Q = \frac{\mathrm{Im}(\omega^2)}{\mathrm{Re}(\omega^2)} \simeq \frac{\rho c_\mathrm{h}}{E\alpha^2 T_0} \frac{1 + (\omega\tau_z)^2}{\omega\tau_z}, \tag{15.35}$$

which is the well-known Zener's formula. More general results can be achieved by considering the standard, fully coupled thermoelastic model. This has been done (Ardito *et al.*, 2008) by implementing the model in a proper finite element program. Since a linear model is considered, the quality factor is estimated after evaluating the modal properties. Owing to the characteristics of microdevices, i.e. very high frequencies involved, high aspect ratios of vibrating beams, extremely low mass and damping, the eigenvalue problem arising from the thermoelastic coupled model presents very ill-conditioned matrices. Standard commercial codes may fail to provide the solution, therefore an *ad-hoc* numerical procedure has been implemented, using the data in Table 15.5.

Assuming $\mathbf{u} = \mathbf{U}(\mathbf{x})\mathrm{e}^{\mathrm{i}\omega t}$, $T = T(\mathbf{x})\mathrm{e}^{\mathrm{i}\omega t}$ and enforcing Equation 15.31 in a weak manner using standard finite elements, a quadratic eigenvalue problem is formulated (Ardito *et al.*, 2008). The eigenvalues are generally complex, the real part giving the oscillatory part of the free-vibration thermoelastic response and the imaginary part influencing the

Table 15.5 Thermal and effective mechanical properties of monocrystalline silicon.

E	150 GPa	ρ	2330 kg/m^3
ν	0.2	c_h	700 J/(kg K)
α	2.6×10^{-6} K^{-1}	k	148 W/(m K)

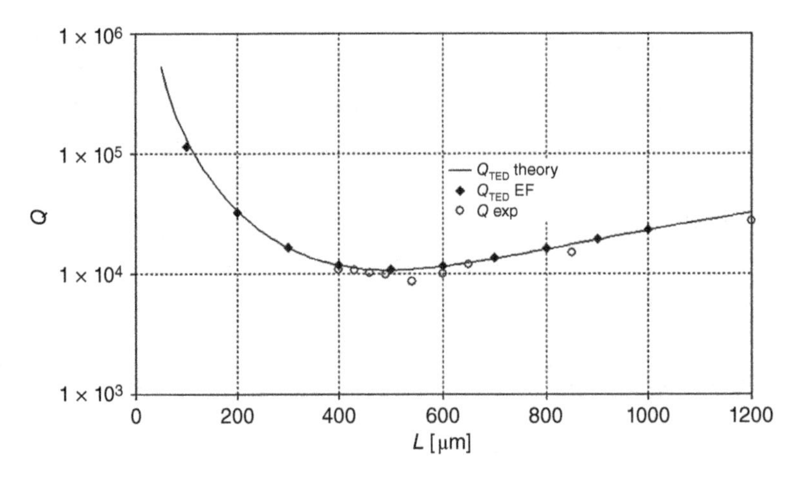

Figure 15.17 Quality factor of single-crystal resonant beams versus beam length, thickness 30 μm: numerical results, Zener's formula and experimental results. Q_{TED}, Q with thermoelastic damping, EF, finite element method; exp, experimental. *Source:* Ardito *et al.* (2008), Figure 2. Reproduced with permission of Springer.

vibration decay. The quality factor Q can be then computed as the ratio between these components.

The numerical results in terms of Q are reported in Figure 15.17, referred to the case of a cantilever beam of thickness 30 μm vibrating in the fist bending mode. The values of Q due to thermoelastic dissipation have been computed numerically for different beam lengths, ranging from 100 μm to 1200 μm (rhombuses in Figure 15.17). The analytical results achieved through Zener's formula are represented in the solid line: good agreement between the finite element method and Zener's solution is noticeable. Moreover, the same figure shows some experimental results (circles) taken from Le Foulgoc *et al.* (2006). It is possible to conclude that thermoelastic dissipation represents the main source of damping for beams in bending with the considered geometry.

Thermoelastic loss is considered a fundamental dissipation mechanism in microbeam resonators. However, classical local thermoelastic analysis is unable to interpret the size effect recently evidenced in resonators when the dimensions become very small, smaller than several micrometres (see e.g. Yasumura *et al.* (2000)), as discussed in Section 15.7.1.

15.6 Solid Damping: Anchor Losses

Anchor losses are due to the scattering of elastic waves from the resonator into the substrate. Since the latter is typically much larger than the resonator itself, it is assumed that all the elastic energy entering the substrate through the anchors is eventually dissipated.

The semi-analytical evaluation of anchor losses has been addressed in several papers with different levels of accuracy (see e.g. the seminal contributions of Cross and Lifshitz (2001) and Jimbo and Itao (1968)). These consider a resonator resting on elastic half-spaces and assume a weak coupling, in the sense that the mechanical mode, as well as the mechanical actions transmitted to the substrate, are those of a rigidly clamped resonator. The displacements and rotations induced in the half-space are provided by suitable Green's functions.

Judge *et al.* (2007) studied analytically the case of a 3D cantilever beam attached either to a semi-infinite space or to a semi-infinite plate of finite thickness. Their results are based on the semi-exact Green's functions established by Bycroft (1956). More recently, Wilson-Rae (2008) generalized all these approaches using the involved framework of radiation tunnelling in photonics. Unfortunately, these contributions provide estimates of quality factors that differ quantitatively. The procedure of Judge *et al.* (2007), which rests on simple mechanical principles, has been recently revisited by Frangi and Cremonesi (2016), starting from the exact Green's functions for the half-space studied by Pak (1987).

15.6.1 Analytical Estimation of Dissipation

Following a rather standard procedure (Jimbo and Itao, 1968; Judge *et al.*, 2007; Photiadis and Judge, 2004), we describe the simplest possible analytical (or semi-analytical) approach based on a decoupling assumption. Let us consider a structure, like the beam of Figure 15.18, attached to semi-infinite elastic spaces and vibrating in one of its fundamental modes with angular frequency ω. The number of anchor points is irrelevant and the procedure must be repeated for each anchor point.

As a consequence, we focus on a specific attachment point and start by considering it as perfectly rigid. Standard theories of structural mechanics permit the expression of concentrated forces and couples exerted by the structure on the support. These generally include a constant component (due, for instance, to pre-stresses or initial deformation) and a sinusoidal varying contribution (see Figure 15.18 for the notation):

axial force: $$n(t) = n_0 + N\mathrm{e}^{\mathrm{i}\omega t} \tag{15.36}$$

shear force: $$r(t) = r_0 + R\mathrm{e}^{\mathrm{i}\omega t} \tag{15.37}$$

bending couple: $$b(t) = b_0 + B\mathrm{e}^{\mathrm{i}\omega t} \tag{15.38}$$

torque: $$\tau(t) = \tau_0 + T\mathrm{e}^{\mathrm{i}\omega t}. \tag{15.39}$$

The shear force and bending couple have, in general, two components which are treated in the same manner.

We now introduce the decoupling assumption, according to which frequencies, forces and couples are not significantly altered if the rigid support is replaced with a deformable

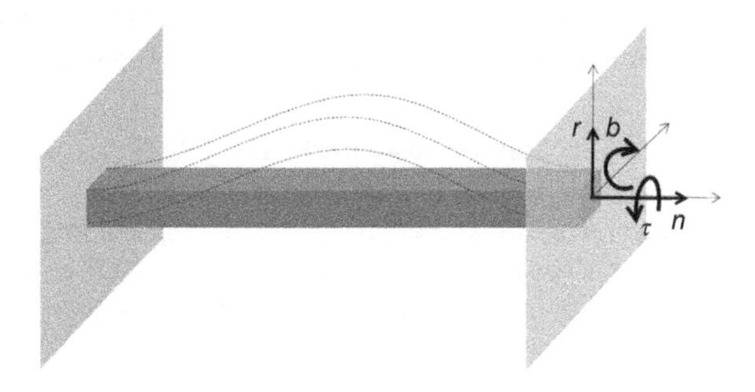

Figure 15.18 Doubly clamped beam. *Source:* Frangi and Cremonesi (2016), Figure 2. Reproduced with permission of Elsevier.

half-space. These concentrated actions induce displacements and rotations:

$$n(t) \quad \rightarrow \quad d(t) = d_0 + D\mathrm{e}^{\mathrm{i}\omega t} \tag{15.40}$$

$$r(t) \quad \rightarrow \quad v(t) = v_0 + V\mathrm{e}^{\mathrm{i}\omega t} \tag{15.41}$$

$$b(t) \quad \rightarrow \quad \phi(t) = \phi_0 + \Phi\mathrm{e}^{\mathrm{i}\omega t} \tag{15.42}$$

$$\tau(t) \quad \rightarrow \quad \psi(t) = \psi_0 + \Psi\mathrm{e}^{\mathrm{i}\omega t} \tag{15.43}$$

where D, V, Φ and Ψ are, in general, complex variables and denote the amplitude of the time-dependent part of, respectively, axial and tangential displacements and bending and torsional rotations. These are known as Green's functions for the half elastic space and it is worth stressing that the real part of these kinematic quantities is, in general, unbounded.

However, the dissipation over one cycle due to the scatter of elastic waves in the infinite half-space is

$$\Delta W = -\pi\left(\mathrm{Im}[D]N + \mathrm{Im}[V]R + \mathrm{Im}[\Phi]B + \mathrm{Im}[\Psi]T\right)$$

and only depends on the bounded imaginary part of the Green's functions.

In particular, setting $k_{\mathrm{T}} = \omega/c_{\mathrm{T}}$, $c_{\mathrm{T}} = \sqrt{\mu/\rho}$, Frangi and Cremonesi (2016) show that:

$$\mathrm{Im}[D] = N\frac{k_{\mathrm{T}}}{\mu}g_N(\nu), \qquad \mathrm{Im}[V] = R\frac{k_{\mathrm{T}}}{\mu}g_R(\nu), \tag{15.44}$$

$$\mathrm{Im}[\Phi] = B\frac{k_{\mathrm{T}}^3}{\mu}g_B(\nu), \qquad \mathrm{Im}[\Psi] = -T\frac{k_{\mathrm{T}}^3}{\mu}\frac{1}{12\pi}, \tag{15.45}$$

where g_N, g_R and g_B are plotted versus the Poisson coefficient ν in Figure 15.19. Most of these functions, rigorously established starting from the work of Pak (1987), are similar to analogous results published by Judge *et al.* (2007), but differ quantitatively. By contrast, the numerical values of the g functions coincide, but for $\mathrm{Im}[\Psi]$, with the expressions given by Wilson-Rae (2008) starting from a totally different perspective.

Finally, summing over all the anchor points:

$$\Delta W = -\pi\sum_i\left(N_i^2\frac{k_{\mathrm{T}}}{\mu}g_N(\nu) + R_i^2\frac{k_{\mathrm{T}}}{\mu}g_R(\nu) + B_i^2\frac{k_{\mathrm{T}}^3}{\mu}g_B(\nu) - T_i^2\frac{k_{\mathrm{T}}^3}{\mu}\frac{1}{12\pi}\right). \tag{15.46}$$

To apply Equation 15.46, only the expressions of R, N, B and T are required. In many cases fully analytical estimates are available, e.g. for a cantilever beam in axial or bending vibrations. More generally, a numerical tool is necessary, as for the case of the buckled beam of interest in this paper, discussed in the following section.

15.6.1.1 Applications: Axial and Bending Modes

Simple applications of these formulae give estimates of the quality factors of cantilever beams of length L and cross-sectional area A, resting on an elastic half-space. For simplicity, the half-space is assumed to be made of the same isotropic material as the beam. In the case of axial vibrations for a cantilever on a rigid support, the axial displacement reads:

$$u = U\sin(k_{\mathrm{B}}x)\mathrm{e}^{\mathrm{i}\omega t}$$

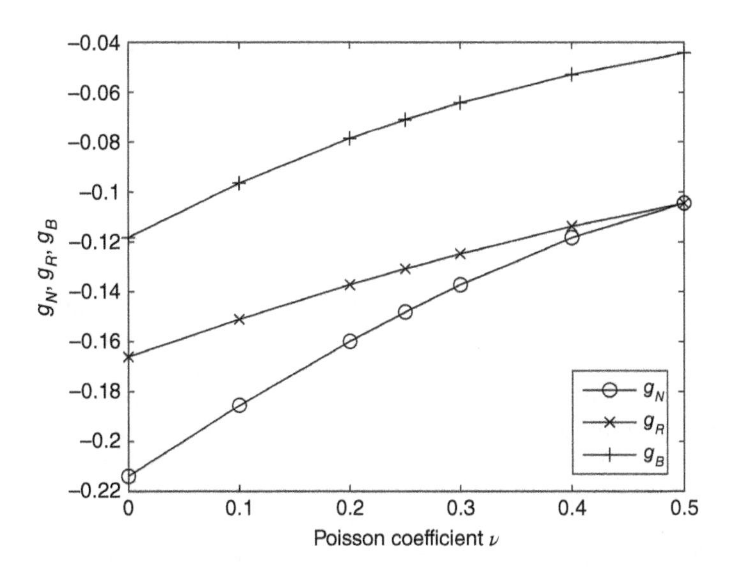

Figure 15.19 Functions g_N, g_R and g_B. *Source:* Frangi and Cremonesi (2016), Figure 3. Reproduced with permission of Elsevier.

with

$$k_{\mathrm{B}} = (1+2m)\frac{\pi}{2L}, \quad \omega = \sqrt{\frac{E}{\rho}}\,k_{\mathrm{B}}.$$

The maximum value of the stored elastic energy is

$$W = \frac{1}{2}U^2 EA k_{\mathrm{B}}^2 \frac{L}{2}$$

and the force exerted on the support $N = EAk_{\mathrm{B}}U$. Assuming that ω, N and W are not significantly altered if the rigid support is replaced with a deformable half-space, N induces a displacement D of the half-space given by Equation 15.40 and the dissipation is

$$\Delta W = -\pi N\,\mathrm{Im}[D] = \pi E^2 A^2 k_{\mathrm{B}}^2 U^2 \frac{1}{\mu} k_{\mathrm{T}}\, g_N(\nu),$$

leading to

$$Q = \frac{L}{EA}\frac{\mu c_{\mathrm{T}}}{g_N(\nu)}\frac{1}{\omega}. \tag{15.47}$$

Similarly, for a bending mode $\psi(x)U$ characterized by a given wave number k_B (e.g. $k_B L = 1.875$ in the first mode) and normalized such that

$$\int_0^L \psi^2 \,\mathrm{d}x = L,$$

the maximum stored energy is $W = (1/2)EILk_B^4U^2$, while the bending couple and shear force read

$$B = 2EIk_B^2U, \qquad R = 2EIk_B^3\beta U, \quad \text{with} \quad \beta = \frac{\sin k_BL - \sinh k_BL}{\cos k_BL + \cosh k_BL}.$$

If only the shear force is considered (the bending dissipation is usually negligible):

$$Q = \frac{L}{4EJk_B^2\beta}\frac{\mu c_T}{g_R(v)}\frac{1}{\omega}, \tag{15.48}$$

where $\omega = \sqrt{EI/(\rho A)}k_B^2$. Equations 15.47 and 15.48 coincide with the formulae given by Wilson-Rae (2008), Table 1.

15.6.2 Numerical Estimation of Anchor Losses

Owing to the presence of complicated geometry or material anisotropy or inhomogeneities, several applications cannot be addressed with the analytical approach described in the previous section and require a more general numerical methodology. Also, the treatment of the substrate as semi-infinite is really most appropriate when characteristic wavelengths of bulk waves in the substrate are small relative to the characteristic dimensions of the substrate; this may not be the case when the waves are induced by low-frequency flexural vibrations.

Dissipative boundary conditions, and the perfectly matched layer approach in particular, are gaining increasing attention in the dedicated literature. The idea of applying the perfectly matched layer technique for microsystems has been proposed by Bindel and Govindjee (2005). Implementing a fully 3D perfectly matched layer analysis for the extraction of the quality factor of a resonator requires, in general, the solution of a large-scale generalized complex symmetric eigenvalue problem, which has possibly hindered its applications in 3D until recently.

Let us consider a cantilever beam resting on a substrate, as depicted in Figure 15.20. Limiting our attention to the case of 'Cartesian' perfectly matched layers, the substrate of infinite extent is truncated at some distance from the resonator by means of dissipative layers orthogonal to the coordinate axes. Every layer is bounded by the two surfaces $x_m = x_m^{(1)}$ and $x_m = x_m^{(2)} = x_m^{(1)} + W_{\mathrm{PML}}$, where W_{PML} is the thickness of the perfectly matched layer. Here, we introduce the 'stretched' transformed complex coordinate y_m:

$$y_m = x_m^{(1)} + \int_{x_m^{(1)}}^{x_m} \lambda_m(s)\mathrm{d}s, \quad \text{with} \quad \lambda_m(s) = 1 - \mathrm{i}\beta\frac{s - x_m^{(1)}}{W_{\mathrm{PML}}}, \tag{15.49}$$

where s is the distance from the plane $x_m = x_m^{(1)}$ and β is a user parameter. Outside the substrate, one has $y_m = x_m$. The same procedure is repeated for all the bounding planes

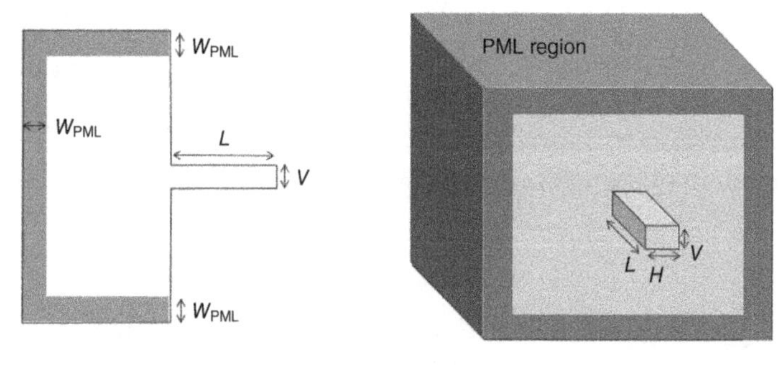

Figure 15.20 Beam on a semi-infinite substrate: geometry and example perfectly matched layers. PML, perfectly matched layer. *Source:* Frangi and Cremonesi (2016), Figure 6. Reproduced with permission of Elsevier.

of the the substrate. At this stage, the principle of virtual power is imposed, as described by Bindel and Govindjee (2005); Frangi *et al.* (2013a).

The modes of vibration now become complex, with complex eigenvalues. It can be shown that, if the structure vibrates in one of these modes, the quality factor is given by the ratio between the real and imaginary parts of the eigenvalue:

$$Q = 2\pi \frac{W}{\Delta W} = \frac{\mathrm{Re}(\omega^2)}{\mathrm{Im}(\omega^2)}.$$

A careful validation of the procedure for cantilever beams has been presented (Frangi *et al.*, 2013a) on the basis of a 3D numerical code developed *ad hoc*. Experimental benchmarks with 3D MEMS resonators have later been described (Frangi *et al.*, 2013b; Segovia-Fernandez *et al.*, 2015), where it has been shown that anchor losses can be estimated with very good accuracy. It should be stressed that these analyses do not require any calibration of constitutive parameters and can be considered exact to within the limits of the assumption that the energy scattered through the anchors is dissipated.

As an example of an application taken from Segovia-Fernandez *et al.* (2015), we analyse the performance of aluminium nitride contour mode resonators with resonance frequency f_r in the range 220 MHz to 1 GHz. The three resonators shown in Figure 15.21 operate at 220 MHz, 370 MHz and 1.05 GHz. These frequencies correspond to three multiples of the first mode of vibration that is displayed in an aluminium nitride plate of 60 μm width and are attained by modifying the electrode pitch. To study the impact of the anchors on Q, the size of the resonator body is kept fixed, whereas the width W_a

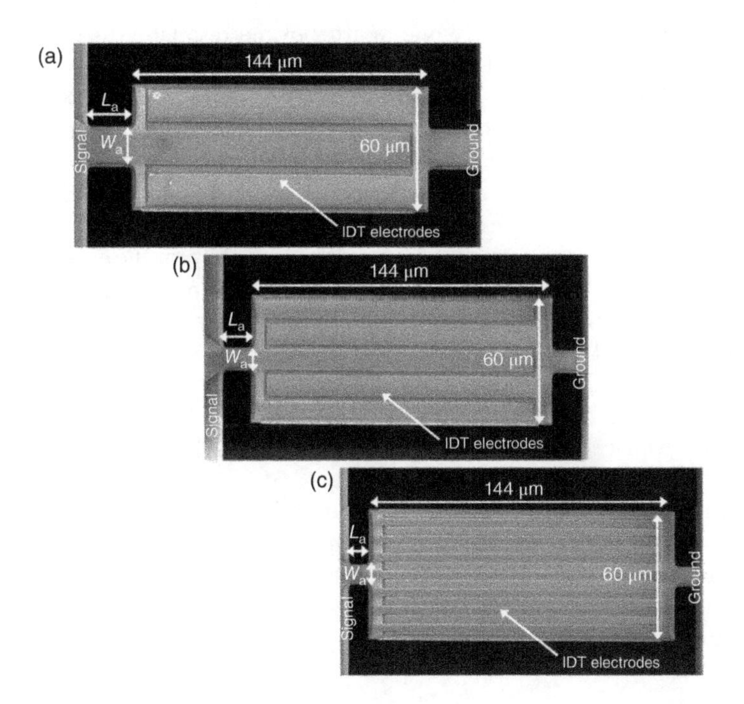

Figure 15.21 SEMs of aluminium nitride contour mode resonators: (a) 220 MHz; (b) 370 MHz; (c) 1.05 GHz. IDT, interdigitated. *Source:* Segovia-Fernandez *et al.* (2015), Figure 1. Reproduced with permission of IEEE.

and length L_a of the anchors are varied as a function of the acoustic wavelength λ for each f_r. Moreover, to better investigate the nature of energy dissipation, we have measured and compared the values of Q exhibited by low (220 MHz) and high (1.05 GHz) frequency resonators at both cryogenic (10 K) and room (300 K) temperatures. Essentially, the device is based on a vibrating aluminium nitride plate sandwiched between two patterned metal layers that are employed to generate an electric field across the thickness of the piezoelectric film h. The resonators used in the experiments belong to the group of lateral field excitations. In this case, the top metal is formed by an array of interdigitated electrodes, whose number is equivalent to the number of fingers n.

The electrode pitch W_f sets the acoustic wavelength of the excited mode (λ is equal to $2W_f$), and hence f_r. For instance, in the idealized case of an isotropic homogeneous 1D resonator unaffected by the presence of anchors, the displacement field would read:

$$u(x,t) = \mathrm{e}^{\mathrm{i}2\pi f_r t} \sin kx, \qquad k = n\frac{\pi}{W} = \frac{\pi}{W_f}, \tag{15.50}$$

where k is the wave number exhibited at f_r and W is the overall width of the resonator ($W = 60$ μm). The bottom metal forms an electrode at a floating potential under the body of the resonator and is used to confine the electric field lines generated by the top electrodes.

For the different resonant frequencies considered (220, 370 and 1050 MHz) these values correspond to three different W_f (20, 12, and 4 μm, respectively) and therefore different values of n (3, 5 and 15). The width ($W = nW_f$) and length (L) of the devices were set to be 60 μm and 144 μm, respectively.

Exploiting the symmetry of the devices, only one-quarter of the resonators is discretized. Figure 15.22 shows the geometrical model and the finite element mesh used in the simulations. A typical mesh for these problems consists of more than 200,000 quadratic tetrahedral or wedge elements. Anchor losses are strongly sensitive to all the macroscopic parameters of the resonators, e.g. W_a, L_a and sidewall angles, but also to geometrical details, such as the fillet radius of the rounded attachments between the anchors and the resonator and substrate (see Figure 15.22).

From Figure 15.23, it appears that anchor losses account very accurately for the experimentally measured dissipation in the devices at 220 MHz and that the numerical tool developed can be used for quantitative predictions. In particular, Q displays a clear peak between $W_a = 10$ μm ($W_a/\lambda = 1/4$) and $W_a = 20$ μm ($W_a/\lambda = 1/2$), corresponding to the cases where the anchor covers the central electrode. A contour plot of the mechanical mode is depicted in Figure 15.24. These figures have been obtained by assembling the contour plots of the u_y displacement component, where y denotes the direction of vibration.

In the contour plot on the left of Figure 15.24, the colour scale has been calibrated so as to show the modal shape in the resonator, while on the right of Figure 15.24, the adopted scale allows appreciation of the pattern of scattered waves in the anchor and in the substrate and their eventual dissipation in the perfectly matched layer regions. High-frequency resonators show a rather different behaviour and are discussed briefly in the following section.

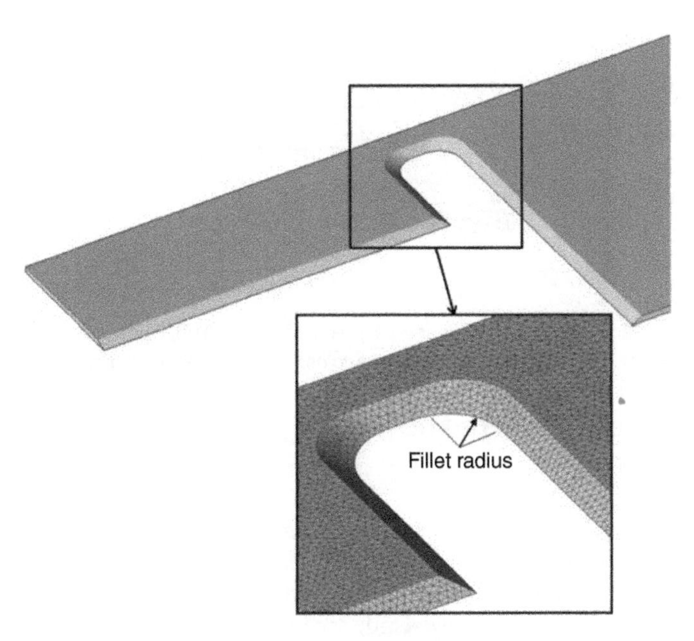

Figure 15.22 Geometric model and detail of a typical mesh. *Source:* Segovia-Fernandez *et al.* (2015), Figure 8. Reproduced with permission of IEEE.

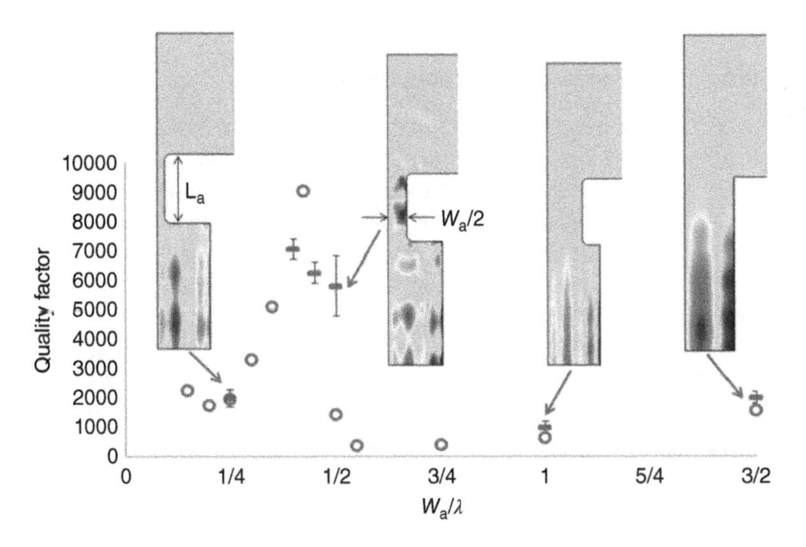

Figure 15.23 220 MHz devices, $L_a/\lambda = 1$. Numerically predicted Q_{anc} (circles) for varying anchor width versus experimental data. Bars denote average experimental data with superposed standard deviations. *Source:* Segovia-Fernandez *et al.* (2015), Figure 9. Reproduced with permission of IEEE. (*See color plate section for the color representation of this figure.*)

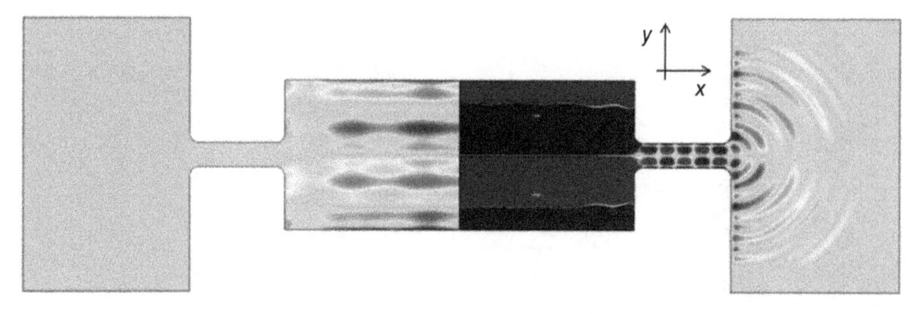

Figure 15.24 u_y for $W_a/\lambda = 1/4$ and $L_a/\lambda = 1$. *Source:* Segovia-Fernandez *et al.* (2015), Figure 11. Reproduced with permission of IEEE. (*See color plate section for the color representation of this figure.*)

15.7 Solid Damping: Additional unknown Sources – Surface Losses

The global picture of solid dissipation is far from being complete and understood, especially when the dimensions of the MEMS decrease towards the nanoscale. In at least two different contexts, briefly described in the following sections, phenomena occurring at the surface of the devices cannot be neglected, especially when the thickness scales down and the surface-to-volume ratio increases. The complicated physics of thin films hence becomes a necessary ingredient, deserving greater attention in the fields of MEMS and NEMS.

15.7.1 Solid Damping: Deviations from Thermoelasticity

According to some recent experimental results (Yang, Ono and Esashi, 2002; Yasumura *et al.*, 2000), classical thermoelastic analysis seems to be unable to interpret the dissipation level demonstrated in resonators with very small thickness. For example, Figure 15.25 reports some experimental results of Yasumura *et al.* (2000) for silicon nitride microbeams. It is evident that, already for a thickness of 1.2 μm, the thermoelastic theoretical approach greatly overestimates the actual quality factor. The discrepancy is even higher for submicrometric beams.

Wang *et al.* (2003) have proved that surface chemistry can play an important role in determining the quality factor of micromechanical silicon resonators, even at room temperature. Indeed, it has been verified that the mechanical energy dissipation in a torsional resonator, with resonant frequencies ranging from 2.2 MHz to 23 MHz, considerably decreases when a 13 Å layer of native silicon oxide (which corresponds to the oxidation of fewer than two silicon bilayers) is replaced by a single monolayer of hydrogen atoms.

15.7.2 Solid Damping: Losses in Piezoresonators

High-frequency piezo-resonators also demonstrate an unexpected contribution of thin layers to the total dissipation. In Section 15.6.2, we have analysed a class of contour mode resonator and shown that losses in low-frequency resonators (220 MHz) are mainly represented by anchor dissipation. Actually, at cryogenic temperatures (10 K) this also seems to hold for 1.05 GHz resonators, as shown in Figure 15.26.

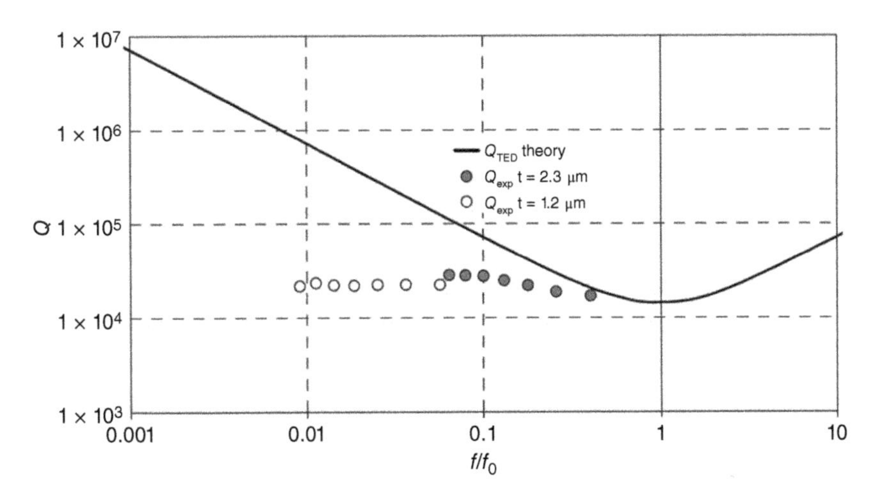

Figure 15.25 Experimental results compared with classical thermoelastic approach, for beam resonators with small thickness t. Q_{TED}, Q with thermoelastic damping, exp, experimental. *Source:* Yasumura *et al.* (2000), Figure 13. Reproduced with permission of IEEE.

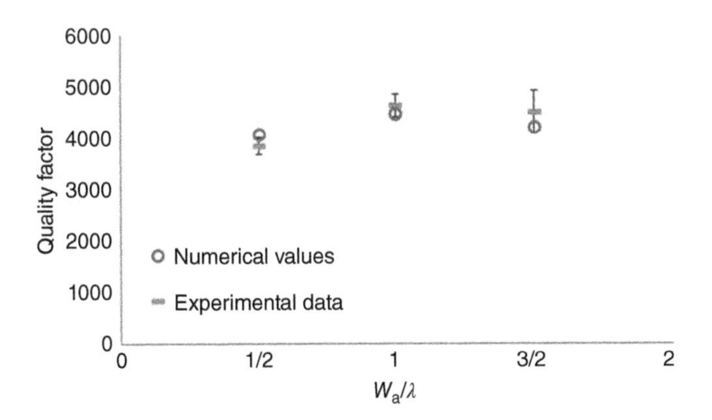

Figure 15.26 1.05 GHz devices, $L_{\mathrm{a}}/\lambda = 1/4$. Numerically predicted Q_{anc} (circles) for varying anchor width versus experimental data at 10 K. *Source:* Segovia-Fernandez *et al.* (2015), Figure 10. Reproduced with permission of IEEE.

However, the situation dramatically changes if we switch to room temperature. Experimental data and estimated anchor losses are collected in Table 15.6. The 1.05 GHz resonator is clearly driven by an additional thermodynamical loss, for which we postulate a quality factor Q_{add} of the form

$$Q_{\mathrm{add}} = \alpha \frac{W}{n}, \tag{15.51}$$

providing the overall numerical prediction $Q_{\mathrm{num}}^{-1} = Q_{\mathrm{anc}}^{-1} + (\alpha W/n)^{-1}$. It is worth stressing that Q_{add} scales proportionally to $1/f_{\mathrm{r}}$, so that its impact on low-frequency resonators is minimal.

Hao and Liao (2010) observed a similar phenomenon for different resonators; the dissipation mechanism has been attributed to microscale viscous phenomena occurring at

Table 15.6 Quality factors for 1.05 GHz resonators.

W_a	L_a	Q_{exp}	Q_{anc}	Q_{num}
4	2	1331	4066	1501
8	2	1586	4475	1553
12	2	1606	4206	1519
4	4	1183	2737	1273
8	4	1219	3194	1363
12	4	1634	4055	1499
4	8	1189	2782	1282
8	8	1496	2938	1314
12	8	1657	4824	1593

the interface between different materials (e.g. between the electrodes and the aluminium nitride). Frangi *et al.* (2013b) proposed a phenomenological law, leading to an expression similar to Equation 15.51. However, the physical origins of this mechanism are still not clear and are likely to be associated with the relatively un-investigated (e.g. thermoelastic) properties of a thin-film piezo.

Assuming the validity of Equation 15.51 as a starting hypothesis, α has been calibrated using all the experimental data available for the high-frequency mode (at 300 K) and estimated by requiring the quadratic error to be minimized. The results of this calibration are listed in the last column of Table 15.6. Further validations have been discussed by Segovia-Fernandez *et al.* (2015) and Frangi *et al.* (2013b).

References

Aoki, K., Kanba, K. and Takata, S. (1997) Numerical analysis of a supersonic rarefied gas flow past a flat plate. *Physics of Fluids*, **9**, 1144–1161.

Ardito, R., Comi, C., Corigliano, A. and Frangi, A. (2008) Solid damping in micro electro mechanical systems. *Meccanica*, **43**, 419–428.

Bao, M., Yang, H., Yin, H. and Sun, Y. (2002) Energy transfer model for squeeze-film air damping in low vacuum. *Journal of Micromechanics and Microengineering*, **12**, 341.

Bhatnagar, P.L., Gross, E.P. and Krook, M. (1954) A model for collision processes in gases. I. Small amplitude processes in charged and neutral one-component systems. *Physical Review*, **94**, 511–525.

Bindel, D.S. and Govindjee, S. (2005) Elastic PMLs for resonator anchor loss simulation. *International Journal for Numerical Methods in Engineering*, **64**, 789–818.

Bird, G.A. (1994) *Molecular Gas Dynamics and the Direct Simulation of Gas Flows*, Clarendon Press.

Bycroft, G.N. (1956) Forced vibrations of a rigid circular plate on a semi-infinite elastic space and on an elastic stratum. *Philosophical Transactions of the Royal Society of London A: Mathematical, Physical and Engineering Sciences*, **248**, 327–368.

Cai, C., Boyd, I.D., Fan, J. and Candler, G.V. (2000) Direct simulation methods for low-speed microchannel flows. *Journal of Thermophysics and Heat Transfer*, **14**, 368–378.

Cercignani, C. (1988) *The Boltzmann Equation and its Applications*, Springer.

Cercignani, C. and Daneri, A. (1963) Flow of a rarefied gas between two parallel plates. *Journal of Applied Physics*, **34**, 3509–3513.

Cercignani, C., Frangi, A., Lorenzani, S. and Vigna, B. (2007) BEM approaches and simplified kinetic models for the analysis of damping in deformable MEMS. *Engineering Analysis with Boundary Elements*, **31**, 451–457.

Chapman, S. and Cowling, T. (1960) *The Mathematical Theory of Non-uniform Gases*, Cambridge University Press.

Christian, R. (1966) The theory of oscillating-vane vacuum gauges. *Vacuum*, **16**, 175–178.

Cross, M.C. and Lifshitz, R. (2001) Elastic wave transmission at an abrupt junction in a thin plate with application to heat transport and vibrations in mesoscopic systems. *Physical Review B*, **64**, 085324.

Frangi, A. (2005) A fast multipole implementation of the qualocation mixed-velocity-traction approach for exterior Stokes flows. *Engineering Analysis with Boundary Elements*, **29**, 1039–1046.

Frangi, A. (2009) A BEM technique for free-molecule flows in high frequency MEMS resonators. *Engineering Analysis with Boundary Elements*, **33**, 493–498.

Frangi, A., Bugada, A., Martello, M. and Savadkoohi, P. (2013a) Validation of PML-based models for the evaluation of anchor dissipation in MEMS resonators. *European Journal of Mechanics – A/Solids*, **37**, 256–265.

Frangi, A. and Cremonesi, M. (2016) Semi-analytical and numerical estimates of anchor losses in bistable MEMS. *International Journal of Solids and Structures*, **92**, 141–148.

Frangi, A., Cremonesi, M., Jaakkola, A. and Pensala, T. (2013b) Analysis of anchor and interface losses in piezoelectric MEMS resonators. *Sensors and Actuators A: Physical*, **190**, 127–135.

Frangi, A., Fedeli, P., Laghi, G. *et al.* (2016) Near vacuum gas damping in MEMS: numerical modeling and experimental validation. *Journal of Microelectromechanical Systems*, **25**, 890–899.

Frangi, A., Frezzotti, A. and Lorenzani, S. (2007) On the application of the BGK kinetic model to the analysis of gas–structure interactions in MEMS. *Computers & Structures*, **85**, 810–817.

Frangi, A., Ghisi, A. and Coronato, L. (2009) On a deterministic approach for the evaluation of gas damping in inertial MEMS in the free-molecule regime. *Sensors and Actuators A: Physical*, **149**, 21–28.

Frangi, A., Ghisi, A. and Frezzotti, A. (2008) Analysis of gas flow in MEMS by a deterministic 3D BGK kinetic model. *Sensors Letters*, **6**, 69–75.

Frangi, A., Spinola, G. and Vigna, B. (2006) On the evaluation of damping in MEMS in the slip-flow regime. *International Journal for Numerical Methods in Engineering*, **68**, 1031–1051.

Gad-el-Hak, M. (ed.) (2002) *The MEMS Handbook*, CRC Press.

Hao, Z. and Liao, B. (2010) An analytical study on interfacial dissipation in piezoelectric rectangular block resonators with in-plane longitudinal-mode vibrations. *Sensors and Actuators A: Physical*, **163**, 401–409.

Hutcherson, S. and Ye, W. (2004) On the squeeze-film damping of micro-resonators in the free-molecule regime. *Journal of Micromechanics and Microengineering*, **14**, 1726.

Jimbo, Y. and Itao, K. (1968) Energy loss of a cantilever vibrator. *Journal of the Horological Institute of Japan*, **47**, 1–15.

Judge, J.A., Photiadis, D.M., Vignola, J.F. *et al.* (2007) Attachment loss of micromechanical and nanomechanical resonators in the limits of thick and thin support structures. *Journal of Applied Physics*, **101**, 013521.

Kádár, Z., Kindt, W., Bossche, A. and Mollinger, J. (1996) Quality factor of torsional resonators in the low-pressure region. *Sensors and Actuators A: Physical*, **53**, 299–303.

Karniadakis, G. and Beskok, A. (2002) *Micro Flows, Fundamentals and Simulation*, Springer.

Le Foulgoc, B., Bourouina, T., Traon, O.L. *et al.* (2006) Highly decoupled single-crystal silicon resonators: an approach for the intrinsic quality factor. *Journal of Micromechanics and Microengineering*, **16**, S45.

Lifshitz, R. and Roukes, M.L. (2000) Thermoelastic damping in micro- and nanomechanical systems. *Physical Review B*, **61**, 5600–5609.

Pak, R.Y.S. (1987) Asymmetric wave propagation in an elastic half-space by a method of potentials. *Journal of Applied Mechanics*, **54**, 121–126.

Park, J.H., Bahukudumbi, P. and Beskok, A. (2004) Rarefaction effects on shear driven oscillatory gas flows: a direct simulation Monte Carlo study in the entire Knudsen regime. *Physics of Fluids*, **16**, 317–330.

Photiadis, D.M. and Judge, J.A. (2004) Attachment losses of high *Q* oscillators. *Applied Physics Letters*, **85**, 482–484.

Segovia-Fernandez, J., Cremonesi, M., Cassella, C. *et al.* (2015) Anchor losses in AlN contour mode resonators. *Journal of Microelectromechanical Systems*, **24**, 265–275.

Sharipov, F. (1996) Rarefied gas flow through a slit: influence of the boundary condition. *Physics of Fluids*, **8**, 262–268.

Sone, Y. (1972) A flow induced by thermal stress in rarefied gas. *Journal of the Physical Society of Japan*, **33**, 232–236.

Veijola, T. (2004) Compact models for squeezed-film dampers with inertial and rarefied gas effects. *Journal of Micromechanics and Microengineering*, **14**, 1109.

Veijola, T., Kuisma, H., Lahdenperä, J. and Ryhänen, T. (1995) Equivalent-circuit model of the squeezed gas film in a silicon accelerometer. *Sensors and Actuators A: Physical*, **48**, 239–248.

Wang, Y., Henry, J.A., Zehnder, A.T. and Hines, M.A. (2003) Surface chemical control of mechanical energy losses in micromachined silicon structures. *Journal of Physical Chemistry B*, **107**, 14270–14277.

Wilson-Rae, I. (2008) Intrinsic dissipation in nanomechanical resonators due to phonon tunneling. *Physical Review B*, **77**, 245418.

Yang, J., Ono, T. and Esashi, M. (2002) Energy dissipation in submicrometer thick single-crystal silicon cantilevers. *Journal of Microelectromechanical Systems*, **11**, 775–783.

Yasumura, K.Y., Stowe, T.D., Chow, E.M. *et al.* (2000) Quality factors in micron- and submicron-thick cantilevers. *Journal of Microelectromechanical Systems*, **9**, 117–125.

Zener, C. (1937) Internal friction in solids. I. Theory of internal friction in reeds. *Physical Review*, **52**, 230–235.

Zook, J., Burns, D., Guckel, H. *et al.* (1992) Characteristics of polysilicon resonant microbeams. *Sensors and Actuators A: Physical*, **35**, 51–59.

16

Surface Interactions

16.1 Introduction

According to Schroeder (2012), Wolfgang Pauli said, 'God made the bulk; surfaces were invented by the devil.' The Austrian physicist, who was also known for his sarcastic humour and his sometimes disrespectful attitude towards his colleagues, has effectively epitomized the inherent difficulties encountered when dealing with surface-related phenomena. A broad spectrum of research activities is connected to surface interactions, whose fundamental features have been studied with reference to the physics and chemistry of elementary interactions (Israelachvili, 2011). The importance of the practical implications led to the development of modern tribology, a whole branch of mechanical engineering pioneered by David Tabor in the 1960s (Bhushan and Gupta, 1997).

In the case of MEMS, surface interactions have an unprecedented importance, for two reasons. First, the surface-to-volume ratio of structures at the microscale is very high: as a matter of fact, such a ratio is roughly proportional to the inverse of the object's leading dimension, so a tremendous increase is expected if the size scales down at the microscopic level (or, *a fortiori*, at the nanoscale). Therefore, the mechanical features that are connected to surface properties are boosted in the case of MEMS and NEMS. Moreover, the surface-based forces are to be compared with restoring elastic forces, which are very small in MEMS. It has been shown in Chapter 2 that the elastic recovery force of a simple cantilever is proportional to the moment of inertia (which in turn is proportional to the fourth power of the object's size) and is inversely proportional to the third power of the beam's length. As a consequence, if the size shrinks homothetically, the elastic restoring forces decreases in a roughly linear fashion.

For these reasons, surface interactions in MEMS have been thoroughly studied since the early age of micro-technology, as witnessed by the review paper of Maboudian (1998) and by the contributions summarized in the Workshop on Tribology Issues and Opportunities in MEMS, which was held in Columbus, OH, in 1997 (Bhushan, 1998). A systematic treatment of micro- and nano-tribology can be found in the classical textbook of Bhushan (1995).

This chapter is focused on one of the most important surface-related phenomena in MEMS, namely the spontaneous adhesion of nominally flat surfaces, which can severely affect the reliability of microdevices. The definition of the problem is followed by an

Mechanics of Microsystems, First Edition. Alberto Corigliano, Raffaele Ardito, Claudia Comi, Attilio Frangi, Aldo Ghisi, and Stefano Mariani.

Companion website: www.wiley.com/go/corigliano/mechanics

overview of possible sources of adhesion. Afterwards, the most prominent experimental works on the topic are summarized, with the purpose of providing a sound basis to the modelling techniques. Finally, some examples are presented and recent trends are summarized.

16.2 Spontaneous Adhesion or Stiction

Spontaneous adhesion is an issue of paramount importance in the field of MEMS, as evidenced by many recent review papers (Maboudian and Howe, 1997; Srikar and Senturia, 2002; Van Spengen, Puers and De Wolf, 2003; Zhao, Wang and Yu, 2003; Zhuang and Menon, 2005). In fact, it plays an important role in both the fabrication process yield and the device reliability during the service life of the device. These scenarios are often addressed in the literature as *stiction failures*. The word *stiction* is a neologism coined from *static friction* (Tas, Gui and Elwenspoek, 2003): this emphasizes the strong correlation with the world of micro-tribology, since friction and wear of contact pairs are tightly connected to adhesive phenomena on contacting surfaces (Bhushan, 2007).

Although adhesion may occur naturally at all scales, it is especially dangerous in MEMS because of their small dimensions, typically large surface-to-volume ratios, small elastic restoring forces due to microbeam slenderness and close proximity of adjacent surfaces. Failure happens when the forces driving decohesion are not sufficient to separate the components, in order to restore the operating configuration. After this catastrophic and irreversible event, the micromachine could be completely unusable and, consequently, should be replaced. The importance of stiction phenomena with respect to MEMS reliability is therefore evident.

Stiction failure may appear in different situations, which can be roughly categorized into two types. The former is the so-called *process stiction* or *release stiction*. In this case, failure occurs during the final step of surface micromachining. After etching away the sacrificial oxide layers in hydrofluoric acid solution, the etchant residues are rinsed with deionized water. During the subsequent drying stage, some liquid bridges are entrapped between the two solids. Depending on the contact angle of the water–solid meniscus, some capillary adhesive forces may be generated. Such forces may induce the sudden collapse of the released surface on the substrate: if the adhesion energy in the contact pair is higher than the elastic energy due to structural deformation, the surface remains stuck to the substrate even after drying. This phenomenon has been studied since the beginning of MEMS history. The pioneering paper of Mastrangelo and Hsu (1993a) contains an interesting study of the capillary effect during the production of micromachines. The authors finally propose a simple index (the so-called elastocapillary number) whose value determines whether the structure reaches contact or not. This number obviously depends on the surface tension of water, the contact angle and the radius of the meniscus and the elastic properties of the structure. Various analytical computations have been performed in the case of beams and plates, both in the small displacement hypothesis and in a nonlinear kinematic regime. Recently, some special fabrication procedures have been suggested to overcome release stiction: in particular, the Sandia group proposed the adoption of supercritical CO_2 drying (De Boer *et al.*, 1998). After rinsing, devices are transferred from water to methanol and placed in a pressure vessel. Herein, gaseous CO_2 is introduced until the supercritical state is reached and then the methanol is displaced. Finally, supercritical CO_2 is vented thus achieving free-standing beams up to 2 mm in length.

The latter type of stiction, often called *in-use stiction* or *post-release stiction*, refers to adhesion that occurs after the release stage, namely during service operations, in storage or because of improper handling. The accidental contact of surfaces may happen, for instance, as a consequence of uncontrolled electrostatic forces, possibly due to static electrification in the handling environment. In the case of sudden drop or impact of the microdevices, inertia forces might be so large as to induce the collapse of a free-standing structure on the substrate. The same effect can be obtained, in the case of unpackaged MEMS, by rapid and strong air flux in the device. While release stiction has a great impact on device yield (which is, of course, very important from the industrial point of view), it is the in-use stiction that poses the greatest threat to MEMS reliability. For this reason, most research examined in the following is concerned with the case of in-use stiction.

Finally, it is important to remember that adhesion is not always a phenomenon to be avoided in MEMS. Some microdevices (for instance, some kinds of microswitch) are properly designed for intermittent or cyclic contact. In these cases, the expression *stiction failure* no longer makes sense. Nevertheless, the adhesion properties of such devices deserve study, since they play a significant role during operation by dictating the forces and power required to operate properly.

16.3 Adhesion Sources

The full description of the physical background of spontaneous adhesion is an awkward task. This section represents an attempt to describe some of the possible physical mechanisms that underlie stiction in MEMS. The approach that is considered herein involves providing a thorough description of surface forces, starting from their essential nature, i.e. intermolecular bonds of different kinds (Israelachvili, 2011). A sharp categorization is introduced in what follows, but it should be kept in mind that all the described phenomena are contemporarily present in adhesion and a strong interference among them is often encountered.

16.3.1 Capillary Attraction

Fluid–solid interactions are directly related to the geometric aspects of contact between liquid drops and solid surfaces. By considering an initially spherical particle, it is necessary to divide its external boundary into a flat area A_f (the one that adheres to the solid) and a curved one A_c. From the geometric point of view, by considering the conservation of the drop volume, the following relationship, which involves the *contact angle* θ, can easily be demonstrated:

$$\frac{dA_c}{dA_f} = \cos\theta. \tag{16.1}$$

The contact angle, which can be easily measured during experiments, is therefore used to evaluate the fluid–solid interaction. It is of great importance in the simulation of adhesive behaviour in wet surfaces: in fact, if θ is less than 90°, an attractive force arises between two wet plates.

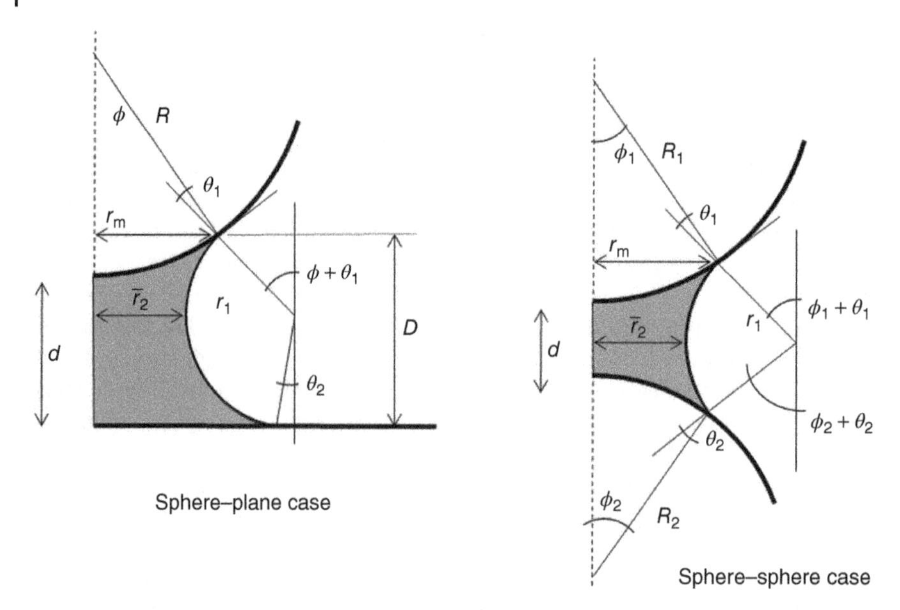

Figure 16.1 Meniscus between spherical asperity and flat surface. *Source*: Ardito *et al*. (2014), Figure 3. Reproduced with permission of Elsevier.

When two surfaces are close to contact, fluids that show a small value of the contact angle (i.e. fluids that *wet* the solid substance) will spontaneously condense from vapour into bulk liquid around surface contact sites (for instance, in cracks, pores or around asperities (Raccurt *et al*., 2004; Van Spengen, Puers and De Wolf, 2002). At equilibrium, the so-called *fluid meniscus* is formed (as depicted in Figure 16.1). The meniscus causes a force that attracts the particles for two reasons: the direct action of the surface tension γ_L of the liquid around the periphery of the meniscus and the pressure inside the meniscus, which is reduced, compared with the outer pressure, by the capillary pressure ΔP.

The computation of both these contributions is subordinated to the evaluation of the shape of the meniscus. The solution of such a problem is a nontrivial task, even in the case of regular asperities.

According to the physical properties of the fluid–surface interface, the meniscus will form the given contact angle θ with the solid surface. The two additional equations describing the thermodynamic equilibrium state of a meniscus are the Young–Laplace equation and the Kelvin equation (Israelachvili, 2011). The former relates the curvature of a liquid interface to the pressure difference ΔP between the fluid and the surrounding gas. If the self-weight of the meniscus is negligible, the enforcement of the mechanical equilibrium of the interface leads to the Young–Laplace equation:

$$\Delta P = \gamma_L \left(\frac{1}{r_1} + \frac{1}{r_2} \right) \equiv \frac{\gamma_L}{r_k} \tag{16.2}$$

In this equation, r_1 and r_2 are the principal radii of curvature that describe the curved shape of the meniscus: the sum of such values, which is invariant with respect to the local reference system, has been defined as r_K, the *Kelvin radius*. In capillarity theory, it is common to consider that the radius is positive if the interface is curved towards the liquid. As a result, a spherical liquid droplet with radius R in a gas has $r_1 = r_2 = 1/R$, whereas a spherical bubble in a liquid has $r_1 = r_2 = -1/R$.

The Kelvin equation relates the relative humidity (RH) to the curvature of the surface of the condensed liquid:

$$r_{\mathrm{K}} = \frac{\gamma_L V_m}{R_u \; T \log RH}, \tag{16.3}$$

where R_{u} is the universal gas constant, T is the absolute temperature and V_{m} is the molar volume of the liquid. For a typical meniscus with a hydrophilic contact angle, $r_1 < 0, r_2 > 0$, $r_2 \gg r_1$ and so $1/r_1 + 1/r_2$ is negative: coherently $RH < 1$ and $\log(RH)$ is negative.

Once the geometric configuration of the meniscus has been obtained, one can compute the area of the wetted region ($A = \pi r_{\mathrm{m}}^2$) and the length of the meniscus' attachment line. The attraction force exerted on the two surfaces is given by the sum of the following actions: (i) the surface tension over the contact length; (ii) the capillary pressure over the wetted area. The key point is represented by the evaluation of the shape of the meniscus. To this purpose, one can follow two different approaches (Payam and Fathipour, 2011), depending on whether or not thermodynamic equilibrium is assumed between the liquid and the surrounding gas. If equilibrium is not assumed a priori, a formulation based on liquid volume conservation must be considered: many researchers refer to this situation (Rabinovich, Esayanur and Moudgil, 2005; Mu and Su, 2007; Payam and Fathipour, 2011); however, it is more realistic to assume thermodynamic equilibrium since, in a typical situation, menisci between MEMS components are caused by capillary condensation and there is no way to evaluate the volume of the liquid bridges (Butt, 2008). If equilibrium is assumed, the mean curvature of the meniscus is constant and the Kelvin equation (16.3) holds, which is solvable either by analytical (Stifter, Marti and Bhushan, 2000; Ardito *et al.*, 2014) or numerical (Pakarinen *et al.*, 2005; Chau *et al.*, 2007) methods.

In the special case of spherical asperities over a flat surface, an approximate solution for the shape of the meniscus can be obtained on the basis of a commonly accepted assumption, namely $|r_2| \gg |r_1|$ and hence $r_{\mathrm{K}} \approx r_1$. In that case, the generating curve for the axisymmetric meniscus will be a circle of radius $|r_{\mathrm{K}}|$. With reference to Figure 16.1, the radius r_{m} of the wetted region can be expressed in terms of the separation d of the sphere from the plane and the height D of the contact point between the meniscus and the sphere. The so-called standard approximation, which is typically adopted in the literature (Israelachvili, 2011), can be applied when the size of the asperity is large, so that $\phi \ll 1$ and $(D - d) \ll R$. In this case, one obtains

$$r_{\mathrm{m}}^2 = 2R(D - d), \quad \text{with} \quad D = |r_{\mathrm{K}}|(\cos\theta_1 + \cos\theta_2). \tag{16.4}$$

The capillary pressure term drops to zero when $d \geq D$; this plays the role of a cut-off distance. The maximum adhesion force is attained when $d = 0$, which means that the sphere is in contact with the flat surface. In that case, the area of the wetted surface is approximately given by

$$A_{\mathrm{L}} \approx 2\pi R|r_{\mathrm{K}}|(\cos\theta_1 + \cos\theta_2). \tag{16.5}$$

In the special case, $\theta_1 = \theta_2 \ll 90°$, one can neglect the effect of the surface tension so that the adhesive force is just obtained by multiplying the Laplace pressure (16.2) by the wetted area (16.5); one finds the well-known formula:

$$F_{\text{asp}} \approx 4\pi\gamma_{\text{L}} R \cos\theta. \tag{16.6}$$

More refined computations can be carried out by relaxing the simplifying hypotheses (Ardito *et al.*, 2014). In any case, it is possible to demonstrate that the adhesion force decreases as the distance d between the contact surface increases, until the cut-off separation is reached. After that threshold, which is directly proportional to the Kelvin radius, see Equation 16.4, the force suddenly drops to zero. From these considerations, it is clear that the adhesion energy, obtained as the work of the adhesive force when the surfaces are separated by an infinite distance, increases as the Kelvin radius increases (in absolute value). This happens, according to Equation 16.3, as the relative humidity increases from 0 to 1 (i.e. from 0% to 100%). As expected, the adhesion due to capillarity is large in the presence of a humid environment.

As a final remark, it is important to underline the dependence of the Kelvin radius on the absolute temperature. Moreover, in reference to Van Spengen, Puers and De Wolf (2003), it has been reported that the water surface tension γ_{L} decreases linearly with temperature:

$$\gamma_{\text{L}} = 121.2 \text{ mJ/m}^2 - 0.167 \text{ mJ/m}^2\text{K} \times T. \tag{16.7}$$

These data are important if the adhesion is to be computed in the presence of a huge temperature variation.

16.3.2 Van der Waals Interactions

The interactions between molecules are strongly dependent on the electrical features of the molecules itself. In the presence of charged molecules (i.e. ions), the main force responsible for the interaction is the Coulomb force. Moreover, interactions between charges, permanent dipoles, quadrupoles, etc. fall into this category. Conversely, in the presence of neutral molecules, there are polarization forces that arise from the dipole moments induced in atoms and molecules by the electric fields of nearby charges and permanent dipoles. These forces contribute to the overall van der Waals interactions, which play a fundamental role in various important phenomena, such as adhesion, surface tension, physical adsorption, wetting, gas, liquids and thin-film properties, strengths of solids, flocculation of particles in liquids and structural properties of condensed macromolecules.

An important contribution to the van der Waals forces, especially in the case of MEMS, is represented by the *dispersion forces*, i.e. forces between nonpolar molecules in the absence of charges and of permanent dipoles. Such forces, whose name is derived from their importance in light dispersion phenomena, are essentially quantum mechanical in origin. A rigorous treatment of dispersion forces is beyond the scope of this book; however, their origin can be intuitively understood as follows. For nonpolar molecules, the time-averaged dipole moment is obviously zero, but the instantaneous position of the electrons with respect to the nucleus gives rise to a finite dipole moment. This dipole influences the surrounding electric field and polarizes any nearby neutral atom, inducing a dipole moment in it. As a consequence of the (induced) dipole–dipole interaction, an instantaneous attractive force is applied to the atoms and the time average of this force is finite.

Some essential aspects of the van der Waals forces deserve attention:

1) They are long-range forces, i.e. their effect is sensible from *large* distance (greater than 10 nm) down to interatomic spacings (≈0.2 nm).
2) These forces are usually attractive but can become repulsive, depending on the material properties.
3) Besides inducing intermolecular forces, the van der Waals interactions tend to align the molecules, thus showing anisotropic behaviour in some cases.
4) The dispersion force between two atoms is affected by the presence of a third atom, from which it follows that the forces are *nonadditive*.

The last observation is of special importance if the focus is on the overall force between two surfaces. In spite of the nonadditivity, the van der Waals force is usually computed by integrating the energies of all the atoms in one body with all the atoms in the other body, thus obtaining the *two-body potential*. The resulting interaction laws for some common geometries have been expressed in terms of the conventional Hamaker constant A, which is proportional to the coefficient that governs the atom–atom interaction and to the number of atoms per unit volume in the two bodies. For silicon, the following value can be adopted (Bergström, 1997):

$$A = 27 \times 10^{-20}\ \mathrm{J}. \tag{16.8}$$

In the application to MEMS rough surfaces, three geometries are important:

$$W = \begin{cases} -\dfrac{A}{d}\dfrac{R_1 R_2}{R_1 + R_2} & \text{two spheres,} \\ -\dfrac{A\,R}{6d} & \text{sphere–flat,} \\ -\dfrac{A}{12\pi d^2} & \text{two flat surfaces,} \end{cases} \tag{16.9}$$

where W is the free energy, A is the Hamaker constant, R is the radius of the sphere and d is the minimum distance between the bodies. The resulting specific force (pressure) can be computed simply by

$$q = \frac{\partial W}{\partial d}. \tag{16.10}$$

As shown by DelRio *et al.* (2005), van der Waals forces could be of great importance in MEMS adhesion, particularly if capillary forces are absent. The authors proposed the use of Equation 16.9 for the evaluation of overall forces between rough surfaces. In the case of almost flat surfaces, the forces can be computed by adopting the last case in Equation 16.9; conversely, if the surfaces are very rough, a significant part of the area is too far apart to contribute to the adhesion, and the interaction would be dominated by the first case, computed for each asperity. In the intermediate cases, i.e. for relatively rough surfaces, evaluation of the surface interaction due to intermolecular forces is a very difficult task. To obtain a reasonable estimate, the so-called *proximity force approximation* (DelRio *et al.*, 2005) is commonly adopted. First, it is necessary to define a set of points, lying on the surface to be considered for adhesion simulation. Each point is endowed with a portion of area and interacts with a corresponding point on the opposite surface. The van der Waals force is then computed by introducing the following hypothesis: the portions of surface around the interacting points are assumed to be flat and parallel to one another. In this way, it is possible to compute the interaction for each pair of points and the overall force is obtained by a summation:

$$F_{\mathrm{vdW}} = \sum_{i=1}^{n_{\mathrm{P}}} S_i \frac{A}{6\pi(d_i + z_0)^3}, \tag{16.11}$$

where S_i is the area of the surface connected with the single point, d_i is the distance between the interacting points and z_0 is the equilibrium separation, namely the distance between two atoms at equilibrium, which for silicon is $z_0 = 0.149$ nm. The introduction of z_0 is important to avoid infinite attractive forces when two points come into contact.

16.3.3 Casimir Forces

As already mentioned, the van der Waals forces considered in the previous section have influence in the distance range 0.2–50 nm. For larger distances, the molecular interaction changes, as demonstrated by Casimir and Polder (Lamoreaux, 2005). The so-called *retarded regime* manifests at such distances, in the following sense. The dispersion forces are generated by the mutual effects of temporary dipoles; if the intermolecular distance is small, the effect of the dipole, which is a perturbation of the electromagnetic field propagating at the speed of light, reaches the other atoms almost instantaneously. Conversely, when two atoms are an appreciable distance apart, the time taken for the electric field to reach the second atom and return is comparable with the period of the fluctuating dipole. As a consequence, the return field will find a dipole oriented in a different manner and less favourably disposed to an attractive interaction. This entails that, by increasing the intermolecular distance, the van der Waals forces decay faster than expected. The retarded van der Waals forces are often referred to as *Casimir forces*.

Historically, Casimir forces have been considered an exotic quantum phenomenon, but now they start to take on a technological importance. In fact, the Casimir force has a major role in modern micro- and nano-technologies, especially with reference to adhesion phenomena, as evidenced by the abundance of literature on the topic (Buks and Roukes, 2001; Ding,Wen and Meng, 2001; Lin and Zhao, 2005). The exact calculation of the Casimir force between surfaces involves very difficult computations, based on quantum mechanics concepts. One relatively simple case is represented by two perfectly conductive flat surfaces with zero thickness. The last feature is important because it allows the analyst to integrate simply on a bi-dimensional domain; the same issue is less relevant with respect to standard van der Waals forces, since they act on a shorter distance. After lengthy computations, one obtains the following expression for the Casimir pressure:

$$q = \frac{\hbar c \pi^2}{240 d^4}, \tag{16.12}$$

where $\hbar = 1.0545 \times 10^{-34}$ Js is the Dirac constant and $c = 2.998 \times 10^8$ m/s is the speed of light in vacuum. By comparing this expression with the standard van der Waals pressure, the different decay rates can easily be understood: Casimir pressure decays as

$1/d^4$, whilst standard van der Waals pressure decays as $1/d^3$. The attractive pressure q lies within 8.12×10^{-7} μN/μm^2 and 3.33×10^{-11} μN/μm^2 when the separation d ranges from 200 nm to 2.5 μm. Such a phenomenon might acquire technological importance in modern micro- and nano-technologies, e.g. in tribology, as demonstrated by Munday, Capasso and Parsegian (2009), who first measured repulsive forces associated with the Casimir effect. It has recently been shown that Casimir forces may hamper the functioning of MEMS and NEMS by inducing a pull-in instability, as explained by Lin and Zhao (2005). A couple of experimental proofs of that fact have been described by Ardito *et al.* (2012) and Zou *et al.* (2013). Attention has also been paid to the positive implications of the attractive Casimir effect, namely its use as an actuation source for micromachines (Bárcenas, Reyes and Esquivel-Sirvent, 2005). Moreover, the aforementioned inversion of sign, which is dependent on the material properties and on the surface shape, could also be favourably used as an anti-stiction provision.

The ideal situation of a perfectly conductive and zero-thickness slab is never encountered in real cases. Some corrections have been studied to consider the finite thickness and the actual material dielectric properties. In the case of silicon, the appropriate reference work is Lambrecht *et al.* (2007), where the following corrected formula is proposed:

$$q_{\mathrm{Si}} = \eta \frac{\hbar c \pi^2}{240 d^4}. \tag{16.13}$$

The corrective factor η is a function of the slab thickness w and of the separation gap d. For instance, in the case of slabs whose thickness is around 10 μm, the corrective factor attains its maximum value $\eta = 0.3$ for gaps in the micrometric range.

The computation of Casimir forces for different geometries should be performed by means of suitable numerical tools: for instance, a formulation based on integral equations has been proposed and exploited by Reid *et al.* (2009). An alternative simplified approach, which has been proven to be accurate in some cases, is the already mentioned proximity force approximation (DelRio *et al.*, 2005), in which Equation 16.12 is applied point-wise and the hypothesis of additivity is enforced. The final expression is similar to what has been obtained for van der Waals forces (16.11): indeed, in some cases the two forces are treated simultaneously, by introducing a suitable *retardation function*, which accounts for variations of the force with separation. Few other contributions in the literature deal with the simplified evaluation of Casimir forces. Emig *et al.* (2001) presented an interesting analytical result. The attractive force between a perfectly flat plate and a parallel plate with organized microasperities in the form of sinusoidal waves is evaluated. A rough surface can be approximated by means of sinusoidal asperities, with a certain pitch λ and amplitude h, assuming that the interaction between two rough surfaces can be described, at least as a first approximation, by the interaction between a flat surface and a surface with asperities twice as deep.

16.3.4 Hydrogen Bonds

The hydrogen bond is a special electrostatic interaction that exists between electronegative atoms (e.g. oxygen, nitrogen, fluorine and chlorine) and hydrogen atoms covalently bound to similar electronegative atoms. This bond is specific of hydrogen atoms because of their tendency to become positively polarized and their uniquely small size, thus enhancing the interaction with electronegative atoms.

The hydrogen bond can be important in MEMS because silicon parts are often covered by thin films endowed with OH terminations (Van Spengen, Puers and De Wolf, 2002). The effect of hydrogen bonds has been studied by Legtenberg *et al.* (1994), who reported the number of OH terminations in a typical microdevice to be about 5 ± 0.5 nm^{-2}. The bond energy is 10–40 kJ/mol, as also reported by Israelachvili (2011), thus achieving a specific energy equal to 0.1–0.3 J/m^2. This value is very high, if compared with some experimental values for surface energy. It must, however, be considered that the range of action of the forces connected to the hydrogen bond is $\approx$0.5 nm (Ripalda, Gale and Jones, 2004), thus far smaller than that reported for the van der Waals pressure. From these considerations, it can be concluded that hydrogen bonds should be considered in MEMS adhesion in the case of small separations (as happens for exceptionally flat surfaces), hydrophilic surfaces (i.e. the presence of OH terminations) and very low relative humidity (in fact, for hydrophilic surfaces in the presence of vapour, the capillary force would dominate).

16.3.5 Electrostatic Forces

The effect of electrostatic forces during adhesion has been deeply studied in the literature (Knapp and de Boer, 2002; Zhang and Zhao, 2004). The research interest is mainly related to the fact that many experimental set-ups, described in the next section, are based on the electrostatic actuation of microbeams: it is therefore important to evaluate the electrostatic field precisely, also accounting for nonlinear behaviour due to beam deflection. Conversely, spontaneous adhesion is usually referred to the case of a movable conductive structure that collapses on a dielectric substrate. This means that no electrostatic forces should be present in normal operating conditions. However, in some special cases, the surface of the dielectric can be capable of localized charge storage for a considerable time. This may induce a parasitic electrostatic field, with consequent increase of the adhesion energy (Van Spengen *et al.*, 2004; Wibbeler, Pfeifer and Hietschold, 1998).

The dielectric charging may happen for many reasons. First, during device testing, scanning electron microscopy is commonly used to inspect the wafer. As reported by Tang *et al.* (1992), the undesirable charging effect originates, in stacked oxide–nitride layers, from the electron beam of the scanning electron microscope.

During MEMS operation, charge accumulation may happen as a consequence of mechanical rubbing of two materials with different work functions. For instance, this is the case of collapsed surfaces after pull-in instability has occurred. Finally, gas discharge within the capacitive air gap is another possible cause of charging. Normally, the voltage level in MEMS is by far less than the breakdown threshold for the air gap. Nevertheless, the following observations must be taken into consideration. First, static electrification in common handling environment can cause a dramatic voltage build-up of several kilovolts. Second, it has been shown by Dhariwal *et al.*, 1994 that Paschen's law, which governs the breakdown voltage, no longer applies for gaps smaller than 4 μm. In particular, for a gap of 1 μm, a breakdown voltage of only 20 V is reported.

If compared with the other causes of adhesion, parasitic electrostatic attraction can be disregarded without introducing any important error. Nevertheless, this phenomenon deserves some special attention in case the extreme conditions referred to in this section are encountered during the MEMS life.

16.4 Experimental Characterization

This section summarizes, in a nonexhaustive way, the most important experimental methods, which have been developed to examine the issue of spontaneous adhesion. A sort of historical perspective is achieved, from pioneering results (obtained in the early 1990s) to procedures at the cutting edge of technology.

16.4.1 Experiments by Mastrangelo and Hsu

One of the earlier attempts in experimental research on adhesive phenomena is represented by the work developed in the Ford Motor Company Scientific Research Laboratory in 1992 (Mastrangelo and Hsu, 1993a,b). This group, as mentioned in Section 16.2, was first interested in the investigation of stiction failure during the rinse–dry phase of the production process. To this purpose, a *contact bound* was determined (Mastrangelo and Hsu, 1993a) with reference to the already defined elastocapillary number. Such a bound was used to predict whether the capillary forces were sufficient to induce pull-in instability of the moving structure and, consequently, contact the underlying layer during the release process.

A different problem is connected with the the so-called *peel bound* (Mastrangelo and Hsu, 1993b), which determines whether the contact can endure permanently because of surface attractive forces. The difference between the two situations is represented by the energy components that come into play. The contact onset involves just the surface energy U_{SL} in the water meniscus and the elastic restoring energy U_E:

$$U_T = U_{SL} + U_E. \tag{16.14}$$

Conversely, after contact has reached, the solid-to-solid surface energy U_{SC} must be added:

$$U_T = U_{SC} + U_{SL} + U_E. \tag{16.15}$$

It is worth noting that, in the considered papers, the solid-to-solid energy U_{SC} was assumed to depend on a free parameter γ_S (the characteristic surface energy, to be identified on the basis of experiments), whose nature was not completely specified. The peel number was used to discriminate the situation of complete detaching with respect to the situation of adhesion to the substrate and was obtained on the basis of equilibrium considerations. If the structure is on the verge of detaching, the peel number attains the unit value. In this situation, one can find an expression of the beam's length l in terms of the Young's modulus E, the specific energy γ_S and the other geometric parameters, the beam's thickness t and the gap with respect to the substrate h (see Figure 16.2). More specifically, the beam's length turns out to be directly proportional to the parameter $(t^3h^2)^{1/4}$.

By using a microscope and a Michelson interferometer attachment, the detachment length was determined for each sample, i.e. the length of the longest beam that did not remain stuck to the substrate. The surface energy of the bond was determined by fitting the experimental detachment length, as obtained in the case of peel number equal to unity. The following results were achieved in terms of surface specific energy γ_S:

$$\gamma_S = \begin{cases} 270 \pm 100 \text{ mJ/m}^2 & \text{for hydrophilic surfaces,} \\ 100 \pm 60 \text{ mJ/m}^2 & \text{for hydrophobic surfaces.} \end{cases} \tag{16.16}$$

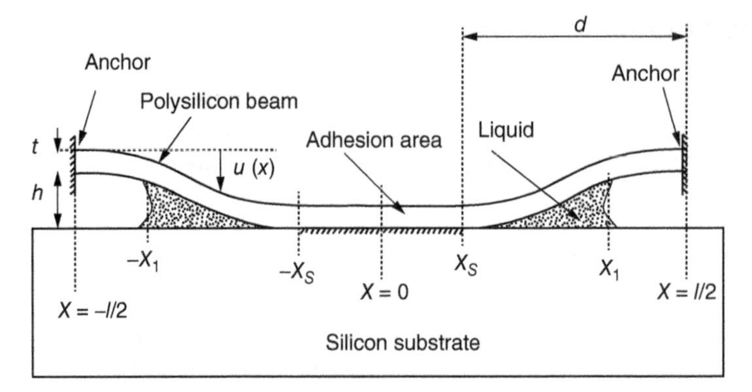

Figure 16.2 Longitudinal cross-section of doubly clamped beam adhering to its substrate. *Source*: Mastrangelo and Hsu (1993b), Figure 1. Reproduced with permission of IEEE.

The experiments were repeated for circular plates, but the results were in some sense inconsistent with those referred to beams. In fact, as explained by Figure 16.3, the previous estimate of surface energy no longer fits the experimental data; moreover, the peel number is in some way inadequate to predict stiction. These errors have been attributed by the authors to the ideal boundary conditions assumed in the calculation (i.e. fully clamped plates).

16.4.2 Experiments by the Sandia Group

The Sandia group started performing experiments on adhesion in the second half of the 1990s. Their research activity was developed on the basis of a thorough examination of the previous work of Mastrangelo and Hsu, who proposed the detachment length as a parameter in order to estimate the surface energy. The critical points were identified as: (i) accuracy in the determination of the adhesive energy; (ii) sensitivity to variations

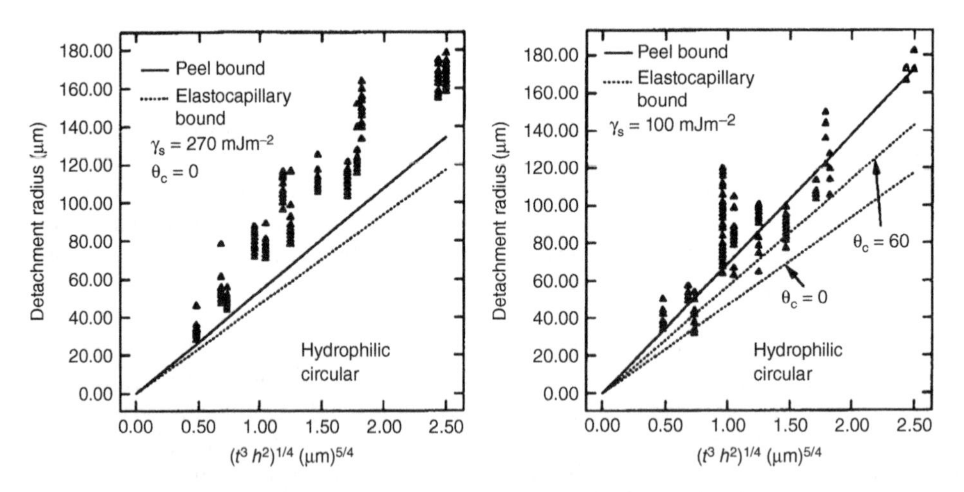

Figure 16.3 Comparisons of theoretical and experimental results for circular plates sticking to the substrate in their central part. *Source*: Mastrangelo and Hsu (1993b), Figure 12. Reproduced with permission of IEEE.

in surface topography; (iii) sensitivity to environmental conditions (namely, to relative humidity).

The first important improvement with respect to Mastrangelo and Hsu was represented by the already mentioned procedure (supercritical CO_2 drying), which can avoid the *release stiction* (De Boer *et al.*, 1998). In this way, a complete array of free-standing beams was obtained. The beams were electrostatically actuated to reach contact with the substrate (see Figure 16.4): attention was paid to the positioning of the electrode outside the landing pad, so that the evaluation of adhesion energy would not include the effect of unwanted electrostatic forces. After the contact configuration was reached, a fringe pattern obtained by interferometry was used to measure the vertical deflection of the beams. As explained by De Boer and Michalske (1999), the fracture mechanics concept of energy release rate was adopted to estimate the adhesion energy on the basis of the detached length s shown in Figure 16.4. The adhered part of the beam is idealized as a crack locus; the adhesion energy can be correlated to the amount of energy that is necessary to induce an advancement of the crack front, which is the energy release rate. At equilibrium, i.e. when the crack stops, the energy release rate is equal to the restoring elastic energy, which, for the case represented in Figure 16.4, can be easily evaluated as

$$\gamma_S = \frac{3}{2} E \frac{h^2 t^3}{s^4}, \tag{16.17}$$

where E is the Young's modulus of silicon and h and t are the initial gap and the beam thickness, respectively. By measuring, via the interferometric technique, the detached length s, the adhesion energy can easily be obtained. It is worth mentioning that the method proposed by the Sandia group seems to be superior to the measuring of the shortest adhered beam for the following reasons: (i) the deflection measurement validates the adhesion value for each beam, (ii) the incremental contact area is well known, (iii) the energy well is very deep. For instance, the foremost results presented by De Boer and Michalske (1999), for uncontrolled humidity, can be summarized as follows:

$$\gamma_S = \begin{cases} 16.5 \pm 8.2 \text{ mJ/m}^2 & \text{for hydrophilic surfaces,} \\ 3.4 \pm 0.5 \text{ mJ/m}^2 & \text{for hydrophobic surfaces.} \end{cases} \tag{16.18}$$

The huge difference with respect to the work of Mastrangelo and Hsu is one important conclusion of the considered paper, related to the length of the tested beams. If the beam is too short, the adhered configuration cannot be represented as in Figure 16.4 (a so-called *S-shaped beam*): the deformed configuration is better described as an *arc-shaped beam*. This configuration should be avoided in experiments, since the very small contact area (which ideally shrinks to a single point) means that the adhesion energy is strongly dependent of local irregularities of the contacting surfaces. Consequently, longer beams should be preferred when designing the experiment.

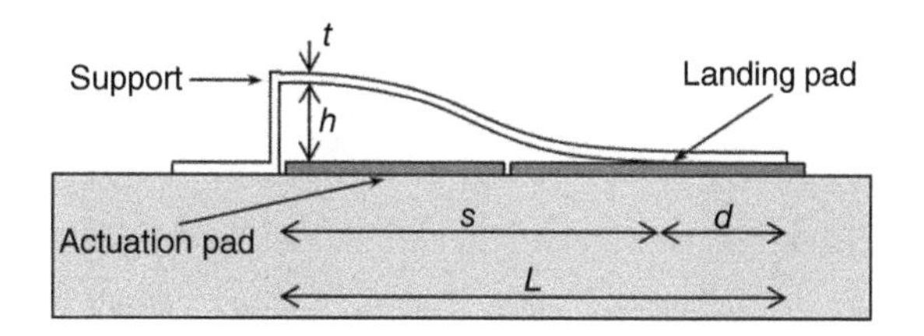

Figure 16.4 Longitudinal cross-section of cantilever beam, as considered in the Sandia experimental set-up. *Source*: De Boer 1998. Reproduced with the permission of Materials Research Society.

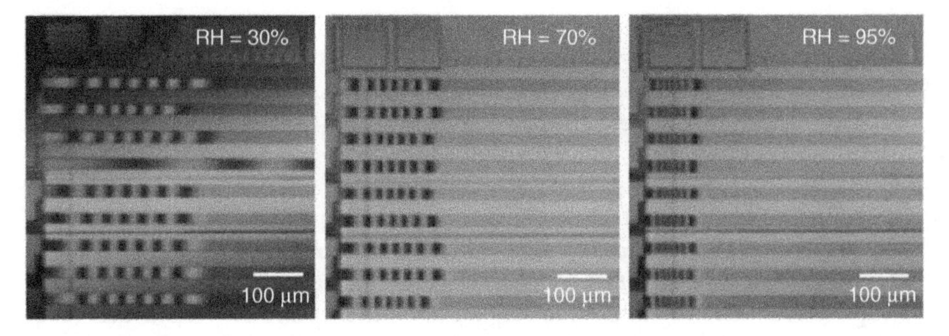

Figure 16.5 Interferogram of detached lengths versus relative humidity (RH), after 40 h exposure. *Source*: De Boer (2007), Figure 2. Reproduced with permission of Springer.

The experimental campaign has been enriched by constructing an environmental interferometric probing station (De Boer *et al.*, 1999; Knapp and de Boer, 2002; De Boer, 2007). The specimens were placed in a humidity-controlled environment and were analysed on the basis of the fracture mechanics concepts. A typical representation of the fringe patterns for different humidity levels is reported in Figure 16.5: the reduction in the detached length with the increase of relative humidity is a clear indicator of the effect of capillary attraction.

The experiments demonstrated that: (i) the results for the as-received samples were erratic, possibly because of thin organic films deposited on the surfaces – the original situation was restored by plasma polishing; (ii) the measured deflection of the beam was in excellent agreement with the theoretical prediction, based on the elastic deformation of a doubly clamped beam subjected to an imposed vertical displacement at the left end support; (iii) the measured adhesion energy (ranging from 1 mJ/m^2 to 50 mJ/m^2, for the smallest and largest relative humidities, respectively) was always smaller than the capillary surface energy, which is assumed to be the main source of adhesion for a hydrophilic surface, such as the one adopted in the tests. A synopsis of the achieved results is represented by the plot of adhesion energy with respect to relative humidity shown in Figure 16.6.

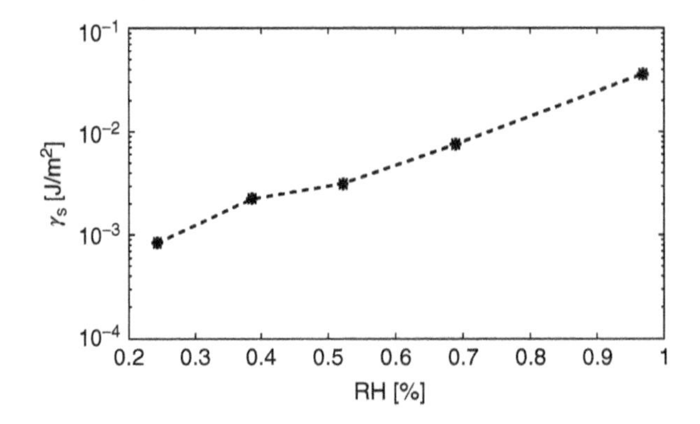

Figure 16.6 Adhesion energy as a function of relative humidity (RH). Data from De Boer *et al.* (1999).

Another important outcome of the Sandia experimental campaign has been represented by the recognition of the roughness level as a paramount parameter that influences the adhesion energy. In a perfectly dry environment (which means that capillary attraction can be ruled out), the adhesion energy can change by one order of magnitude if the roughness increases from 2.6 nm rms to 10.3 nm rms (DelRio *et al.*, 2005): the flatter the surface, the higher the adhesion energy. Clearly, the specific adhesion energy is always referred to the nominal area of the contacting surfaces. A more complex behaviour is encountered for a humid environment. For low humidity, the dry conditions are recovered; for relative humidities greater than 90%, the capillary condensation is so widespread that the effect of roughness is overwhelmed by the fluid film. In the transition regime, some interesting situations may appear, see e.g. the results shown in De Boer *et al.* (1999) (Figure 16.6) compared with the later results reported in DelRio *et al.* (2007). In the latter case, which is referred to higher roughness, an abrupt transition of adhesion energy is observed for relative humidities around 65%. This is because, in the presence of bigger asperities, capillary condensation may happen only if the relative humidity is greater than a certain threshold.

16.4.3 Experiments by the Virginia Group

A group of researchers from the University of Virginia has reported some interesting experimental results (Jones, Begley and Murphy, 2003; Jones, Murphy and Begley, 2003).

The experimental configuration was very similar to that proposed by the Sandia group. In fact, Jones, Begley and Murphy (2003) claimed that microcantilever beams offer two essential advantages: (i) ease of fabrication and (ii) ability to vary the ratio between elastic restoring energy and surface adhesive energy over many orders of magnitude by simply changing the beam dimensions. Conversely, they criticized the Sandia approach, as regards actuation and measurement procedures. In particular, the interferometric technique involves a rather high intrinsically level of uncertainty; since the energy release rate depends on the fourth power of the detached length s, see Equation 16.17, the measurement errors may have a strong impact on the estimate of the surface energy.

The proposed improvement of the measurement technique was the adoption of mechanical actuation by means of a micro-indenter, as shown in Figure 16.7.

The indenter was used to push the beam on the substrate by applying a certain force; contemporarily, the indenter displacement was recorded, and an estimate of the system stiffness was provided. After the elimination of the spurious effect due to the indenter and the contact stiffness, the detached length was computed by comparing the experimental stiffness with the computation performed with reference to the model of an elastic doubly clamped beam. Finally, the fracture mechanics approach, already introduced by the Sandia group, yielded the estimate of the surface energy.

Preliminary studies (Jones, Begley and Murphy, 2003) confirmed that computation of the effective stiffness (for small loads or displacements of the indenter) was clearly sensitive to the detached length. This indicated that good resolution is possible in evaluating the adhesion energy without viewing the sample, by slightly perturbing the system with small loads and calculating the system's stiffness. The agreement was better if the load was applied at locations far from the support post; for loading near the built-in end, the compliance of the support influenced the measured response and increased the

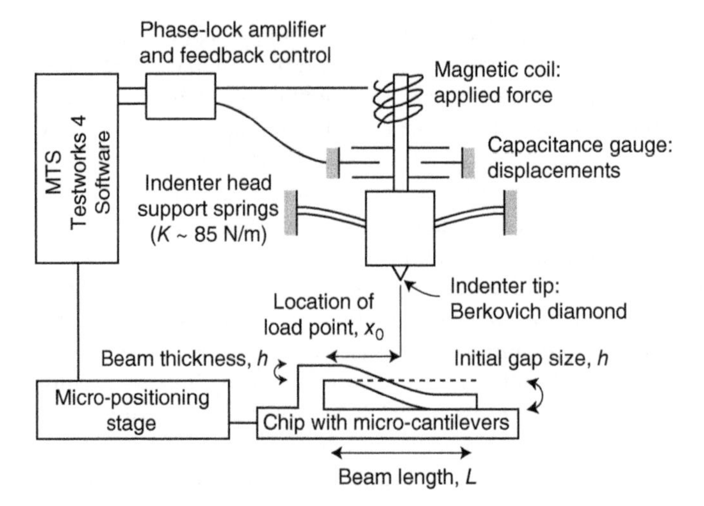

Figure 16.7 Experimental set-up proposed by the Virginia group. *Source*: Jones, Murphy and Begley (2003), Figure 1. Reproduced with permission of Springer.

discrepancy. The experimental outcomes furnished an estimate of the surface energy, with reference to a low-humidity environment. The estimated value was

$$\gamma_S = 20.6 \pm 1.22 \text{ mJ/m}^2. \tag{16.19}$$

This value is approximately 20 times greater than those reported in the literature for the same humidity level and beam fabrication process (De Boer *et al.*, 1999). A possible explanation of such a discrepancy, as proposed by the authors, is the different stiction mechanism for just-released structures (such as those tested by De Boer *et al.* (1999)) and aged specimens, considered by the Virginia group. As a matter of fact, several phenomena, which may happen during ageing in ambient air, are responsible for an increase of the adhesive energy: surfaces may become more hydrophilic because of hydrocarbon surface contamination; parasitic charges may accumulate, thus increasing the electrostatic attraction; etc.

An important result obtained by this group (Jones, Begley and Murphy, 2003; Jones, Murphy and Begley, 2003) is related to the cyclic behaviour of adhesion. The particular actuation procedure is able to perform a loading–unloading cycle, with changes in the adhered length. It has been clearly shown that there is an initial *shakedown* period until the beam reaches a stationary conditions. As depicted in Figure 16.8, the *shakedown* period lasts for five cycles; afterwards, the cycles follow the same loading–unloading hysteresis loop.

The cyclic behaviour could be partly explained by considering local plasticization around the indenter tip. It has also been shown that the stuck area permanently increased during the first five cycles, thus suggesting that the adhesive energy also changes during repeated loading and unloading. This is of paramount importance in the modelling of stiction, since it could be explained by considering the plastic deformation that happens, at the microscopic level, in correspondence to surface asperities.

Some possible improvement could be envisaged for the Virginia group testing procedure. In particular, it would be very interesting to reverse the sign of the applied

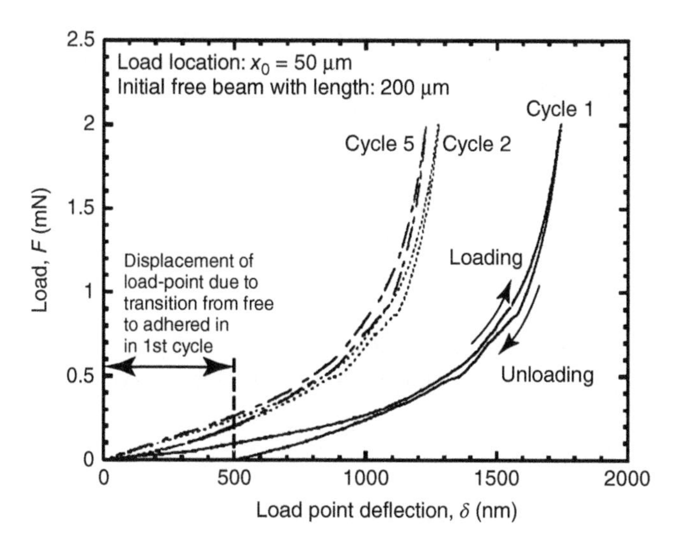

Figure 16.8 Experimental load–displacement curve for the indentation point: cyclic behaviour. *Source*: Jones, Begley and Murphy (2003), Figure 10. Reproduced with permission of Elsevier.

load, in order to perform some sort of *peel test*. Obviously, this cannot be done if the micro-indenter is adopted in the configuration depicted in Figure 16.7.

16.4.4 Peel Experiments

The aforementioned peel test has been performed, with reference to a different experimental arrangement, by Leseman, Carlson and Mackin (2007). The cantilever beams can be freely moved with respect to the substrate, so that they are rigidly brought into contact with a flat surface and are subsequently gently peeled off. During the test, the microcantilever beams were moved using an external piezoactuator; the deformed configuration of the beams was recorded using the same interferometric technique discussed in the previous sections. The tested beams were characterized by two lengths: 1000 μm and 1500 μm. It was shown that, in dry conditions, the adhered configuration was always of the arc-shaped type: this is possibly due to the low surface energy with respect to the restoring elastic force, which in turn depends on the beam length. Since, for arc-shaped beams, it is difficult to apply the fracture mechanics approach, it was chosen to characterize stiction using the average pull-off force, without attempting to calculate the adhesion energy. Unfortunately, this prevents comparison with other results in the literature.

In case of wet stiction, the adhesion energy clearly increased and S-shaped beams were achieved. The energy release rate method was applied to find an estimate of the adhesion energy, which is in good agreement with results previously reported (De Boer and Michalske, 1999).

Leseman, Carlson and Mackin (2007) reported that the results for wet stiction were less disperse, in a statistical sense, than those for dry stiction. This can easily be explained by considering the drawbacks of arc-shaped beams, already discussed in Section 16.4.2. The experiments might be easily improved by introducing some

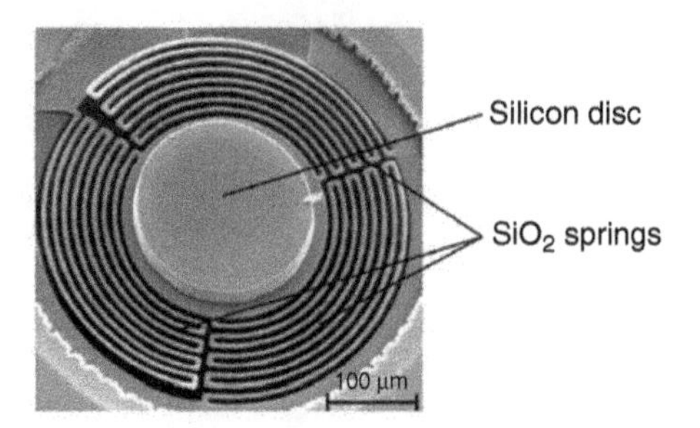

Figure 16.9 Micrograph of a structure tested by the ETH group. *Source*: Bachmann, Kühne and Hierold (2006), Figure 2. Reproduced with permission of Elsevier.

loading–unloading cycles, in order to achieve comparative results to those presented by the Virginia group (Jones, Begley and Murphy, 2003; Jones, Murphy and Begley, 2003).

16.4.5 Pull-in Experiments

One of the most relevant pull-in experiments was proposed by researchers from ETH Zurich (Bachmann, Kühne and Hierold, 2006; Bachmann and Hierold, 2007). The experimental device, depicted in Figure 16.9, was characterized by many silicon discs suspended by means of SiO_2 folded springs; different elastic restoring energies could be achieved by simply changing the stiffness of the folded springs. The choice of different materials was dictated by fabrication reasons but was also fruitfully exploited in obtaining a weaker suspension with respect to the stiffness of the disc: in fact, the Young's modulus of SiO_2 is more or less half that of silicon.

In the experimental campaign, 12 different designs were tested, considering 60 identical structures for each design: a statistically meaningful sample was thus obtained. The discs were electrostatically actuated in the transverse direction in order to collapse on the substrate. After voltage release, the sample was inspected and the sticking frequency evaluated, i.e. the percentage of discs, for each design, that remained stuck after release was recorded.

The surface energy was calculated by a fracture mechanics approach. Instead of using the energy release rate concept, the ETH group proposed to adopt a statistical Weibull approach. The sticking probability P is expressed as:

$$P = \exp\left[-\left(\frac{W_{\text{mech}}}{\gamma A_{\text{eff}}}\right)^m\right], \tag{16.20}$$

where W_{mech} is the elastic energy (known and different for the various devices). The total surface energy γA_{eff} and the Weibull modulus m were evaluated by a least squares curve fit of the experimental data. The results are summarized in Figure 16.10.

One important issue is represented by the computation of the specific surface energy. To this purpose, it is necessary to compute the effective contact area A_{eff}. This would be a very simple task, if the disc surfaces were regular. Unfortunately, the particular fabrication process yielded slightly curved discs. The curvature is induced by the presence of a thin SiO_2 layer on the upper surface of the discs; the unavoidable residual tensile stress in SiO_2 and the absence of a symmetric layer on the lower surface induce an uncontrolled curvature of the disc. The effective contact surface is represented by an annular region, whose area can be estimated with a high level of uncertainty. For this reason, the value for the specific adhesion energy is smaller than other values reported in the literature.

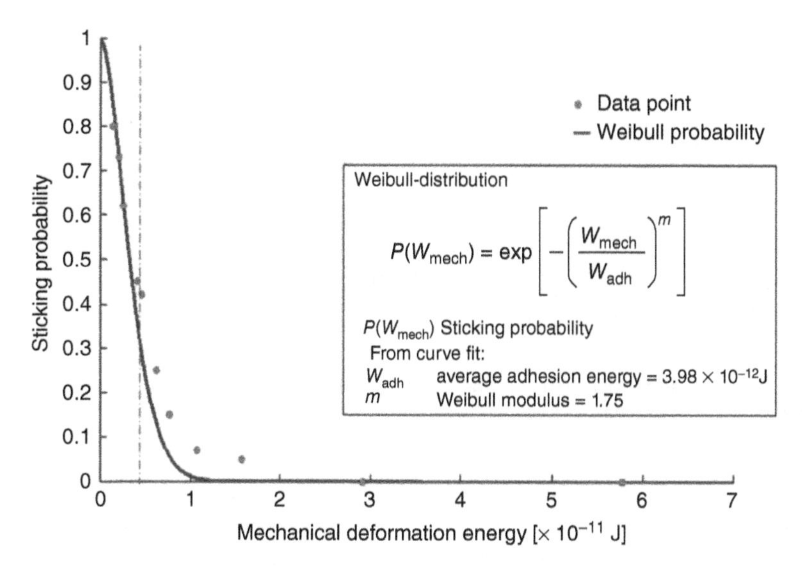

Figure 16.10 Results of experimental tests after application of the Weibull approach. *Source:* Bachmann, Kühne and Hierold (2006), Figure 8. Reproduced with permission of Elsevier.

In spite of this criticism, the idea of testing single contact pairs of regular shapes instead of microbeams seems to be very interesting, especially if it could be exploited in loading–unloading cycles. This kind of test, in fact, is directly focused on the interfacial behaviour between contacting surfaces and could be of great help in assessing the load–displacement behaviour of the adhered surfaces.

A different, and very important, version of the pull-in test has its origins in the works of Yu *et al.* (2007), of the group from the Indian Institute of Technology (Basu, Prabhakar and Bhattacharya, 2007) and of the group at the Politecnico di Milano (Ardito *et al.* 2013).

To be precise, these tests should be named *pull-in–pull-off*. The experimental devices were represented by deformable structures that had been electrostatically attracted towards the substrate until the pull-in voltage was reached and the structure suddenly collapsed on the substrate. Afterwards, the applied voltage was progressively reduced until the elastic force restored the unstuck configuration of the structure. A theoretical plot of current versus applied voltage is sketched in Figure 16.11, along with actual experimental data. It is evident that the pull-off voltage is reduced by the presence of adhesive forces, which conversely do not influence the pull-in value. The amount of adhesive energy can be estimated indirectly on the basis of the measured pull-off voltage. The differences between the theoretical and experimental curves can be justified by considering that the contact interface is not perfect from the electrical point of view.

In the study of Yu *et al.* (2007), the reference microdevice consisted of tilting elements, connected to a torsion hinge suspended above the surface by supports. The elements were electrostatically actuated and could land on the substrate if the electrostatic torque were higher than the hinge restoring torque. Afterwards, the elements were released by reducing the applied voltage. The experiments were performed on a very large array of microdevices (about 3000), whose position was controlled through

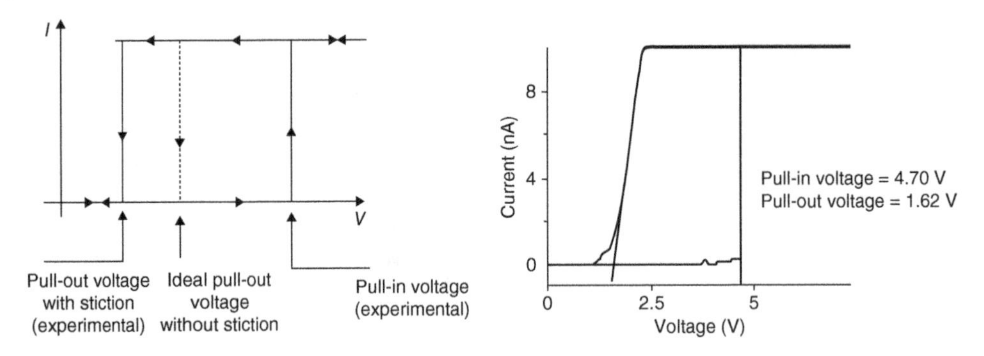

Figure 16.11 Theoretical (left) and experimental (right) plots of current versus applied voltage during a pull-in–pull-off test. *Source*: Bhattacharya *et al.* (2005), Figures 4 and 5a. Reproduced with permission of Wiley.

an optical system (which included a microscope with ring illumination, a CCD camera and a frame grabber) and subsequent digital image processing. A situation of aggressive wear was simulated by introducing a high-frequency horizontal actuation in the landed configuration.

In view of the statistical significance of the tested sample, the authors drew some very important conclusions:

1) Worn surfaces show higher adhesion strength with respect to the original situation. This has been demonstrated by the left shift of the release curve, which indicates lower pull-off voltage and higher surface energy. This behaviour can easily be understood by considering the effect of asperities on adhesion, already discussed by the Sandia group and further investigated in Section 16.5.
2) Hermetically packaged devices are less prone to stiction than are de-lidded MEMS. This result, mainly due to the presence of environmental humidity, is increasingly evident for highly scrubbed devices.

In the work by the Indian Institute of Technology group (Basu, Prabhakar and Bhattacharya, 2007), the pull-in–pull-off tests were performed on different microstructures, namely on cantilever beams. The pull-off voltage was correlated to the stiction force and some comparative results were shown with reference to various porosity levels of the silicon surface. The paper by Basu, Prabhakar and Bhattacharya (2007) includes a detailed description of the electromechanical model, which was adopted to evaluate the stiction force on the basis of the pull-off voltage. First, an analytical model was developed for the solution of Laplace's equation governing the electrostatic field. The model was compared with numerical outcomes obtained using the commercial code CoventorWare: good qualitative agreement was shown, even though there was no quantitative comparison. The mechanical model is very simple, but nevertheless is able to provide a good estimate of the stiction force. The most important conclusion of this work is that the formation of porous silicon on the surface, by means of stain etching, reduces the stiction force; this is an expected conclusion, because the effect of greater porosity should be more or less the same as that of an increase of roughness.

Ardito *et al.* (2013) presented a new experimental methodology with the main purpose of allowing direct measurement of adhesion energy in real-life MEMS. In particular, no optical measurement was employed; the adhesion energy was measured

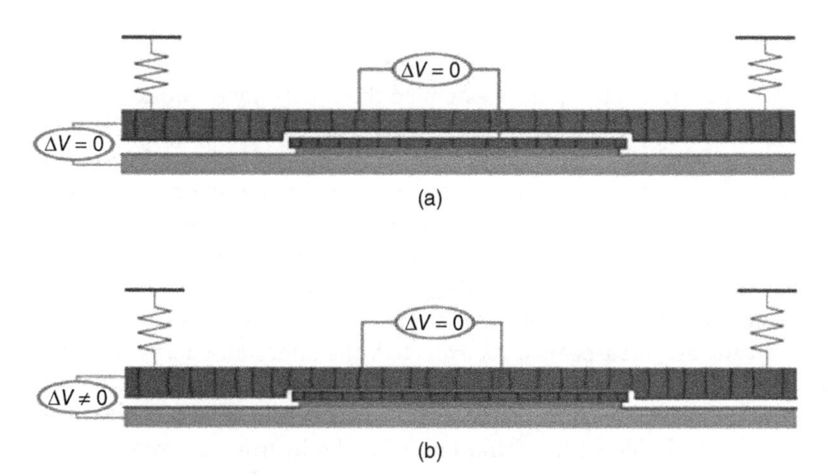

Figure 16.12 Cross-section of stuck–unstuck device proposed by Ardito *et al.* (2013) to study adhesion energy by means of electrostatic actuation and capacitive read-out. *Source*: Ardito *et al.* (2013), Figure 1. Reproduced with permission of Springer.

in the actual operating conditions of packaged and unpackaged MEMS and no unstable pull-in was encountered. For this reason, the proposed method should be referred to as *stuck–unstuck* rather than *pull-in–pull-off*. The experimental device, whose cross-section is depicted in Figure 16.12, is constituted of a movable mass connected to a set of (linear) elastic springs. The actuation is electrostatic and occurs at a limited portion of the device, whereas the central part lands progressively onto the substrate, which is electrically grounded, along with the movable mass. The typical stuck–unstuck curve, as depicted in Figure 16.11, is measured on the basis of a capacitive read-out. The key point of the method is the detaching branch of the curve. In the unstuck position, the separation between surfaces is large enough to assume that the adhesion forces are negligible. For this reason, and in view of the conservative nature of the forces at play, one can define the adhesion energy as the increase in elastic and electric energy from the stuck to the unstuck configuration. The final result is referred to a low-humidity environment with a fairly high surface roughness (about 10 nm rms); the obtained specific adhesion energy is

$$\gamma_S = 13.4 \pm 0.89\ \mu\text{J/m}^2, \tag{16.21}$$

which is in good agreement with the data reported, e.g., by the Sandia group in a similar environment.

Similar concepts have been exploited by the groups at Berkeley (Shavezipur *et al.*, 2012), and IMEC (De Coster *et al.*, 2013). In both cases, the energy concept is abandoned and the focus is moved to the *adhesion force*, which can be computed from the difference between the restoring force and the electrostatic force at the pull-off voltage. An interesting point, which is raised by Shavezipur *et al.* (2012), is that the adhesion force is just *weakly* correlated to the nominal contact area. In view of the very nature of the adhesive forces, one would expect the adhesion forces to be roughly proportional to the contact area. Nevertheless, the presence of some *leading asperities* in the rough surface may lead to a substantial independence of the force with respect to the contact area. This concept is further examined in the next section.

16.4.6 Tests for Sidewall Adhesion

Some researchers of the Sandia group have extended the experimental work to somewhat different problems. An interesting topic is represented by the sidewall adhesion, investigated in Ashurst *et al.* (2003) in cooperation with researchers from the University of California, Berkeley. While adhesion to the substrate is the reference problem for most studies, stiction on the lateral walls of the micromachines deserves the same attention. The experimental set-up consisted of an array of coupled beams which, after electrostatic actuation, could stick to each other on a vertical surface. The adhesion energy was computed by adopting a method that is very similar to the one described previously for vertically actuated beams. As expected, the adhesion energy was very different with respect to in-plane stiction. Nevertheless, the results were not fully understood. For instance, it was not clear why the sidewall adhesion energy was higher than in-plane values for hydrophobic surfaces and lower for the hydrophilic case. Further studies should be also devoted to the elimination of spurious effects due to residual stresses, which induce horizontal bending of the beams in their reference configuration.

Starting from the work of Timpe and Komvopoulos (2005), the gold standard in the evaluation of sidewall adhesion has been represented by micromachines that push two nominally parallel surfaces one against the other. Two important advantages can be envisaged: (i) the inaccuracy of tests based on beam deformation in the horizontal plane can be completely circumvented; (ii) the experimental method is fully representative of realistic situations, i.e. the contact between movable masses and the lateral stopper. A typical scheme of this kind of experiment is depicted in Figure 16.13. The contact between the vertical surfaces is obtained by means of electrostatic actuation and the adhesion forces are measured by considering the equilibrium equation at the moment of sudden separation. The displacement of the shuttle is monitored by a stereo-microscope, to detect the moment of detachment. Several tests have been performed, for different contact area and different environmental conditions: the final outcomes were that adhesion force increases with contact area, relative humidity and contact load.

A similar experimental device was exploited by Van Spengen, Bakker and Frenken (2007), with the important difference of adopting the pull-in–pull-off procedure, which was already considered for out-of-plane adhesion. The adhesion force was evaluated by means of electrostatic actuation and capacitive read-out, thus avoiding the necessity of optical measurements. Friedrich, Raudzis and Müller-Fiedler (2009) and Heinz *et al.* (2014) used the same concept, with some interesting novelty with the purpose of achieving a linear response with respect to the applied voltage in the actuation phase. The authors repeated the experiments several times, for different contact areas and different contact loads: an intriguing point was represented by the fact that the adhesion force seemed to be independent of area, while it seemed to increase for higher contact loads.

More recently, Dellea *et al.* (2016) conducted a broad experimental campaign, based on an automatic procedure which is able to obtain the adhesion force (in the usual pull-in–pull-off conditions) for a huge number of cycles with different kinetic energy at the impact. Figure 16.14 shows the designed device, which is compact and was studied to provide inline monitoring of adhesion forces for MEMS in realistic operating conditions. The results have been thoroughly examined and some conclusion have been reached: the adhesion force on the pristine device appears to be independent of the stopper area; during collision cycles, an initial growth of the adhesion force occurs,

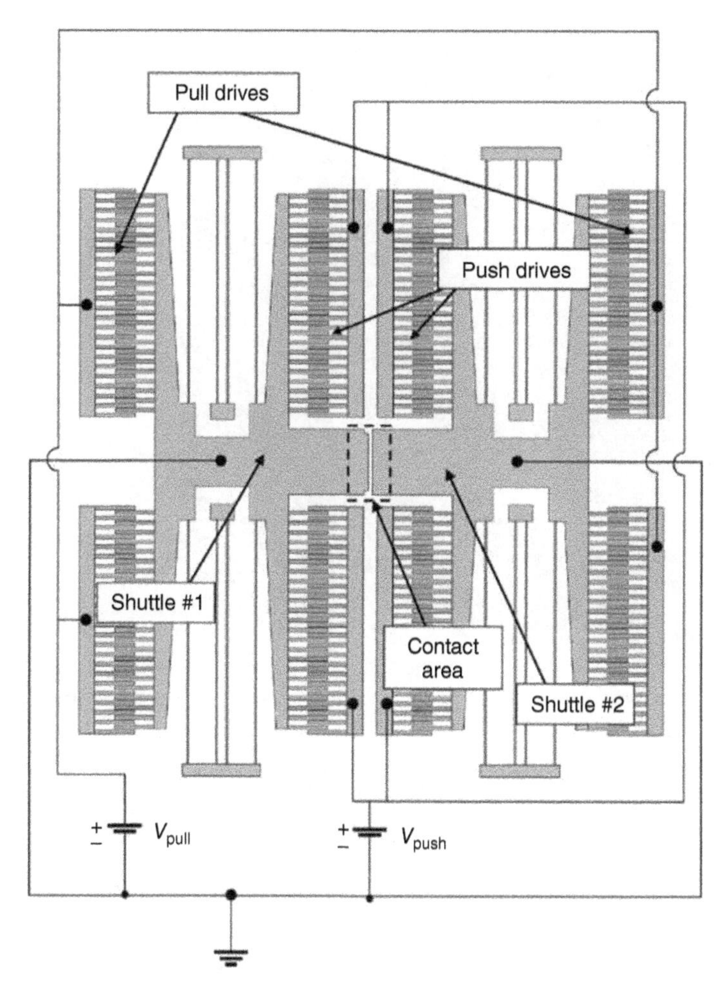

Figure 16.13 Typical layout for evaluation of sidewall adhesion force. *Source*: Timpe and Komvopoulos (2005), Figure 2. Reproduced with permission of IEEE.

followed by a stabilization; adhesion force is linearly related to the kinetic impact energy; a dependence of the adhesion force on the specimen area is shown (~1.5× larger adhesion for a stopper with 2× larger contact area). All the evidence obtained suggest the following considerations and interpretation of the phenomena. First, in the pristine samples, a dominant asperity model may explain the independence of the specimen area. Second, the underlying phenomenon, responsible for slow adhesion force growth during collision cycles, is probably not only related to plastic deformation of dominant microasperities. One hypothesis to explain the observed slow evolution is the occurrence of a wear phenomenon, resulting from the complex interaction of plastic deformation and mechanical deterioration due to surface fatigue. Post-mortem visual analyses have provided some partial confirmation of the proposed explanation. In fact, SEMs of the lateral surface revealed the presence of a *top scallop*, which is by far more pronounced than the other asperities. It has been noticed that the aforementioned

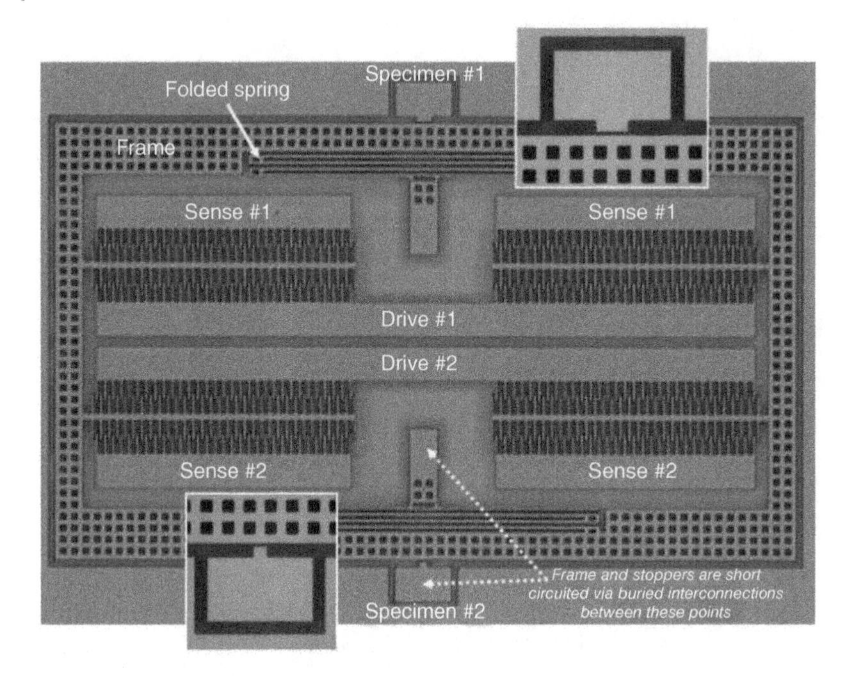

Figure 16.14 Device for inline monitoring of sidewall adhesion force, proposed by the group at the Politecnico di Milano. *Source*: Dellea *et al.* (2016), Figure 2. Reproduced with permission of IEEE.

scallops is substantially reduced on the edges that were subject to impact cycles. Such an edge consumption would be compatible with a fatigue-like mechanical effect and could explain the increase and the stabilization of adhesion force during the impact cycles.

16.5 Modelling and Simulation

16.5.1 Lennard-Jones Potential

As explained in the previous section, the evaluation of the surface interaction due to intermolecular forces is very difficult. To obtain a reasonable estimate, it is common to describe the interaction by means of some surface energy potentials, which typically depend on the distance d between contacting surfaces. The most widely used potential is the Lennard-Jones potential (Israelachvili, 2011), which is referred in its original form to the interaction between two isolated molecules. This potential captures the mild attraction (6th power) of surfaces as they approach one another, basically due to the van der Waals forces, and the strong repulsion (12th power), or steric forces, when they come too close to one another. Such a potential, sketched in Figure 16.15, has the advantage of taking into account most dominant adhesive forces, including long-range van der Waals forces and short-range Born repulsion.

A common mathematical form of the Lennard-Jones law is:

$$p(d) = 4\,\epsilon \left[\left(\frac{\sigma}{d} \right)^{12} - \left(\frac{\sigma}{d} \right)^{6} \right], \tag{16.22}$$

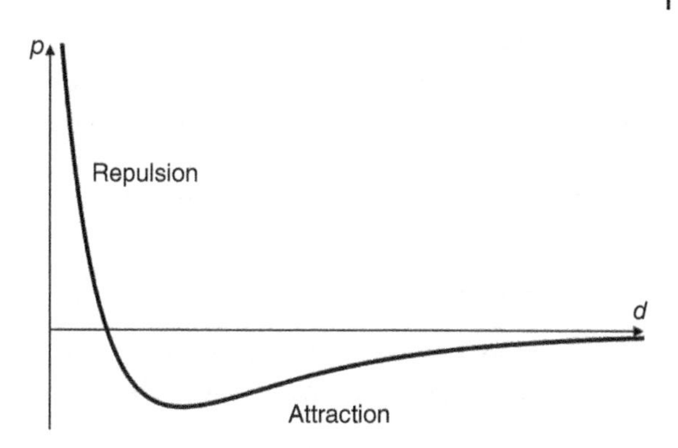

Figure 16.15 Lennard-Jones potential for contacting molecules.

where two parameters have been introduced, namely the maximum attractive energy ϵ (to be inferred from the quantum mechanical properties of the interacting molecules) and the characteristic separation σ (i.e. the distance at which the potential is equal to zero). This parameter is usually different from the molecule radius.

The minimum of the potential (i.e. the maximum attractive energy) is reached for $d = 1.12\sigma$. At this separation, equilibrium is reached and the van der Waals energy is $p(d) = -2\epsilon$, while the repulsive contribution is ϵ. Thus, the repulsive contribution decreases the strength of the binding energy by 50%. This can be compared with a different contact model, where each atom is represented as a rigid sphere (*hard* contact). In the latter case the equilibrium separation is equal to the molecule radius and the corresponding energy is simply the van der Waals energy at contact.

16.5.2 Tribological Models: Hertz, JKR, DMT

The experimental results reported in Section 16.4 and the physical interpretations summarized in Section 16.3 demonstrate the great importance of surface roughness on the adhesive behaviour. The computation of surface interactions between contacting asperities can be faced in a simplified way by idealizing the asperities by an approximate model of contacting spheres, characterized by the local principal curvature radii of the surfaces. For this reason, it is essential to consider the most important models to simulate the contact behaviour of spheres.

In 1881, Hertz developed his classical theory for local elastic deformations of smooth contacting bodies under external loads. Two spheres of radii R_1 and R_2 and Young's modulus E are pressed by a force F. The Hertz theory yields the following results:

$$\begin{cases} a = \dfrac{R\,F}{E} & \text{contact radius,} \\ \delta = \dfrac{F\,a}{E} & \text{displacement,} \\ q(\rho) = \dfrac{3F\sqrt{1-\rho^2}}{2\pi a^2} & \text{pressure,} \end{cases} \tag{16.23}$$

where $R = R_1R_2/(R_1 + R_2)$ and $\rho = r/a$ is the dimensionless distance from the contact axis.

The Hertz model does not include adhesion between surfaces. In the 1970s, Johnson, Kendall and Roberts extended the model by achieving the equilibrium situation through an energy balance considering elastic energy, mechanical potential energy and surface energy (JKR model (Johnson, Kendall and Robert, 1971)). The results, in terms of contact radius, displacement and pressure, are dependent on the specific surface energy γ and reduce to the Hertz outcomes if the particular value $\gamma = 0$ is considered. An important conclusion of the JKR model is the total pull-off force, which for a sphere on a flat surface reads:

$$F_s = -3\pi R\gamma. \tag{16.24}$$

A nonsmooth transition between contact and noncontact situation appears: separation occurs abruptly once the contact radius has fallen to

$$a_s = 0.63a = 0.63\sqrt[3]{\frac{12\pi R^2\gamma}{E}}. \tag{16.25}$$

This formula indicates that the critical radius is equal to 63% of the contact radius at equilibrium.

Some difficulties connected with the JKR model were pointed out by various authors in the 1970s. In particular, the fact that infinite stress is predicted at the edge of the contact circle is questionable. This is a strict consequence of the implicit assumption of attractive forces, which act over an infinitesimally small distance. This unphysical situation has been overcome by including a finite-range force, described by an extension of the Lennard-Jones potential from the case of atomistic interaction to the representation of adhesive forces between macrospheres. This model, developed by Derjaguin, Muller and Toporov (DMT model (Derjaguin, Muller and Toporov, 1975)), is derived for two interacting spheres (or an equivalent sphere and a flat surface), which, in the current model, represent a single asperity contact. The DMT model gives the attractive pressure outside the contact region according to the modified Lennard-Jones potential, thus ignoring adhesion from within the contact region. The separation force predicted by the DMT model is

$$F_s = -4\pi R\gamma. \tag{16.26}$$

The difference between the JKR and DMT approaches, as demonstrated by Equations 16.24 and 16.26, is the source of a long-standing controversy. The development of a self-consistent numerical solution has been strongly encouraged, in order to finally settle the issue. The general conclusion of earlier attempts is that DMT theory is valid for hard solids with small adhesion and the JKR model becomes increasingly applicable in case of soft solids with large adhesion.

Interest in the efficient numerical solution for the problem of adhesive contact still remains. Many contributions have recently been presented with different computational approaches (Attard, 2000; Feng, 2000). Without entering into technical detail, it is worth mentioning that the surface forces are modelled by considering a pressure distribution derived from the Lennard-Jones potential:

$$q(r) = \frac{A}{6\pi d^3}\left[\left(\frac{z_0}{d}\right)^6 - 1\right], \tag{16.27}$$

where $d(r)$ is the point-wise separation, which depends on the distance from the contact axis r, A is the Hamaker constant and z_0 is the equilibrium spacing. For silicon spheres

$A = 27 \times 10^{-20}$ J and $z_0 = 0.149$ nm. The potential model in Equation 16.27 implicitly entails that the surface energy, used in the JKR and DMT models, is given by

$$\gamma = \frac{A}{16\pi z_0^2}. \tag{16.28}$$

The pressure in Equation 16.27 is combined with the equilibrium equation and the governing relationship for the sphere deformation, which depends on the complete elliptic integral of the first kind. Eventually, one obtains a nonlinear integral equation, which can be solved numerically by a sort of collocation method. Numerical results confirm the ambit of applicability of JKR and DMT model, but show also some interesting aspects of adhesive contact, e.g. elastic instability at surface detachment and hysteretic behaviour in the case of cyclic loading and unloading.

16.5.3 Computation of Adhesion Energy

In the previous section, the tribological model was described with reference to the problem of contacting spheres, considered as a representative model of single asperity contact in rough MEMS surfaces. These models could be directly applied to the problem of adhered contact in MEMS, as in Decuzzi and Srolovitz (2004), by introducing the simplistic hypothesis that the surfaces are endowed with a regular pattern of asperities. If a realistic estimate of the adhesion energy is to be computed, taking account of the stochastic nature of the actual rough surfaces is unavoidable (Bhushan, 1996).

The first important issue when dealing with random surfaces is the correct evaluation of the elastic and plastic behaviour during contact. This problem has been studied for more than 50 years, but there is still a big controversy over the validity of the available formulations. One of the most popular models was proposed by Greenwood and Williamson in 1966 (GW model (Greenwood and Williamson, 1966)). In this model, the two surfaces are represented with a combined rough surface in contact with a smooth rigid plane, as described in Figure 16.16. By introducing the Gaussian hypothesis for the distribution of asperity heights, the GW model is able to predict the deformed configuration of the system. Despite the fact that such models have been used for many years, the same original authors have recently exposed serious doubts on its validity (Greenwood and Wu, 2001).

The GW model has been improved by Chang, Etsion and Bogy in the late 1980s (CEB model (Chang, Etsion and Bogy, 1987)). Some of the hypotheses of the GW asperity model have been maintained, namely the equivalence represented in Figure 16.16 and the assumption of noninteracting neighbouring asperities. The deformation at the contacting regions can be elastic, plastic or elastoplastic, depending on the nominal pressure, surface roughness and material properties. The overall contact force is analytically evaluated by integrating over the whole spectrum of asperity heights, having defined a Gaussian distribution of this stochastic quantity. The rather complex formula is not reported herein, for the sake of simplicity, but can be found in Chang, Etsion and Bogy (1987). Such a formula is constituted by the sum of two integrals, referred to the elastically and plasticly deformed asperities, respectively. The contact force is accompanied by the adhesive force, which is analytically computed using a similar procedure. The total adhesion can be separated into two components, corresponding to interactions between noncontacting asperities and between contacting asperities; both terms are represented by properly modified Lennard-Jones potentials.

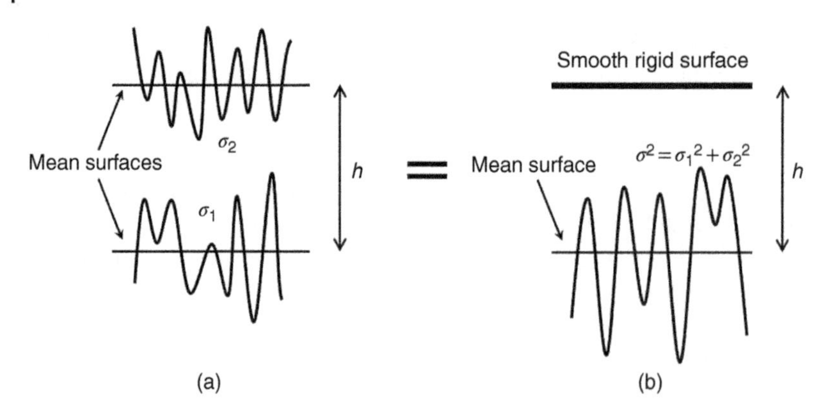

Figure 16.16 Contact between rough surfaces: equivalence between original situation (a) and simplified model (b) in the GW approach; σ_1 and σ_2 represent the rms roughness of the original surfaces.

In the case of wet stiction, the CEB model has been modified to include the capillary force, thus obtaining the so-called sub-lubrication method (SBL (Suh and Polycarpou, 2003)). This model is, in some sense, alternative to meniscus modelling, introduced in Section 16.3.1. The SBL model considers the presence of a lubricant film characterized by extremely small thickness and strong adhesion to the underlying solids. However, the meniscus model (valid only in static conditions) is more useful when the lubricant is relatively thick and abundant. As a matter of fact, the latter hypothesis seems more close to reality for MEMS in a humid environment. The SBL model has been applied to polysilicon MEMS by Suh and Polycarpou (2003), showing sufficient agreement with respect to experimental measurements.

With the purpose of improving the reproduction of experimental results, a different probability density function has been assumed for the asperity height. Tayebi and Polycarpou (2006) have adapted the SBL model and introduced a Pearson's (instead of a Gaussian) distribution. In this way, it has been possible to study the sensitivity of adhesion forces with respect to the degree of asymmetry and sharpness of the asperity distribution (statistically measured by the skewness and kurtosis indices, respectively). Comparison with experiments is encouraging, but not yet completely satisfactory.

A totally different approach has been proposed by Komvopoulos (2003), who characterized the elastoplastic deformation forces in terms of the geometrical properties of a fractal model of the asperities. Adhesion was introduced by properly modelling the van der Waals, electrostatic and capillary forces for the fractal geometry. Such an accurate model is, however, unable to provide a precise and reliable prediction of adhesion force.

Another model, recently proposed by researchers from the University of Toronto (Hariri, Zu and Ben Mrad, 2006, 2007), is based on a modification of the GW model. The difference is basically constituted by a new model for the geometric description of asperities (the so-called *n-point asperity model*, as opposed to the *three-point asperity model* in the GW approach). The elastoplastic contact force and the adhesive actions are modelled on the basis of the geometry of contact between rough surfaces. Capillarity is introduced through the meniscus model. The analytical results have been compared with experimental values of surface energy. Figure 16.17 shows two curves for the new model, obtained for different values of the λ^* parameter (correlation length), which

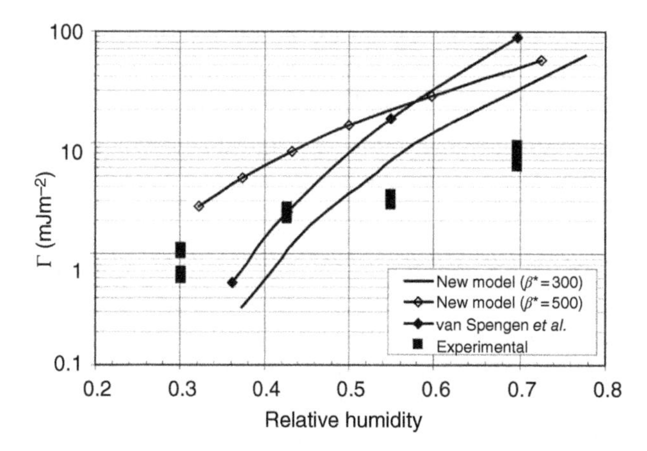

Figure 16.17 Specific adhesion energy versus relative humidity for wet stiction. Experimental results are from De Boer *et al.* (1999), 'new model' refers to results of Hariri, Zu and Ben Mrad (2007). Results of Van Spengen, Puers and De Wolf (2002) are also reported. *Source*: Hariri *et al.* (2007), Figure 10. Reproduced with permission of IEEE.

governs the autocorrelation function. In fact, the rough surface is assumed to be a broadly stationary and ergodic process, characterized by an autocorrelation function of the exponential form. In spite of the good qualitative agreement for one curve, no exact match has yet been achieved.

Figure 16.17 also reports some computational results obtained by Van Spengen, Puers and De Wolf (2002). These authors have adopted a drastically simplified approach. The GW model is assumed as the basis for a description of rough surface contact but elastic deformation is completely neglected. This sort of rigid-plastic behaviour led to very simple formulae for the evaluation of repulsive and adhesive forces (the latter include capillary and van der Waals actions). Comparison with experimental results shows that such a simplified method is not reliable for a silicon surface. Moreover, the same authors have published a corrigendum for the hardness parameter of silicon: such a correction would induce further deviation from the actual adhesive behaviour.

Among recent simulation attempts, DelRio, Dunn and de Boer (2008) have presented a comprehensive numerical procedure in which the two surfaces are represented as a plane contacting spherical surfaces with equal radii and an equivalent Gaussian distribution of elevations. They extended previous models (the GW and CEB models) to explore the effects of surface topographical correlations, plasticity and disjoining pressure on capillary adhesion due to adsorbed water layers. In particular, the presence of adsorbed water layers on the hydrophilic polysilicon surfaces only slightly modifies the asperity geometry but seems to have an important role in the capillary adhesion process, which is highly sensitive to the form of asperities.

From the aforementioned considerations, it is evident that further research is needed for the correct prediction of adhesive forces. In particular, computer modelling of the actual contact between stochastic rough surfaces would be desirable. In this way, it could be possible to simulate precisely elastic and plastic deformation, the generation of microscopic interactions, and cyclic behaviour in the case of repeated loading and unloading.

16.6 Recent Advances

16.6.1 Finite Element Analysis of Adhesion between Rough Surfaces

In the previous section, some computational models were presented to study the adhesion forces for simple configurations, e.g. a sphere over a flat surface. On the other hand, it has been shown (Cho and Park, 2004) that the problem of an adhesive sphere could be solved in a genuine finite element environment, by modelling elastic parts through conventional finite elements and performing contact analysis. More recently, the Lennard-Jones interatomic potential was used in finite element analysis to obtain an innovative formulation of the frictionless contact problem (Sauer and Wriggers, 2009) and of frictional and adhesive behaviour (Wriggers and Reinelt, 2009). A recent proposal, presented by Ardito, Corigliano and Frangi (2013), focuses on the computational study of adhesion of rough surfaces by means of finite element analysis. An automatic procedure is adopted to generate random surfaces with predefined stochastic properties and to obtain a 3D finite element mesh. Adhesive forces are modelled in a simplified way and are applied as surface tractions in a nonlinear finite element analysis, which is referred to a representative portion of a MEMS surface. In this way, it is possible to evaluate the elastic and plastic deformation of random asperities and to account for their effects on adhesion in a loading–unloading cycle.

16.6.1.1 Artificial Rough Surfaces

The roughness of a surface is commonly defined as the height variation of the surface compared with a reference plane. It is measured on a single linear profile or on a surface portion. Generally, the roughness is characterized by a set of statistical parameters, which are now defined for the simple case of a surface whose height z is described in a discrete way (i.e. the height values are picked at a finite number N of positions with a constant sampling interval). The centre line or mean line is defined as the line that equalizes the area subtended by the profile over this line to the area under this line:

$$m = \frac{1}{N} \sum_{i=1}^{N} z_i. \tag{16.29}$$

The basic statistic parameter is the so-called roughness root mean square:

$$R_q = \sqrt{\frac{1}{N} \sum_{i=1}^{N} z_i^2}. \tag{16.30}$$

Other important parameters are the variance σ, the skewness (Sk) and the kurtosis (Ku); the latter two can be used to compare the height distribution to the Gaussian case, which is characterized by $Sk = 0$, $Ku = 3$:

$$\sigma^2 = \frac{1}{N} \sum_{i=1}^{N} (z_i - m)^2, \tag{16.31}$$

$$Sk = \frac{1}{\sigma^3 N} \sum_{i=1}^{N} (z_i - m)^3, \tag{16.32}$$

$$Ku = \frac{1}{\sigma^4 N}\sum_{i=1}^{N}(z_i - m)^4. \tag{16.33}$$

A 2D rough surface can be interpreted as an ergodic stochastic process, governed by the spatial autocorrelation function. The autocorrelation function of a casual function can be interpreted as the measure of how precisely future values of the function can be predicted through the last observations. Such a concept can easily be extended to the case of a rough surface. It has been observed that many technically interesting surfaces have an exponential autocorrelation function. The measure of how quickly the exponential function decays is termed the correlation length. For spatial separations greater than this length, the autocorrelation function reduces to a small fraction of its original value, generally 10%. A Gaussian and isotropic surface can be entirely statistically described by two parameters: the rms and the correlation length λ^*. In this case, the exponential autocorrelation function reads:

$$C(x_i, y_i) = R_q^2 \exp\left\{-2.3\left[\left(\frac{x_i}{\lambda^*}\right)^2 + \left(\frac{y_i}{\lambda^*}\right)^2\right]\right\}. \tag{16.34}$$

The artificial rough surface can be obtained by adopting a digital filtering technique, as suggested by Hu and Tonder (1992). The starting point is represented by a sequence of independent random numbers $h(r, s)$, extracted by a Gaussian distribution with zero mean and predefined standard deviation. The application of a digital finite impulse filter, denoted $f(k, l)$, allows the desired distribution of heights to be obtained. The coefficients of the filter can be obtained by working in the frequency domain: in fact, from signal theory it is known that the power spectral density of heights is equal to the Fourier transform of the autocorrelation function, which is given by Equation 16.34. The final distribution of heights can be obtained in two ways: (i) the filter coefficients are obtained by the inverse Fourier transform, then the convolution equation is applied; (ii) convolution is transformed in a simple product in the frequency domain and the values are obtained from an inverse Fourier transform of the result. After many numerical tests, it has been concluded that the second option is by far preferable: the obtained surface is smoother and the computing time is about ten times shorter than for the first option. Excellent results have been obtained, in terms of statistical properties of the generated surfaces. Finally, the finite element mesh for a rough surface has been obtained: a typical example of a finite element model is given in Figure 16.18, which refers to the case of a roughness of 10 nm rms and a correlation length of 300 nm.

16.6.1.2 Details of the Finite Element Model and Results

Mechanical analyses at the microscale are based on a representative portion of the adhered surfaces, which is, for simplicity, a square shape. To obtain significant results, the representative area must include a certain number of asperities, in order to consider the actual behaviour of the whole surface. It is necessary to include the substrate until a critical depth, after which the mechanical effects are negligible, because of the dependence of the adhesive phenomena on the deformation of the surfaces. For this reason, Ardito, Corigliano and Frangi (2013) used a 3D model with fine meshes, small enough to describe well the geometry of asperities. A large-scale problem can be envisaged. The computational burden is enhanced by the strongly nonlinear mechanical model. A frictionless contact model is introduced, including a

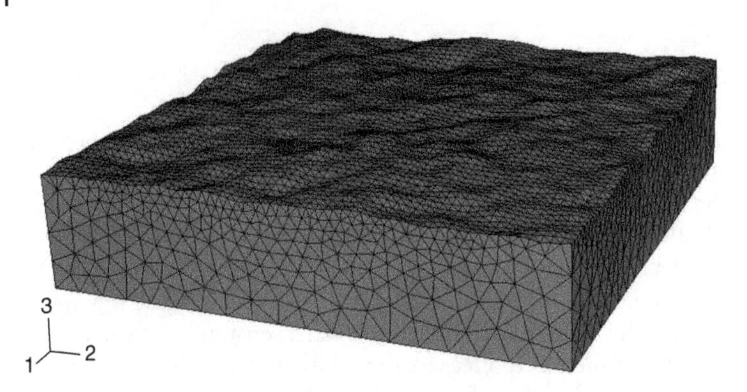

Figure 16.18 Finite element model for typical artificially generated rough surface. *Source*: Ardito *et al.* (2013), Figure 4. Reproduced with permission of Elsevier.

hard pressure (p)–overclosure (h) relationship. The constitutive model for silicon at the microscale is assumed to be elastic-perfectly plastic, in spite of the fact that such a material exhibits, at the macroscale, a fragile behaviour. However, it must be considered that the presence of plastic deformation is customarily accepted in order to explain the silicon response to nanoindentation tests (Bhushan, 1995).

Adhesive forces are introduced in the finite element model as distributed pressure on the adhered surface. To model these forces, the proximity force approximation has been used, in the sense that each material point on one surface interacts only with the corresponding point on the other surface. For this reason it is necessary to identify the pairs of points to be considered in the pressure computation. The adhesive forces, which are represented by capillary attraction and van der Waals interaction, depend on the relative distance between the interacting points. At each time increment, a scan of the current geometric configuration is executed and the force magnitude is computed. Strictly speaking, this approach is in contrast with the backward difference scheme, since the force is computed at the beginning of the time step and the constitutive model is enforced at the end. However, as the time increment is sufficiently small, the effect of such inconsistency can be neglected.

As an example, a critical comparison between the computational procedure proposed by Ardito, Corigliano and Frangi (2013) and the experimental data reported by DelRio *et al.* (2005) is considered. It is worth noting that the results in Figure 16.19 have been obtained without any fitting parameters: the only parameters that influence the comparative analyses are represented by standard literature data for silicon (namely, the Hamaker constant and equilibrium separation). In view of the specific experimental procedure adopted therein, it has been chosen to consider the loading phase only, thus neglecting the occurrence of plastic strain. This assumption seems quite reasonable, in view of the smooth landing of microcantilevers on the substrate during experiments and of the limited force that is exerted after contact. Figure 16.19 shows the comparison, which confirms the consistency of the proposed model: excellent agreement is obtained for the higher roughness level and both the trend and the absolute values of adhesion energy for the other surfaces are described with sufficient accuracy by the numerical simulations, performed in the absence of capillary forces.

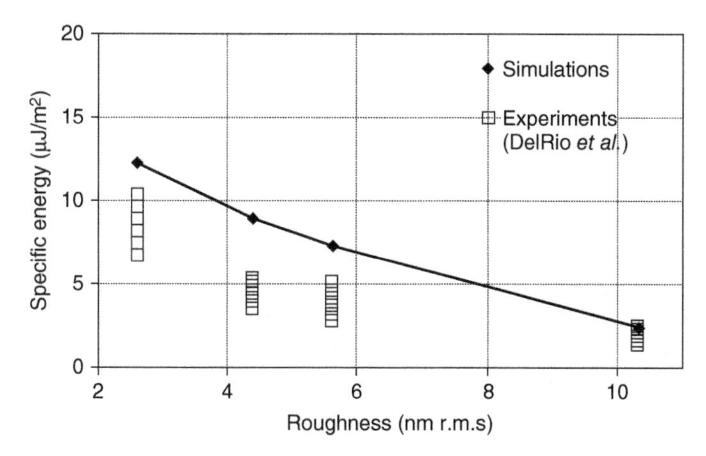

Figure 16.19 Comparison of numerical results with the experimental results of DelRio *et al.* (2005), for the specific case of perfectly dry conditions (van der Waals forces only). *Source*: Ardito, Corigliano and Frangi (2013), Figure 11. Reproduced with permission of Elsevier.

16.6.2 Accelerated Numerical Techniques

A major issue of the procedure proposed by Ardito, Corigliano and Frangi (2013) was the computational burden, which impeded the study of significant portions of the rough surface. Ardito *et al.* (2016) proposed a new algorithm to obtain an equivalent surface made of spherical caps with the same statistical properties of the real situation. A further advantage of using spherical caps is represented by the possibility of applying analytical formulae to simulate the mechanical behaviour of asperities. In this way, it was possible to develop a new simulation procedure, which combined the elastic–plastic deformation of the asperities and the formulae that represent adhesive behaviour. The proposed procedures were able to obtain the traction–separation curve, even for large portions of the rough surfaces, with negligible computational effort.

16.6.2.1 Rough Surface Represented by Spherical Caps

The artificial rough surface, generated with the Hu–Tonder method, can be used with the proximity force approximation approach but it doesn't allow one to apply the analytical models for capillarity and van der Waals interaction, as described in Section 16.3, which can be used only in the case of spherical asperities. Consequently, a different procedure for obtaining rough surfaces, constituting a set of spherical caps, has been implemented. The novelty of the proposal, with respect to previous contributions pioneered by Greenwood and Williamson (1966), is the achievement of a physical representation of the rough surface, paying attention to statistical analysis of the obtained results. The surface is constituted of spherical caps, whose radii and positions are suitably varied in order to match the statistical parameters (listed in Equations 16.30 to 16.33) of the real surface. Virtual prototyping of the rough surface allows for a precise description of the mechanical and adhesive behaviour of each pair of contacting spheres, thus obtaining a physical model for adhesion in the absence of any fitting parameter. This advantage is shared by the finite-element-based technique described by Ardito,

Corigliano and Frangi (2013), but the new procedure, which allows for the adoption of analytical formulae for describing the mechanical and adhesive behaviour, is by far more effective in terms of computational burden. This means that one can easily analyse large representative portions of the rough surface (i.e. sides longer than 10 μm) and that the analyses can be repeated several times to obtain a statistically significant result. Conversely, it is worth noting that the proposed algorithm can be applied only in the case of *moderately* rough surfaces, as happens in the case of silicon: in fact, in the presence of very sharp asperities or abrupt changes of texture, the proposed algorithm would yield surfaces that would not fully represent the real situation.

The new algorithm starts from the definition of a uniform grid, where the centres of the spheres are positioned. The spacing is based on the desired correlation length: after several parametric analyses, it is possible to figure out that the optimal choice is one-half of the correlation length. The radii of the spheres are initialized to a uniform value (bias). Such a value, for each sphere, is added to a random number extracted from a Gaussian distribution, with zero mean and standard deviation in agreement with the desired R_q value. In the random extraction, it is necessary to rule out the values that exceed, in absolute value, the threshold $z_p = 2\sigma$: such a provision aims at avoiding the presence of too large spherical caps, which might overwhelm the contiguous sphere, thus yielding a corrupted result. Conversely, the presence of too small spheres might endanger the algorithmic process, since the desired outcome should be characterized by a moderate overlap between adjacent spheres. The surface is finally composed of spheres with slightly different radius, which seems more realistic than the previous approaches of DelRio, Dunn and de Boer (2008), who considered identical spheres with different alignments of their centres. In the subsequent algorithmic steps, the bias is iteratively updated, by multiplication with a small quantity $1 < \alpha < 1.1$, until the surface is endowed with the correct R_q. By increasing the bias, in fact, one obtain an increase in the overlap between the spherical caps. Consequently, there is a decrease in R_q, which is computed, as for the other statistical parameters, taking into account all the points that constitute the *semi-continuous* distribution that defines the texture. It is worth pointing out that the choice of the initial value for the bias $R_{\mathrm{b}}^{(0)}$ represents a nontrivial task. If the initial bias is too small, one might obtain a surface with the desired R_q that is simply a flat surface with isolated small half-spheres. Conversely, if the initial radius is too large, the procedure could be nonconvergent.

The final outcome of the proposed algorithm is depicted in Figure 16.20: the statistical features of such an artificial rough surfaces turn out to be in excellent agreement with the desired values.

16.6.2.2 Numerical Outcomes and Comparison with Experiments

Numerical models for adhesive forces have been applied to representative artificial rough surfaces generated employing the techniques described in the previous section. Both the capillarity models and the van der Waals formula are applied to each pair of spheres, located on facing surfaces. The overall attractive force is obtained by simply adding the contributions from all the pairs of spheres.

The analyses are performed as follows: the upper and lower rough surfaces are modelled by means of spherical caps and are initially placed at a sufficient distance, which is next incrementally decreased. At each step, attractive forces are computed and contact is monitored. An adhesion energy is estimated by integrating the net attractive forces

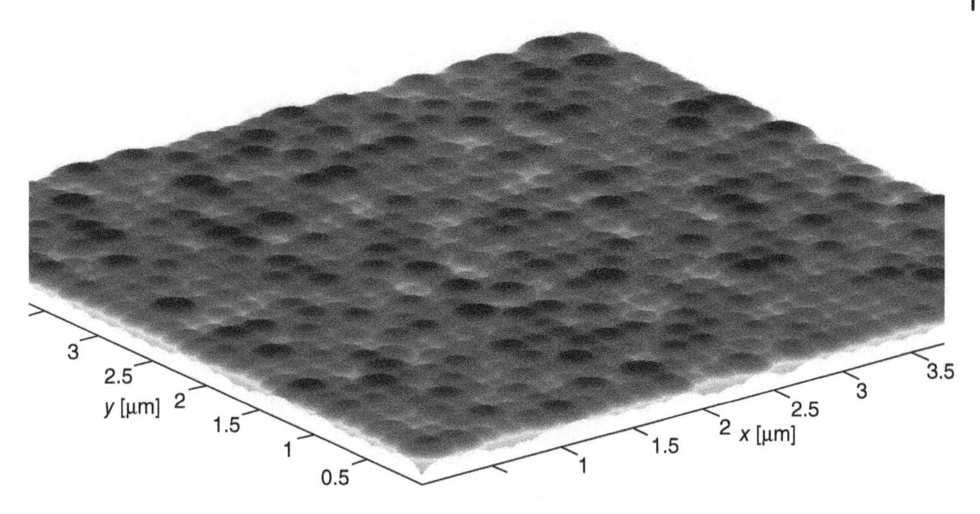

Figure 16.20 Detailed view of numerically generated rough surface. The same length scale is adopted for both in-plane and out-of-plane dimensions. *Source*: Ardito *et al.* (2016), Figure 5a. Reproduced with permission of Elsevier.

over distance, from infinity to the equilibrium position where the global force vanishes (contact forces balance attractive forces). We will refer to this energy as 'load adhesion energy'. Next, the distance between the two surfaces is further decreased until the average separation is zero. In this phase, most asperities undergo strong plastic deformation. The process is next reverted, incrementally augmenting the separation, and a second 'unload adhesion energy' is obtained, integrating the power of net attractive forces from the new equilibrium distance to infinity.

The proposed method has been compared with experiments that included the effect of humidity. Figure 16.21 shows the adhesion energy for two surfaces of roughness 2 nm rms: numerical outcomes are compared with the experiments reported in De Boer (2007) and with a numerical model presented by Van Spengen, Puers and De Wolf (2002). Numerical predictions are always the mean over 30 different repetitions of the analysis on different surface realizations and present both 'load' and 'unload' estimates of the adhesion energy. This comparison is a strong validation of the proposed model. Indeed, experiments lie within the expected lower bound (load adhesion energy) and upper bound (unload adhesion energy). The two bounds, moreover, are rather close, which is reasonable, since the asperities here are small.

An additional experimental benchmark is represented by the data presented by DelRio *et al.* (2007), which considered surfaces with roughness equal to 2.6 nm rms for the landing pad and 2.4 nm rms for the upper surface. In this case, the experimental data are compared with the 'load adhesion energy' only, which is more representative of the experimental conditions declared by DelRio *et al.* (2007). Figure 16.22 shows the computational results for the complete model (i.e. capillary force modelled by the analytical approach plus van der Waals attraction) and the case of van der Waals forces only, for small relative humidities. Satisfactory agreement with experiment is obtained in both cases, providing additional validation of the proposed method. The sudden decrease in experimental adhesion energy for relative humidity of 0.65–0.7 is not caught by the present model. In fact, the transition between capillary attraction

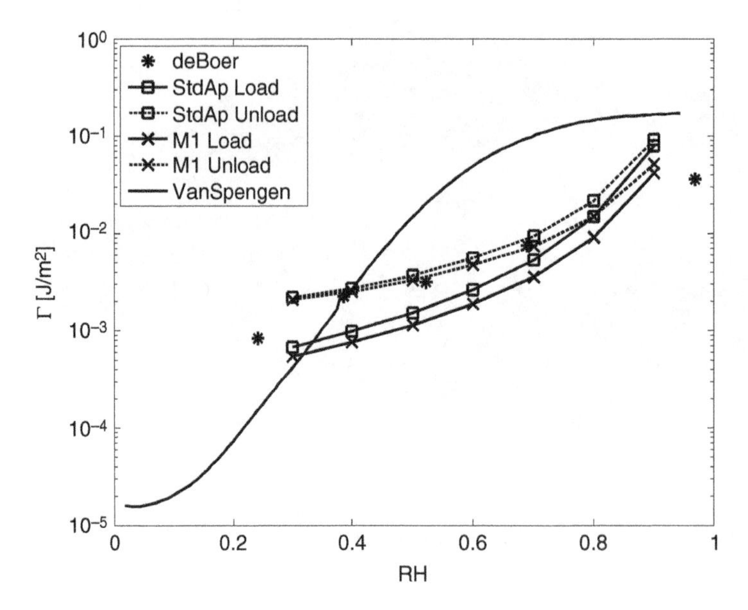

Figure 16.21 Adhesion energy versus relative humidity (RH) for a surface of roughness 2 nm rms: comparison of experimental data from De Boer (2007) and numerical outcomes. M1, model 1; StdAp, standard approximation. *Source*: Ardito *et al.* (2016), Figure 9. Reproduced with permission of Elsevier.

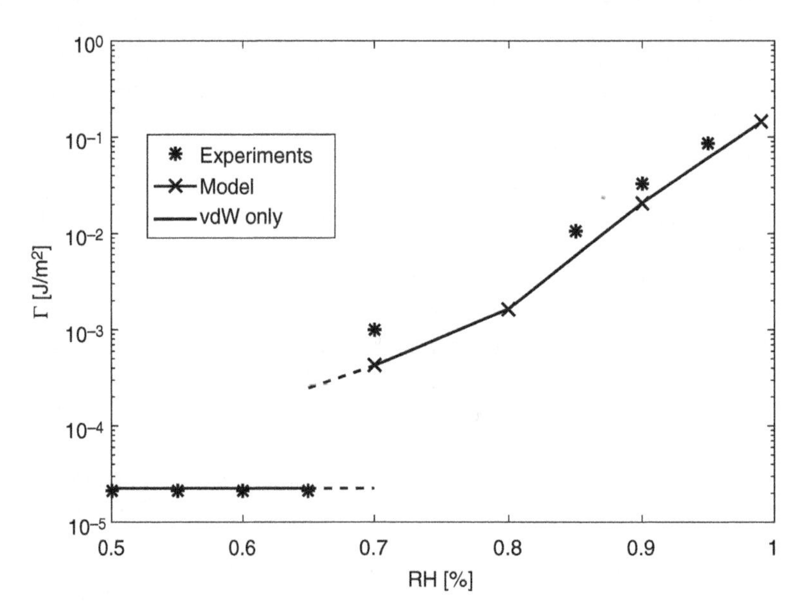

Figure 16.22 Adhesion energy vs. relative humidity (RH): comparison of experimental data, from DelRio *et al.* (2007), and numerical outcomes. vdW, van der Waals. *Source*: Ardito *et al.* (2016), Figure 11. Reproduced with permission of Elsevier.

and van-der-Waals-dominated attraction involves complex phenomena of rupture of the meniscus, which can be influenced by the presence of adsorbed water layers on the hydrophilic polysilicon surface, (DelRio, Dunn and de Boer, 2008). Such features are not included in the present version of the proposed model and will represent an interesting development.

References

Ardito, R., Baldasarre, L., Corigliano, A. *et al.* (2013) Experimental evaluation and numerical modeling of adhesion phenomena in polysilicon MEMS. *Meccanica*, **48** (8), 1835–1844.

Ardito, R., Corigliano, A. and Frangi, A. (2013) Modelling of spontaneous adhesion phenomena in micro-electro-mechanical systems. *European Journal of Mechanics – A/Solids*, **39**, 144–152.

Ardito, R., Corigliano, A., Frangi, A. and Rizzini, F. (2014) Advanced models for the calculation of capillary attraction in axisymmetric configurations. *European Journal of Mechanics – A/Solids*, **47**, 298–308.

Ardito, R., Frangi, A., Corigliano, A. *et al.* (2012) The effect of nano-scale interaction forces on the premature pull-in of real-life micro-electro-mechanical systems. *Microelectronics Reliability*, **52** (1), 271–281.

Ardito, R., Rizzini, F., Frangi, A. and Corigliano, A. (2016) Evaluation of adhesion in microsystems using equivalent rough surfaces modeled with spherical caps. *European Journal of Mechanics – A/Solids*, **57**, 121–131.

Ashurst, W.R., De Boer, M.P., Carraro, C. and Maboudian, R. (2003) An investigation of sidewall adhesion in MEMS. *Applied Surface Science*, **212–213**, 735–741.

Attard, P. (2000) Interaction and deformation of elastic bodies: origin of adhesion hysteresis. *Journal of Physical Chemistry B*, **104** (45), 10635–10641.

Bachmann, D. and Hierold, C. (2007) Determination of pull-off forces of textured silicon surfaces by AFM force curve analysis. *Journal of Micromechanics and Microengineering*, **17** (7), 1326–1333.

Bachmann, D., Kühne, S. and Hierold, C. (2006) Determination of the adhesion energy of MEMS structures by applying Weibull-type distribution function. *Sensors and Actuators, A: Physical*, **132**, 407–414.

Bárcenas, J., Reyes, L. and Esquivel-Sirvent, R. (2005) Scaling of micro- and nanodevices actuated by Casimir forces. *Applied Physics Letters*, **87** (26), 263106.

Basu, S., Prabhakar, A. and Bhattacharya, E. (2007) Estimation of stiction force from electrical and optical measurements on cantilever beams. *Journal of Microelectromechanical Systems*, **16** (5), 1254–1262.

Bergström, L. (1997) Hamaker constants of inorganic materials. *Advances in Colloid and Interface Science*, **70** (1–3), 125–169.

Bhattacharya, E., Rani, H.A., Babu, U.V. *et al.* (2005) Effect of porous silicon formation on stiction in surface micromachined MEMS structures. *Physica Status Solidi (A) Applications and Materials Science*, **202** (8), 1482–1486.

Bhushan, B. (1995) *Handbook of Micro/Nano Tribology*, CRC Press.

Bhushan, B. (1996) Methodology for roughness measurement and contact analysis for optimization of interface roughness. *IEEE Transactions on Magnetics*, **32**, 1819–1825.

Bhushan, B. (ed.) (1998) *Tribology Issues and Opportunities in MEMS – Proceeding of the Workshop*, Springer.

Bhushan, B. (2007) Nanotribology and nanomechanics of MEMS/NEMS and BioMEMS/BioNEMS materials and devices. *Microelectronic Engineering*, **84** (3), 387–412.

Bhushan, B. and Gupta, B.K. (1997) *Handbook of Tribology: Materials, Coatings, and Surface Treatments*, Krieger Publishing Company.

Buks, E. and Roukes, M.L. (2001) Stiction, adhesion energy, and the Casimir effect in micromechanical systems. *Physical Review B – Condensed Matter and Materials Physics*, **63** (3), 334021–334024.

Butt, H.J. (2008) Capillary forces: influence of roughness and heterogeneity. *Langmuir*, **24** (9), 4715–4721.

Chang, W.R., Etsion, I. and Bogy, D.B. (1987) An elastic–plastic model for the contact of rough surfaces. *Journal of Tribology*, **109**, 257–263.

Chau, A., Rignier, S., Delchambre, A. and Lambert, P. (2007) Three-dimensional model for capillary nanobridges and capillary forces. *Modelling and Simulation in Materials Science and Engineering*, **15** (3),305.

Cho, S.S. and Park, S. (2004) Finite element modeling of adhesive contact using molecular potential. *Tribology International*, **37** (9), 763–769.

De Boer, M.P. (2007) Capillary adhesion between elastically hard rough surfaces. *Experimental Mechanics*, **47**, 171–183.

De Boer, M.P., Clews, P.J., Smith, B.K. and Michalske, T.A. (1998) Adhesion of polysilicon microbeams in controlled humidity ambients. *MRS Online Proceedings Library Archive*, **518**, 131–136.

De Boer, M.P., Knapp, J.A., Mayer, T.M. and Michalske, T.A. (1999) The role of interfacial properties on MEMS performance and reliability. *Proceedings of SPIE*, **3825**, 2–15.

De Boer, M.P. and Michalske, T.A. (1999) Accurate method for determining adhesion of cantilever beams. *Journal of Applied Physics*, **86** (2), 817–827.

De Coster, J., Ling, F., Witvrouw, A. and De Wolf, I. (2013) *Dedicated Test Structure for the Measurement of Adhesion Forces Between Contacting Surfaces in MEMS Devices*, Transducers and Eurosensors XXVII: The 17th International Conference on Solid-State Sensors, Actuators and Microsystems, June 16–20, 2013, Barcelona, Spain, IEEE.

Decuzzi, P. and Srolovitz, D.J. (2004) Scaling laws for opening partially adhered contacts in MEMS. *Journal of Microelectromechanical Systems*, **13** (2), 377–385.

Dellea, S., Ardito, R., De Masi, B. *et al.* (2016) Sidewall adhesion evolution in epitaxial polysilicon as a function of impact kinetic energy and stopper area. *Journal of Microelectromechanical Systems*, **25** (1), 134–143.

DelRio, F.W., De Boer, M.P., Knapp, J.A. *et al.* (2005) The role of van der Waals forces in adhesion of micromachined surfaces. *Nature Materials*, **4** (8), 629–634.

DelRio, F.W., Dunn, M.L. and de Boer, M.P. (2008) Capillary adhesion model for contacting micromachined surfaces. *Scripta Materialia*, **59** (9), 916–920.

DelRio, F.W., Dunn, M.L., Phinney, L.M. *et al.* (2007) Rough surface adhesion in the presence of capillary condensation. *Applied Physics Letters*, **90** (16), 163104.

Derjaguin, B.V., Muller, V.M. and Toporov, Y.P. (1975) Effect of contact deformations on the adhesion of particles. *Journal of Colloid And Interface Science*, **53** (2), 314–326.

Dhariwal, R.S., Milne, N., Yang, S.J. *et al.* (1994) *Breakdown Electric Field Strength Between Small Electrode Spacings in Air*, Proceedings Micro System Technologies, October 19–21, 1994, Berlin, Germany, VDE-Verlag.

Ding, J.N., Wen, S.Z. and Meng, Y.G. (2001) Theoretical study of the sticking of a membrane strip in MEMS under the Casimir effect. *Journal of Micromechanics and Microengineering*, **11** (3), 202–208.

Emig, T., Hanke, A., Golestanian, R. and Kardar, M. (2001) Probing the strong boundary shape dependence of the Casimir force. *Physical Review Letters*, **87** (26), 2604021–2604024.

Feng, J.Q. (2000) Contact behavior of spherical elastic particles: a computational study of particle adhesion and deformations. *Colloids and Surfaces A: Physicochemical and Engineering Aspects*, **172** (1–3), 175–198.

Friedrich, T., Raudzis, C. and Müller-Fiedler, R. (2009) Experimental study of in-plane and out-of-plane adhesions in microelectromechanical systems. *Journal of Microelectromechanical Systems*, **18** (6), 1326–1334.

Greenwood, J.A. and Williamson, J.B.P. (1966) Contact of nominally flat surfaces. *Proceedings of the Royal Society of London A*, **295**, 300–319.

Greenwood, J.A. and Wu, J.J. (2001) Surface roughness and contact: an apology. *Meccanica*, **36** (6), 617–630.

Hariri, A., Zu, J.W. and Ben Mrad, R. (2006) Modeling of dry stiction in micro electro-mechanical systems (MEMS). *Journal of Micromechanics and Microengineering*, **16** (7), 1195–1206.

Hariri, A., Zu, J.W. and Ben Mrad, R. (2007) Modeling of wet stiction in microelectromechanical systems (MEMS). *Journal of Microelectromechanical Systems*, **16** (5), 1276–1285.

Heinz, D.B., Hong, V.A., Ng, E.J. *et al.* (2014) *Characterization of Stiction Forces in Ultra-clean Encapsulated MEMS Devices*, Proceedings of the IEEE International Conference on Micro Electro Mechanical Systems (MEMS), January 26–30, 2014, San Francisco, CA, IEEE.

Hu, Y.Z. and Tonder, K. (1992) Simulation of 3-D random rough surface by 2-D digital filter and Fourier analysis. *International Journal of Machine Tools and Manufacture*, **32** (1–2), 83–90.

Israelachvili, J. (2011) *Intermolecular and Surface Forces*, Academic Press.

Johnson, K.L., Kendall, K. and Robert, A.D. (1971) Surface energy and the contact of elastic solids. *Proceedings of the Royal Society of London A*, **324**, 301–303.

Jones, E.E., Begley, M.R. and Murphy, K.D. (2003) Adhesion of micro-cantilevers subjected to mechanical point loading: modeling and experiments. *Journal of the Mechanics and Physics of Solids*, **51** (8), 1601–1622.

Jones, E.E., Murphy, K.D. and Begley, M.R. (2003) Mechanical measurements of adhesion in microcantilevers: transitions in geometry and cyclic energy changes. *Experimental Mechanics*, **43** (3), 280–288.

Knapp, J.A. and de Boer, M.P. (2002) Mechanics of microcantilever beams subject to combined electrostatic and adhesive forces. *Journal of Microelectromechanical Systems*, **11** (6), 754–764.

Komvopoulos, K. (2003) Adhesion and friction forces in microelectromechanical systems: mechanisms, measurement, surface modification techniques, and adhesion theory. *Journal of Adhesion Science and Technology*, **17** (4), 477–517.

Lambrecht, A., Pirozhenko, I., Duraffourg, L. and Andreucci, P. (2007) The Casimir effect for silicon and gold slabs. *Europhysics Letters*, **77** (4).

Lamoreaux, S.K. (2005) The Casimir force: background, experiments, and applications. *Reports on Progress in Physics*, **68** (1), 201–236.

Legtenberg, R., Tilmans, H.A.C., Elders, J. and Elwenspoek, M. (1994) Stiction of surface micromachined structures after rinsing and drying: model and investigation of adhesion mechanisms. *Sensors and Actuators: A. Physical*, **43** (1–3), 230–238.

Leseman, Z.C., Carlson, S.P. and Mackin, T.J. (2007) Experimental measurements of the strain energy release rate for stiction-failed microcantilevers using a single-cantilever beam peel test. *Journal of Microelectromechanical Systems*, **16** (1), 38–43.

Lin, W.H. and Zhao, Y.P. (2005) Casimir effect on the pull-in parameters of nanometer switches. *Microsystem Technologies*, **11** (2–3), 80–85.

Maboudian, R. (1998) Surface processes in MEMS technology. *Surface Science Reports*, **30** (6), 207–269.

Maboudian, R. and Howe, R.T. (1997) Critical review: adhesion in surface micromechanical structures. *Journal of Vacuum Science and Technology B: Microelectronics and Nanometer Structures*, **15** (1), doi: 10.1116/1.589247.

Mastrangelo, C.H. and Hsu, C.H. (1993a) Mechanical stability and adhesion of microstructures under capillary forces – Part I: Basic theory. *Journal of Microelectromechanical Systems*, **2** (1), 33–43.

Mastrangelo, C.H. and Hsu, C.H. (1993b) Mechanical stability and adhesion of microstructures under capillary forces – Part II: Experiments. *Journal of Microelectromechanical Systems*, **2** (1), 44–55.

Mu, F. and Su, X. (2007) Analysis of liquid bridge between spherical particles. *China Particuology*, **5** (6), 420–424.

Munday, J.N., Capasso, F. and Parsegian, V.A. (2009) Measured long-range repulsive Casimir–Lifshitz forces. *Nature*, **457** (7226), 170–173.

Pakarinen, O.H., Foster, A.S., Paajanen, M. *et al.* (2005) Towards an accurate description of the capillary force in nanoparticle–surface interactions. *Modelling and Simulation in Materials Science and Engineering*, **13** (7), 1175–1186.

Payam, A.F. and Fathipour, M. (2011) A capillary force model for interactions between two spheres. *Particuology*, **9** (4), 381–386.

Rabinovich, Y.I., Esayanur, M.S. and Moudgil, B.M. (2005) Capillary forces between two spheres with a fixed volume liquid bridge: theory and experiment. *Langmuir*, **21** (24), 10992–10997.

Raccurt, O., Tardif, F., D'Avitaya, F. and Vareine, T. (2004) Influence of liquid surface tension on stiction of SOI MEMS. *Journal of Micromechanics and Microengineering*, **14** (7), 1083–1090.

Reid, M.T.H., Rodriguez, A.W., White, J. and Johnson, S.G. (2009) Efficient computation of Casimir interactions between arbitrary 3D objects. *Physical Review Letters*, **103** (4), 040401.

Ripalda, J.M., Gale, J.D. and Jones, T.S. (2004) Hydrogen-bridge bonding on semiconductor surfaces: density-functional calculations. *Physical Review B – Condensed Matter and Materials Physics*, **70** (24), 1–4.

Sauer, R.A. and Wriggers, P. (2009) Formulation and analysis of a three-dimensional finite element implementation for adhesive contact at the nanoscale. *Computer Methods in Applied Mechanics and Engineering*, **198** (49–52), 3871–3883.

Schroeder, M.R. (2012) *Fractals, Chaos, Power Laws: Minutes from an Infinite Paradise*, Courier Corporation.

Shavezipur, M., Gou, W., Fisch, M. *et al.* (2012) Inline measurement of adhesion force using electrostatic actuation and capacitive readout. *Journal of Microelectromechanical Systems*, **21** (4), 768–770.

Srikar, V.T. and Senturia, S.D. (2002) The reliability of microelectromechanical systems (MEMS) in shock environments. *Journal of Microelectromechanical Systems*, **11** (3), 206–214.

Stifter, T., Marti, O. and Bhushan, B. (2000) Theoretical investigation of the distance dependence of capillary and van der Waals forces in scanning force microscopy. *Physical Review B*, **62**, 13667–13673.

Suh, A.Y. and Polycarpou, A.A. (2003) Adhesion and pull-off forces for polysilicon MEMS surfaces using the sub-boundary lubrication model. *Journal of Tribology*, **125** (1), 193–199.

Tang, W.C., Lim, M.G. and Howe, R.T. (1992) Electrostatic comb drive levitation and control method. *Journal of Microelectromechanical Systems*, **1** (4), 170–178.

Tas, N.R., Gui, C. and Elwenspoek, M. (2003) Static friction in elastic adhesion contacts in MEMS. *Journal of Adhesion Science and Technology*, **17** (4), 547–561.

Tayebi, N. and Polycarpou, A.A. (2006) Adhesion and contact modeling and experiments in microelectromechanical systems including roughness effects. *Microsystem Technologies*, **12** (9), 854–869.

Timpe, S.J. and Komvopoulos, K. (2005) An experimental study of sidewall adhesion in microelectromechanical systems. *Journal of Microelectromechanical Systems*, **14** (6), 1356–1363.

Van Spengen, W.M., Bakker, E. and Frenken, J.W.M. (2007) A 'nano-battering ram' for measuring surface forces: obtaining force-distance curves and sidewall stiction data with a MEMS device. *Journal of Micromechanics and Microengineering*, **17** (7), S91–S97.

Van Spengen, W.M., Puers, R. and De Wolf, I. (2002) A physical model to predict stiction in MEMS. *Journal of Micromechanics and Microengineering*, **12** (5), 702–713.

Van Spengen, W.M., Puers, R. and De Wolf, I. (2003) On the physics of stiction and its impact on the reliability of microstructures. *Journal of Adhesion Science and Technology*, **17** (4), 563–582.

Van Spengen, W.M., Puers, R., Mertens, R. and De Wolf, I. (2004) A comprehensive model to predict the charging and reliability of capacitive RF MEMS switches. *Journal of Micromechanics and Microengineering*, **14** (4), 514–521.

Wibbeler, J., Pfeifer, G. and Hietschold, M. (1998) Parasitic charging of dielectric surfaces in capacitive microelectromechanical systems (MEMS). *Sensors and Actuators, A: Physical*, **71** (1–2), 74–80.

Wriggers, P. and Reinelt, J. (2009) Multi-scale approach for frictional contact of elastomers on rough rigid surfaces. *Computer Methods in Applied Mechanics and Engineering*, **198** (21–26), 1996–2008.

Yu, T., Ranganathan, R., Johnson, N. *et al.* (2007) *In situ* characterization of induced stiction in a MEMS. *Journal of Microelectromechanical Systems*, **16** (2), 355–364.

Zhang, Y. and Zhao, Y.P. (2004) Static study of cantilever beam stiction under electrostatic force influence. *Acta Mechanica Solida Sinica*, **17** (2), 104–112.

Zhao, Y.P., Wang, L.S. and Yu, T.X. (2003) Mechanics of adhesion in MEMS – a review. *Journal of Adhesion Science and Technology*, **17** (4), 519–546.

Zhuang, Y.X. and Menon, A. (2005) On the stiction of MEMS materials. *Tribology Letters*, **19** (2), 111–117.

Zou, J., Marcet, Z., Rodriguez, A.W. *et al.* (2013) Casimir forces on a silicon micromechanical chip. *Nature Communications*, **4**, 1845.

Index

Mechanics of Microsystems, First Edition. Alberto Corigliano, Raffaele Ardito, Claudia Comi, Attilio Frangi, Aldo Ghisi, and Stefano Mariani.

Companion website: www.wiley.com/go/corigliano/mechanics